DSP for Embedded and Real-Time Systems

Expert Guide

DSP for Embedded and Real-Time Systems

Expert Guide

Robert Oshana

AMSTERDAM • BOSTON • HEIDELBERG • LONDON • NEW YORK • OXFORD
PARIS • SAN DIEGO • SAN FRANCISCO • SINGAPORE • SYDNEY • TOKYO

Newnes is an imprint of Elsevier

Newnes is an imprint of Elsevier
The Boulevard, Langford Lane, Kidlington, Oxford, OX5 1GB
225 Wyman Street, Waltham, MA 02451, USA

First published 2012

British Library Cataloguing in Publication Data
A catalogue record for this book is available from the British Library

Library of Congress Number: 2012938003

ISBN: 978-0-12-386535-9

Contents

Author Biographies

Kia Amiri

kiaa@alumni.rice.edu

Kia Amiri received his M.S. and Ph.D. degrees in Electrical and Computer Engineering at Rice University, Houston, TX, in 2007 and 2010, where he has been a member of the Center for Multimedia Communication (CMC) lab. His research focus is in the area of physical layer design and hardware architecture for wireless communication. He is the co-inventor of 7 (pending) patents in this area. During the summer and fall of 2007, Kia worked on developing and implementing MIMO algorithms as part of the Advanced Systems Technology Group in Xilinx, San Jose, CA.

Arokiasamy I

arokia@freescale.com

Arokiasamy I is presently working as Engineering Manager for LTE Layer-1 with Freescale, with over 25 years of Software Architecture and Development experience covering, Industrial Automation Systems for Hydro/Power Stations, Realtime Operating systems(RTOS), compiler backend for MIPS platform as part of CodeWarrior tool, WiMAX PHY for Freescale MSC8144 platform etc. Arokiasamy I received his Bachelors of Engineering in Electronics and Communication Engineering from Indian Institute Of Science, Bangalore, India.

Dr Michael C. Brogioli

Dr Michael C. Brogioli is currently a computer engineering and software consultant in Austin, TX, USA, as well as an adjunct professor of computer engineering at Rice University. In addition, he also serves on the advisory boards of a number of early stage technology companies in the Austin and New York areas. Prior to this, he was a Senior Member Technical Staff and Lead Architect of compilers and build tools in their Technology Solutions Organization. During his time at Freescale, he also worked in conjunction with the hardware development teams on next generation DSP platforms. Prior to his time at Freescale, Dr Brogioli worked with Texas Instruments Advanced Architecture and Chip Technologies group in Stafford, TX, USA and Intel's Microprocessor Research Labs in Santa Clara, CA,

USA. He has authored over a dozen peer reviewed publications as well as co-authored three books in the embedded computing and signal processing space. He holds a Ph.D. and M.S. in Computer Engineering from Rice University in Houston, TX, USA, as well as a B.S. in Electrical Engineering from Rensselaer Polytechnic Institute in Troy, NY, USA.

Cristian Caciuloiu

Cristian Căciuloiu is a Software Engineer in Freescale Semiconductor Romania. He received BSc/MSc in Computer Science in 1999 from the University POLITEHNICA of Bucharest, Romania. Cristian joined Freescale (Motorola's Semiconductor Products Sector at that time) in 2000 and worked in various embedded software projects for several DSP and PowerPC platforms. From 2005 onwards, he held senior technical leadership and managerial positions, the most recent being the Software Architect for the VoIP Media Gateway Software Products from the Signal Processing Software and Systems team in Freescale Romania.

Joseph R. Cavallaro
cavallar@rice.edu

Joseph R. Cavallaro is a Professor in the Electrical and Computer Engineering Department and Director of the Center for Multimedia Communication at Rice University, Houston, Texas. He is also a Docent in the Centre for Wireless Communications at the University of Oulu in Finland. He received a B.S. from the University of Pennsylvania (1981), an M.S. from Princeton University (1982) and a Ph.D. from Cornell University (1988), all in Electrical Engineering. He has been engaged in research since 1981 after joining AT&T Bell Laboratories as a Member of Technical Staff and his research interests include VLSI signal processing with applications to wireless communication systems.

Stephen Dew
stephendew@yahoo.com

Stephen Dew is currently a Compiler Engineer at Freescale Semiconductor. Previously, he held positions in Applications Engineering at Freescale and StarCore, LLC. He received a Masters of Science in Electrical and Computer Engineering from the Georgia Institute of Technology in 2001.

Dr Chris Dick
chris.dick@xilinx.com

Dr Chris Dick is the DSP Chief Scientist at Xilinx and the engineering manager for the Xilinx Wireless Systems Engineering team. Chris has worked with signal processing technology for two decades and his work has spanned the commercial, military and academic sectors. Prior to joining Xilinx in 1997 he was a professor at La Trobe University, Melbourne Australia for

13 years and managed a DSP Consultancy called Signal Processing Solutions. Chris' work and research interests are in the areas of fast algorithms for signal processing, digital communication, MIMO, OFDM, software defined radios, VLSI architectures for DSP, adaptive signal processing, synchronization, hardware architectures for real-time signal processing, and the use of Field Programmable Arrays (FPGAs) for custom computing machines and real-time signal processing. He holds a bachelor's and Ph.D. degrees in the areas of computer science and electronic engineering.

Melissa Duarte

megduarte@gmail.com

Melissa Duarte was born in Cucuta, Colombia. She received the B.S. degree in Electrical Engineering from the Pontificia Universidad Javeriana, Bogota, Colombia, in 2005, and the M.S and Ph.D. degrees in Electrical and Computer Engineering from Rice University, Houston, TX, in 2007 and 2012 respectively. Her research is focused on the design and implementation of architectures for next-generation wireless communications, full-duplex systems, feedback based Multiple Input Multiple Output (MIMO) antenna systems, and the development of a Wireless Open Access Research Platform (WARP) for implementation and evaluation of algorithms for wireless communications. Melissa was a recipient of a Xilinx Fellowship at Rice University 2006–2007, a TI Fellowship at Rice University 2007–2008, and a Roberto Rocca Fellowship 2009–2012.

Michelle Fleischer

Michelle.fleischer@freescale.com

Michelle Fleischer grew up in Southeastern Michigan and majored in electrical engineering at Michigan Technological University in Houghton Michigan where she earned her BSEE (1992) and MSEE (1995) degrees. In 1992 she started working on a research grant for the Keweenaw research center performing acoustic modeling experiments for NASA and active noise control experiments for the US Army Tank Command (TACOM). In 1995 she started work at Texas Instruments in Dallas Texas. In 2006 she joined Freescale in Chicago Illinois as a Field Applications Engineer, supporting key accounts such as Motorola and RIM for the wireless group and key networking accounts for the Networking and Multimedia group. Ms. Fleischer has authored numerous application notes, design documents, software specifications, and software designs throughout her career.

Umang Garg

Umang.Garg@gmail.com

Umang Garg is currently an Engineering Manager for LTE Layer 2 software development in Networking and Multimedia Solutions Group at Freescale He has 10 years of experience in software development for DSP and Communication Processors. His current work interest is in the area of leveraging multi-core architectures for LTE software development while previous work interests have included Audio, Speech, Image Processing and VoIP technologies. Umang has a B.E.(Honors) in Computer Systems Engineering from University of Sussex, U.K. and a Master of Science in Advanced Computing from Imperial College, U.K.

Vatsal Gaur

vatsal@freescale.com

Vatsal Gaur is a DSP Software Engineer with over 7 years of experience in wireless baseband domain. He has worked on Layer-1 software development for various technologies such as GSM, GPRS, EDGE, WiMAX and LTE. His current work includes algorithm development and system design to enable various features of LTE eNodeB. Before joining Freescale, Vatsal had worked as a technical lead at Hughes software systems (now Aricent Technologies). Vatsal received his Bachelors of Engineering in Electrical and Electronics from Birla Institute of Technology, India.

Mircea Ionita

Mircea.Ionita@freescale.com

Mircea Ioniță is the manager of the Networking and Signal Processing Software and Systems Team within Freescale Semiconductor Romania since 2007. Mircea joined Freescale/ Motorola in 2001 where he held senior technical and managerial positions until 2007 when he was appointed to lead the Networking and Signal Processing Software and Systems Team. Mircea received his Bachelors of Science, Masters of Science and Doctorate degrees in Electronic Engineering from the University POLITEHNICA of Bucharest, Romania.

Nitin Jain

nitin19@gmail.com

Nitin Jain received his B.E. degree specializing in Electronics and Communications from Visvesvaraya Technological University, Belgaum, India, in 2003. From 2004 to 2007, he worked with MindTree's Research group and was responsible for developing Speech and Audio algorithms, and integrating them within different Bluetooth and mobile products. Nitin joined Freescale in 2008 as a Senior Engineer in the Networking group and has been involved in WiMAX and LTE physical layer development for Macro and Femto product segments on Freescale's baseband devices. As a Lead Engineer, his recent assignment is to perform LTE Femto eNB integration and verification with commercial-grade higher layers and user equipments. His work has appeared in renowned magazines and conference in Embedded and DSP domain.

Michael Kardonik
michael.kardonik@gmail.com

Michael Kardonik's career begun at Motorola Semiconductor Israel at 1999 where he worked on board support packages for PowerQUICC II processors. Later, Michael led engineering team that designed and implemented SmartDSP OS for Motorola's StarCore DSP. After his relocation to Austin, US, Michael served as DSP Applications Engineer, Tools engineer responsible for advanced debugging technologies. Today, Michael is part of Architecture team at Freescale Semiconductor working on current and next generation of QorIQ. Michael received his Bachelors of Arts in Economics and Computer Science degree from Ben Gurion University of Negev Israel and Masters of Science in Software Engineering degree from the University of Texas at Austin.

Robert Krutsch

Robert Krutsch is a DSP Software Engineer at Freescale Semiconductor with a primary focus in medical and baseband applications. He received a BSc/MSc degree specializing in automatics and computer science from the Polytechnic University of Timisoara; a BSc/Msc degree from the Automation Institute at the University of Bremen.

Tai Ly
tai.ly@ni.com

Tai Ly obtained his PhD in High Level Synthesis in 1993 from the University of Alberta, Canada, and has since worked in Research and Development in a number of companies including Synopsys, 0-In Design Automation, Mentor Graphics and Synfora Inc, He holds 14 Patents in the fields of High Level Synthesis, Assertion Verification, Formal Verification, and Clock Domain Crossing Verification. He is currently a Principal Engineer in National Instrument.

Akshitij Malik
akshitij.malik@freescale.com

Akshitij Malik is a Software Engineer with over 11 years of experience in the Telecom/Wireless industry. Akshitij worked as a Senior Principal Engineer at Hughes Systique and Hughes Software Systems LLC before moving on to his current role as a Lead Software Engineer at Freescale Semiconductor. Past works include development of L3 software for WCDMA RNCs, L2 software for WiMAX Base-Stations amongst other topics. Akshitij received his Bachelors of Engineering in Computer Engineering from the University of Pune, India.

Robert Oshana

robert.oshana@freescale.com

Robert Oshana 30 years of experience in the software industry, primarily focused on embedded and real-time systems for the defense and semiconductor industries. He has BSEE, MSEE, MSCS, and MBA degrees and is a Senior Member of IEEE. Rob is a member of several Advisory Boards including the Embedded Systems group, where he is also an international speaker. He has over 100 presentations and publications in various technology fields and has written several books on Embedded software technology. Rob is an adjunct professor at Southern Methodist University where he teaches graduate software engineering courses. He is a Distinguished Member of Technical Staff and Director of Global Software R&D for Networking and Multimedia at Freescale Semiconductor.

Raghu Rao

raghu.rao@xilinx.com

Raghu Mysore Rao is a Principal Engineer and System Architect with the Communications Signal Processing Group, Communications Business Unit, Xilinx Inc. He and his team are currently working on signal processing and digital communication algorithms for FPGAs. He is an IEEE senior member. He has a Ph.D. in wireless communications from University of California, Los Angeles. In the past, he has worked at Texas instruments India, Exemplar Logic Inc. and Mentor Graphics Inc. His interests are in digital communication and signal processing algorithms and architectures for their efficient implementation on FPGAs.

Ashutosh Sabharwal

ashu@rice.edu

Ashutosh Sabharwal is a Professor of Electrical and Computer Engineering at Rice University. His research interests are in information theory, network protocols and high-performance wireless platforms. He is the founder of Rice Wireless Open-Access Research Platform (http://warp.rice.edu), which is in use at 100+ organizations worldwide.

Yang Sun

ysunrice@gmail.com

Yang Sun received the B.S. degree in Testing Technology and Instrumentation in 2000, and the M.S. degree in Instrument Science and Technology in 2003, both from Zhejiang University, Hangzhou, China. He received the Ph.D. degree in electrical and computer engineering from Rice University, Houston, Texas, in 2010. He is currently a senior engineer in Qualcomm Inc. His research interests include parallel algorithms and VLSI architectures for wireless communication systems, digital signal processing systems, multimedia systems, and general purpose computing systems.

Adrian Susan
Adrian.Susan@freescale.com

Adrian Susan is the technical manager of L1 DSP Software Components Team in Freescale Romania located in Bucharest. Adrian is a Software Engineer, having graduated from the "Politehnica" University of Timioara with a B.Sc. in Computer Science. Adrian worked for Freescale Romania since 2004 where he was a member of the team in charge with the development of Freescale VoIP multimedia gateway.

Andrew Temple
temple.andrewr@gmail.com

Andrew Temple is an application engineer with over 10 years of experience in the semiconductor industry. Andrew worked as an applications engineer at Texas Instrument and StarCore LLC before moving on to his current role as a DSP Applications Engineer at Freescale Semiconductor. Past works include application notes on power consumption, bus interfacing, deadlock arbitration, Serial Rapid IO usage, Ethernet performance and connectivity, amongst other topics. Andrew received his Bachelors of Science in Computer Engineering and Masters of Science in Software Engineering degrees from the University of Texas at Austin.

Guohui Wang
gw2@rice.edu

Guohui Wang received his B.S. degree in EE from Peking University, Beijing, China, and M.S. degree in CS from Institute of Computing Technology, Chinese Academy of Sciences, Beijing, China. Since 2008, he has been pursuing a Ph.D. degree in Department of Electrical and Computer Engineering, Rice University, Houston, Texas. His research interests include mobile computing, VLSI signal processing for wireless communication systems and parallel signal processing on GPGPU.

Bei Yin
by2@rice.edu

Bei Yin received his B.S. degree in Electrical Engineering from Beijing University of Technology, Beijing, China, in 2002, and his M.S. degree in Electrical Engineering from Royal Institute of Technology, Stockholm, Sweden, in 2005. From 2005 to 2008, he worked at ARCA Technology Co. as an ASIC/SoC design engineer. He is currently a Ph.D. student in the Department of Electrical and Computer Engineering at Rice University, Houston, Texas. His research interests include VLSI signal processing and wireless communications.

DSP in Embedded Systems: A Roadmap

DSP Software development for embedded systems follows the standard hardware/software co-design model for embedded systems as shown in Figure 1.

The development process can be divided into six phases;

- Phase 1; Product Specification
- Phase 2; Algorithmic Modeling
- Phase 3; HW/SW Partitioning
- Phase 4; Iteration and Selection
- Phase 5; Real-Time SW Design
- Phase 6; HW/SW Integration

This book will cover each of these important phases of DSP Software Development.

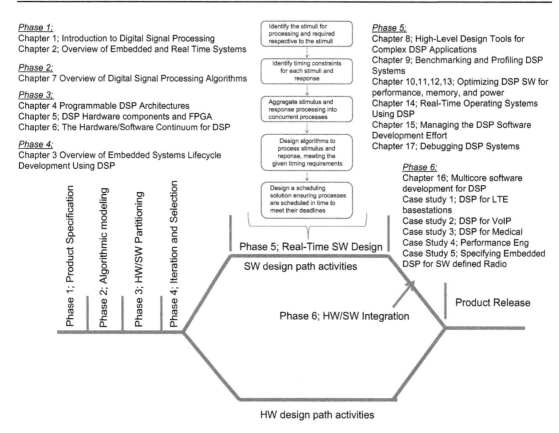

Figure 1:
DSP Software Development Following the Embedded HW/SW Co-design Model

Phase 1 — Product Specification

Phase 1 is an overview of embedded and real time systems and introduces the reader to the unique aspects of this type of software development.

There are several embedded systems challenges that we need to comprehend before we can have a discussion on digital signal processing. These include a significant degree of complexity of environments as well as interactions between systems, an increasing percentage of software within embedded components, the need for software code reuse and rapid reengineering, fast innovation and release cycles driven by shifting market demands, numerous real time demands and the need for requirements management, and the increasing focus on quality and process maturity. Chapters 1 and 2 (page 1 and 15) will provide an overview of DSP as well as embedded systems in general, and review the key differences that exist in embedded systems in general and DSP specifically.

Phase 2 — Algorithmic Modeling

Phase 2 focuses on the understanding of the fundamental algorithmic nature of signal processing. Digital signal processing is about the representation of discrete time signals using a sequence of numbers or symbols and the subsequent processing of these signals. DSP includes areas like audio and speech signal processing, sonar and radar signal processing, statistical signal processing, digital image processing, communications, system control, biomedical signal processing, and many other areas. DSP algorithms are used to process these digital signals. There are a set of fundamental algorithms used in signal processing such as fourier transforms, digital filters, convolution, and correlations. Chapter 7 (page 113) will introduce and explain some of the most important and fundamental DSP algorithms in order to set the base for many of the topics later in the book.

Phase 3 — Hardware/Software Partitioning

An important phase of any embedded development project is the partitioning of the system into hardware and software components.

Much of DSP is programmable. Programmable architectures for digital signal processing take a number of forms, each having their own tradeoffs in terms of cost, power consumption, performance and flexibility. At one end of the spectrum, digital signal processing system designers can achieve extremely high levels of efficiency and performance via the use of proprietary assembly language implementations of their application. At the other end of the spectrum, system developers can implement digital signal processing software stacks using ordinary ANSI C or C++ or other domain specific languages, executing the resulting algorithm on commercial desktop computers. Chapter 4 (page 63) details tradeoffs in implementations at varying points on a continuum, with one end being utmost digital signal processing performance and the other end of the continuum being flexibility and portability of a software implementation. Tradeoffs for each solution are detailed along the way, with the goal of guiding the digital signal processing system developer to the solution that meets their particular use case needs.

DSP is also implemented using Field Programmable Gate Arrays (FPGA) . As an example of this, in chapter 5 (page 75), we discuss the architectural challenges associated with spatial multiplexing and diversity gain schemes and introduce FPGA architectures and report experimental results for the FPGA realization of these systems. Chapter 5 (page 75) will introduce a flexible architecture and implementation of a spatial multiplexing MIMO detector, Flex-sphere, and its FPGA implementation. We also present a hardware architecture for beamforming in a WiMAX system as a way to enhance the diversity and performance in the next-generation wire- less systems.

Hardware platforms for digital signal processing systems come in varying designs, each with inherent programmability, power and performance tradeoffs. What is suitable for one system designer may not be suitable for another. Chapter 6 (page 103) details various digital signal processing platform designs as they pertain to system configurability and programmability. At one end of the spectrum, application specific integrated circuits (ASICs) are detailed as a high performance, low configurability solution. At the other end of the spectrum, general purpose embedded microprocessors with SIMD style extensions are presented as a highly configurable solution with robust software programmability support. Various design points are presented throughout such as reconfigurable field programmable gate array (FPGA) based solutions, and high performance application specific integrated processors (ASIP) with varying degrees of software programmability. Chapter 6 (page 103) will present tradeoffs for each system design, as a way of guiding the system developer in choosing the right digital signal processing hardware platform and component for their immediate and future system deployment needs.

Phase 4 — Iteration and Selection

Another key concern for DSP developers is managing the embedded lifecycle. This starts with selection of the solution space for the DSP problem you are trying to solve, and defining an embedded system that meets performance as well as cost, time to market and other important system constraints. As mentioned earlier, an embedded system is a specialized computer system that is integrated as part of a larger system. Many embedded systems are implemented using digital signal processors. The DSP will interface with the other embedded components to perform a specific function. The specific embedded application will determine the specific DSP to be used. For example, if the embedded application is one that performed video processing, the system designer may choose a DSP that is customized to perform media processing including video and audio processing. Chapter 3 (page 29) will discuss the embedded lifecycle and the various options for DSP and how to determine overall system performance and capability.

Phase 5 — Real-Time Software Design

Real-time software design follows the five step process outlines in Figure 1;

1. Identify the stimuli for processing and required responses to the stimuli
2. Identify timing constraints for each stimuli and response
3. Aggregate stimulus and response processing into concurrent processes
4. Design algorithms to process stimulus and response, meeting the given timing requirements

5. Design a scheduling solution ensuring processes are scheduled in time to meet their deadlines

We will discuss details of this process in this phase of the process.

1. Part 1; Identify the stimuli for processing and required responses to the stimuli

Firstly, we need to identify the system stimuli for signal processing as well as the response to these stimuli. This must be done regardless of whether we are using hardware or software.

We introduce a simple practical but very powerful specification technique in Case Study 2: DSP for medical devices (page 493) to give the developer a set of guidelines for this level of specification. The focus is on DSP development process focusing on specifying the behavior of embedded systems using DSP. The criticality of correct, complete, testable requirements is a fundamental tenet in software engineering. Both functional and financial success is affected by the quality of requirements. This starts with good requirements. Requirements may range from a high-level abstract statement of a service or of a system constraint to a detailed mathematical functional specification. Requirements are needed to Specify external system behavior, specify implementation constraints, serve as reference tool for maintenance, record forethought about the life cycle of the system i.e. predict changes, and to characterize responses to unexpected events. This case study introduces a rigorous behavioral specification technique that can greatly reduce risk by exposing ambiguities in requirements and making explicit otherwise tacit information. Such an external, or "black box" specification can be developed from behavioral requirements in a systematic manner through the process of sequence enumeration. This process results in an arguably complete, consistent, and traceable specification of external system behavior. Sequence abstraction provides a powerful means to manage and focus the enumeration process. A simple cell phone is used to reinforce these concepts.

There are some more advanced techniques that can be used for both hardware and software. Chapter 8 (page 133) focuses on a system level methodology of designing and synthesizing real-time digital signal processing systems. This is another challenge in the area of DSP development. Current hardware designs and implementations for DSP systems have a huge time gap between the development of algorithms for new DSP applications and their hardware implementation. The high level design and synthesis tools create application-special DSP accelerators from high abstraction-level for complex DSP processing hardware, which greatly reduces the design cycle while still maintaining area and power efficiency. This chapter presents two high level design methodologies, 1) C-to-RTL high level synthesis (HLS) for ASIC/FPGA implementation of the DSP systems, and 2) System Generator for FPGA implementation of the DSP systems. In the case studies, we will present three complex DSP accelerator designs using high level design tools: 1) Low-density parity-check (LDPC) decoder accelerator design using PICO C, 2) Matrix multiplication accelerator design using Catapult C, and 3) QR decomposition accelerator design using System Generator.

2. Identify timing constraints for each stimuli and response

A key part of optimizing DSP software is being able to properly profile and benchmark a DSP kernel and a DSP system. With a solid benchmark and profiling of a DSP kernel, both best case and in system performance can be assessed. Proper profiling and benchmarking can often be an art form. It is often the case that an algorithm is tested in nearly ideal conditions, and that performance is then used within a performance budget. Truly understanding the performance of an algorithm requires being able to model system effects along with understanding an algorithms best-case performance. System effects can include changes such as a running operating system, executing out of memories with different latencies, cache overhead, and managing coherency with memory. All of these effects require a carefully crafted benchmark, which can model these behaviors in a standalone fashion. If modeled correctly the standalone benchmark can very closely replicate a DSP kernel's execution, as it would behave in a running system. Chapter 9 (page 157) discusses how to perform this kind of benchmark.

3. Aggregate stimulus and response processing into concurrent processes

Once we understand what goes in and what comes out, we need to design and development the DSP software solution. Software development using DSPs is subject to many of the same constraints and development challenges that other types of software development face. These include a shrinking time to market, tedious and repetitive algorithm integration cycles, time intensive debug cycles for real-time applications, integration of multiple differentiated tasks running on a single DSP, as well as other real-time processing demands. Up to 80% of the development effort is involved in analysis, design, implementation, and integration of the software components of a DSP system.

Chapter 15 (page 335), will cover several topics related to managing the DSP software development effort. The first section of this chapter will discuss DSP system engineering and problem framing issues that relate to DSP and real-time software development. High level design tools are then discussed in the context of DSP application development. Integrated development environments and prototyping environments are also discussed. Specific challenges to the DSP application developer such as code tuning, profiling and optimization are reviewed. At the end of the chapter, there are several white papers that discuss related topics to DSP and real-time system development, integration, and analysis.

4. Design algorithms to process stimulus and response, meeting the given timing requirements

Writing DSP software to meet real-time constraints is challenging, so we dedicate several chapters to this area. Historically, DSP software was written in assembly languages. However, with the advent of the modern compiler, writing efficient DSP code using the C language is standard. However, since most DSPs have features that cannot be expressed fully in the C language, various alternatives exist to augment the standard languages of C and C++ such as extensions, higher-level languages and auto-vectorizing compilers. Chapter 10 (page 169) explains the high-level languages and programming models available for writing DSP software.

Optimization is the process of transforming a piece of code to make more efficient (either in terms of time, space, or power) without changing its output or side-effects. The only difference visible to the code's user should be that it runs faster and/or consumes less memory or power. Optimization is really a misnomer in the sense that the name implies one is trying to find an "optimal" solution— in reality, optimization aims to improve, not perfect, the result.

DSP is all about optimization, and this includes optimization for performance, memory, and power.

Chapter 11 (page 181) focuses on code optimization for performance. This is a critical step in the development process as it directly impacts the ability of the system to do it's intended job. Code that executes faster means more channels, more work performed and competitive advantage. This chapter is intended to help programmers write the most efficient code possible. It starts with an introduction to using the tool chain, covers the importance of knowing the Digital Signal Processor architecture before optimization, then moves on to cover wide range of optimization techniques. Techniques are presented which are valid on all programmable DSP architectures — C-language optimization techniques and general loop transformations. Real-world examples are presented throughout. To illustrate the concepts, DSPs from both Texas Instruments and Freescale are discussed.

Chapter 12 (page 217) focuses on memory optimization. Optimizing application code around memory system performance can often yield impressive gains in both executable code size, as well as runtime performance. This chapter illustrates methods that applications developers can utilize to improve both the static size of their executables in the presence of all too often resource constrained embedded systems. In addition, the chapter illustrates code optimization techniques that can either provide explicit performance gains of the application code, or implicit gains by improving the software build tools' ability to optimize code at compilation, assembly and link time.

Chapter 13 (page 241) focuses on optimization for power. This chapter is intended to be a resource for programmers needing to optimize a DSP for power consumption using strictly software. In order to provide the most comprehensive source for DSP software power optimization, this chapter provides a basic introduction into power consumption background, measurement techniques, and then goes into the details of power optimization. The chapter will focus on three main areas: algorithmic optimization, software controlled hardware power optimization (low power modes, clock control, and voltage control), and data flow optimization with a discussion into the functionality and power considerations when using fast SRAM type memories (common for cache) and DDR SDRAM.

5. Design a scheduling solution ensuring processes are scheduled in time to meet their deadlines

DSP applications are very demanding in terms of data rates and real-time computational requirements. These applications also can have very diverse real-time requirements. The DSP designer must understand the real-time design issues or order to obtain maximum performance. In addition, the resulting complexity requires the use of a real-time operating system. Key characteristics of a DSP RTOS include, extremely fast context switch times, very low interrupt latency times, optimized scheduler and interrupt handler, task, event, message, circular queue, and timer management, resource and semaphore processing, fixed block and memory management, and full pre-emption and ability to also have cooperative and time slice scheduling. Chapter 14 (page 291) will cover DSP RTOS in depth in order to be able to effectively utilize a DSP RTOS in application development.

This entire phase of real-time software design and development requires the engineer to build and debug the system in iterative steps using software development tools. DSP debug technology includes both hardware and software technology. Debug hardware consists of functionality on the DSP chip, which enables the collection of data. This data provides state behavior and other system visibility. Hardware is also required to extract this information from the DSP device at high rates and format the data. Debug software provides additional higher level control and an interface with the host computer, usually in terms of an interface called a debugger. The debugger provides the development engineer an easy migration from the compilation process (compiling, assembling, and linking an application) to the execution environment. The debugger takes the output from the compilation process and loads the image into the target system. The engineer then uses the debugger to interact with the emulator to control and execute the application and find and fix problems. These problems can be hardware as well as software problems. The debugger is designed to be a complete integration and test environment. Chapter 17 (page 381) discusses a sequence of activities and techniques to debug complex DSP systems.

Case Study 1 (page 423) will pull all of this together with an excellent case study on performance engineering for DSP systems. The focus is on a case study related to Performance Engineering. Expensive disasters can be avoided when system performance evaluation takes place relatively early in the software development lifecycle. Applications will generally have better performance when alternative designs are evaluated prior to implementation. Software Performance Engineering (SPE) is a set of techniques for gathering data, constructing a system performance model, evaluating the performance model, managing risk of uncertainty, evaluating alternatives, and verifying the models and results. SPE also includes strategies for the effective use of these techniques. Software performance engineering concepts have been used on programs developing digital signal processing applications concurrently with a next generation DSP-based array processor. Algorithmic performance and an efficient implementation were driving criteria for the program. As the processor was being developed concurrently with the software application a significant amount of the system and software development would be completed prior to the availability of physical hardware. This led to incorporation of SPE techniques into the development life-cycle. The techniques were incorporated cross-functionally into both the systems engineering organization responsible for developing the signal processing algorithms and the software and hardware engineering organizations responsible for implementing the algorithms in an embedded real-time system.

Phase 6 — Hardware/Software Integration

The integration phase is where the system is integrated into a fully functional real time system. This is where we choose to describe several cases studies that reinforce many of the points discussed in the preceding chapters. We will cover the more useful and important aspects of system integration as it relates to DSP systems. Topical areas in this phase include;

- Integration of Multicore DSP systems
- Integration of DSP systems for base stations
- Integration of DSP systems for medical
- Integration of DSP systems for Voice over IP (VoIP)
- Integration of DSP systems for Software Defined Radio

Multicore Digital Signal Processors have grown in importance in recent years mainly because of the emergence of data-intensive applications, such as video and high-speed Internet browsing on mobile devices. These applications demand significant computational performance and at lower cost and power consumption. In Chapter 16 (page 361), we discuss these topics and use the example of porting a single core application to a multicore environment, which entails the consideration of complex programming and process scheduling in addition to performing the required process algorithms. This chapter discusses two basic multicore programming methods: multiple-single-cores and true-multiple-cores. The true-multiple-cores model is used to port a motion JPEG application to a multicore DSP device and serves to illustrate the concepts presented. The chapter also addresses some of the concerns that arise when porting applications to a multi-core environment along with proposed solutions to address these issues.

As more and more DSP algorithm components are developed for a growing number of signal processing centric applications, the need for a programming model and standard is also emerging. Like other coding standards, a DSP programming standard can improve engineering efficiency and improve integration time and effectiveness. It can also make integration of DSP components from multiple vendors more effective.

Phase 6 also includes a case study describing a Long Term Evolution (LTE) basestation design of layer 1 and layer 2 software and some of the techniques used to develop a multicore implementation of this application (Case Study 1, page 423). It brings together many points raised earlier in the book. This case study summarizes the various challenges and their resolution when migrating single-core embedded application software to multi-core SoCs (System-on-a-Chip). The migration of a LTE eNodeB protocol stack is taken as the subject of this case-study.

This case study describes the essential software engineering practices which could be used by engineering teams for enhancing the viability of projects that involve software development over complex embedded platforms.

Thereafter the case study describes through a well defined 3-Step process, the steps required to develop the embedded applications for multicore SoCs with the help of associated examples at each step. Each of the steps is further broken down into sub steps, which seek to ensure that there is a measurable progress at the end of each phase of development. The various technical challenges and their resolution are described in detail, along with various suggestions, using the example of a high performance multicore SoC.

There is a case study on DSP for Voice over IP (VoIP) systems (Case Study 3, page 523). VoIP offers many cost related advantages over legacy copper-based telephony, on both the physical side (the equipment and wiring materials), as well as the logical side (considering the flexibility to add services and to differentiate the cost model). This case study gives an introduction on how the DSP made possible the VoIP ubiquity in the last decade.

There are many packet-based voice technologies that reshaped the telecommunication network, once the migration from twisted-pair and E1/T1 trunk wiring made room for Ethernet LAN and optical backbones. One of the workhorse systems that made the infrastructure transition possible was the VoIP Gateway. Segments of the network were replaced with IP-based technologies but still the service had to be operational when the legacy and new technologies collided. A VoIP Media Gateway handled the voice and signaling information that is required to interface two technologically different sides of the network, for example the TDM network to a packet network. Full-duplex real-time voice or fax/modem communications are compressed by the DSP engine and encoded into IP packets and then sent to the IP network. Packets received from the network are decoded and then decompressed towards TDM side. A dynamic jitter buffer automatically compensates for network delay variation enabling real-time voice communication. Voice processing includes echo cancellation, voice compression, voice-activity detection, and voice packetization. Other functions included are signaling detection and relay and fax/modem relay. DSP technology enabled all of this underlying technology. The architecture of a Media Gateway will be described in this case study.

There is also a case study on the use of DSP in the medical field (Case Study 2, page 493). The example used is a real time ultrasound system. Real time ultrasound systems have been available for more than forty years. During this time the architecture of such systems has changed significantly, bringing new capabilities, in terms of quality and operation modes.

In this case study we present an overview of some of the most common operation modes concentrating on an engineering approach to answer some of the frequent design questions. The text is focused on implementation of beamforming and B-Mode on modern DSP architecture, discussing tradeoffs and hardware capabilities.

Today's generation of DSP bring a lot of processing power, making them more suitable for medical ultrasound applications. The use of cases and examples in this chapter (Case Study 2, page 493) will show what could be implemented on DSP, where the bottlenecks are and what are the benefits.

There is a case study on Software Defined Radio (SDR) (Case Study 6, page 599); Digital signal processing has become a powerful premise for the world of wireless communications. All the latest technologies, including OFDM, CDMA, SC-FDMA that represent the main foundation of the 3 G-4 G networks, such as HSPA, LTE or WiMAX, are now made possible by the high density of digital algorithms that can be squeezed in a small, low-power chip. On top of that, additional signal processing techniques, such as beamforming or spatial multiplexing, have contributed to the achievement of throughputs of hundreds Mbps and spectral efficiencies of tens of b/s/Hz. Also, some other complex algorithms that can be found in channel estimators, equalizers and bit decoders allow all these techniques to operate in a tough radio environment, including in non-line of sight communications with high mobility, even at large distance.

Large-scale integration allows a handheld device to include a multi-standard terminal, compliant with a large variety of wireless standards. Think about a smart phone, that now conveniently chooses the best technology for data and voice transmission, according to its service needs and to the channel conditions. This is why your smart phone can connect to the BTS through GSM, EDGE, UMTS or soon, LTE. At the same time, it has capabilities for Wi-Fi, Bluetooth, GPS connections, all these in a very small terminal, with decent battery life. In the future, this service selection and multiplexing will only be defined through software. Without digital signal processing, relying solely on analog components, there would be no possible way to achieve such performance. Right now, the analog part in a transceiver is losing its pace in front of the digital part. More and more operations are performed in the digital part, including filtering, up/down conversion, compensations of the distortions produced in the analog chain, at a smaller price and with better performances. Moreover, it is expected that the analog part will be completely conquered in the future, with the help of high-speed A/D and D/A converters, so that the digital transceiver will be glued directly to the antenna. This part is called *front-end* in a transceiver and while some 10 years ago it used to be completely analog, it is becoming more and more digital. This case study (Case Study 6, page 599) attempts to reveal some of the digital signal processing existing in the front end of a digital transceiver.

DSP systems require accurate specification of requirements for processing for many applications. In this book we have a case study involving the specification steps for a cell phone application utilizing DSP technology. Detailed and accurate specification techniques lead to systems that meet customer requirements and perform well in the field. This case study will walk the reader through these interesting and practical steps to specify software systems.

Finally, there are case studies that overview software performance engineering (SPE). SPE is a systematic, quantitative approach that leads to cost-effective development of software systems to meet system performance requirements. SPE is a software-oriented approach that focuses on architecture, design, and implementation choices. There is an excellent case study focusing on the use of SPE for a large DSP application with hard real-time deadlines.

CHAPTER 1

Introduction to Digital Signal Processing

Robert Oshana

Chapter Outline

What is digital signal processing?

Digital signal processing (DSP) is the method of processing signals and data in order to enhance, modify, or analyze those signals to determine specific information content. It involves the processing of real-world signals that are converted to, and represented by, sequences of numbers. These signals are then processed using mathematical techniques, in order to extract certain information from the signal, or to transform the signal in some preferably beneficial way.

The term 'digital' in DSP requires processing using discrete signals to represent the data in the form of numbers that can be easily manipulated. In other words, the signal is represented numerically. This type of representation implies some form of quantization of one or more properties of the signal, including time.

This is just one type of digital data; other types include ASCII numbers and letters that have a digital representation as well.

The term 'signal' in DSP refers to a variable parameter. This parameter is treated as information as it flows through an electronic circuit. The signal usually starts out in the analog world as a constantly changing piece of information.[1] Examples of real-world signals include:

- Air temperature
- Sound
- Humidity
- Speed
- Position
- Flow
- Light
- Pressure
- Volume

The signal is essentially a voltage that varies among a theoretically infinite number of values. This represents patterns of variation of physical quantities. Other examples of signals are sine waves, the waveforms representing human speech, and the signals from a conventional television. A signal is a detectable physical quantity. Messages or information can be transmitted based on these signals.

A signal is called one dimensional (1-D) when it describes variations of a physical quantity as a function of a single independent variable. An audio/speech signal is one dimensional because it represents the continuing variation of air pressure as a function of time.

Finally, the term 'processing' in DSP relates to the processing of data using software programs as opposed to hardware circuitry. A DSP is a device or a system that performs signal processing functions on signals from the real (analog) world, primarily using software programs to manipulate the signals. This is an advantage in the sense that the software program can be changed relatively easily to modify the performance or behavior of the signal processing. This is much harder to do with analog circuitry.

Since DSPs interact with signals in the environment, the DSP system must be 'reactive' to the environment. In other words the DSP must keep up with changes in the environment. This is the concept of 'real-time' processing and we will talk about this shortly.

Advantages of DSP

There are many advantages of using a digital signal processing solution over an analog solution. These include:

---

[1] Usually because some signals may already be in a discrete form. An example of this would be a switch which is represented discretely as being either open or closed.

- Changeability; it is easy to reprogram digital systems for other applications, or to fine tune existing applications. A DSP allows for easy changes and updates to the application.
- Repeatability; analog components have characteristics that may change slightly over time or temperature variances. A programmable digital solution is much more repeatable due to the programmable nature of the system. Multiple DSPs in a system, for example, can also run the exact same program and be very repeatable. With analog signal processing, each DSP in the system would have to be individually tuned.
- Size, weight, and power; a DSP solution that requires mostly programming means the DSP device itself consumes less overall power than a solution using all hardware components.
- Reliability; analog systems are reliable to the extent to which the hardware devices function properly. If any of these devices fail due to physical condition, the entire system degrades or fails. A DSP solution implemented in software will function properly as long as the software is implemented correctly.
- Expandability; to add more functionality to an analog system, the engineer must add more hardware. This may not be possible. Adding the same functionality to a DSP involves adding software, which is much easier.

DSP systems

The signals that a DSP processor uses come from the real world. Because a DSP must respond to signals in the real world, it must be capable of changing based on the changes it sees in the real world. We live in an analog world in which the information around us changes, sometimes very quickly. A DSP system must be able to process these analog signals and respond back to the real world in a timely manner. A typical DSP system (Figure 1-1) consists of the following:

- Signal source; something that is producing the signal such as a microphone, a radar sensor, or a flow gauge.
- Analog signal processing (ASP); circuitry to perform some initial signal amplification or filtering.
- Analog-to-digital conversion (ADC); an electronic process in which a continuously variable signal is changed, without altering its essential content, into a multi-level (digital) signal. The output of the ADC has defined levels or states. The number of states is almost always a power of two — that is, 2, 4, 8, 16, etc. The simplest digital signals have only two states, and are called binary.
- Digital signal processing (DSP); the various techniques used to improve the accuracy and reliability of modern digital communications. DSP works by clarifying, or standardizing, the levels or states of a digital signal. A DSP system is able to

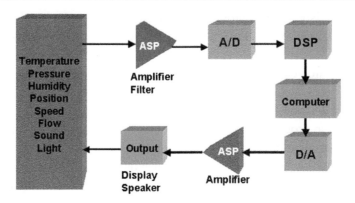

Figure 1-1:
A DSP system.

differentiate, for example, between human-made signals, which are orderly, and noise, which is inherently chaotic.

- Computer; if additional processing is required in the system, additional computing resources can be applied, if necessary. For example, if the signals being processed by the DSP are to be formatted for display to a user, an additional computer can be used to perform these tasks.
- Digital-to-analog conversion (DAC); the process in which signals having a few (usually two) defined levels or states (digital) are converted into signals having a theoretically infinite number of states (analog). A common example is the processing, by a modem, of computer data into audio-frequency (AF) tones that can be transmitted over a twisted pair telephone line.
- Output; a system for realizing the processed data. This may be a terminal display, a speaker, or another computer.

Systems operate on signals to produce new signals. For example, microphones convert air pressure to electrical current, and speakers convert electrical current to air pressure.

Analog-to-digital conversion

The first step in a signal processing system is getting the information from the real world into the system. This requires transforming an analog signal to a digital representation suitable for processing by the digital system. This signal passes through a device called an analog-to-digital converter (A/D or ADC). The ADC converts the analog signal to a digital one by sampling or measuring the signal at a periodic rate. Each sample is assigned a digital code (Figure 1-2). These digital codes can then be processed by the DSP. The number of different codes or states is almost always a power of two (2, 4, 8, 16, etc.). The simplest digital signals have only two states. These are referred to as binary signals.

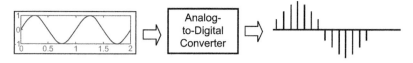

Figure 1-2:
Analog-to-digital conversion for signal processing.

Examples of analog signals are waveforms representing human speech and signals from a television camera. Each of these analog signals can be converted to digital form using ADC and then processed using a programmable DSP.

Digital signals can be processed more efficiently than analog signals. Digital signals are generally well-defined and orderly which makes them easier for electronic circuits to distinguish from *noise*, which is chaotic. Noise is basically unwanted information. Noise can be anything from the background sound of an automobile engine, to a scratch on a picture that has been converted to digital format. In the analog world noise can be represented as electrical or electromagnetic energy that degrades the quality of signals and data. Noise, however, occurs in both digital and analog systems. Sampling errors (we'll talk more about this later) can degrade digital signals as well. Too much noise can degrade all forms of information including text, programs, images, audio and video, and telemetry. Digital signal processing provides an effective way to minimize the effects of noise by making it easy to filter this 'bad' information out of the signal.

As an example, assume that the analog signal in Figure 1-2 needs to be converted into a digital signal for further processing. The first question to consider is how often to *sample* or measure the analog signal in order to represent that signal accurately in the digital domain. The sample rate is the number of samples of an analog event (like sound) that are taken per second to represent the event in the digital domain. Let's assume that we are going to sample the signal at a rate of T seconds. This can be represented as:

Sampling period (T) = 1 / Sampling frequency (fs)

where the sampling frequency is measured in hertz (Hz).[2]

If the sampling frequency is 8 kilohertz (kHz), this would be equivalent to 8000 cycles per second. The sampling period would then be:

T = 1 / 8000 = 125 microseconds = 0.000125 seconds

This tells us that, for this signal being sampled at this rate, we would have 0.000125 seconds to perform all the processing necessary before the next sample arrived (remember, these

---

[2] Hertz is a unit of frequency (of change in state or cycle in a sound wave, alternating current, or other cyclical waveform) of one cycle per second. The unit of measure is named after Heinrich Hertz, the German physicist.

samples arrive on a continuous basis and we cannot fall behind in processing them). This is a common restriction for *real-time systems* which we will discuss shortly.

If we now know the time restriction, we can then determine the processor speed required to keep up with this sampling rate. Processor speed is measured not by how fast the clock rate is for the processor, but how fast the processor executes instructions. Once we know the processor instruction cycle time, we can determine how many instructions we have available to process the sample:

Sampling period (T) / Instruction cycle time = number of instructions per sample

For a 100 MHz processor that executes one instruction per cycle, the instruction cycle time would be:

1/100MHz = 10 nano seconds
125 us / 10 ns = 12,500 instructions per sample
125 us / 5 ns = 25,000 instructions per sample (for a 200 MHz processor)
125 us / 2 ns = 62,500 instruction per sample (for a 500 MHz processor)

As this example demonstrated, the higher the processor instruction cycle execution, the more processing we can do on each sample. If it were this easy, we could just choose the highest processor speed available and have plenty of processing margin. Unfortunately, it is not as easy as this. Many other factors including cost, accuracy, and power limitations must also be considered. Embedded systems have many constraints such as these, as well as size and weight (important for portable devices). For example, how do we know how fast we should sample the input analog signal to accurately represent it in the digital domain? If we do not sample often enough, the information we obtain will not be representative of the true signal. If we sample too much, we may be 'over-designing' the system, and overly constraining ourselves.

The Nyquist criteria

One of the most important rules of sampling is called the Nyquist Theorem[3], which states that the highest frequency which can be accurately represented is one-half of the sampling rate. The Nyquist rate specifies the minimum sampling rate that fully describes a given signal. Adhering to the Nyquist rate enables accurate reconstruction of the original signal from the samples. The actual sampling rate required to reconstruct the original signal must be somewhat higher than the Nyquist rate, because of various quantization errors introduced by the sampling process.

For example, human hearing ranges from 20 Hz to 20,000 Hz, so to imprint sound to a CD, the frequency must be sampled at a rate of 40,000 Hz to reproduce the 20,000 Hz signal. The CD standard is to sample 44,100 times per second, or 44.1 kHz.

---

[3] Named in 1933 after scientist Harry Nyquist.

If a signal is not sampled at the Nyquist rate, the data sampled will not accurately represent the true signal. Consider the sine wave below:

The dashed vertical lines are sample intervals. The dots are the crossing points on the signal. These represent the actual samples taken during the conversion process (for example, an analog-to-digital converter). If the sampling rate in Figure 1-3 is below the required Nyquist frequency, this causes a problem. If the signal is reconstructed, the resultant waveform, as shown in Figure 1-4, can occur.

This signal looks nothing like the input. This undesirable feature is referred to as 'aliasing'. Aliasing is the generation of a false (or alias) frequency, along with the correct one, when performing frequency sampling in a given signal.

Aliasing manifests itself differently, depending on the signal affected. Aliasing shows up images as a jagged edge or stair-step effect. Aliasing affects sound by producing a 'buzz'. In order to reduce or eliminate this noise, the output from the ADC process is usually low-pass filtered to remove the higher frequency signals above the Nyquist frequency. Low-pass

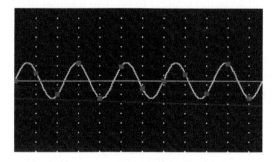

Figure 1-3:
A signal sampled at a rate below the Nyquist rate will not fully represent the true signal.

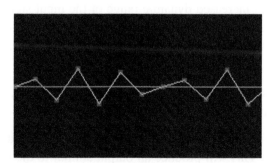

Figure 1-4:
A reconstructed waveform showing the problem when the Nyquist theorem is not followed.

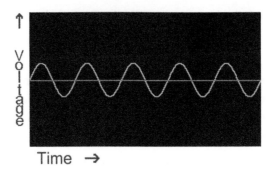

Figure 1-5:
Analog data plotted over time.

filtering also eliminates unwanted high-frequency noise and interference introduced prior to the sampling phase. We will talk more about filtering algorithms in coming chapters.

Let's assume we want to convert an analog audio into a digital signal for further processing. We use the term 'analog' in an audio context to refer to a signal that represents sound waves traveling through the air. A simple audio tone, such as a sine wave, will cause the air to form evenly spaced ripples of alternating high and low pressure. When these signals enter a microphone (or an eardrum for that matter), they cause sensors to move evenly back and forth, at the same rate, producing a voltage. The measured voltage coming from the microphone, plotted over time, will look similar to that shown in Figure 1-5.

If we want to edit, manipulate, or otherwise transmit this signal over a communication link, the signal must be digitized first. The incoming analog voltage levels are converted into binary numbers using analog-to-digital conversion. The two important constraints when performing this operation are sampling frequency (how often the voltage is measured) and resolution (the size of digital numbers used to measure the voltage, specifically the size or width of the A/D converter.

Larger ADCs allow for an increased dynamic range of the input signal. When an analog waveform is digitized we are essentially taking continuous 'snapshots' of the waveform at a certain interval, called the sampling frequency, and storing those snapshots as binary codes or numbers.

When the waveform is reconstructed from the sequence of numbers, the result will be a 'stair-step' approximation of what we started with (Figure 1-6).

To convert this digital data back into analog voltages, the stair-step approximation must be 'smoothed' using a filter. This will produce an output similar to the input (assuming no additional processing has been performed). However, if the sampling frequency, signal resolution, or both, are too low, the reconstructed waveform will be of lower quality. Failing

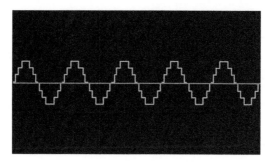

Figure 1-6:
Resultant analog signal reconstructed from the digitized samples.

to process a signal at the sample rate (or higher) is as bad as a miscalculation in a hard real-time system such as a CD player. This can be generalized to:

(Number of instructions to process * Sample rate) < Fclk * Instructions/cycle (MIPS)

where Fclk is the clock frequency of the DSP device.

The required sampling rate depends on the application. There is a wide range of sampling rates, from radar and signal processing applications on the high end, where the sampling rate may be up to 1 gigahertz and beyond to control and instrumentation which require a much lower sampling rate, in the order of 10 to 100 hertz. Algorithm complexity must also be taken into consideration. In general, the more complex the algorithm, the more instruction cycles are required to compute the result, and the lower the sampling rate must be to accommodate the time for processing these complex algorithms.

Digital-to-analog conversion

In many applications, a signal must be sent back out to the real world after being processed, enhanced and/or transformed while inside the DSP. Digital-to-analog conversion (DAC) is a process in which digital signals having a few (usually two) defined levels or states are converted into analog signals having a very large number of states.

Both the DAC and the ADC are of significance in many applications of digital signal processing. The fidelity of an analog signal can often be improved by converting the analog input to digital form using a DAC, clarifying or enhancing the digital signal, and then converting the enhanced digital impulses back to analog form using an ADC (a single digital output level provides a DC output voltage).

Figure 1-7 shows a digital signal passing through a digital-to-analog (D/A or DAC) converter which transforms the digital signal into an analog signal and outputs that signal to the environment.

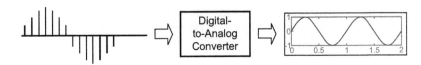

Figure 1-7:
Digital-to-analog conversion.

Applications for DSPs

In this section, we will explore some common applications for DSPs. Although there are many different DSP applications, I will focus on three categories:

- Low cost, good performance DSP applications
- Low power DSP applications
- High performance DSP applications

Low cost DSP applications

DSPs are becoming an increasingly popular choice as low cost solutions in a number of different areas. One popular area is electronic motor control. Electric motors exist in many consumer products, from washing machines to refrigerators. The energy consumed by the electric motor in these appliances is a significant portion of the total energy consumed by the appliance.

Controlling the speed of the motor has a direct effect on the total energy consumption of the appliance. In order to achieve the performance improvements necessary to meet energy consumption targets for these appliances, manufacturers use advanced three-phase variable speed drive systems. DSP-based motor control systems have the bandwidth required to enable the development of more advanced motor drive systems for many domestic appliance applications.

Application complexity has continued to grow as well, from basic digital control to advanced noise and vibration cancellation applications. As the complexity of these applications has grown, there has also been a migration from analog-to-digital control. This has resulted in an increase in reliability, efficiency, flexibility, and integration, leading to overall lower system costs.

Many of the early control functions used what is called a microcontroller as the basic control unit. As the complexity of the algorithms in motor control systems increased, the need also grew for higher performance and more programmable solutions. Digital signal processors provide much of the bandwidth and programmability required for such applications. DSPs are now finding their way into some of the more advanced motor control technologies:

- Variable speed motor control
- Sensorless control

- Field oriented control
- Motor modeling in software
- Improvements in control algorithms
- Replacement of costly hardware components with software routines

Motor control is one example of a low cost DSP application. In this example, the DSP is used to provide fast and precise PWM switching of the converter. The DSP also provides the system with fast, accurate feedback of the various analog motor control parameters such as current, voltage, speed, temperature, etc. There are two different motor control approaches; open-loop control and closed-loop control. The open-loop control system is the simplest form of control. Open-loop systems have good steady state performance and the lack of current feedback limits much of the transient performance. A low cost DSP is used to provide variable speed control of the three-phase induction motor, providing improved system efficiency.

A closed-loop solution is more complicated. A higher performance DSP is used to control current, speed, and position feedback, which improves the transient response of the system and enables tighter velocity/position control. Other, more sophisticated, motor control algorithms can also be implemented in the higher performance DSP.

There are many other applications using low cost DSPs. Refrigeration compressors, for example, use low cost DSPs to control variable speed compressors which dramatically improve energy efficiency. Low cost DSPs are used in many washing machines to enable variable speed control, which has eliminated the need for mechanical gearing. DSPs also provide sensorless control for these devices, which eliminates the need for speed and current sensors. Improved off-balance detection and control enable higher spin speeds, which get clothes dryer with less noise and vibration. Heating, ventilating, and air conditioning (HVAC) systems use DSPs in variable speed control of the blower and inducer, which increases furnace efficiency and improves comfort.

Power efficient DSP applications

We live in a portable society. From cell phones to personal digital assistants (PDAs), we work and play on the road! These systems are dependent on the batteries that power them. The longer the battery life can be extended, the better. So it makes sense for the designers of these systems to be sensitive to processor power. Having a processor that consumes less power enables longer battery life, and makes these systems and applications possible.

As a result of reduced power consumption, systems dissipate lower heat. This results in the elimination of costly hardware components like heat sinks to dissipate the heat effectively. This leads to overall lower system cost, as well as smaller overall system size, because of the reduced number of components. Continuing along this same line of reasoning, if the system can be made less complex with fewer parts, designers can bring these systems to market more quickly.

Low power devices also give the system designer a number of new options, such as potential battery back-up to enable uninterruptible operation, as well as the ability to do more with the same power (as well as cost) budget, to enable greater functionality and/or higher performance.

There are several classes of systems that are suitable for low power DSPs. Portable consumer electronics use batteries for power. Since the average consumer of these devices wants to minimize the replacement of batteries, the longer they can go on the same batteries, the better off they are. This class of customer also cares about size. Consumers want products they can carry with them, clip onto their belts, or carry in their pockets.

Certain classes of system require designers to adhere to a strict power budget. These include those that have a fixed power budget, such as systems that operate on limited line power, battery back-up, or with a fixed power source. For these classes of systems, designers aim to deliver functionality within the constraints imposed by the power supply. Examples include many defense and aerospace systems. These systems also have very tight size, weight, and power restrictions. Low power processors give designers more flexibility under all three of these important constraints.

Another important class of power-sensitive systems include high density systems. These systems are often high performance, or multi-processor systems. Power efficiency is important for these types of systems, not only because of the power supply constraints, but also because of heat dissipation concerns. These systems contain very dense boards with a large number of components per board. There may also be several boards per system in a very confined area. Designers of these systems are concerned about reduced power consumption, as well as heat dissipation. Low power DSPs can lead to higher performance and higher density. Fewer heat sinks and cooling systems enable lower cost systems that are easier to design. The main concerns for these systems are:

- Creating more functions per channel
- Achieving more functions per square inch
- Avoiding cooling issues (heat sinks, fans, noise)
- Reducing overall power consumption

Power is the limiting factor in many systems today. Designers must optimize the system design for power efficiency at every step. One of the first steps in any system design is the selection of the processor. A processor should be selected based on an architecture and instruction set optimized for power efficient performance. For signal processing intensive systems, a common choice is a DSP.

As an example of a low power DSP solution, consider a solid state audio player. This system requires a number of DSP-centric algorithms to perform the signal processing necessary to produce high fidelity quality music sound. A low power DSP can handle the decompression, decryption, and processing of audio data. This data may be stored on external memory devices which can be interchanged like individual CDs. These memory devices can be reprogrammed as well. The user interface functions can be handled by a microcontroller. The memory device which holds the audio data may be connected to the micro which reads it and transfers it to the DSP. Alternatively, data might be downloaded from a PC or internet site and played directly, or written onto blank memory devices. A digital-to-analog (DAC) converter translates the digital audio output of the DSP into an analog form to eventually be played on user headphones. The entire system must be powered from batteries, (for example, two AA batteries).

For this type of product, a key design constraint would be power. Customers do not like replacing the batteries in their portable devices. Thus, battery life, which is directly related to system power consumption, is a key consideration. By not having any moving parts, a solid state audio player uses less power than previous generation players (such as tapes and CDs). Since this is a portable product, size and weight are obviously also key concerns. Solid state devices, such as the one described here, are also size efficient because they contain fewer parts in the overall system.

To the system designer, programmability is a key concern. With a programmable DSP solution, this portable audio player can be updated with the newest decompression, encryption, and audio processing algorithms instantly from the world wide web, or from memory devices. A low power DSP-based system solution like the one described here could have system power consumption as low as 200mW. This will allow the portable audio player to have three times the battery life of a CD player on the same two AA battery supply.

High performance DSP applications

At the high end of the performance spectrum, DSPs utilize advanced architectures to perform signal processing at high rates. Advanced architectures such as Very Long Instruction Word (VLIW) use extensive parallelism and pipelining to achieve high performance. These advanced architectures take advantage of other technologies, such as optimizing compilers, to achieve this performance. There is a growing need for high performance computing. Applications include:

- DSL modems
- Base station transceivers
- Wireless LAN
- Multimedia gateways

- Professional audio
- Networked cameras
- Security identification
- Industrial scanners
- High speed printers
- Advanced encryption

Conclusion

Though analog signals can also be processed using analog hardware (i.e., electrical circuits containing active and passive elements), there are several advantages to digital signal processing:

- Analog hardware is usually limited to linear operations; digital hardware can implement nonlinear operations.
- Digital hardware is programmable, which allows for easy modification of the signal processing procedure in both real-time and non real-time modes of operation.
- Digital hardware is less sensitive than analog hardware to variations such as temperature.

These advantages lead to lower cost, which is the main reason for the ongoing shift from analog to digital processing in wireless telephones, consumer electronics, industrial controllers, and numerous other applications.

The discipline of signal processing, whether analog or digital, consists of a large number of specific techniques. These can be roughly categorized into two families:

- Signal-analysis/feature-extraction techniques which are used to extract useful information from a signal. Examples include speech recognition, location, and identification of targets from radar signals, and detection and characterization of changes in meteorological or seismographic data.
- Signal filtering/shaping techniques which are used to improve the quality of a signal. Sometimes this is done as an initial step before analysis or feature extraction. Examples of these techniques include the removal of noise and interference using filtering algorithms, separating a signal into simpler components, and other time- and frequency-domain averaging.

A complete signal processing system usually consists of many components and incorporates multiple signal processing techniques.

CHAPTER 2

Overview of Real-time and Embedded Systems

Robert Oshana

Chapter Outline

Real-time systems

A real-time system is a system that is required to react to stimuli from the environment (including the passage of physical time) within time intervals dictated by the environment. The Oxford dictionary defines a real-time system as:

> *Any system in which the time at which output is produced is significant.*

This is usually because the input corresponds to some movement in the physical world, and the output has to relate to that same movement. The lag from input time to output time must be sufficiently small for acceptable timeliness. Another way of thinking of real-time systems is that they can consist of any information processing activity or system which has to respond

DSP for Embedded and Real-Time Systems. DOI: 10.1016/B978-0-12-386535-9.00002-0

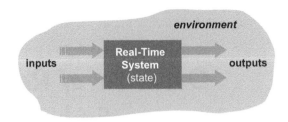

Figure 2-1:
A real-time system reacts to inputs from the environment and produces outputs that affect the environment.

to externally generated input stimuli within a finite and specified period. Generally, real-time systems are systems that maintain a continuous and timely interaction with their environment (Figure 2-1).

Soft and hard real-time systems

Correctness of a computation depends not only upon its results but also upon the time at which its outputs are generated. A real-time system must satisfy response-time constraints or suffer significant system consequences. If the consequences consist of a degradation of performance, but not outright failure, the system is referred to as a soft real-time system. If the consequences are system failure, then the system is referred to as a hard real-time system (e.g., an anti-lock braking system in an automobile).

A system function (hardware, software, or a combination of both) is considered hard real-time if, and only if, it has a hard deadline for the completion of an action or task. This deadline must always be met, otherwise the task has failed. The system may have one or more hard real-time tasks, as well as other non-real-time tasks. This is acceptable, as long as the system can properly schedule these tasks in such a way that the hard real-time tasks always meet their deadlines. Hard real-time systems are also commonly embedded systems. Embedded systems are specialized systems that are part of a larger system. We will look in more detail at embedded systems later in the chapter.

Differences between real-time and time-shared systems

Real-time systems are different from time-shared systems in three fundamental areas (Table 2-1). These include:

- System capacity
- Responsiveness
- Overload

Real-time systems offer a high degree of schedulability, where timing requirements of the system must be satisfied at high degrees of resource usage.

Characteristic	Time-shared systems	Real-time systems
System capacity	High throughput	Schedulability and the ability of system tasks to meet all deadlines
Responsiveness	Fast average response time	Ensured worst case latency which is the worst-case response time to events
Overload	Fairness to all	Stability; when the system is overloaded important tasks must meet deadlines while others may be starved

They also ensure worst case latency, guaranteeing that the system will still operate under worst case response time to events. Real-time systems also offer stability under transient overload — when the system is overloaded by events and it is impossible to meet all deadlines, the deadlines of selected critical tasks must still be guaranteed.

DSP systems are hard real-time

DSP systems usually qualify as hard real-time systems. As an example, assume that an analog signal is to be processed digitally. The first question to consider is how often to sample or measure the analog signal in order to represent that signal accurately in the digital domain. As we discussed in Chapter 1, the sample rate is the number of samples of an analog event (such as sound) that are taken per second to represent the event in the digital domain. We now know that the signal must be sampled at a rate equal to at least twice the highest frequency that we wish to preserve. If this signal contains important components at 4 kHz, then the sampling frequency would need to be at least 8 kHz. The sampling period would then be:

$T = 1 / 8000 = 125$ microseconds $= 0.000125$ seconds

This tells us that, for this signal being sampled at this rate, we would have 0.000125 seconds to perform *all* the processing necessary before the next sample arrived. When samples are arriving on a continuous basis, and the system cannot fall behind in processing them, and must still produce correct results — this is a *hard real-time system*.

Hard real-time systems

The collective timeliness of the hard real-time tasks is binary — so they will either meet their deadlines (in a correctly functioning system), or they will not (in which case the system is infeasible). In all hard real-time systems, collective timeliness is deterministic. This determinism does not imply that the actual individual task completion times, or the task execution ordering, are necessarily known in advance.

A computing system being hard real-time says nothing about the magnitudes of the deadlines. They may be microseconds or weeks. There is a bit of confusion with regard to the usage of the term 'hard real-time.' Some relate hard real-time to response-time magnitudes below some arbitrary threshold, such as 1 msec. This is not correct, however. Many of these systems actually happen to be soft real-time. These systems would be more accurately termed 'real fast,' or perhaps 'real predictable,' but certainly not hard real-time.

The feasibility and costs (e.g., in terms of system resources) of hard real-time computing depend on how well known, *a priori*, are the relevant future behavioral characteristics of the tasks and execution environment. These task characteristics include:

- Timeliness parameters, such as arrival periods or upper bounds
- Deadlines
- Worst case execution times
- Ready and suspension times
- Resource utilization profiles
- Precedence and exclusion constraints
- Relative importance

There are also important characteristics relating to the execution environment:

- System loading
- Resource interactions
- Queuing disciplines
- Arbitration mechanisms
- Service latencies
- Interrupt priorities and timing
- Caching

Deterministic collective task timeliness in hard (and soft) real-time computing requires that the future characteristics of the relevant tasks and execution environment be deterministic, i.e., known absolutely in advance. The knowledge of these characteristics must then be used to pre-allocate resources so that all deadlines will always be met.

Usually, the future characteristics of the tasks and execution environment must be adjusted to enable scheduling and resource allocation which meets all deadlines. Different algorithms or schedules which meet all deadlines are evaluated with respect to other factors. In many real-time computing applications it is common that the primary factor is maximizing processor utilization.

Allocation for hard real-time computing has been performed using various techniques. Some of these techniques involve conducting an off-line enumerative search for a static schedule which will deterministically always meet all deadlines. Scheduling algorithms include the use of priorities that are assigned to the various system tasks. These priorities can be assigned either off-line by application programmers, or online by the application or operating system software. The task priority assignments may either be static (fixed), as with rate monotonic algorithms, or dynamic (changeable), as with the earliest-deadline-first algorithm.

Real-time event characteristics

Real-time events fall into one of three categories:

- *Asynchronous events* are entirely unpredictable. An example of this is a cell phone call arriving at a cellular base station. As far as the base station is concerned, the action of making a phone call cannot be predicted.
- *Synchronous events* are predictable events and occur with precise regularity. For example, the audio and video in a camcorder take place in a synchronous fashion.
- *Isochronous events* occur with regularity within a given window of time. For example, audio data in a networked multimedia application must appear within a certain window of time when the corresponding video stream arrives. Isochronous is a sub-class of asynchronous.

In many real-time systems, task and future execution environment characteristics are hard to predict. This makes true hard real-time scheduling infeasible. In hard real-time computing, deterministic satisfaction of the collective timeliness criterion is the driving requirement. The necessary approach to meeting that requirement is static (i.e., *a priori*) scheduling of deterministic task and execution environment characteristic cases. The requirement for advance knowledge about each of the system tasks and their future execution environment to enable off-line scheduling and resource allocation significantly restricts the applicability of hard real-time computing.

Efficient execution and the execution environment

Real-time systems are time critical and the efficiency of their implementation is more important than in other systems. Efficiency can be categorized in terms of processor cycles, memory, or power. This constraint may drive everything from the choice of processor to the choice of the programming language. One of the main benefits of using a higher level language is to allow the programmer to abstract away implementation details and concentrate on solving the problem. This is not always true in the embedded system world. Some higher level languages have instructions that are an order of magnitude slower than the assembly language. However, higher level languages can be effectively used in real-time systems using the right techniques. We will be discussing much more about this topic in the chapter on optimizing source code for DSPs.

Resource management

A system operates in real time as long as it completes its time-critical processes with acceptable timeliness. 'Acceptable timeliness' is defined as part of the behavioral or 'non-functional' requirements for the system. These requirements must be objectively quantifiable and therefore measureable (stating that the system must be 'fast,' for example, is not quantifiable). A system is said to be real-time if it contains some model of real-time resource management (these resources must be explicitly managed for the purpose of operating in real time). As mentioned earlier, resource management may be performed statically off-line or dynamically online.

Real-time resource management comes at a cost. The degree to which a system is required to operate in real time cannot necessarily be attained solely by hardware over-capacity (e.g., high processor performance using a faster CPU). There must also be some form of real-time resource management to be cost effective. Systems which must operate in real time consist of both real-time resource management and hardware resource capacity. Systems which have interactions with physical devices require higher degrees of real-time resource management. These computers are referred to as embedded systems which we spoke about earlier. Many of these embedded computers use very little real-time resource management. The resource management that is used is usually static and requires analysis of the system prior to it executing in its environment. In a real-time system, physical time (as opposed to logical time) is necessary for real-time resource management, in order to relate events to the precise moments of occurrence. Physical time is also important for action time constraints as well as measuring costs incurred as processes progress to completion. Physical time can also be used for logging history data.

All real-time systems make trade-offs of scheduling costs versus performance in order to reach an appropriate balance for attaining acceptable timeliness between the real-time portion of the scheduling optimization rules and the off-line scheduling performance evaluation and analysis.

Reactive and embedded real-time systems

There are two types of real-time systems: reactive and embedded. Reactive real-time systems involve constant interaction with the environment (such as a pilot controlling an aircraft). An embedded real-time system is used to control specialized hardware that is installed within a larger system (such as a microprocessor that controls anti-lock brakes in an automobile).

Challenges in real-time system design

Designing real-time systems poses significant challenges to the designer. One of the most significant challenges comes from the fact that real-time systems must interact with the environment. The environment is complex and changing, and these interactions can therefore become very complicated. Many real-time systems don't just interact with one, but many different entities in the environment, each with different characteristics and rates of interaction. A cell phone base station, for example, must be able to handle calls from literally thousands of cell phone subscribers at the same time. Each call may have different requirements for processing. All of this complexity must be managed and coordinated.

Response time

Real-time systems must respond to external interactions in the environment within a predetermined amount of time. These systems must produce the correct result and produce it in a timely way. This implies that response time is as important as producing correct results. Real-time systems must be engineered to meet these response times. Hardware and software must be designed to support response-time requirements for these systems. Optimal partitioning of the system requirements into hardware and software is also important.

Real-time systems must be architected to meet system response-time requirements. Using combinations of hardware and software components, engineering makes architecture decisions, such as inter-connectivity of the system processors, system link speeds, processor speeds, memory size, and I/O bandwidth. Key questions to be answered include:

- Is the architecture suitable? To meet the system response-time requirements, the system can be architected using one powerful processor or several smaller processors. Can the application be partitioned among the several smaller processors without imposing large communication bottlenecks throughout the system? If the designer decides to use one powerful processor, will the system meet its power requirements? Sometimes a simpler architecture may be the better approach — more complexity can lead to unnecessary bottlenecks which cause response-time issues.
- Are the processing elements powerful enough? A processing element with high utilization (greater than 90%) will lead to unpredictable run-time behavior. At this utilization levels lower priority tasks in the system may get starved. As a general rule, real-time systems that are loaded at 90% take approximately twice as long to develop due to the cycles of optimization and integration issues with the system at these utilization rates. At 95% utilization, systems can take three times longer to develop due to these same issues. Using multiple processors will help but the inter-processor communication must be managed.
- Are the communication speeds adequate? Communication and I/O is a common bottleneck in real-time embedded systems. Many response-time problems come not from the processor being overloaded, but in latencies in getting data into and out of the system. In other cases, overloading a communication port (greater than 75%) can cause unnecessary queuing in different system nodes and this causes delays in messages passing throughout the rest of the system.
- Is the right scheduling system available? In real-time systems tasks that are processing real-time events must take higher priority. But how do you schedule multiple tasks that are all processing real-time events? There are several scheduling approaches available and the engineer must design the scheduling algorithm to accommodate the system priorities in

order to meet all real-time deadlines. Because external events may occur at any time, the scheduling system must be able to pre-empt currently running tasks to allow higher priority tasks to run. The scheduling system (or real-time operating system) must not introduce a significant amount of overhead into the real-time system.

Recovering from failures

Real-time systems interact with the environment which is inherently unreliable. Therefore real-time systems must be able to detect and overcome failures in the environment. Also, since real-time systems are also embedded into other systems and may be hard to get at (such as a space craft or satellite) these systems must also be able to detect and overcome internal failures as well (there is no 'reset' button in easy reach of the user!). Also, since events in the environment are unpredictable, it is almost impossible to test for every possible combination and sequence of events in the environment. This is a characteristic of real-time software that makes it somewhat non-deterministic in the sense that it is almost impossible in some real-time systems to predict the multiple paths of execution based on the non-deterministic behavior of the environment.

Examples of internal and external failures that must be detected and managed by real-time systems include:

- Processor failures
- Board failures
- Link failures
- Invalid behavior of the external environment
- Inter-connectivity failure

Distributed and multi-processor architectures

Real-time systems are becoming so complex that applications are executed on multi-processor systems that are distributed across some communication systems. This poses challenges to the designer that relate to the partitioning of the application in a multi-processor system. These systems will involve processing on several different nodes. One node may be a DSP, while another may be a more general purpose processor. Some may even specialize in hardware processing elements. This leads to several design challenges for the engineering team.

Initialization of the system

Initializing a multi-processor system can be very complicated. In most multi-processor systems the software load file resides on the general purpose processing node. Nodes that are directly connected to the general purpose processor, for example a DSP, will initialize first. After these nodes complete loading and initialization, other nodes connected to it may then go through this same process until the system completes initialization.

Processor interfaces

When multiple processors must communicate with each other, care must be taken to ensure that messages sent along interfaces between the processors are well defined and consistent with the processing elements. Differences in message protocol including endianness, byte ordering, and other padding rules can complicate system integration, especially if there is a system requirement for backwards compatibility.

Load distribution

As mentioned earlier, multiple processors lead to the challenge of distributing the application, and possibly developing the application to support an efficient partitioning of the application among the processing elements. Mistakes in partitioning the application can lead to bottlenecks in the system and this degrades the full entitlement of the system by overloading certain processing elements and leaving others under-utilized. Application developers must design the application to be efficiently partitioned across the processing elements.

Centralized resource allocation and management

In a system of multiple processing elements, there is still a common set of resources including peripherals, cross bar switches, and memory that must be managed. In some cases the operating system can provide mechanisms such as semaphores to manage these shared resources. In other cases there may be dedicated hardware to manage the resources. Either way, important shared resources in the system must be managed in order to prevent more system bottlenecks.

Embedded systems

An embedded system is a specialized computer system that is usually integrated as part of a larger system. An embedded system consists of a combination of hardware and software components to form a computational engine that will perform a specific function. Unlike desktop systems which are designed to perform a general function, embedded systems are constrained in their application. Embedded systems often perform in reactive and time-constrained environments as described earlier. A rough partitioning of an embedded system consists of the hardware which provides the performance necessary for the application

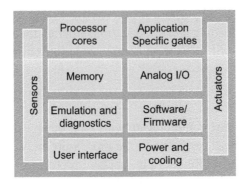

Figure 2-2:
Typical embedded system components.

(and other system properties such as security) and the software which provides the majority of the features and flexibility in the system. A typical embedded system is shown in Figure 2-2.

Some of the features of typical embedded system components include:

- Processor core — at the heart of the embedded system is the processor core(s). This can range from a simple inexpensive 8 bit microcontroller to a more complex 32 or 64 bit microprocessor. The embedded system designer must select the most cost sensitive device for the application that can meet all of the functional and non-functional (timing) requirements.
- Analog I/O — D/A and A/D converters are used to get data from the environment and pass it back out to the environment. The embedded designer must understand the type of data required from the environment, the accuracy requirements for that data, and the input/ output data rates, in order to select the right converters for the application. The external environment drives the reactive nature of the embedded system. Embedded systems have to be at least fast enough to keep up with the environment. This is where analog information such as light or sound pressure, or acceleration, are sensed and input into the embedded system.
- Sensors and Actuator — sensors are used to sense analog information from the environment. Actuators are used to control the environment in some way.
- Embedded systems also have user interfaces — these interfaces may be as simple as a flashing LED or can be sophisticated, such as a cell phone or digital still camera interface.
- Application-specific gates — hardware acceleration such as ASIC or FPGA is used for accelerating specific functions in the application that have high performance requirements. The embedded designer must be able to map or partition the application appropriately using available accelerators to gain maximum application performance.
- Software is a significant part of embedded system development. Over the last several years, the amount of embedded software has grown faster than Moore's law, with the amount

doubling approximately every 10 months. Embedded software is usually optimized in some way (performance, memory, or power). More and more embedded software is written in a high level language like C/C++ with some of the more performance critical pieces of code still written in assembly language.

- Memory is an important part of an embedded system and embedded applications can run out of either RAM or ROM depending on the application. There are many types of volatile and non-volatile memory used for embedded systems and we will talk more about this later in the chapter.

- Emulation and diagnostics — many embedded systems are difficult to see or to get to. There needs to be a way to interface to embedded systems to debug them. Diagnostic ports such as a JTAG (Joint Test Action Group) are used to debug embedded systems. On chip emulation is used to provide visibility into the behavior of the application. These emulation modules provide sophisticated visibility into the run-time behavior and performance, in effect replacing external logic analyzer functions with on-board diagnostic capability.

Embedded systems are reactive systems

A typical embedded system responds to the environment via sensors and controls the environment using actuators (Figure 2-3). This imposes a requirement on embedded systems to achieve performance consistent with that of the environment. This is why embedded systems are referred to as reactive systems. A reactive system must use a combination of hardware and software to respond to events in the environment within defined constraints. Complicating the matter is the fact that these external events can be either periodic and predictable, or aperiodic and hard to predict. When scheduling events for processing in an embedded system, both periodic and aperiodic events must be considered and performance must be guaranteed for worst case rates of execution. This can be a significant challenge.

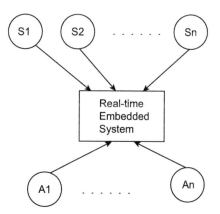

Figure 2-3:
A model of sensors and actuators in embedded systems.

There are several key characteristics of embedded systems:

- Monitoring and reacting to the environment — embedded systems typically get input by reading data from input sensors. There are many different types of sensors that monitor various analog signals in the environment including temperature, sound pressure, and vibration. This data is processed using embedded system algorithms. The results may be displayed in some format to a user or simply used to control actuators (like deploying the airbags and calling the police)
- Controlling the environment — embedded systems may generate and transmit commands that control actuators such as airbags, motors, etc.
- Processing of information — embedded systems process the data collected from the sensors in some meaningful way, such as data compression/decompression, side impact detection, etc.
- Application specifics — embedded systems are often designed for applications such as airbag deployment, digital still cameras, or cell phones. Embedded systems may also be designed for processing control laws, finite state machines, and signal processing algorithms. Embedded systems must also be able to detect and react appropriately to faults in the internal computing environment as well as the surrounding systems.

Summary

Many of the items that we interface with or use on a daily basis contain an embedded system. An embedded system is a system that is 'hidden' inside the item we interface with. Systems such as cell phones, answering machines, microwave ovens, VCRs, DVD players, video game consoles, digital cameras, music synthesizers, and cars all contain embedded processors. A late model automobile can contain up to 80 embedded microprocessors. These embedded processors keep us safe and comfortable by controlling such tasks as anti-lock braking, climate control, engine control, audio system control, and airbag deployment.

Embedded systems have the added burden of reacting quickly and efficiently to the external 'analog' environment. That may include responding to the push of a button, or a sensor, to trigger an air bag during a collision, or the arrival of a phone call on a cell phone. Simply put, embedded systems have deadlines which can be hard or soft. Given the 'hidden' nature of embedded systems, they must also react to, and handle, unusual conditions without the intervention of a human.

DSPs are useful in embedded systems principally for signal processing. Its ability to perform complex signal processing functions in real time gives DSP a key advantage over other forms of embedded processing. DSPs must respond in real time to analog signals from the environment, convert them to digital form, perform value added processing to those digital signals and, if required, convert the processed signals back to analog form to send back out to the environment.

We still need to discuss the special architectures and techniques that allow DSP to perform these real-time embedded tasks so quickly and efficiently. These topics are discussed in the coming chapters.

Programming embedded systems requires an entirely different approach from that used in desktop or mainframe programming. Embedded systems must be able to respond to external events in a very predictable and reliable way. Real-time programs must not only execute correctly, but they must also execute on time. A late answer is a wrong answer. Because of this requirement, we will be looking at issues such as concurrency, mutual exclusivity, interrupts, hardware control, multi-tasking, and processing later in the book.

CHAPTER 3

Overview of Embedded Systems Development Lifecycle Using DSP

Robert Oshana

Chapter Outline

Embedded systems

As we have seen in Chapter 2, an embedded system is a specialized computer system that is integrated as part of a larger system. Many embedded systems are implemented using digital signal processors (DSPs). The DSP will interface with the other embedded components to perform a specific function. The specific embedded application will determine the corresponding DSP to be used. For example, if the embedded application is one that performs video processing, the system designer may choose a DSP that is customized to perform media processing including video and audio processing. This device contains dual channel video

DSP for Embedded and Real-Time Systems. DOI: 10.1016/B978-0-12-386535-9.00003-2

ports that are software configurable for input or output, as well as video filtering and automatic horizontal scaling and support of various digital TV formats such as HDTV, multi-channel audio serial ports, multiple stereo lines, and an ethernet peripheral to connect to IP packet networks. It is obvious that the choice of DSP system depends on the embedded application.

In this chapter we will discuss the basic steps involved in developing an embedded application using DSP.

The embedded system lifecycle using DSP

In this section we will give an overview of the general embedded system lifecycle using DSP. There are many steps involved in developing an embedded system — some are similar to other system development activities and some are unique. We will step through the basic process of embedded system development focusing on DSP applications.

Step 1 — Examine the overall needs of the system

Comparing and choosing a design solution is a difficult process. Often the choice comes down to emotion or attachment to a particular vendor or processor, or to inertia based on prior choosing and comfort level. The embedded designer must take a positive logical approach to comparing solutions, based on well-defined selection criteria. For DSP, there is a set of specific selection criteria that must be discussed. Many signal processing applications will require a mix of several system components in the overall design solution including:

- Human interface
- Signal processing chain
- I/O interface
- Control code
- Glue logic

What is a DSP solution?

A typical DSP product design uses the is digital signal processor itself, analog/mixed signal functions, memory, and software; all is designed with a deep understanding of overall system function. In the product, the analog signals of the real world, signals representing anything from temperature to sound or images, are translated into digital bits — zeros and ones — by an analog/mixed signal device. Then the digital bits or signals are processed by the DSP. Digital signal processing is much faster and more precise than traditional analog processing. This type of processing speed is needed for today's advanced communications devices where information requires instantaneous processing, and in many portable applications that are connected to the internet.

There are many selection criteria for embedded DSP systems. These include, but are not limited to:

- Price
  - System costs
  - Tools
- Time to market
  - Ease of use
  - Existing algorithms
  - Reference designs
  - RTOS and debug tools
- Performance
  - Sampling frequency
  - Number of channels
  - Signal processing requirements
  - System integration
- Power
  - System power
  - Power analysis tools

These are the major selection criteria defined by Berkeley Design Technology Incorporated (bdti.com). Other selection criteria may be 'ease of use' which is closely linked to 'time to market' and also 'features.' Some of the basic rules to consider in this phase are:

- For a fixed cost, maximize performance
- For a fixed performance, minimize cost

Step 2 — Select the hardware components required for the system

In many systems, a general purpose processor (GPP), field programmable gate array (FPGA), microcontroller (uC), or DSP is not used as a single-point solution. This is because designers often combine solutions, maximizing the strengths of each device (Figure 3-1).

One of the first decisions that designers often make when choosing a processor is whether they would like a software programmable processor in which functional blocks are developed in software using C or assembly, or a hardware processor in which functional blocks are laid out logically in gates. Both FPGAs and application specific integrated circuits (ASICs) may integrate a processor core (very common in ASICs).

Hardware gates

Hardware gates are logical blocks laid out in a flow; therefore any degree of parallelization of instructions is theoretically possible. Logical blocks have very low latency, and therefore

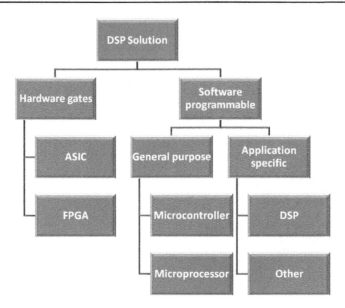

Figure 3-1:
Many applications, multiple solutions.

FPGAs are more efficient for building peripherals than 'bit-banging' using a software device.

If a designer chooses to design in hardware, he or she may design using either FPGA or ASIC. FPGAs are termed 'field programmable' because their logical architecture is stored in a non-volatile memory and booted into the device. Thus FPGAs may be reprogrammed in the field simply by modifying the non-volatile memory (usually FLASH or EEPROM). ASICs are not field programmable. They are programmed at the factory using a mask which cannot be changed. ASICs are often less expensive and/or lower power. They often have sizable non-recurring engineering (NRE) costs.

Software programmable

In this model, instructions are executed from memory in a serial fashion (i.e. one per cycle). Software programmable solutions have limited parallelization of instructions. However, some devices can execute multiple instructions in parallel in a single cycle. Because instructions are executed from memory in the CPU, device functions can be changed without having to reset the device. Also, because instructions are executed from memory, many different functions or routines may be integrated into a program without the need to lay out each individual routine in gates. This may make a software programmable device more cost effective for implementing very complex programs with a large number of subroutines.

If a designer chooses to design in software, there are many types of processors available to choose from. There are a number of general purpose processors but, in addition, there are processors that have been optimized for specific applications. Examples of such application-specific processors are graphics processors, network processors, and digital signal processors (DSPs). Application-specific processors usually offer higher performance for a target application, but are less flexible than general purpose processors.

General purpose processors

Within the category of general purpose processors (GPPs) are microcontrollers (uC) and microprocessors (uP) (see Figure 3-2).

Microcontrollers usually have control-oriented peripherals. They are usually lower cost and lower performance than microprocessors. Microprocessors usually have communications-oriented peripherals. They are usually higher cost and higher performance than microcontrollers.

Note that some GPPs have integrated MAC units. It is not particularly a strength of GPPs to have this capability, because all DSPs have MACs, but it is worth noting because a student might mention it. Regarding performance of the GPP's MAC, it is different for each one.

Microcontrollers

A microcontroller is a highly integrated chip that contains many or all of the components comprising a controller. This includes a CPU, RAM and ROM, I/O ports, and timers. Many

Key Strengths	• Familiar design environment (tools, SW, emulator) • Robust communication peripherals • Ability to use operating systems for control code • Great for compiling generic (non-tuned code)
Signal processing effectiveness	Fair to good
Key application focus	PC and some hand held

Figure 3-2:
General purpose processor solutions.

Key Strengths	• Good control peripherals
	• May have ability to use mid range operating systems
	• Very low cost
	• Integrated FLASH
	• Low power
Signal processing effectiveness	Poor to fair
Key application focus	Embedded control, home appliances

Figure 3-3:
Microcontroller solutions.

general-purpose computers are designed the same way. But a microcontroller is usually designed for very specific tasks in embedded systems. As the name implies, the specific task is to control a particular system, hence the name microcontroller. Because of this customized task, the device parts can be simplified, which makes these devices very cost effective solutions for these types of applications (see Figure 3-3).

Some microcontrollers can actually do a multiply and accumulate (MAC) in a single cycle. But that does not necessarily make it DSP-centric. True DSP can allow two 16×16 MACs in a single cycle, including bringing the data in over the busses. It is this that truly makes the application a DSP. So, devices with hardware MACs might get a 'fair' rating. Others get a 'poor' rating. In general, microcontrollers can do DSP but they will generally do it slower.

FPGA solutions

An FPGA is an array of logic gates that are hardware-programmed to perform a user-specified task. FPGAs are arrays of programmable logic cells interconnected by a matrix of wires and programmable switches. Each cell in an FPGA performs a simple logic function. These logic functions are defined by an engineer's program. FPGAs contain large numbers of these cells (1000–100,000) available to use as building blocks in DSP applications. The advantage of using FPGAs is that the engineer can create special purpose functional units that can perform limited tasks very efficiently. FPGAs can be reconfigured dynamically as well (usually 100–1,000 times per second, depending on the device). This makes it possible to optimize FPGAs for complex tasks at speeds higher than that which can be achieved using a general

Key Strengths	• Very fast computation • Very good design support tools • PLD usually required in design • Ability to synthesize many peripherals • Easy to develop with • Flexible features, field programmable
Signal processing effectiveness	Excellent for high speed and parallel signal processing
Key application focus	Glue logic, radar, sensor arrays

Figure 3-4:
FPGA solutions for DSP.

purpose processor. The ability to manipulate logic at the gate level means it is possible to construct custom DSP-centric processors that efficiently implement the desired DSP function. This is possible by simultaneously performing all of the algorithm's subfunctions. This is where the FPGA can achieve performance gains over a programmable DSP processor.

The DSP designer must understand the trade-offs when using FPGA (see Figure 3-4). If the application can be done in a single programmable DSP, that is usually the best way to go, since talent for programming DSP is usually easier to find than FPGA designers. Also, software design tools are common, cheap, and sophisticated, which improves development time and cost. Most of the common DSP algorithms are also available in well-packaged software components. It is harder to find these same algorithms implemented and available for FPGA designs.

FPGA is worth considering, however, if the desired performance cannot be achieved using one or two DSPs, or when there may be significant power concerns (although DSP is also a power efficient device − benchmarking needs to be performed), or when there may be significant programmatic issues when developing and integrating a complex software system.

Typical applications for FPGA include radar/sensor arrays, physical system and noise modeling and any high I/O and high bandwidth application.

Digital signal processors

A DSP is a specialized microprocessor used to efficiently perform calculations on digitized signals that are converted from the analog domain. One of the big advantages of DSP is the programmability of the processor, which allows important system parameters to be easily changed to accommodate the application. DSPs are optimized for digital signal manipulations.

DSPs provide ultra-fast instruction sequences, such as shift and add, and multiply and add. These instruction sequences are common in many math-intensive signal processing applications. DSPs are used in devices where this type of signal processing is important, such as sound cards, modems, cell phones, high-capacity hard disks, and digital TVs (Figure 3-5).

In reality, the choice is not that clear, as DSP functions, for example, are making their way onto microcontrollers and vice versa. The automotive segment is a good example of this. High performance power train applications for six cylinder and higher require significant signal processing. Microcontrollers are used for this purpose (see Figure 3-6).

The MPC5554 in this example uses the zen 6 processor core with 32K of cache and memory management unit. It includes 2MB of Flash and 111K SRAM distributed as Data RAM, eTPU coprocessor RAM, and Cache. This microcontroller has general communications peripherals common to automotive and industrial applications such as CAN, SPI, Serial, and LIN. The MPC5554 has dual analog-to-digital converters supporting 40 channels.

This microcontroller is highly flexible and can be programmed to perform complex scatter/gather operations. But the key feature relevant to this discussion is the signal processing engine (SPE) located inside the core.

The SPE is a SIMD architecture processing engine — Single Instruction Multiple Data (see Figure 3-7). This allows multiple data elements to be acted upon by a single common operation.

Key Strengths	• Architecture optimized for computing DSP algorithms
	• Excellent MIP/mW/$$ trade-off
	• Efficient compilers, can program entire app in C
	• RTOS for task scheduling
	• Low power
Signal processing effectiveness	Good to excellent
Key application focus	Cell phone, telecom infrastructure, multimedia

Figure 3-5:
DSP processor solutions.

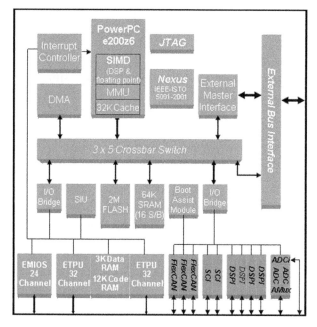

Figure 3-6:
The MPC5554 is used in the automotive industry specifically addressing high performance power train applications 6 cylinder and above.

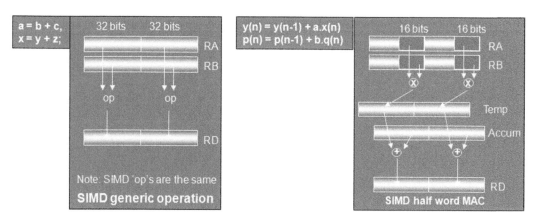

Figure 3-7:
Architecture of the signal processing engine (SPE).

This improves performance on algorithms that have tight inner loops and operate on large sets of data. Signal processing algorithms are examples of this type of algorithm.

The SPE utilizes the existing components of the core to bring improved performance with minimum additional complexity of the core and easier integration with existing tools and software. From a user perspective the most visible aspect of the core with SPE APU is that the registers are shared with the basic core and extended to 64 bits.

The pipeline for instruction and data fetch are also extended to 64 bits for optimal transfer of data through the SIMD. This is essentially an arithmetic processing unit (APU) that provides signal processing capabilities aimed specifically at DSP operations, such as filters and FFTs.

SPE engines like this provide significant DSP capability, such as SIMD functionality, full set of arithmetic, logical and floating point operations, multiple Multiply and Multiply-Accumulate Instructions, 16 and 32 bit loads and stores and data movement instructions, FFT address calculation through bit-reverse increment instruction, 64-bit accumulator, etc.

SPE can be used to accelerate DSP algorithms. Let's take a look at an FIR filter. Figure 3-8 shows a basic FIR filter. In Figure 3-9 we take a look at what a similar code set written for the SPE might look like.

First there are a set of creation intrinsics — these intrinsics create new generic 64-bit opaque data types from the given inputs passed by value. More specifically, they are created from the following inputs:

- 1 signed or unsigned 64-bit integer
- 2 single-precision floats
- 2 signed or unsigned 32-bit integers
- 4 signed or unsigned 16-bit integers

The intrinsic 'evmwlumiaaw' stands for Vector, Multiply, Word, Low, Unsigned, Modulo, Integer, and Accumulate in Words.

For each word element in the accumulator, the corresponding word unsigned integer elements in **rA** and **rB** are multiplied. The least significant 32 bits of each product are added to the

```
void fir(uint32_t *y, unit32_t n, unit32_t *x)
{
    unit32_t i, j;
    unit32_t tap[16] = {3, 5, 6, 22, 2, 3, 3, 6,
                        15, 1, 23, 261, 31451, 7345, 123445, 15};
    for(i = 0; i < n; i++) {
        y[i] = 0;
        for (j = 0; j < 16; j++) {
            y[i] += tap[j] * x[j+i];
        }
    }
}
```

Figure 3-8:
Basic FIR filter.

```
void fir(uint32_t *y, uint32_t n, uint32_t *x)
{
    uint32_t i, j;
    __ev64_opaque__ y0, t;
    __ev64_u32__ tap[16] = {3, 3, 5, 5, 6, 6, 22, 22, 2, 2, 3, 3, 3, 3,
                            6, 6, 15, 15, 1, 1, 23, 23, 261, 261,
                            31451, 31451, 7345, 7345, 123445, 123445, 15, 15};
    for(i = 0; i < n ; i+=2 )
    {
        // clear accumulator
        __ev_set_acc_u64( 0ULL );
        for(j = 0; j < 16; j++)
        {
            t  = __ev_create_u32(x[i+j], x[i+j+1]);
            y0 = __ev_mwlumiaaw(tap[j], t);
        }
        y[i]   = __ev_get_upper_u32(y0);
        y[i+1] = __ev_get_lower_u32(y0);
    }
}
```

Figure 3-9:
FIR filter using SPE intriniscs.

contents of the corresponding accumulator word and the result is placed into **rD** and the accumulator.

The 'Get_Upper/Lower' intrinsics specify whether the upper or lower 32 bits of the 64-bit opaque data type are returned. Only signed/unsigned 32-bit integers or single-precision floats are returned.

To initialize the accumulator, the 'evmra' instruction is used.

DSP acceleration units like SPE are used to help cost reduce applications from having to use multiple devices. Figure 3-10 shows how a custom ASIC can be removed from this engine

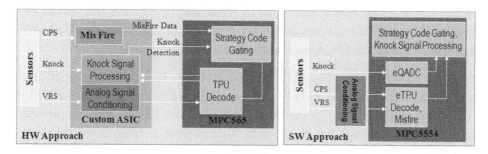

Figure 3-10:
SPE are DSP capabilities inside the microcontroller that can be used to reduce the overall number of components in a DSP system.

control application by using the SPE inside the microcontroller, thus cost reducing the device. In summary, although we can assess DSP capability by discrete processor type, such as micro, DSP, ASIC, etc., many times the solution is a hybrid of several of these processing elements.

A general signal processing solution

The solution shown in Figure 3-10 allows each device to perform the tasks that it is best at, achieving a more efficient system in terms of cost/power/performance. For example, in Figure 3-10, the system designer may put the system control software as follows: state machines and other communication software on the general purpose processor or microcontroller, the high performance, single dedicated fixed functions on the FPGA, and the high I/O signal processing functions on the DSP.

When planning the embedded product development cycle there are multiple opportunities to reduce cost and/or increase functionality using combinations of GPP/uC, FPGA, and DSP. This becomes more of an issue in higher-end DSP applications. These are applications which are computationally intensive and performance critical. These applications require more processing power and channel density than can be provided by GPPs alone. For these high-end applications there are software/hardware alternatives which the system designer must consider. Each alternative provides different degrees of performance benefits and must also be weighed against other important system parameters including cost, power consumption, and time to market.

The system designer may decide to use an FPGA in a DSP system for the following reasons:

- decision to extend the life of a generic, lower-cost microprocessor or DSP by offloading computationally intensive work to an FPGA
- decision to reduce or eliminate the need for a higher-cost, higher performance DSP processor
- increased computational throughput

If the throughput of an existing system must increase to handle higher resolutions, or larger signal bandwidths, an FPGA may be an option. If the required performance increases are computational in nature, an FPGA may again be an option.

Since the computational core of many DSP algorithms can be defined using a small amount of C code, the system designer can quickly prototype new algorithmic approaches on FPGAs before committing to hardware or other production solutions like an ASIC implementing 'glue' logic; various processor peripherals and other random or 'glue' logic are often consolidated into a single FPGA. This can lead to reduced system size, complexity, and cost.

By combining the capabilities of FPGAs and DSP processors, the system designer can increase the scope of the system design solution. A combination of fixed hardware and programmable processors is a good model for enabling flexibility, programmability, and computational acceleration of hardware for the system. An example of this is shown in Figure 3-11. This multicore DSP device has six DSP cores and fixed hardware gates (accelerators) to perform specialized functions. The MAPLE accelerator performs signal processing functions like FFT and Viterbi decoding. The QUICC accelerator performs network protocol functions. The SEC accelerator performs security protocol processing.

DSP acceleration decisions

In DSP system design, there are several things to consider when determining whether a functional component should be implemented in hardware or software:

- Computational complexity; the system designer must analyze the algorithm to determine if it maps well onto the DSP architecture and programming model, or whether it maps better onto a hardware model to exploit certain forms of parallelism that cannot be found in a von Neumann or Harvard architecture of a programmable DSP.
- Signal processing algorithm parallelism; modern processor architectures have various forms of instruction level parallelism (ILP). Modern DSPs have very long instruction word (VLIW) architecture. These DSPs exploit ILP by grouping multiple instructions (adds, multiplies, loads, and stores) for execution in a single processor cycle. For DSP algorithms that map well to this type of instruction parallelism, significant performance gains can be realized. But not all signal processing algorithms exploit such forms of parallelism. Filtering algorithms such as finite impulse response (FIR) algorithms are recursive and are sub-optimal when mapped to programmable DSPs. Data recursion prevents effective parallelism and ILP. As an alternative, the system designer can build dedicated hardware engines in an FPGA.
- Computational complexity; depending on the computational complexity of the algorithms, these may run more efficiently on an FPGA instead of a DSP. It may make sense, for certain algorithmic functions, to implement in a FPGA and free up programmable DSP cycles for other algorithms. Some FPGAs have multiple clock domains built into the fabric which can be used to separate different signal processing hardware blocks into separate clock speeds based on their computational requirements. FPGAs can also provide flexibility by exploiting data and algorithm parallelism using multiple instantiations of hardware engines in the device.
- Data locality; the ability to access memory in a particular order and granularity is important. Data access takes time (clock cycles) due to architectural latency, bus contention, data alignment, direct memory access (DMA) transfer rates, and even the type of memory being used in the system. For example, static RAM (SRAM), which is very fast, but much more expensive than dynamic RAM (DRAM), is often used as cache memory due to its speed. Synchronous DRAM (SDRAM), on the other hand, is directly

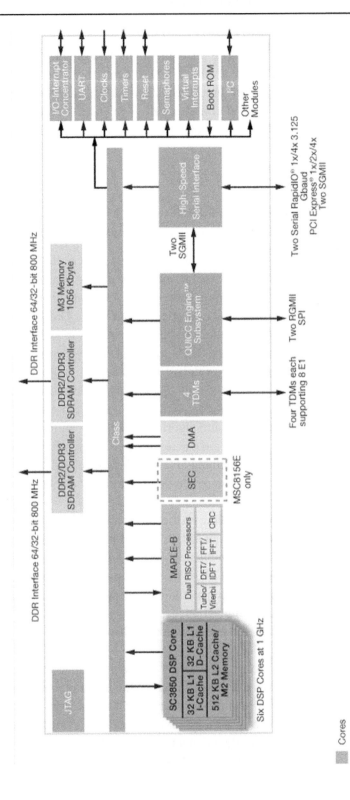

Figure 3-11:
General signal processing solution.

dependent on the clock speed of the entire system (that's why they call it synchronous). It basically works at the same speed as the system bus. The overall performance of the system is driven in part by which type of memory is being used. The physical interfaces between the data unit and the arithmetic unit are the primary drivers of the data locality issue.

- Data parallelism; many signal processing algorithms operate on data that is highly capable of parallelism, such as many common filtering algorithms. Some of the more advanced high-performance DSPs have single instruction multiple data (SIMD) capability in the architectures and/or compilers that implement various forms of vector processing operations. FPGA devices are also good at this type of parallelism. Large amounts of RAM are used to support high bandwidth requirements. Depending on the DSP processor being used, an FPGA can be used to provide this SIMD processing capability for certain algorithms that have these characteristics.

A DSP-based embedded system could incorporate one, two or all three of these devices depending on various factors:

- Number of signal processing tasks/challenges
- Sampling rate
- Memory/peripherals needed
- Power requirements
- Availability of desired algorithms
- Amount of control code
- Development environment
- Operating system
- Debug capabilities
- Form factor
- System cost

The trend in embedded DSP development is moving more toward programmable solutions. There will always be a trade-off, depending on the application.

'Cost' can mean different things to different people. Sometimes, the solution is to go with the lowest 'device cost.' However, if the development team then spends large amounts of time re-doing work, the project may be delayed and the 'time to market' window may extend, which, in the long run, costs more than the savings of the low cost device.

The first point to make is that a purely 100% software or hardware solution is usually the most expensive option. A combination of the two is the best. In the past, more functions were done in hardware and less in software. Hardware was faster, cheaper (ASICs) and good C compilers for embedded processors just weren't available. However, today, with better compilers, and faster and lower-cost processors available, the trend is toward more of a software programmable solution. A software-only solution is not (and most likely never will be) the best overall option. Some hardware will still be required. For example, let's say you have 10 functions to perform and two of them require extreme speed. Do you purchase a very fast processor (which costs 3—4 times the speed you need for the other 8 functions) or do you spend 1 times the price on a lower-speed processor and purchase an ASIC or FPGA to do only those two critical functions? It's probably best to choose the combination.

Cost can be defined by a combination of the following:

- Device cost
- NRE
- Manufacturing cost
- Opportunity cost
- Power dissipation
- Time to market
- Weight
- Size

A combination of software and hardware always gives the lowest cost system design.

Step 3 — Understand DSP basics and architecture

One compelling reason to choose a DSP processor for an embedded system application is performance. Three important questions to understand when deciding on a DSP are:

- What makes a DSP a DSP?
- How fast can it go?
- How can I achieve maximum performance without writing in assembly?

In this section we will begin to answer these questions. So what does make a DSP a DSP? A DSP is really just an application-specific microprocessor. They are designed to do a certain thing, signal processing, very efficiently. For example, consider the signal processor algorithms in Figure 3-12.

Notice the common structure of each of the algorithms:

- They all accumulate a number of computations
- They all sum over a number of elements
- They all perform a series of multiplies and adds

These algorithms all share some common characteristics; they perform multiplies and adds over and over again. This is generally referred to as the sum of products (SOP).

DSP designers have developed hardware architectures that allow the efficient execution of algorithms to take advantage of this algorithmic specialty in signal processing. For example,

Algorithm	Equation
Finite Impulse Response Filter	$y(n) = \displaystyle\sum_{k=0}^{M} a_k\, x(n-k)$
Infinite Impulse Response Filter	$y(n) = \displaystyle\sum_{k=0}^{M} a_k\, x(n-k) + \sum_{k=1}^{N} b_k\, y(n-k)$
Convolution	$y(n) = \displaystyle\sum_{k=0}^{N} x(k) h(n-k)$
Discrete Fourier Transform	$X(k) = \displaystyle\sum_{n=0}^{N-1} x(n)\, \exp[-j(2\pi/N)nk]$
Discrete Cosine Transform	$F(u) = \displaystyle\sum_{x=0}^{N-1} c(u).f(x).\cos\!\left[\dfrac{\pi}{2N}u(2x+1)\right]$

Figure 3-12:
Typical DSP algorithms.

$$Y = \sum_{i=1}^{count} a_i * x_i$$

```
for (i = 1; i < count; i++) {
    sum += a[i] * x[i] }
```

Figure 3-13:
Architectural features of DSPs accommodate the algorithmic structure.

some of the specific architectural features of DSPs accommodate the algorithmic structure described in Figure 3-13.

As an example, consider the FIR diagram in Figure 3-14. This is a DSP algorithm which clearly shows the multiply/accumulate and demonstrates the need for doing MACs very fast, along with reading at least two data values. As shown, the filter algorithm can be implemented using a few lines of C source code. The signal flow diagram in Figure 3-8 shows this

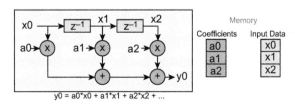

Figure 3-14:
Signal flow graph for FIR filter.

algorithm in a more visual context. Signal flow diagrams are used to show overall logic flow, signal dependencies, and code structure. They make a nice addition to code documentation.

To execute at top speed, a DSP needs to:

- Read *at least* two values from memory
- Multiply coeff * data
- Accumulate (+) answer (an * xn) to running total…
- …and do all of the above in a <u>single cycle</u> (or less)

DSP architectures support the requirements above via:

- High-speed memory architectures supporting multiple accesses/cycle
- Multiple read busses allowing 2 (or more) data reads/cycle from memory
- The processor pipeline overlaying CPU operations allowing 1-cycle execution

All of these things work together to result in the highest possible performance when executing DSP algorithms.

Other DSP architectural features are summarized below:

- Circular buffers; automatically wrap pointer at end of the data or coefficient buffer
- Repeat single, and repeat block; execute next instruction or block of code with zero loop overhead
- Numerical issues; handles fixed or floating point math issues in hardware (e.g., saturation, rounding, overflow)
- Unique addressing modes; address pointers have their own ALU which is used to auto- increment and auto-decrement pointers, and create offsets with no cycle penalty
- Instruction parallelism; executes up to eight instructions in a single cycle.

Models of DSP processing

There are two types of DSP processing models, single sample model and block processing model. In a single sample model of signal processing (Figure 3-15a) the output must result before the next input sample. The goal is minimum latency (in-to-out time). These systems tend to be interrupt intensive, as interrupts drive the processing for the next sample. Example DSP applications include motor control and noise cancellation.

In the block processing model (Figure 3-15b) the system will output a buffer of results before the next input buffer fills. DSP systems like this use the DMA to transfer samples to the buffer. There is increased latency in this approach as the buffers are filled before processing. However, these systems tend to be computationally efficient. The main types of DSP applications that use block processing include cellular telephony, video, and telecom infrastructure.

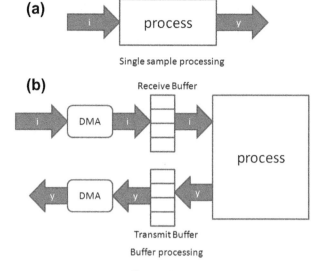

Figure 3-15:
(a) Single sample and (b) block processing models of DSP.

An example of stream processing is averaging data sample. A DSP system that must average the last three digital samples of a signal together, and output a signal at the same rate as that which is being sampled, must do the following:

- Input a new sample and store it
- Average the new sample with the last two samples
- Output the result

These three steps must complete before the next sample is taken. This is an example of stream processing. The signal must be processed in real time. A system that is sampling at 1000 samples per second has one-thousandth of a second to complete the operation, in order to maintain real-time performance.

Block processing, on the other hand, accumulates a large number of samples at a time and processes those samples while the next buffer of samples is being collected. Algorithms such as the fast fourier transform (FFT) operate in this mode.

Block processing (processing a block of data in a tight inner loop) can have a number of advantages in DSP systems:

- If the DSP has an instruction cache, this cache will optimize instructions to run faster the second (or next) time through the loop.
- If the data access adheres to a locality of reference (which is quite common in DSP systems) the performance will improve. Processing the data in stages means the data in any

given stage will be accessed from fewer areas, and is therefore less likely to thrash the data caches in the device.

- Block processing can often be done in simple loops. These loops have stages where only one kind of processing is taking place. In this manner there will be less thrashing from registers to memory and back. Most, if not all, of the intermediate results can be kept in registers or in a level one cache.
- By arranging data access to be sequential, even data from the slowest level of memory (DRAM) will be much faster because the various types of DRAM assume sequential access.

DSP designers will use one of these two methods in their system. Typically, control algorithms will use single-sample processing because they cannot delay the output very long, such as in the case of block processing. In audio/video systems, block processing is typically used — because there can be some delay tolerated from input to output.

Input/output options

DSPs are used in many different systems including motor control applications, performance-oriented applications, and power-sensitive applications. The choice of a DSP processor is dependent on not just the CPU speed or architecture, but also on the mix of peripherals or I/O devices used to get data in and out of the system. After all, much of the bottleneck in DSP applications is not in the compute engine, but in getting data in and out of the system. Therefore, the correct choice of peripherals is important in selecting the device for the application. Example I/O devices for DSP include:

- GPIO; A flexible parallel interface that allows a variety of custom connections.
- UART; universal asynchronous receiver-transmitter. This is a component that converts parallel data to serial data for transmission and also converts received serial data to parallel data for digital processing.
- CAN; controller area network. The CAN protocol is an international standard used in many automotive applications.
- SPI; serial peripheral interface. A 3-wire serial interface developed by Motorola.
- USB; universal serial bus. This is a standard port that enables the designer to connect external devices (digital cameras, scanners, music players, etc.) to computers. The USB standard supports data transfer rates of 12Mbps (million bits per second).
- HPI; host port interface. This is used to download data from a host processor into the DSP.

Calculating DSP performance

Before choosing a DSP processor for a specific application, the system designer must evaluate three key system parameters:

1. Maximum CPU performance; what is the maximum number of times the CPU can execute your algorithm (maximum number of channels)?
2. Maximum I/O performance; can the I/O keep up with the maximum number of channels?
3. Available high speed memory; is there enough high speed internal memory?

With this knowledge, the system designer can scale the numbers to meet the application's needs and then determine:

- CPU load (% of maximum CPU)
- At this CPU load, what other functions can be performed.

The DSP system designer can use this process for any CPU they are evaluating. The goal is to find the 'weakest link' in terms of performance, so that you know what the system constraints are. The CPU might be able to process numbers at sufficient rates, but if the CPU cannot be fed with data fast enough, then having a fast CPU doesn't really matter. The goal is to determine the maximum number of channels that can be processed, given a specific algorithm, and then work that number down based on other constraints (maximum input/output speed, and available memory).

As an example, consider the system shown in Figure 3-16. The goal is to determine the maximum number of channels that this specific DSP processor can handle, given a specific algorithm. To do this, we must first determine the benchmark of the chosen algorithm (in this case, a 200-tap FIR filter). The relevant documentation for an algorithm like this (from a library of DSP functions) gives us the benchmark with two variables: nx (size of buffer) and nh (# coeffs) — these are used for the first part of the computation.

This FIR routine takes about 106K cycles per frame. Now, consider the sampling frequency. A key question to answer at this point is, 'How many times is a frame FULL per second?' To answer this, divide the sampling frequency (which specifies how often a new data item is sampled) by the size of the buffer. Performing this calculation determines that we fill about 47 frames per second.

CPU		
	FIR benchmark	$(nx/2)(nh+7) = 128 * 207 =$ 26496 cycles/frame
	# times frame full/sec	(samp freq / frame size) = 48000/256 = 187.5 frames/second
	MIP calculation	(frames/sec) (cycles/frame) = 187.5 * 26496 = 4.97M cycles/second
	Conclusion	FIR takes approx 5 MIPS on this DSP
	Max # channels	60 @ 300 Mhz

Figure 3-16:
Computing the number of channels possible for a DSP system.

Next is the most important calculation — how many MIPS does this algorithm require of a processor? We need to find out how many cycles this algorithm will require per second. Now we multiply frames/second by cycles/frame and perform the calculation using this data to get a throughput rate of about 5 MIPS. Assuming this is the only computation being performed on the processor, the channel density (how many channels of simultaneous processing can be performed by a processor) is a maximum of $300/5 = 60$ channels. This completes the CPU calculation. This result cannot be used in the I/O calculation.

Algorithm: 200-tap (nh) low-pass FIR filter

Frame size: 256 (nx) 16-bit elements

Sampling frequency: 48 kHz

The next question to answer is 'Can the I/O interface feed the CPU fast enough to handle 60 channels?' Figure 3-17 shows this. Step one is to first calculate the 'bit rate' required of the serial port. To do this, the required sampling rate (48 kHz) is multiplied by the maximum channel density (60). This is then multiplied by 16 (assuming the word size is 16 — which it is, given the chosen algorithm). This calculation yields a requirement of 46 Mbps for 60 channels operating at 48 kHz. In this example what can the DSP serial port support? The specification says that the maximum bit rate is 50 Mbps (1/2 the CPU clock rate up to 50 Mbps). This tells us that the processor can handle the rates we need for this chosen application. Can the DMA move these samples from the McBSP to memory fast enough? Again, the specification tells us that this should not be a problem.

The next step considers the issue of required data memory.

Assume that all 60 channels of this application are using different filters — i.e., 60 different sets of coefficients and 60 double buffers (this can be implemented using a ping pong buffer on both the receive and transmit sides. This is a total of 4 buffers per channel, hence the *4 +

I/O		
	Required I/O Rate	48K samples/sec * # channels = 48000 * 16 * 60 = 48.08 Mbps
	DSP SP rate	Serial port is full dupllex = 50 Mbps
	DMA rate	(2 * 16 bit transfers/cycle) * 300 Mhz = 9600 Mbps
	Required Data Memory	(60 * 200) + (60 * 4 * 256) + (60 * 2 * 199) = 97K * 16 bit
	Available internal memory	32K * 26 bit

Figure 3-17:
Computing maximum number of channels and required memory and I/O.

the delay buffers for each channel (only the receive side has delay buffers, so the algorithm becomes:

Number of channels * 2 * delay buffer size

$= 60 * 2 * 199$

This is extremely conservative and the system designer could save some memory if this is not the case. But this is a worst case scenario. So, we'll have 60 sets of 200 coefficients, 60 double buffers (ping and pong on receive and transmit, hence the *4) and we'll also need a delay buffer of the number of coefficients -1, which is 199 for each channel. So, the calculation is:

(#Channels * #coefficients) + (#Channels * 4 * frame size) + (#Channels * #delay_buffers * delay_buffer_size)

$= (60 * 200) + (60 * 4 * 256) + (60 * 2 * 199)$

This results in a requirement of 97K of memory. The DSP only has 32K of on-chip memory, so this is a limitation. Again, you can re-do the calculation assuming only one type of filter is used, or look for another processor.

Once you've done these calculations, you can 'back off' the calculation to the exact number of channels your system requires, determine an initial theoretical CPU load that is expected, and then make some decisions about what to do with any additional bandwidth that is left over.

There are two sample cases that help drive discussion on issues related to CPU load. In the first case, the entire application only takes 20% of the CPU's load. What do you do with the extra bandwidth? The designer can add more algorithmic processing, increase the channel density, increase the sampling rate to achieve higher resolution or accuracy, or decrease the clock/voltage so that the CPU load goes up and you save lots of power. It is up to the system designer to determine the best strategy here based on the system requirements.

The second example application is the other side of the fence — where the application takes more processing power than the CPU can handle. This leads the designer to consider a combined solution. The architecture of this again depends on the application's needs.

DSP software

DSP software development is primarily focused on achieving the performance goals of the system. It is more efficient to develop DSP software using a higher level language like C or C++ but it is not uncommon to see some of the high performance, MIPS-intensive algorithms written, at least partially, in assembly language. When generating DSP algorithm code, the designer should use one or more of the following approaches:

- Find existing algorithms (free code).

- Buy or license algorithms from vendors. These algorithms may come bundled with tools or may be classes of libraries for specific applications.
- Write the algorithms in house. If using this approach, implement as much of the algorithm as possible in C/C++. This usually results in faster time to market and requires a common skill found in the industry.

It is much easier to find a C programmer than a DSP assembly language programmer. DSP compiler efficiency is fairly good and significant performance can be achieved using a compiler with the right techniques. There are several tuning techniques used to generate optimal code and this will be discussed in later chapters.

To tune your code and get the highest efficiency possible, the system designer needs to know three things:

- The architecture
- The algorithms
- The compiler

There are several ways to help the compiler generate efficient code. Compilers are pessimistic by nature, so the more information that can be provided about the system algorithms, where data is in memory, etc., the better. The DSP compiler can achieve 100% efficiency versus hand-coded assembly if the right techniques are used. There are pros and cons to writing DSP algorithms in assembly language as well, so if this must be done, these must be understood from the beginning.

The compiler is generally part of a more comprehensive integrated development environment (IDE) that includes a number of other tools. Figure 3-18 shows some of the main components of a DSP IDE including real-time analysis, simulation, user interfaces, and code generation capabilities.

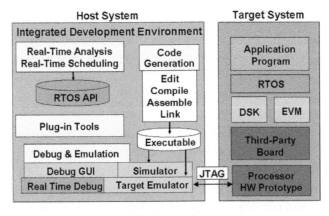

Figure 3-18:
Main components of a DSP IDE.

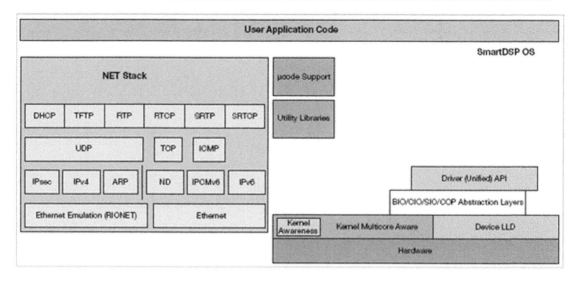

Figure 3-19:
DSP RTOS component architecture.

DSP IDEs also provide several components for use on the target application side including a real-time operating system, network stacks, low level drivers (LLDs), APIs, abstraction layers, and other support software as shown in Figure 3-19. This allows DSP developers to get systems up and running quickly.

Code tuning and optimization

One of the main differentiators between developers of non-real-time systems and real-time systems is the phase of code tuning and optimization. It is during this phase that the DSP developer looks for 'hot spots' or inefficient code segments and attempts to optimize those segments. Code in real-time DSP systems is often optimized for speed, memory size, or power. DSP code build tools (compilers, assemblers, and linkers) are improving to the point where developers can write the majority, if not all, of their application in high-level languages like C or C++. Nevertheless, the developer must provide help and guidance to the compiler in order to get the technology entitlement from the DSP architecture.

DSP compilers perform architecture-specific optimizations and provide the developer with feedback on the decisions and assumptions that were made during the compile process. The developer must iterate in this phase, to address the decisions and assumptions made during the build process, until the performance goals are met. DSP developers can give the DSP compiler specific instructions using a number of compiler options. These options direct the compiler as to the level of aggressiveness to use when compiling the code, whether to focus

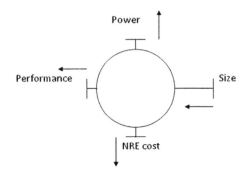

Figure 3-20:
Optimization trade-offs between size, power, performance, and cost.

on code speed or size, whether to compile with advanced debug information, and many other options.

Given the potentially large number of degrees of freedom in compile options and optimization axes (speed, size, power), the number of trade-offs available during the optimization phase can be enormous (especially considering that every function or file in an application can be compiled with different options (see Figure 3-20)). Profile-based optimization can be used to graph a summary of code size versus speed options. The developer can choose the option that meets the goals for speed and power and have the application automatically built with the option that yields the selected size/speed trade-off selected.

Typical DSP development flow

DSP developers follow a development flow that takes them through several phases:

- Application Definition — it's during this phase that the developer begins to focus on the end goals for performance, power, and cost.
- Architecture Design — during this phase, the application is designed at a systems level using block diagram and signal flow tools if the application is large enough to justify using these tools.
- Hardware/ Software mapping — in this phase a target decision is made for each block and signal in the architecture design
- Code Creation — this phase is where the initial development is done, prototypes are developed, and mockups of the system are performed.
- Validate /Debug — functional correctness is verified during this phase
- Tuning /Optimization — this is the phase where the developer's goal is to meet the performance goals of the system

- Production and Deployment − release to market
- Field testing

Developing a well-tuned and optimized application involves several iterations between the validate phase and the optimize phase. Each time through the validate phase the developer will edit and build the modified application, run it on a target or simulator, and analyze the results for functional correctness. Once the application is functionally correct, the developer will begin a phase of optimization on the functionally correct code. This involves tuning the application toward the performance goals of the system (speed, memory, power, for example), running the tuned code on the target or simulator to measure the performance, and evaluation where the developer will analyze the remaining 'hot spots' or areas of concern that have not yet been addressed, or are still outside the goals of performance for that particular area (see Figure 3-21).

Once the evaluation is complete, the developer will go back to the validate phase where the new, more optimized code is run to verify functional correctness has not been compromised. If the performance of the application is within acceptable goals for the developer, the process stops. If a particular optimization has broken the functional correctness of the code, the developer will debug the system to determine what has broken, fix the problem, and continue with another round of optimization. Optimizing DSP applications inherently leads to more complex code, and the likelihood of breaking something that used to work increases the more aggressively the developer optimizes the application. There can be many cycles in this process, continuing until the performance goals have been met for the system.

Generally, a DSP application will initially be developed without much optimization. During this early period, the DSP developer is primarily concerned with functional correctness of the application. Therefore, the 'out of box' experience from a performance perspective is not that impressive, even when using the more aggressive levels of optimization in the DSP compiler. This initial view can be termed the 'pessimistic' view, in the sense that there are no aggressive

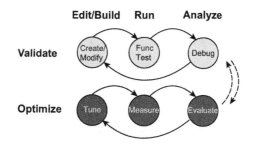

Figure 3-21:
DSP developers iterate through a series of optimize-and-validate steps until the goals for performance are achieved.

assumptions made in the compiled output, there is no aggressive mapping of the application to the specific DSP architecture, and there is no aggressive algorithmic transformation to allow the application to run more efficiently on the target DSP.

Significant performance improvements can come quickly by focusing on a few critical areas of the application:

- Key tight loops in the code with many iterations
- Ensuring critical resources are in on-chip memory
- Unrolling key loops where applicable

The techniques to perform these optimizations are discussed in the chapter on optimizing DSP software. If these few key optimizations are performed, the overall system performance goes up significantly. As Figure 3-22 shows, a few key optimizations on a small percentage of the code really leads to significant performance improvements. Additional phases of optimization get more and more difficult as the optimization opportunities are reduced, as well as the cost/benefit of each additional optimization. The goal of the DSP developer must be to continue to optimize the application until the performance goals of the system are met, not until the application is running at its theoretical peak performance. The cost/benefit does not justify this approach.

After each optimization, the profiled application can be analyzed for where the majority of cycles or memory is being consumed by the application. DSP IDEs provide advanced profiling capabilities that allow the DSP developer to profile the application and display useful information about the application such as code size, total cycle count, and number of

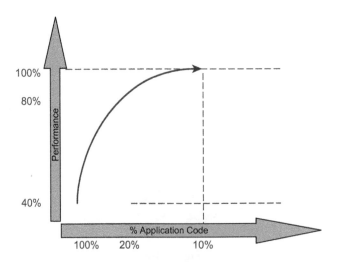

Figure 3-22:
Optimizing DSP code takes time and effort to reach the desired performance goals.

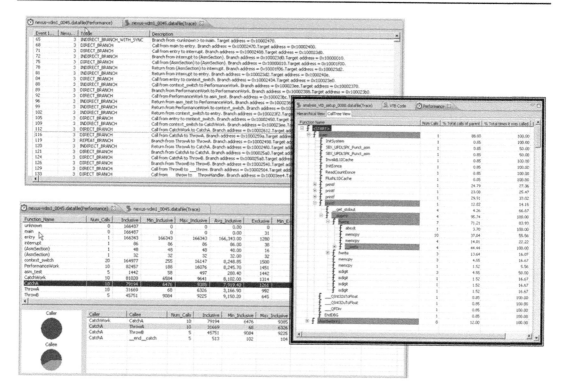

Figure 3-23:
Profiling and determining hot spots in DSP applications is a common exercise, and tools can aid in this process.

times the algorithm looped through a particular function. This information can then be analyzed to determine which functions are optimization candidates (Figure 3-23).

An optimal approach to profiling and tuning a DSP application is to attack the right areas first. These represent those areas where the greatest performance improvement can be gained with the smallest effort. A Pareto ranking of the biggest performance areas will guide the DSP developer towards those areas where the most performance can be gained.

Generally, the top eight to ten performance intensive algorithms in the DSP system are measured, analyzed, and then optimized to achieve full system performance goals. Figure 3-24 is an example of an actual set of benchmarks used to assess full system performance goals. Although there were over 200 algorithms in the system, none were chosen due to the fact that they consumed a majority of the cycles. This then became the suite of benchmarks on which to focus the optimization process.

In order to get a good system view of where the cycles are being consumed, on-chip instrumentation in the form of counters, timers, etc., are used to enable the extraction of

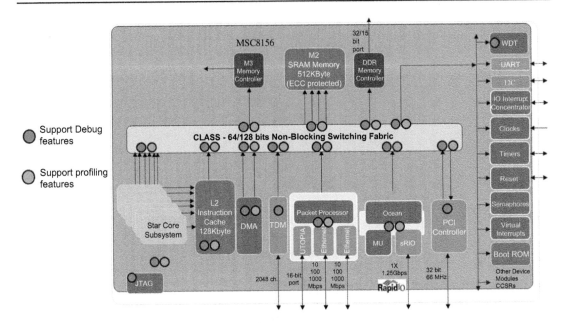

Figure 3-24:
Advanced DSP SoCs have peripherals, communication busses, cores, and accelerators instrumented with counters and timers to support both profiling and debug.

various types of profiling and debug data. Advanced DSP SoCs have peripherals, communication busses, cores, and accelerators instrumented with counters and timers to support both profiling and debug. This is shown in Figure 3-25 for a six core multicore DSP device. By having this support on chip, software tools, profilers, and even the running

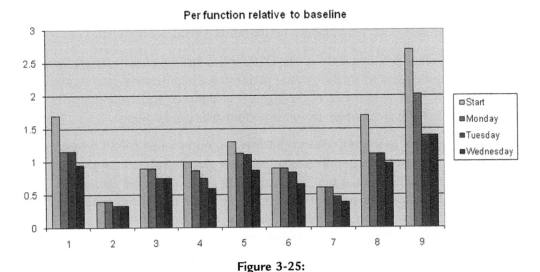

Figure 3-25:
Generally the top eight to ten performance intensive algorithms in the DSP system are measured, analyzed, and then optimized to achieve full system performance goals.

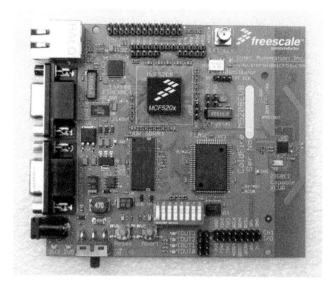

Figure 3-26:
Evaluation board for DSP.

application software can take advantage of this information, to make decisions about optimization, debug, etc. This allows the DSP developer to visualize debug and profiling information at various points in the DSP system including the cores, the communication fabric, the accelerators and peripherals, and other important processing elements in the SoC, which allows for full system analysis (Figure 3-27).

A DSP starter kit (Figure 3-26) is easy to install and allows the developer to get started writing code very quickly. The starter kits usually come with daughter card expansion slots, the target hardware, software development tools, a parallel port interface for debug, a power supply, and the appropriate cables.

Putting it all together

There are five major stages to the DSP development process:

- System concept and requirements — this phase includes the elicitation of the system level functional and non-functional (sometimes called 'quality') requirements. Power requirements, quality of service (QoS), performance and other system level requirements are elicited. Modeling techniques such as signal flow graphs are constructed to examine the major building blocks of the system.
- System algorithm research and experimentation — during this phase, the detailed algorithms are developed based on the given performance and accuracy requirements. Analysis is first done on floating point development systems to determine if these performance and accuracy requirements can be met. These systems are then ported,

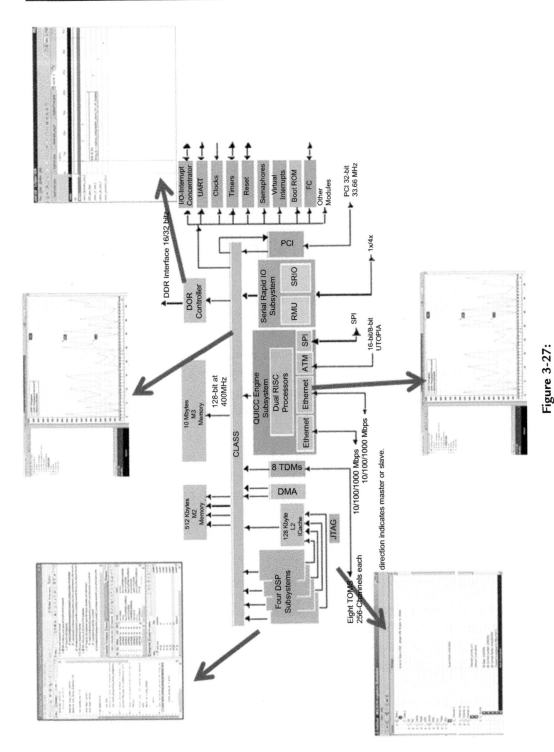

Figure 3-27:

Visibility must be provided throughout the DSP SoC.

if necessary, to fixed point processors for cost reasons. Inexpensive evaluation boards are used for this analysis.

- System design — during the design phase, the hardware and software blocks for the system are selected and/or developed. These systems are analyzed using prototyping and simulation to determine whether the right partitioning has been performed and whether the performance goals can be realized using the given hardware and software components. Software components can be custom developed or reused, depending on the application.
- System implementation — during the system implementation phase, inputs from system prototyping, trade-off studies, and hardware synthesis options are used to develop a full system co-simulation model. Software algorithms and components are used to develop the software system. Combinations of signal processing algorithms and control frameworks are used to develop the system.
- System integration — during the system integration phase, the system is built, validated, tuned, if necessary, and executed in a simulated environment or in hardware in the loop simulation environment. The scheduled system is analyzed and potentially re-partitioned if performance goals are not being met.

In many ways, the DSP system development process is similar to other development processes. Given the increased amount of signal processing algorithms, early simulation-based analysis is required more for these systems. The increased focus on performance requires the DSP development process to focus more on real-time deadlines and iterations of performance tuning.

CHAPTER 4

Programmable DSP Architectures

Mike Brogioli

Chapter Outline

Common features of programmable DSP architectures

Most modern digital signal processing (DSP) architectures are significantly more irregular than their general purpose computing counterparts, and certainly more complex than most 8/16 bit microcontroller architectures. This is due, in part, to the need to pack high levels of computational power into the processor design to handle the computational demands of many DSP workloads. At the same time, due to the rigid power demands of many portable devices, such as cellular handsets and portable media players, many architectural constructs that are common in general purpose computing, such as out-of-order execution of instructions within the processor pipeline, cannot be afforded. At the same time, one long standing and ever increasing goal for modern DSP architectures is to provide a platform which can be programmed in C and C++ -like programming languages. In doing this, customers can reduce their time to market and increase the portability of their software base, should they decide to use another DSP architecture as their platform further down the road. The following section describes the features of many popular DSP architectures, many of which contribute to significant numerical processing performance gains over their general purpose processor or embedded microcontroller counterparts.

DSP core and ISA features

Modern DSP cores contain a number of features that enhance the number crunching ability of the core to meet the ever demanding needs of DSP applications. At the same time, these features are added to maximize performance while meeting the strict power consumption demands of the overall system and environment. This presents an interesting challenge for both the user and the programming environment. While the user often wants to maximize the performance of their algorithm on the target platform of choice, given the often irregular nature of the underlying DSP architecture the programming language of choice may not be able to sufficiently represent the underlying architecture. This provides a challenge for compilers, assemblers, and linkers when targeting such DSP architectures. Often, a programmer may choose to write certain DSP kernels, or portions of kernels, at the assembly language level, or use highly powerful, target architecture-specific 'intrinsic functions' to access various architectural constructs. The following section identifies and discusses such features specific to the programmable DSP space.

Features of the programmable DSP space

Most modern DSP architectures contain application domain specific instructions to accelerate performance of kernels. Examples of this include multiply-accumulate instructions, commonly referred to as 'MAC' instructions. Because it is often common in DSP kernels to multiply values together and accumulate the result in specific accumulator value, dedicated instructions are made available to perform such computation, rather than issue multiple addition and multiply instructions with corresponding load and store instructions. DSP kernels that commonly use MAC instructions include Fast Fourier Transforms, Finite Impulse Response Filtering, and many others. Most modern DSPs will offer different permutations of MAC instructions that operate on varying types of operands, such as 16-bit input values, that accumulate into 40-bit output values. Additionally, the computation may be saturating or non-saturating in nature, dependent on instructions within the DSP instruction set, or various mode bits that can be configured within the processor at runtime.

Due to the large amounts of instruction level parallelism, and data level parallelism within most DSP workloads, it is common for modern DSP architectures to use a Very Long Instruction Word (VLIW) architecture, as described previously. Due to large amounts of data parallelism that often exist in performance critical DSP kernels, many DSP architectures utilize single instruction multiple data (SIMD) extensions in their instruction set. SIMD permits a single atomic instruction to perform the same computation over multiple sets of input operands on a single ALU within the processor's pipeline. In doing this, the density of computation to instruction ratio within the program is increased, as the instructions now execute over short vectors of multiple input operands. An example of such operations would be a multimedia kernel that loads multiple pixel values from memory, adjusts each of the red, green, and blue values by some constant value, and then writes the values back to memory. Rather than load each value at runtime separately, perform the desired computation, and then write the value back to memory, multiple input operands can be loaded in parallel within a single atomic instruction, computed upon in parallel, and written back to memory in parallel.

Issues surrounding use of SIMD operations

There are a number of challenges with utilizing SIMD operations in a DSP environment. In some DSP architectures, loads from memory must be aligned to certain memory boundaries. It may be the programmer's responsibility to ensure that data is aligned on proper memory boundaries to ensure correctness, which is often done via 'pragma' statements that are inserted into the code base. Additionally, not all computation can be vectorized across SIMD ALUs within the pipeline. Code with significant amounts of control flow, such as if-then-else statements, may be of significant challenge to the compiler in vectorizing across SIMD engines. As a result, optimally vectorizing code across SIMD engines within the DSP pipeline is often a manually intensive programming effort in the event that highly optimizing vectorizing compilers are not available for the target processor. In such a case, as an alternative, programmers often map performance critical sections of computations across the SIMD engines manually via assembly language, or specify the vector computation using DSP architecture specific intrinsic functions. Some vendors supply native keywords such as '_vector' to denote vectors of computation that the compiler may assume are safe to map to the various SIMD ALUs within the pipeline. Lastly, in certain cases the compiler may auto vectorize C-language code onto the SIMD ALUs if the computation can be proven safe at compile time for certain use cases.

DSP kernels quite often perform their computation over buffers of input values that exhibit temporal and/or spatial reuse. It is not uncommon for a DSP application programmer to put data into a local fast memory location, perhaps in SRAM, for low latency high bandwidth access to the computation. Rather than require the application programmer to manually check whether or not the index into the memory buffer has reached the end boundary and requires overlap checking, DSP instruction sets often contain hardware modulo addressing. This type of addressing implements circular buffers in memory that can be used as temporary storage without the need to check the buffer overrun and wrap conditions. In essence, the circular buffer with hardware modulo addressing provides an efficient means for a first-in, first-out data structure. Circular buffers are quite often used to store the most recent values of a continuously updated signal. An example of this is illustrated in Figure 4-1.

Since DSPs are commonly used in portable devices that operate on battery power, system resources are often at a premium. Program memory is one such resource that often comes at a premium. In the case of mobile cellular handsets, it is often desirable for the compiler to reduce the size of the program executable as much as possible to fit in the limited program memory resources. At the same time, performance of the DSP kernels is critical and often performance cannot be sacrificed for the sake of program size in memory. Therefore, most DSPs often have multiple encoding schemes. Examples of this include the Arm Thumb2 encoding scheme, as well as the Freescale StarCore SC3850 DSP architecture. In these architectures, instructions are classified into 'premium' groupings, whereby powerful

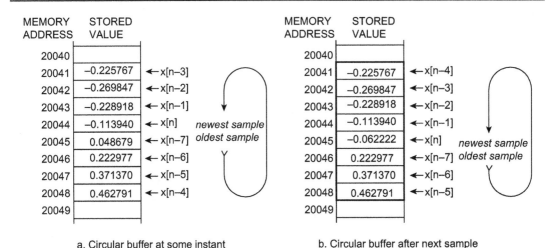

Figure 4-1:
Circular buffers are quite often used to store the most recent values of a continuously updated signal.

instructions that are commonly used within an application are given both a standard 32-bit encoding, as well as a premium 16-bit encoding. When the compiler is compiling for program size optimization, it will attempt to select those instructions that have a premium encoding, requiring fewer bits, in order to conserve program memory requirements of the executable. It should be noted, however, that instructions may be limited in their functionality when they are implemented using premium encoding. Since the instruction is represented with fewer bits in memory, the number of variants that can be encoded in the limited bitset is often reduced. As such, premium encodings of instructions may limit the set of registers that may be used as input and output operands for the instructions. In some cases, there may be implicit couplings between register operands as well; for instance, input registers may be required to be contiguous such as register pairs R0, R1, or R2, R3. Lastly, such premium encodings may reduce the total number of operands and instead overwrite input operands with the output value. Examples of this would be absolute value instructions that are represented as 'ABS R0, R1' in a normal encoding, where R0 is the input value, and R1 is the output value. In a premium encoding, the instruction would be represented as 'ABS R0', where R0 is the input operand, and the value is destroyed after execution by being overwritten with the output operand. As such, while the encoding may be smaller and reduce program memory requirements, performance of code may deteriorate by some degree due to extra copy operations that may be required to preserve the input value for later use in computation.

Most DSP architectures contain deep processor pipelines requiring multiple clock cycle execution of multiply instructions, and containing rather large branch delay slots on change of flow instructions. For example, the Texas Instruments C6400 series DSPs have a four cycle branch delay slot, which can be costly to program performance if not filled with computation by the compiler, or manually by the assembly programmer. This problem is only exacerbated by the lack of out-of-order execution hardware on most modern DSP architectures as well.

Predicated execution

In order to circumvent costly change of flow instructions required to process control code commonly containing if-then-else statements, many modern DSP architectures contain predicated instructions. Predicated execution is a mechanism whereby the execution of a given instruction is predicated with data values within the processor. In doing this, short branches within the program that may be utilized to implement if-then-else conditions within a C-program, are avoided and bubbles associated with the delay slots of the branch instructions do not deteriorate program performance at runtime. Rather than skip over and branch around the short sections of code in the 'then' or 'else' conditions, predicated processors simply execute the code contained in all clauses of the if-then-else statement. This effectively permits the processor to execute both branch code paths simultaneously without the performance penalties associated with executing the branch instructions. Control flow in the program is converted into data dependencies within the program via the predication bits associated with each instruction in the various clauses of the control flow statement.

Typically, predicated execution is handled behind the scenes from the application writer by the optimizing compiler. The compiler will often convert control flow dependences within the program to data dependences on predicate registers, or predicate bits within the architecture. As such, costly branching instructions can be avoided at runtime. In predicated DSP architectures, with the appropriate hardware support, the behavior of an operation changes according to the value of the previously computed value or 'predicate.' In using predication, multiple regions of the control flow graph are converted into a single region containing predicated code, or instructions. In effect, what were previously control dependences are converted into data dependences. However, given an optimizing compiler, the user does not need to worry about this in their input source code.

In the case of full instruction set predication, instructions take additional operands that determine at runtime whether they should execute or simply be ignored when executed through the DSP's pipeline. The additional operands are predicate operands, or guards. In the case of partial instruction set predication, special operations, such as conditional move instructions, achieve similar results.

There are a number of compiler techniques available to support predicated execution. They include if-conversion, logical reductions of predicate registers, reverse if-conversion, and hyperblock-based scheduling. Modulo scheduling also falls into this category.

In many embedded systems, predicates are implemented using the general scalar registers of the machines. In other system designs, the predicates are implemented using status bits within control registers. In either case, predicate register real estate is a precious and limited commodity within the system. As such, often the compiler will attempt to reduce the number of predicates needed for a given region of code

An example of a modern predicated architecture used for DSP workloads is the ARM instruction set, which includes a 4-bit set of condition code predicates. The Texas Instruments C6x architecture also supports full predication, with five of the general purpose registers used for predicate values.

Figure 4-2 illustrates a segment of control flow, where the runtime behavior of the program is dependent on whether or not the variable STATUS is set to ACTIVE. If STATUS equals ACTIVE, then the if-statement executes; otherwise the else-statement executes.

Figure 4-3 shows the pseudo-assembly code generated in the standard case under which predicated execution is not used. As can be seen, the test for equivalence to ACTIVE is performed by the CMP.NEQ instruction, followed by a branch to the relevant control flow destination. The else-statement also contains a branch instruction to the join point labeled 'JOIN.'

Figure 4-4 shows the same source code now compiled with support for predicated execution. Notice the use of registers P00 and R01 as predicates. The CMP.EQ instruction is used to set the values of P00 and P01 to true and false, respectively, and these are used as guards on subsequent instructions that were previously part of the if-statements control flow. In this case, both the then-clause and else-clause of the

```
IF(STATUS == ACTIVE)

{

        N_ACTIVE_ITEMS++;

        TOTAL_SUMMARY += RUNNING_COUNT;

}

ELSE

{

        N_INACTIVE_ITEMS++;

}
```

Figure 4-2:
Runtime behavior of the program is dependent on whether or not the variable STATUS is set to ACTIVE.

```
        {
                CMP.NEQ PO1 = RS,ACTIVE// CMP STATUS TO ACTIVE
                (PO1==TRUE) BR ELSE
        }
    .LABEL THEN
        {
                ADD RT = RT,RP          // SUM TOTAL_SUMMARY + RUNNING COUNT
                ADD RA = RA,1           // INCREMENT N_ACTIVE_ITEMS
                BR JPIN                 // BRANCH TO JOIN
        }
    .LABEL ELSE
        {
                ADD RI = RI,1           // INCREMENT N_INACTIVE_ITEMS

    .LABEL JOIN
```

Figure 4-3:
Pseudo-assembly code generated in the standard case under which predicated execution is not used.

```
        {
                CMP.EQ PO0,PO1 = RS,ACTIVE      // CMP STATUS TO ACTIVE
        }
        {
                (PO0==TRUE) ADD RT = RT,RP
                (PO0==TRUE) ADD RA = RA,1       // INCREMENT N_ACTIVE_ITEMS
                (PO0==TRUE) ADD RI = RI,1       // INCREMENT N_INACTIVE_ITEMS
        }
```

Figure 4-4:
Source code compiled with support for predicated execution.

if-statement are executed in parallel since the instructions from each are guarded by PO0 and PO1, effectively compressing the critical path length of the control-flow statement, and removing all branch instructions and associated delay slots from the control flow statements.

Memory architecture

Most programmable DSP cores today utilize specialized memory architectures that can fetch multiple data and instruction elements per clock cycle. It is often desirable to have architectures for which load and store instructions are the only ones that may access memory. In doing so, it is left up to the compiler to arrange the rest of the work as regards how to use them. If an architecture is allowed memory operands, this will create barriers to instruction level parallelism for the compiler, due to the latency of memory operations and possible data dependences through memory. As such, most DSPs vary in terms of addressing modes, access sizes and alignment restrictions on access to memory.

The addressing modes of the DSP architecture form the basis of the address generation unit. The most common addressing modes for programmable DSP architectures are as follows:

- Register addressing — a register entry within the register file is used as the address pointer.
- Direct or absolute addressing — the address is part of the instruction itself, and the programmer specifies the address within the instruction.
- Register post increment — the address pointer will take a register value and increment this value by a default 'step' value.
- Register post decrement — the address pointer will take a register value and decrement this value by a default 'step' value.
- Segment plus offset — the address pointer will take a register value added with a defined offset from the instruction.
- Indexed addressing — the address pointer will take a register value and add an indexed register value.

Additionally, there are two other modes of addressing that are common, not only to VLIW architectures, but specifically to DSP architectures. Modulo addressing mode is an addressing mode, mentioned previously, that forces the memory to be used as a FIFO or cyclic ring buffer. Finite Impulse Response filters are typically the most common use of modulo addressing modes, whereby FIFO structures are implemented directly in memory and make use of the modulo addressing capabilities of the DSP core.

Bit reversed addressing mode is another addressing mode that is common to DSP architectures. Fast Fourier Transform is another key kernel that requires high performance in DSPs.

Access sizes

Most modern DSP architectures are capable of accessing memory at varied data widths that are native to the machine's computation. For example, many DSP architectures can access 8/16/32 bit data operands. In addition, the architectures can also often access 20-bit and

40-bit operands as well, should the architecture provide such data resolution. This is often the case for operations such as multiply accumulate, whereby the bit width resolution of the output operands is higher than that of the input operands. Lastly, due to the VLIW nature of most DSP architectures, they are often capable of accessing packed data elements from memory. For instance, a given DSP architecture that has 128-bit load-store bandwidth, may be able to access 16, 16-bit operands in parallel from memory in one SIMD vector. This is often desirable for computational kernels with high levels of instruction level parallelism, whereby multiple instructions can execute across the multiple VLIW ALUs in parallel.

Alignment issues

As was mentioned above, many VLIW architectures are capable of executing load and store operations of multiple data elements in parallel from memory. This is very useful where there is significant instruction level parallelism. By accessing multiple elements in parallel, and computing across them with SIMD instructions, the ratio of instructions in the program to computation performed is decreased, thereby yielding higher performance.

One challenge in achieving such performance scenarios is the underlying memory system of the DSP architecture. Often, DSP architectures have memory alignment restrictions which ensure that a given memory access exists on certain memory alignment boundaries. Structures may or may not align cleanly to word boundaries. Consider, for example, a structure containing an array of bytes! Additionally, it is common for a program to operate on randomly selected sections within an array, which again causes alignment problems at runtime. A common solution to the alignment restriction of many embedded systems is for the build tools to perform 'versioning' at compilation time. In essence, a given loop will have two versions generated in the executable. A switch test is performed at runtime to test whether or not the data is aligned, and if so, the optimized version of the loop is executed. In the event that the switch test determines that data is not aligned at runtime, the non-optimized version of the loop is executed. While this can result in performance improvements at runtime on systems with alignment restrictions that cannot be guaranteed to be met at runtime, there are the obvious trade-offs in executable code size. Another common solution to this is to provide the programmer with 'pragma' statements that communicate to the build tools that alignment restrictions are guaranteed to be enforced.

Data operations

Most modern DSP architectures contain a mix of arithmetical operations to meet the varied needs of application developers. It is common for almost all programmable DSP cores to support integer computation. While this portion of the DSP's instruction set may be used for the signal processing portion of the application being run, it is also used to map the semantics of the C-like programming language onto the DSP cores.

The integer portion of the instruction set is often supplemented with other forms of computation targeted at the DSP portion of the application. Many DSPs include native floating point operations. It should be noted that full IEEE compliance and double precision are rarely characteristic of embedded applications, rather relying on single precision floating point operations. It is often sufficient to omit various IEEE rounding modes and exception handling on floating point operations for the needs of embedded computing. Furthermore, many devices simply opt to emulate floating point computation via software libraries, if native floating point is not a driving requirement. This is only a viable solution when floating performance is not on the critical path of the application, as emulation of floating point instructions in software can often take hundreds of clock cycles on the native VLIW host processor.

It is common for many DSP architectures to support fixed point arithmetic instead of floating point arithmetic. Fixed point data formats are not quite the same as integer data formats. Fixed point format is used to represent numbers that lie between zero and one, with the decimal point assumed to reside after the most significant bit. The most significant bit in this case contains the sign of the number. The size of the fraction represented by the smallest bit is the precision of the fixed point format. The size of the largest number that can be represented by the fixed point format is the dynamic range of the format.

To efficiently use the fixed point format, the programmer must explicitly manage how the data is handled at program runtime. If the fixed point operand becomes too large for the native word size, the programmer will have to scale the number down by shifting to the right. In doing this, lower precision bits may be lost. If the fixed point operand becomes too small, the number of bits actually representing the number is reduced as well. As such, the programmer may decide to scale the number up so as to benefit from an increased number of bits representing the number. In both cases, it is the responsibility of the programmer to manage how many bits are being used to represent the number.

One last data format worth mentioning is the use of saturating arithmetic in DSP architectures. The C programming language semantics imply wraparound behavior for integer data types. However, in many embedded application spaces, such as audio and video processing, it is desirable to have the operands saturate at a maximum or a minimum value. Hardware support for this has both its pros and cons. While an instruction set that has native support for saturating arithmetic can efficiently pipeline such operations, supporting equivalent functionality via min() and max() operations in software can be performed as well. While the compute time is longer to use explicit min() and max() instructions, this may be a sufficient solution for architectures which do not rely on saturating arithmetic for the majority of their computation.

One caveat to supporting saturating arithmetic, without a corresponding non-saturating variant within the instruction set is that, for programmers, the resulting functionality can be confusing and error prone. Consider, for example, computation that is in wraparound mode, and reordered by the build tools at optimization time. Since wraparound mode instructions preserve their arithmetic properties, such as commutativity and associativity at runtime, perceived behavior is not altered when the program is optimized. In the case where the saturating variant of the instructions is used, however, the build tools may either be prohibited from optimizing such computation, or the user may see variance in the output at the bit compatibility level, due to reorderings on the computation performed at optimization time. This can often be difficult to manage for both the end programmer as well as the tools developers.

CHAPTER 5

FPGA in Wireless Communications Applications

Kiarash Amiri[1], Melissa Duarte[1], Joe Cavallaro[1], Chris Dick[2], Raghu Rao[2], Ashutosh Sabharwal[1]

[1]*Rice University, ECE Dept., Houston, TX,* [2]*Xilinx Inc., San Jose, CA*

Chapter Outline

Introduction

In the past decade we have witnessed explosive growth in the wireless communications industry with over 4 billion subscribers worldwide. While first and second generation systems focused on voice communications, third generation networks (3GPP and 3GPP2) embraced code division multiple access (CDMA) and had a strong focus on enabling wireless data services. As we reflect on the rollout of 3G services, the reality is that first generation 3G

DSP for Embedded and Real-Time Systems. DOI: 10.1016/B978-0-12-386535-9.00005-6

systems did not entirely fulfill the promise of high-speed transmission, and the rates supported in practice were much lower than those claimed in the standards. Enhanced 3G systems were subsequently deployed to address the deficiencies. However, the data rate capabilities and network architecture of these systems were insufficient to address the insatiable consumer and business sector demand for the nomadic delivery of media and data-centric services to an increasingly rich set of mobile platforms.

With these considerations in mind, the move to 4G technologies like 3G LTE (long term evolution) [1] and WiMAX [2] is proceeding at an extremely rapid rate. The goal of next generation systems is to provide high-data rate, low-latency, and high reliability (minimizing outages and connection drops), employing packet-optimized radio access technology, supporting flexible bandwidth allocation. Additional key objectives are to drive down the cost of infrastructure equipment and consumer terminals and to employ a more efficient modulation scheme than the CDMA technology used in 3G systems, in order to make more optimal use of precious communication bandwidth.

To meet all of these requirements a significant re-structuring of both the physical layer (PHY) and network architecture is required. Increased spectral efficiency is delivered in all 4G cellular broadband systems through the use of orthogonal frequency division multiplexing (OFDM) as the preferred modulation scheme. The technology center-piece for delivering improved data rates and communication link robustness lies in more aggressive use of the spatial dimension of communication systems through multiple-input multiple-output (MIMO) techniques [3]. 4G wireless systems will employ MIMO processing to equip next generation systems with spatial multiplexing (improved data rates) and diversity (improved reliability) capabilities. MIMO systems (Figure 5-1) can increase the data rate and provide multiplexing gain by transmitting different data on different antennas [4], also known as spatial multiplexing. MIMO systems can also improve the reliability and error performance in the receiver through diversity gain, i.e., by providing the receiver with multiple copies of the transmitted signal. One practical way to realize such diversity gains is to beamform a message across multiple transmit antennas [5, 6]. The beamforming procedure utilizes limited information about the channel state information; this channel

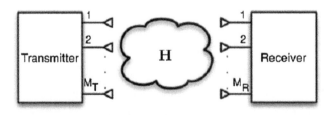

Figure 5-1:
Multiple antenna (MIMO) system.

information is usually provided through a limited feedback link from the receiver to the transmitter.

The implementation of 4G infrastructure equipment has several challenges. The first is related to the compute requirements of MIMO-OFDM systems. The more sophisticated MIMO-OFDM advanced receiver architectures have compute requirements that are several orders of magnitude greater than those of 3G CDMA systems. A second requirement is related to bridging the gap between legacy standards (e.g., 3GPP) and next generation 4G protocols and systems. In many cases 4G infrastructure equipment will need to have multi-mode capability and support not only the MIMO-OFDM PHY of WiMAX and 3G LTE systems, but also the W-CDMA (wideband CDMA) PHY of 3GPP-based networks. The silicon technologies deployed in next generation systems will not only need to supply enormous compute capability (many hundreds of giga-operations/second), but they will also require significant flexibility in order to support multi-mode base station infrastructure equipment.

Field programmable gate arrays (FPGAs) with their inherently parallel structure are increasingly the technology of choice for addressing the compute and flexibility requirements of next generation systems. One of the fundamental areas in many academic and industrial research organizations is that of hardware realization of advanced MIMO receivers.

Spatial multiplexing MIMO systems

We assume a MIMO OFDM system with M_T transmit antennas and $M_R \geq M_T$ receive antennas. We denote the K subcarriers by the subscript $k = 1, \ldots, K$ throughout this chapter. The input-output model is captured by:

$$\tilde{y}_k = \tilde{H}_k \tilde{s}_k + \tilde{n}_k \tag{1}$$

where $\tilde{H}_k$ is the complex-valued $M_R \times M_T$ channel matrix, $\tilde{s}_k = [\tilde{s}_{1,k}, \tilde{s}_{2,k}, \ldots \tilde{s}_{M_T,k}]_T$ is the M_T-dimensional transmitted vector, where each $\tilde{s}_{j,k}, j = 1, \ldots, M_T$, is chosen from a complex-valued constellation Ω_j of the order $w_{j,k} = |\Omega_{j,k}|$, $\tilde{n}_k$ is the circularly symmetric complex additive white Gaussian noise vector of size M_R, and $\tilde{y}_k = [\tilde{y}_{1,k}, \tilde{y}_{2,k}, \ldots \tilde{y}_{M_R,k}]_T$ is the M_R-element received vector. Note that we do not restrict all the parallel M_T streams to use the same modulation order; rather, each stream, which corresponds to one of the antennas of one of the users, may be using either the 4-, 16-, or 64-QAM modulation. Also, note that even though we focus on one transmitter with multiple antennas, the results in this section can be extended to multiple users such that the sum of the number of antennas of all of them equals M_T.

We assume that the complete channel information, i.e., the channel matrix coefficients for each subcarrier, are known in the receiver. Moreover, since the detection procedure for each subcarrier is performed independently of other subcarriers, we will drop the k subscript unless it is needed.

Therefore, the original MIMO model can be simplified to $\tilde{y} = \tilde{H}\,\tilde{s} + \tilde{n}$.

The preceding MIMO equation can be further decomposed into real-valued numbers as follows [7]:

$$y = Hs + n \tag{2}$$

corresponding to:

$$\begin{pmatrix} \Re(\tilde{y}) \\ \Im(\tilde{y}) \end{pmatrix} = \begin{pmatrix} \Re(\tilde{H}) & -\Im(\tilde{H}) \\ \Im(\tilde{H}) & \Re(\tilde{H}) \end{pmatrix} \begin{pmatrix} \Re(\tilde{s}) \\ \Im(\tilde{s}) \end{pmatrix} + \begin{pmatrix} \Re(\tilde{n}) \\ \Im(\tilde{n}) \end{pmatrix} \tag{3}$$

with $M = 2M_T$ and $N = 2M_R$ presenting the dimensions of the new model. We call the ordering in (2) the conventional ordering. Using the conventional ordering, all the computations can be performed in real values, which would simplify the implementation complexity. Note that after real-valued decomposition, each s_i, $i = 1, \ldots, M$, in s is chosen from a set of real numbers, Ω_i', with $p_i' = \sqrt{p_i}$ elements. For instance, for a 64-QAM modulation, each s_i can take any of the values in the set $\Omega' = \{\pm7, \pm5, \pm3, \pm1\}$. The general optimum detector for such a system is the maximum-likelihood (ML) detector which minimizes $\| y - Hs \|2$ over all the possible combinations of the s vector. Notice that for high order modulations and large number of antennas, this detection scheme incurs an exhaustive exponentially growing search among all the candidates, and is not practically feasible in a MIMO receiver. However, it has been shown that using the QR decomposition of the channel matrix, the distance norm can be simplified [8, 9, 10] as follows:

$$D(s) = \|y - Hs\|^2$$
$$= \|Q^H y - Rs\|^2 = \sum_{i=M}^{1} \left| y_i' - \sum_{j=i}^{M} R_{i,j} s_j \right|^2 \tag{4}$$

where $H = QR$, $QQ_H = I$, and $y' = Q^H y$. Note that the transition in (4) is possible through the fact that R is an upper triangular matrix.

The norm in (4) can be computed in M iterations starting with $i = M$. When $i = M$, i.e., the first iteration, the initial partial norm is set to zero, $T_{M+1}(s_{(M+1)}) = 0$. Using the notation of [11], at each iteration the Partial Euclidean Distances (PEDs) at the next levels are given by

$$T_i(s^{(i)}) = T_{i+1}(s^{(i+1)}) + \left| e_i(s^{(i)}) \right|^2 \tag{5}$$

With $s_{(i)} = [s_i, s_{i+1}, \ldots, s_M]_T$, and $i = M, M-1, \ldots, 1$, where

$$\left| e_i(s^{(i)}) \right|^2 = \left| y_i{}' - R_{i,i}s_i - \sum_{j=i+1}^{M} R_{i,j}s_j \right|^2 \tag{6}$$

$$= \left| J_{i+1}(s^{(i+1)}) - R_{i,i}s_i \right|^2. \tag{7}$$

One can envision this iterative algorithm as a tree traversal with each level of the tree corresponding to one i value, and each node having $p'i$ children.

The tree traversal can be performed in either a breadth-first or a depth-first manner. In the depth-first tree search [12, 11, 13], only one node is expanded at any time and once the end of the tree is reached, the traversal continues by visiting the new nodes in the higher levels of the tree. Therefore, each level can be visited more than one time.

In the breadth-first tree search [14, 15], however, each level is only visited once, and more than one node per level is expanded. Once the end of the tree, or the leaves level, is reached, the minimum candidate is chosen. A typical breadth-first tree search is the K-best detector. In the K-best detector, at each level, only the best K nodes, i.e. the K nodes with the smallest T_i, are chosen for expansion. Note that such a detector requires sorting a list of size $K \times p'$ to find the best K candidates. For instance, for a 16-QAM system with $K = 10$, this requires sorting a list of size $K \times p' = 10 \times 4 = 40$ at most of the tree levels. This introduces a long delay for the next processing block in the detector unless a highly parallel sorter is used. Highly parallel sorters, on the other hand, consist of a large number of compare-select blocks, and result in dramatic area increase.

Flex-Sphere detector

In order to simplify the sorting step, which significantly reduces the delay of the detector, a sort-free strategy can be utilized. Moreover, we discuss how using a new modified real-valued decomposition ordering (M-RVD) scheme can help in designing a flexible architecture. Finally, we discuss the design and implementation of the flexible sphere detector, Flex-Sphere [16, 17], that can support a range of modulation orders and number of antennas.

Tree Traversal for Flex-Sphere detection

Using the sort-free technique, the long sorting operation is effectively simplified to a minimum-finding operation [16, 18]. The detailed steps of this algorithm are described in the Flex-Sphere Tree Traversal algorithm.

Algorithm 1: Flex-Sphere Tree Traversal

Input: $\mathbf{R}, \mathbf{y}'$
$T_{M+1}(s^{(M+1)}) = 0$
$\mathcal{L} \leftarrow \emptyset$
$\mathcal{L}' \leftarrow \emptyset$
$i \leftarrow M$
\\ Full expansion of the first level:
- Compute T_i with (5),
- $\mathcal{L} \leftarrow \{(s^{(i)}, T_i(s^{(i)}))_j | j = 1, ..., p'\}$
- $i \leftarrow i - 1$
\\ Full expansion of the second level:
- **for** each $(s^{(i+1)}, T_{i+1}(s^{(i+1)})) \in \mathcal{L}$, **repeat**
    - compute $(s^{(i)}, T_i(s^{(i)}))_j$ children pairs, $j = 1, ..., p'$
    - $\mathcal{L}' \leftarrow \mathcal{L}' \cup \{(s^{(i)}, T_i(s^{(i)}))_j | j = 1, ..., p'\}$
- **end**
- $\mathcal{L} \leftarrow \mathcal{L}'$
- $\mathcal{L}' \leftarrow \emptyset$
\\ Minimum-based expansion of the next levels:
- **for** $i = M - 2$ down to $i = 1$, **repeat**
    - **for** each $(s^{(i+1)}, T_{i+1}(s^{(i+1)})) \in \mathcal{L}$, **repeat**
        - compute $(s^{(i)}, T_i(s^{(i)}))_j$ children pairs, $j = 1, ..., p'$
        - $(s^{(i)}, T_i(s^{(i)}))_{min} \leftarrow \arg \min_{\{(s^{(i)}, T_i(s^{(i)}))_j | j=1,...,p'\}} T_i(s^{(i)})$
        - $\mathcal{L}' \leftarrow \mathcal{L}' \cup \{(s^{(i)}, T_i(s^{(i)}))_{min}\}$
    - **end**
    - $\mathcal{L} \leftarrow \mathcal{L}'$
    - $\mathcal{L}' \leftarrow \emptyset$
    - $i \leftarrow i - 1$
- **end**

- $(s^{(i)}, T_i(s^{(i)}))_{detected} \leftarrow \arg \min_{\mathcal{L}} T_i(s^{(i)})$

An example of this algorithm is illustrated in Figure 5-2 for a 4×4, 64-QAM system. Note that, as described above, the first two levels are fully expanded to guarantee high performance; whereas for the following levels, only the best candidate in the children list of a parent node is expanded. In other words, after passing the first two levels, p_{M_T} nodes are expanded, and for each of those p_{M_T} nodes, the best child node among its p'_M children nodes is selected as the survived node. Therefore, the new node list would contain p_{M_T} nodes in the third level. These p_{M_T} nodes are expanded in a similar way to the fourth level, and this procedure continues until the very last level, where the minimum-distance node is taken as the detected node.

Moreover, from the Schnorr-Euchner (SE) ordering [19], we know that finding

$$(s^{(i)}, T_i(s^{(i)}))_{min} \leftarrow \arg \min_{\{(s^{(i)}, T_i(s^{(i)}))_j | j=1,...,p'\}} T_i(s^{(i)})$$

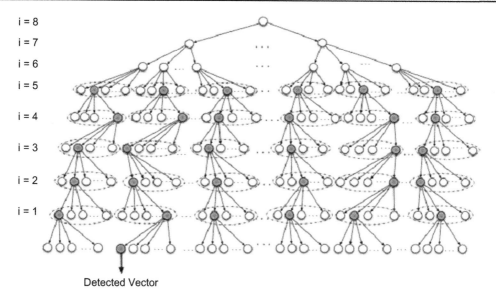

Figure 5-2:
Flex-Sphere algorithm for a 64-QAM, 4 × 4 system. The topmost two levels are fully expanded. The nodes marked with black are the minimum in their own set, where each set is denoted by a dashed line. Note that because of the real-valued decomposition, each node has only $M = 2 \times M_T = 8$.

basically corresponds to finding the real-valued constellation point closest to

$$\frac{1}{R_{ii}} J_{i+1}(s^{(i+1)})$$

See equation (7). Thus, the long sorting of K-best is avoided.

Modified real-valued decomposition (M-RVD) ordering

For the sort-free detector described in the preceding section, we can use a modified real-valued decomposition (M-RVD) ordering which improves the BER performance compared to the ordering given in equation (2). The new decomposition is summarized as [20]:

$$\hat{y} = \hat{H}\hat{s} + \hat{n} \tag{8}$$

or,

$$
\begin{pmatrix}
\Re(\tilde{y}_1) \\
\Im(\tilde{y}_1) \\
\Re(\tilde{y}_2) \\
\Im(\tilde{y}_2) \\
\vdots \\
\Re(\tilde{y}_{M_R}) \\
\Im(\tilde{y}_{M_R})
\end{pmatrix}
= \hat{H}
\begin{pmatrix}
\Re(\tilde{s}_1) \\
\Im(\tilde{s}_1) \\
\Re(\tilde{s}_2) \\
\Im(\tilde{s}_2) \\
\vdots \\
\Re(\tilde{s}_{M_T}) \\
\Im(\tilde{s}_{M_T})
\end{pmatrix}
+
\begin{pmatrix}
\Re(\tilde{n}_1) \\
\Im(\tilde{n}_1) \\
\Re(\tilde{n}_2) \\
\Im(\tilde{n}_2) \\
\vdots \\
\Re(\tilde{n}_{M_R}) \\
\Im(\tilde{n}_{M_R})
\end{pmatrix}
\tag{9}
$$

where $\hat{H}$ is the permuted channel matrix of equation (3) whose columns are reordered to match the other vectors of the new decomposition ordering in equation (8). It is worth noting that since the difference between RVD and M-RVD is the grouping of the signals, there is no extra computational cost associated with this modified ordering. We will see in the following sections how this ordering could reduce the latency for Flex-Sphere.

FPGA design of the configurable detector for SDR handsets

In this section, the main features of the architecture and the FPGA implementation of the SDR handset detector are presented. We use the Xilinx System Generator [21] to implement the proposed architecture. In order to support all the different numbers of antenna/user and modulation orders, the detector is designed for the maximal case, i.e., $M_T \times M_R$, 64-QAM case, and configurability elements are introduced in the design to support different configurations.

PED computations

Computing the norms in (7) is performed in the PED blocks. Depending on the level of the tree, three different PED blocks are used: The PED in the first real-valued level, PED, corresponds to the root node in the tree, $i = M = 2M_T = 8$. The second level consists of $64 = 8$ parallel PED2 blocks, which compute 8 PEDs for each of the 8 PEDs generated by PED_1, thus, generating 64 PEDs for the $i = 7$ level. Followed by this level, there are 8 parallel general PED computation blocks, PEDg, which compute the closest node PED for all 8 outputs of each of the PED2s. The next levels will also use PEDg. At the end, the Min Finder unit detects the signal by finding the minimum of the 64 distances of the appropriate level. The block diagram of this design is shown in Figure 5-3.

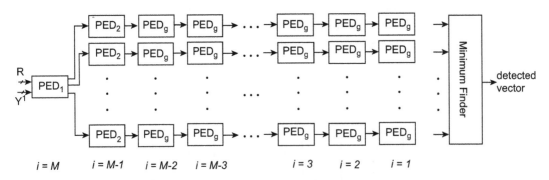

Figure 5-3:
The block diagram of the Flex-Sphere. Note that there are M parallel PEDs at each level. The input to the Minimum Finder is fed from the appropriate PED block.

Configurable design

In order to ensure the configurability of the Flex-Sphere, it needs to support different M_T as well as different modulation orders for different users. The configurability of the detector is achieved through two input signals, M_T and $q_{(i)}$, which control the number of antennas and the modulation order, respectively. These two inputs can change based on the system parameters at any time during the detection procedure. Therefore, this configurability is a real-time operation.

Number of Antennas

The M_T determines the number of detection levels, and it is set through M_T input to the detector, which in turn, would configure the Min Finder appropriately. Therefore, the minimum finder can operate on the outputs of the corresponding level, and generate the minimum result. In other words, the multiplexers in each input of the Min Finder block choose which one of the four streams of data should be fed into the Min Finder. Therefore, the inputs to the Min Finder would be coming from the $i = 5$, 3 or 1, if M_T is 2, 3, or 4, respectively, (see Figure 5-3).

The M_T input can change on-the-fly; thus, the design can shift from one mode to another mode based on the number of streams it is attempting to detect at any time. Moreover, as will be shown later, the configurability of the minimum finder guarantees that less latency is required for detecting a smaller number of streams.

Modulation Order

In order to support different modulation orders per data stream, the Flex-Sphere uses another input control signal $q_{(i)}$ to determine the maximum real value of the modulation order of the i-th level. Thus, $q_{(i)} \in \{1, 3, 7\}$. Moreover, since the modulation order of each level is changing, a simple comparison-thresholding can not be used to find the closest candidate for Schnorr-Euchner [19] ordering. Therefore, the following conversion is used to find the closest SE candidate:

$$\tilde{s} = g\left(2\left[\frac{J+1}{2}\right] - 1\right) \tag{10}$$

where [.] represents rounding to the nearest integer, $J = (1/R_{ii}) \cdot J_{i+1}$ of equation (7), and $g(.)$ is:

$$g(x) = \begin{cases} -q^{(i)} & x \leq -q^{(i)} \\ x & -q^{(i)} \leq x \leq q^{(i)} \\ q^{(i)} & x \geq q^{(i)} \end{cases} \tag{11}$$

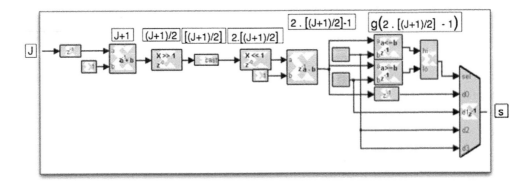

Figure 5-4:
The pipelined System Generator block diagram for equation (10) in the PED$_g$ to support different modulation orders.

All of these functions can be readily implemented using the available building blocks of the Xilinx System Generator (see Figure 5-4). Note that the multiplications/divisions are simple one-bit shifts.

For the first two levels, which correspond to the in-phase and quadrature components of the last antenna, the PED of the out-of-range candidates are simply overwritten with the maximum value; thus, they will be automatically discarded during the minimum-finding procedure.

Modified real-valued decomposition (M-RVD)

Using the real-valued decomposition, the two extra adders that are required for each complex multiplication can be avoided, thus avoiding the unnecessary FPGA slices on the addition operations. Moreover, while using the complex-valued operations requires the SE ordering of [11], which would be a demanding task given the configurable nature of the detector, with the real-valued decomposition, the SE ordering can be implemented more efficiently and simply for the proposed configurable architecture as described earlier. Also, note that even though some of the multiplications can be replaced with shift-adds in an area-optimized ASIC design, for an FPGA implementation, the appropriate design choice is to use the available embedded multipliers, commonly known as XtremeDSP and DSP48E in Virtex-4 and Virtex-5 devices.

It is noteworthy that if the conventional real-valued decomposition of (3) were employed, then the results for a 2×2 system would have been ready only after going through all the in-phase tree levels and the first two quadrature levels. However, with the modified real-valued decomposition (M-RVD), each antenna is isolated from other antennas in two consecutive levels of the tree. Therefore, there is no need to go through the latency of the unnecessary levels. Thus, using the M-RVD technique offers a latency reduction compared to the conventional real-valued decomposition.

Timing analysis

Each of the PED_g blocks are responsible for expanding 8 nodes; thus, the folding factor of the design is $F = 8$. In order to ensure a high maximum clock frequency, several pipelining levels are introduced inside each of the PED computation blocks. The latency of the PED_1, PED_2, and PED_g blocks are 7, 17, and 22, respectively. Note that the larger latency of the PED_g blocks is due to more multiplications required to compute the PEDs of the later levels. The Min Finder block has a latency of 8.

As mentioned earlier, different values of M_T require different numbers of tree levels, which incurs different latencies. The latencies of the three different configurations of M_T are presented in Table 5-1. In computing the latencies, an initial 8 cycles are required to fill up the pipeline path.

Table 5-1: Latency for different values of M_T.

M_T	Latency
$M_T = 2$	$8 + PED_1 + PED_2 + 2 \cdot PED_g + Min\_Finder = 84$
$M_T = 3$	$8 + PED_1 + PED_2 + 4 \cdot PED_g + Min\_Finder = 128$
$M_T = 4$	$8 + PED_1 + PED_2 + 6 \cdot PED_g + Min\_Finder = 172$

Xilinx FPGA implementation results for $M_T = 3$

Table 5-2 presents the System Generator implementation results of the Flex-Sphere on a Xilinx Virtex-4 FPGA, xc4vfx100-10ff1517 [21] for 16-bits precision. The maximum number of detectable streams is set to $M_T = 3$. The maximum achievable clock frequency is

Table 5-2: FPGA resource utilization summary of the proposed Flex-Sphere for the Xilinx Virtex-4, xc4vfx100-10ff1517 device.

No. of Antennas	2, 3
Modulation Order	{4, 16, 64}-QAM
Max. Data Rate	562.5 Mbps
Number of Slices	18,825/42,176 (44%)
Number of Slice FFs	23,961/84, 352 (28%)
Number of LUTs	30,297/84, 352 (35%)
Number of DSP48E	129/160 (80%)
Max. Freq.	250 MHz

250 MHz. Since the design folding factor is set to F $= 8$, the maximum achievable data rate, i.e., $M_T = 3$ and $p_i = 64$, is:

$$D = \frac{M_T \cdot \log \omega}{F} \cdot f_{max} = 562.5 \text{ [Mbps]}. \tag{12}$$

Xilinx FPGA implementation results for $M_T = 4$

Table 5-3 presents the System Generator implementation results of the Flex-Sphere on a Xilinx Virtex-5 FPGA, xc5vsx95t-3ff1136 [21] for 16-bits precision and $M_T = 4$. The maximum achievable clock frequency is 285.71 MHz. Since the design folding factor is set to F $= 8$, the maximum achievable data rate, i.e., $M_T = 4$ and $p_i = 64$, is:

$$D = \frac{M_T \cdot \log \omega}{F} f_{max} = 857.1 \text{ [Mbps]}. \tag{13}$$

The Flex-Sphere can support different numbers of antennas and modulation orders, and achieves high data rate requirements of various wireless standards. Table 5-4 summarizes the data rates for all of the different scenarios of the $M_T = 4$, Virtex-5 implementation.

Table 5-3: FPGA resource utilization of the proposed Flex-Sphere.

Device	XC5VSX95
No. of Antennas	2, 3, 4
Modulation Order	{4, 16, 64}-QAM
Max.Data Rate	857.1 Mbps
BER $= 10^{-4}$ @ SNR $=$	$= 25$ dB
Number of Slices	11,604/14,720 (78%)
Number of Registers/FFs	27, 115/58,880 (46%)
Number of Slice LUTs	33, 427/58,880 (56%)
Number of DSP48E/Multipliers	321/640 (50%)
Number of block RAMs	0(0%)
Max. Freq.	285.71 MHz

Table 5-4: Data rate for different configurations of the 4 × 4.

	4-QAM	16-QAM	64-QAM
$M_T = 2$	142.7 Mbps	285.7 Mbps	428.4 Mbps
$M_T = 3$	214.1 Mbps	428.4 Mbps	642.7 Mbps
$M_T = 4$	285.7 Mbps	571.4 Mbps	857.1 Mbps

Simulation results

In this section, we present the simulation results for the Flex-Sphere, and compare the performance of the FPGA fixed-point implementation with that of the optimum floating-point maximum-likelihood (ML) results. The Xilinx System Generator implementation of the Flex-Sphere detector is shown in Figure 5-5. Prior to the M-RVD, introduced earlier, we employ the channel ordering of [22] to further close the gap to ML. Also, we make the assumption that all the streams are using the same modulation scheme. We assume a Rayleigh fading channel model, i.e., complex-valued channel matrices with the real and imaginary parts of each element drawn from the normal distribution.

In order to ensure that all the antennas in the receiver have similar average received SNR, and none of the users' messages are suppressed with other messages, a power control scheme is employed. Figure 5-6 shows the simulation results for the maximal 4×4 configuration. As can be seen, the proposed hardware architecture implementation performs within, at most, 1 dB of the optimum maximum-likelihood detection.

Beamforming for WiMAX

Multiple Input Multiple Output (MIMO) antenna systems can be used to increase the data rate (multiplexing gain), improve reliability (diversity gain) or both, in certain combinations [23]. The previous section presented schemes for achieving multiplexing gain. This section explains how full diversity gain is achieved in a WiMAX system via beamforming techniques. Implementation challenges of a beamforming WiMAX system are analyzed and results of experiments performed on an FPGA-based testbed are presented.

MIMO schemes that achieve full diversity gain can be either open loop or closed loop. In a closed loop system, the transmitter has some knowledge of the instantaneous channel realization and if the forward and reverse channels are not reciprocal, the transmitter acquires channel state information from feedback sent by the receiver. Transmit beamforming is an example of a closed loop scheme that achieves full diversity. In an open loop system, the transmitter does not require knowledge of the channel state information; Space Time Codes (STC) are an example of open loop full diversity achieving schemes [24].

The potential for more reliable communication at higher data rates has made closed loop transmit beamforming techniques part of the physical layer of standards for future wireless communications, for example, 802.11n, WiMAX, and 3GPP [25]. Transmit beamforming has been considered in WiMAX Frequency Division Duplexing (FDD) mode, where the forward and feedback channels are not on the same frequency band, hence they are not reciprocal.

Beamforming in wideband systems

The physical layer of the WiMAX standard achieves wideband communication via Orthogonal Frequency Division Multiplexing (OFDM). OFDM is a multicarrier scheme in

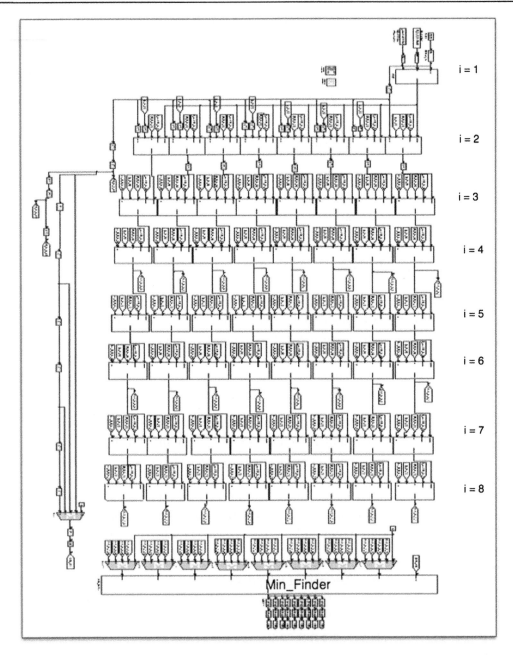

Figure 5-5:
Xilinx System Generator implementation of the Flex-Sphere detector.

which subcarriers occupy orthogonal narrowband channels. Beamforming is a narrowband scheme that can be easily extended to wideband OFDM systems by applying beamforming techniques per subcarrier.

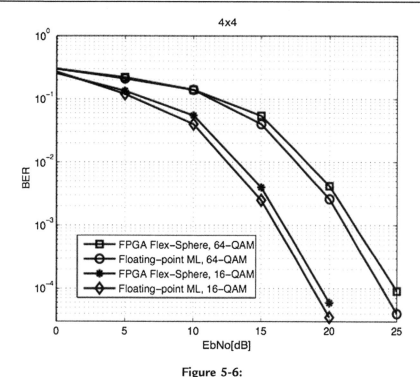

Figure 5-6:
BER plots comparing the performance of the floating-point maximum-likelihood (ML) with the FPGA implementation. Note that the channel preprocessing of [22] is employed to improve performance.

Figure 5-7 shows a beamforming OFDM system where channel state information is sent from the receiver to the transmitter via a feedback channel. The system has M_T transmitter antennas, M_R receiver antennas, and K data subcarriers. $\tilde{s}_k$ is used to represent the complex symbol transmitted on subcarrier k, $1 \leq k \leq K$. The time span of the channel impulse response is assumed to be shorter than the time span of the cyclic prefix (this is true for any well designed OFDM system); the baseband relationship between the complex symbol transmitted on subcarrier k and the corresponding received signal y_k is given by:

$$y_k = z_k^H \tilde{H}_k w_{b_k} \tilde{s}_k + z_k^H \tilde{n}_k. \tag{14}$$

The $M_T \times 1$ vector used for beamforming subcarrier k is represented by w_{bk}. The K beamforming vectors $w_{b1}, \ldots, w_{bK}$ are chosen from a beamforming codebook W of cardinality $|W| = 2_B$, hence, $W = \{w_1, w_2, \ldots, w_{|W|}\}$, and index b_k specifies the index of the beamforming vector chosen to beamform subcarrier k, $1 \leq b_k \leq |W|$. The channel matrix corresponding to subcarrier k is represented by the $M_R \times M_T$ matrix $\tilde{H}_k$, the entries of $\tilde{H}_k$

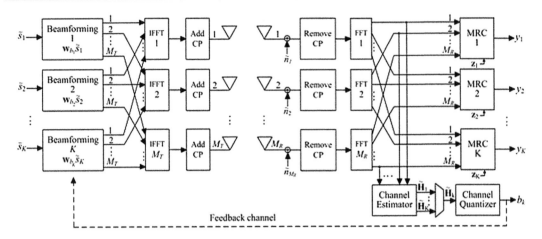

Figure 5-7:
Baseband representation of a beamforming MIMO-OFDM system. The FFT and IFFT blocks compute a Fast Fourier Transform and Inverse Fast Fourier Transform respectively. CP is used to denote cyclic prefix.

are i.i.d., and each entry is a circularly symmetric complex Gaussian random variable with zero mean and unit variance per complex dimension. To simplify the analysis, it is assumed that the receiver has perfect knowledge of the channel matrix $\tilde{H}_k$. The $M_R \times 1$ vector z_k is the Maximum Ratio Combining (MRC) vector for subcarrier k. The $M_R \times 1$ noise vector $\tilde{n}_k$ has i.i.d. entries where each entry is assumed to be circularly symmetric complex Gaussian with zero mean and variance N_0 per complex dimension. Each subcarrier is transmitted with the same average power E_s; this constraint is met by setting $E[|\tilde{s}_k|2] = E_s$ and $\|w_{b_k}\| = 1$. The WiMAX standard specifies beamforming codebooks for different numbers of transmitter antennas and codebook cardinalities; all the beamforming vectors in the WiMAX codebooks satisfy $\|w_{b_k}\| = 1$. (Note that $\|\cdot\|$ is used to denote the vector two-norm and $(\cdot)_H$ denotes the conjugate transposition of a matrix.)

The SNR for subcarrier k is given by:

$$\text{SNR}_k = \frac{E_s \left| z_k^H \tilde{H}_k w_{b_k} \right|^2}{\|z_k\|^2 N_0}, \tag{15}$$

the MRC vector and the index of the beamforming vector that maximize SNR_k are [5]

$$z_k = \frac{\tilde{H}_k w_{b_k}}{\left\| \tilde{H}_k w_{b_k} \right\|} \tag{16}$$

and

$$b_k = \arg \max_{1 \leq i \leq |w|} \left\| \tilde{\mathbf{H}}_k w_i \right\|^2 \tag{17}$$

respectively.

In a beamforming FDD system like the one considered in the WiMAX standard, the beamforming codebook W is known to both the transmitter and the receiver. The K indices b_1, $b_2,\ldots,b_K$ computed by the channel quantizer at the receiver are fedback to the transmitter. Based on these indices, the transmitter beamforms subcarriers 1, 2, . . . , K using vectors w_{b1}, $w_{b2}, \ldots, w_{bK}$ respectively. The K indices b_1, $b_2,\ldots,b_K$ computed by the channel quantizer are also used to compute the MRC vectors as shown in (16). Since the codebook has cardinality $|W| = 2^B$, each index b_k is represented using B bits. The total amount of feedback bits is equal to KB, which corresponds to B bits of feedback information per subcarrier. The WiMAX standard specifies different possible mechanisms to send the KB bits of feedback information from the receiver to the transmitter; in some configurations (e.g. adjacent subcarrier permutations [26]) the amount of feedback bits can be reduced by clustering subcarriers.

The process of quantizing each subcarrier channel $\tilde{\mathbf{H}}_k$ into B bits takes place at the channel quantizer. Since each subcarrier goes through an independent channel, each subcarrier channel must be quantized independently. In order to reutilize hardware resources the quantizer processes one subcarrier at a time, as shown in Figure 5-7. The input-output relationship of the channel quantizer is given by equation (17) which is computed by exhaustive searching over the elements in the codebook W.

The system depicted in Figure 5-7 achieves full diversity if $|W| \geq M_T$ and the beamforming codebook is designed based on the Grassmannian beamforming criterion [5]. The WiMAX standard specifies codebooks for 2, 3, and 4 transmit antennas. For 2 transmit antennas the WiMAX standard specifies a codebook of size $|W| = 8$; for 3 and 4 transmitter antennas the standard specifies one codebook of size $|W| = 8$ and one codebook of size $|W| = 64$. The WiMAX codebooks seem to be designed based on the Grassmannian criterion [27] and it can be verified via Monte Carlo simulation that WiMAX codebooks achieve full diversity.

Computational requirements and performance of a beamforming system

Implementation of the beamforming and MRC blocks in Figure 5-7 is straightforward; the most intensive computation in each of the blocks is a vector multiplication. Also, the K beamforming and K MRC blocks can be reduced to one beamforming block and one MRC block by processing one subcarrier at a time.

Implementation of the channel quantizer in Figure 5-7 can require a large amount of resources depending on the number of transmitter antennas and codebook size. The entries of $\tilde{\mathbf{H}}_k$ are complex numbers and the WiMAX standard specifies that the entries of w_i are complex numbers rounded to four decimal places. Hence, computing $||\tilde{H}_k w_i||2$ for all the $|W|$ codewords requires $|W| M_T M_R$ complex multiplications, $|W| M_T M_R - |W| M_R$ complex additions, $2|W| M_R$ real multiplications, and $2|W| M_R - |W|$ real additions. After computing $||\tilde{H}_k w_i||_2$ for all $|W|$ codewords the channel quantizer searches for the codeword with the largest $||\tilde{H}_k w_i||_2$. Implementing a tree search requires $|W|-1$ relational blocks that compare two inputs and then output the greater of the two.

The total amount of resources required for channel quantization for one subcarrier in different WiMAX configurations is shown in Table 5-5. Notice that as the number of antennas and codebook cardinality increases, the number of resources required increases dramatically. The maximum number of embedded multipliers in an FPGA is 512 for the Xilinx Virtex-4 family [28] and 1056 for the Xilinx Virtex-5 family [29]. Hence, implementing the channel quantizer can become a bottleneck for the implementation of WiMAX systems.

The amount of resources required for channel quantization can be reduced by reutilizing resources, but resource reutilization would increase the latency of the channel quantizer. Reutilization is possible as long as the timing constraints are met. In a WiMAX system the timing constraints can be quite tight, specially in high mobility scenarios and in frequency division duplexing (FDD) mode. In a high mobility scenario the channel coherence interval decreases and feedback information must be sent before it becomes stale and in FDD mode feedback information can potentially be sent as soon as it is available. In both FDD mode and high mobility the faster the feedback information is sent the more throughput the system will have, since more payload data will be sent during a coherence interval.

Another alternative to reduce the amount of resources required for channel quantization is to use a mixed codebook scheme as proposed in [30]. In a mixed codebook scheme, the WiMAX codebook is used at the transmitter for beamforming and at the receiver for MRC, and a mapped version of the WiMAX codebook is used at the receiver's channel quantizer. The entries in the mapped version of the WiMAX codebook belong to $\{0,+1,-1,+j,-j\}$; hence, all the complex multiplications required for channel quantization can be implemented via simple changes and swapping of the real and imaginary parts of the channel matrix entries.

Table 5-5: Resources required for channel quantization.

	$M_T = 2, M_R = 1$ $\|w\| = 8$	$M_T = 3, M_R = 3$ $\|w\| = 8$	$M_T = 4, M_R = 4$ $\|w\| = 64$
Complex Multipliers	16	72	1024
Complex Adders	8	48	768
Real Multipliers	16	48	512
Real Adders	8	40	448
Relational	7	7	63

Table 5-6: Resources required for channel quantization using a WiMAX codebook and a mapped WiMAX codebook. Results correspond to a system with 4 transmitter and 4 receiver antennas and a codebook of 64 codewords.

	WiMAX Codebook	Mapped WiMAX Codebook
Complex Multipliers	1024	0
Complex Adders	768	768
Real Multipliers	512	576
Real Adders	8	448
Negators	0	2048
5-input Multiplexer	0	2048
Relational	63	63

Table 5-6 compares the amount of resources required for channel quantization when using a mapped WiMAX codebook with the amount of resources required when using a WiMAX codebook. The mapped version of the WiMAX codebook can be obtained via the vector mapping proposed in [30]. Using a mixed codebook scheme will affect performance because the codebook used for beamforming at the transmitter and MRC at the receiver is different from the codebook used for channel quantization at the receiver. However, the performance loss is very small, since by construction, the mapped WiMAX codebook has quantization regions similar to the quantization regions of the WiMAX codebook from which it is obtained [30]. Notice that the mixed codebook scheme remains WiMAX compliant because the mapped WiMAX codebook is only used for channel quantization.

Observe from Table 5-6 that when using a mapped WiMAX codebook for channel quantization, the number of multiplications required reduces to zero. This reduction is obtained by increasing the number of five input multiplexers and negators (a negator block is a very simple block that computes the arithmetic negation or two's complement of its input). For both ASIC and FPGA implementations, 2048 five input multiplexers, 2048 negators, and 64 real multipliers require fewer resources than implementing 1024 complex multipliers.

Figure 5-8 shows the performance of different beamforming WiMAX configurations. All simulations were run for a one subcarrier model, and the results obtained represent the per subcarrier behavior in an OFDM system. The results in Figure 5-8 correspond to the best case scenario of perfect channel estimate at the receiver, noiseless and zero delay feedback, and floating-point processing. The results in Figure 5-8 show that in a feedback-based beamforming system, such as FDD WiMAX, few bits of feedback can achieve performance close to the ideal case of infinite feedback. Also, observe that using a mixed codebook scheme (labeled MC in the figure) results in small performance degradation, and as shown in Table 5-6, the amount of resources can be significantly reduced by using a mixed codebook scheme. Figure 5-8 also shows the performance of the Alamouti STC [31], which is an open

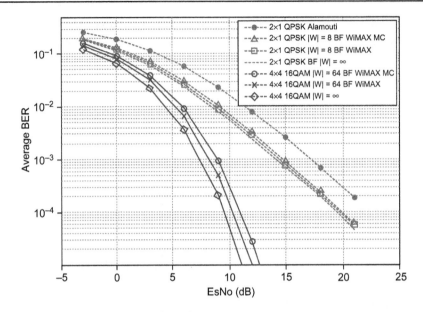

Figure 5-8:
Performance of a $M_T \times M_R$ beamforming WiMAX system for different values of M_T, M_R, and $|W|$. BF is used to denote beamforming, MC is used to denote mixed codebook scheme, and $|W| = \infty$ is used to denote an infinite feedback scenario.

loop scheme. Observe that the closed loop beamforming scheme outperforms the open loop scheme.

Beamforming experiments using WARPLab

WARPLab is a framework for rapid prototyping of physical layer algorithms. The WARPLab framework combines the ease of MATLAB with the capabilities of the Wireless Open Access Research Platform (WARP) developed at Rice University [32]. This section describes in detail the WARPLab framework and presents experiment results that show the performance gains that can be obtained with beamforming in a WiMAX system.

WARPLab framework

WARP provides a unique platform to develop, implement, and test advanced wireless communication algorithms. The platform architecture consists of four main components: custom hardware, platform support packages, open-access repository, and research applications.

The custom hardware allows implementation and scalability of sophisticated signal processing algorithms and provides extensible peripheral options for radios and user interface. The main component of the WARP hardware is a Xilinx Virtex II-Pro FPGA, a new

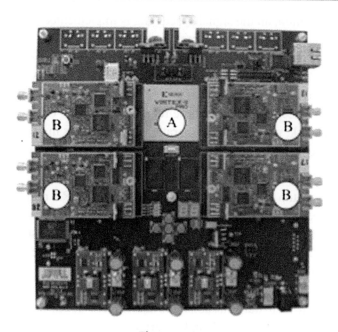

Figure 5-9:
WARP board with a radio board in each of the four daughtercard slots. A: Xilinx Virtex-II Pro FPGA.
B: Radio board.

version of the WARP hardware will soon be released, and in this new version, the Xilinx Virtex II Pro FPGA has been replaced by a more powerful Xilinx V-4 FPGA. The WARP board, shown in Figure 5-9, has four daughter card slots and each slot is connected to a dedicated bank of I/O pins on the FPGA, providing a flexible, high-throughput interface. As shown in Figure 5-9, the four daughter card slots can be used to connect the FPGA to four different radio boards so that up to a 4×4 MIMO system can be built. The radio boards have been designed by Rice University students and these boards are capable of targeting both the 2.4 GHz and 5 GHz ISM bands. They are intended for wide band applications, such as OFDM, with a bandwidth up to 40 MHz.

The platform support packages facilitate seamless use of the WARP hardware by researchers working at all layers of wireless network design. The open-access repository [33], accessible from the Internet, is the central archive for all source codes, models, platform support packages, application building blocks, research applications, design documents, and hardware design files associated with WARP. The contents of the repository are verified by the project administrator at Rice University.

WARP allows clean-slate design and prototyping of physical layer algorithms for wireless communications via two possible design flows: 1) real time implementation and 2) WARPLab framework that allows real-time RF transmission and offline processing on a host PC.

In real-time implementation, all the signal processing is implemented in the FPGA, allowing implementation of a complete end-to-end real-time system. However, many physical layer researchers are interested in rapid prototyping and over-the-air testing of new algorithms without having to deal with the details of FPGA implementation, and without having to implement mechanisms that belong to higher layers, like carrier sensing, contention resolution protocols, and packet detection. To meet these requirements, Rice University has developed the WARPLab framework, which allows generation of waveforms in MATLAB and provides simple m-code functions that allow over-the-air transmission of these waveforms using the WARP hardware.

The basic WARPLab setup is shown in Figure 5-10. Two WARP nodes are connected to a host PC via an Ethernet switch. Up to 16 WARP nodes can be connected to the switch and controlled from the host PC. A user first uses MATLAB to construct waveforms that their algorithm would transmit. These waveforms are loaded into the assigned WARP transmit nodes via Ethernet links that are controlled by custom code on both the PC and WARP nodes. The host PC then triggers the beginning of the experiment by telling the transmit nodes to begin their transmissions and the receive nodes to begin capturing data from the radio. Once transmission and capture are completed, the captured waveforms are passed to the host PC via the Ethernet links. The user can then use MATLAB to process the received waveforms and determine the effects of real radios and wireless channels on their novel algorithm.

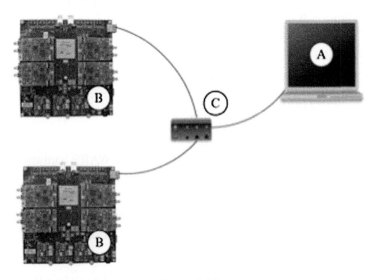

Figure 5-10:
Basic WARPLab setup. A: Host PC, B: WARP node, C: Switch. The switch is connected to the WARP nodes and the host PC via Ethernet links.

The WARPLab framework provides the software necessary for easy interaction with the WARP nodes directly from the MATLAB workspace. The software consists of FPGA code and m-code functions which are all available in the WARP repository [34].

Experiment setup and results

This section shows experiment results for a beamforming system with two transmitter antennas and one receiver antenna. The system was designed and tested using the WARPLab framework. Experiments were performed over a wireless channel emulated using Spirent's SR-5500 wireless channel emulator. The experiment setup is shown in Figure 5-11. Two radios at the transmitter node are connected to the two RF inputs of the channel emulator which emulates two independent RF wireless channels. The two RF outputs of the channel emulator are added to emulate a 2×1 system; the output of the RF adder is connected to one radio at the receiver node. The channel emulator is connected to the host computer via Ethernet; the host computer controls the two WARP nodes and the channel emulator via the Ethernet links.

The system implemented is a single subcarrier system, the results representing the per subcarrier behavior in an OFDM system. Table 5-7 summarizes the experiment conditions; the two emulated RF channels were both set using exactly the same parameters. Since the delay spread is much smaller than the symbol period, the transmitted signal goes through a flat fading channel.

Transmission of the payload data was done in packets of 110 symbols. The number of symbols per packet was limited to 110 due to the characteristics of the transmitted signal

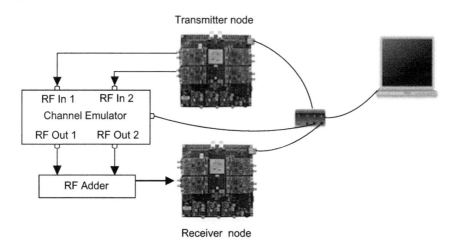

Figure 5-11:
Experiment setup. The basic WARPLab setup was connected to a channel emulator and an RF adder to emulate a system with two transmitter antennas and one receive antenna.

Table 5-7: Experimental conditions.

Parameter	Value
Number of transmitter antennas	2
Number of receiver antennas	1
Carrier frequency	2.4 GHz
Number of subcarriers	1
Bandwidth	625 kHz
Sampling frequency	40 MHz
Pulse shaping filter	Squared Root Raised Cosine (SRRC)
SRRC roll-off factor	1
Symbol time	3.2 μs
Modulation	16 QAM
Coding Rate	1 (No error correction code)
Energy per symbol	-20 dBm at input of channel emulator
Paths per emulated RF channel	3
Envelope per path	Rayleigh flat fading all 3 paths
Fading Doppler per path	0.1 Hz (0.04 km/h) in all 3 paths emulates block fading channel
Delay per path	Path 1=0 μs, Path 2=0.05 μs, Path 3=0.1 μs
Relative path loss	Path 1=0 dB, Path 2=3.6 dB, Path 3=7.2 dB

(symbol time of 3.2 μs and sampling frequency of 40 MHz) and the maximum number of samples that can be stored per receiver radio in a node, which is limited to 2_{14} samples [34]. As shown in Figure 5-12, two pilot sequences were sent before transmission of payload. The first pilot sequence was used for computation of the channel estimate used for channel quantization, and the total pilot energy was set equal to twice the energy per symbol. The second pilot sequence was a beamformed pilot sequence, which was used to obtain an estimate of the channel, times the beamforming vector. This estimate was used for MRC at the receiver and the total pilot energy was equal to the total energy per symbol. The error-free feedback channel was implemented in the host PC (the host PC is connected to both transmitter and receiver). The feedback delay was approximately 60 ms, which is the time it took to send training, estimate the channel, quantize the channel estimate, and start transmitting the beamformed signal.

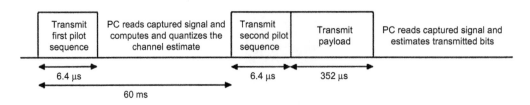

Figure 5-12:
Time diagram of transmitted signal and PC processing for beamforming experiment.

In order to compare the performance of a closed loop scheme like beamforming and an open loop scheme like Alamouti, a 2 × 1 Alamouti scheme was also implemented and tested using the WARPLab framework. In the Alamouti scheme, transmission of the payload data was also done in packets of 110 symbols. For the Alamouti implementation, only one pilot sequence was sent and payload was sent immediately after the pilot sequence. The total pilot energy was equal to the total energy per symbol.

Simulation results in Figure 5-8 showed that the beamforming scheme has better performance than the Alamouti scheme and that the mixed codebook scheme for efficient implementation results in small performance degradation. These two observations are verified by the experiment results shown in Figure 5-13. The results also show that, as expected, few feedback bits have performance close to the performance of infinite feedback. Observe that the plots of the experiment results are not as smooth as the plots of the simulation results. This is very likely due to hardware non-linearities and the fact that simulations were run more for total number of bits than were experiments because simulations ran faster.

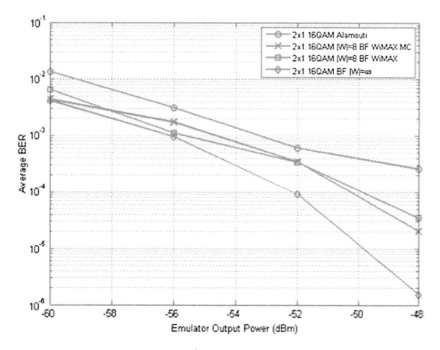

Figure 5-13:
Experiment results for a 2 × 1 beamforming system and a 2 × 1 Alamouti system. BF is used to denote beamforming, MC is used to denote mixed codebook scheme, and $|W| = \infty$ is used to denote an infinite feedback scenario.

Conclusion

In this chapter, we discussed the architectural challenges of spatial multiplexing and diversity gain schemes, and further, introduced FPGA-centric architectures and experiments for these systems. We introduced a flexible architecture and implementation of a spatial multiplexing MIMO detector, Flex-sphere, and its FPGA implementation. We also presented a hardware architecture for beamforming for WiMAX, as a way to enhance the diversity and performance in the next-generation wireless systems. Finally, we showed how we utilized the WARP platform for studying the effects of MIMO systems in real-world scenarios.

References

[1] [Online]. Available: http://www.3gpp.org/
[2] [Online]. Available: http://wirelessman.org/
[3] G. Foschini, Layered space-time architecture for wireless communication in a fading environment when using multiple antennas, Bell Labs. Tech. Journal 2 (1996).
[4] G.D. Golden, G.J. Foschini, R.A. Valenzuela, P.W. Wolniansky, Detection algorithms and initial laboratory results using V-BLAST space-time communication architecture, Electronics Letters 35 (1999) 14–15.
[5] D.J. Love, R.W. Heath, T. Strohmer, Grassmannian beamforming for multiple-input multiple-output wireless systems, IEEE Trans. Inform. Theory 49 (Oct 2003) 2735–2747.
[6] K.K. Mukkavilli, A. Sabharwal, E. Erkip, B. Aazhang, On beamforming with finite rate feedback in multiple-antenna systems, IEEE Trans. Inform. Theory 49 (Oct 2003) 2562–2579.
[7] Z. Guo, P. Nilsson, A 53.3 Mb/s 4×4 16-QAM MIMO decoder in 0.35µm CMOS, IEEE Int. Symp. Circuits Syst. 5 (May 2005) 4947–4950.
[8] M.O. Damen, H.E. Gamal, G. Caire, On maximum likelihood detection and the search for the closest lattice point, IEEE Trans. on Inf. Theory 49 (no 10) (Oct 2003) 2389–2402.
[9] U. Fincke, M. Pohst, Improved methods for calculating vectors of short length in a lattice, including a complexity analysis, Math. Computat 44 (no 170) (Apr 1985) 463–471.
[10] B. Hochwald, S. ten Brink, Achieving near-capacity on a multiple-antenna channel, IEEE Trans. on Comm. 51 (Mar 2003) 389–399.
[11] A. Burg, M. Borgmann, M. Wenk, M. Zellweger, W. Fichtner, H. Bolcskei, VLSI implementation of MIMO detection using the sphere decoding algorithm, IEEE Journal of Solid-State Circuits 40 (no 7) (July 2005) 1566–1577.
[12] K. Amiri, J.R. Cavallaro, FPGA implementation of dynamic threshold sphere detection for MIMO systems, 40th Asilomar Conf on Signals, Systems and Computers (Nov 2006).
[13] D. Garrett, L. Davis, S. ten Brink, B. Hochwald, G. Knagge, Silicon complexity for maximum likelihood MIMO detection using spherical decoding, IEEE JSSC 39 (no 9) (Sep 2004) 1544–1552.
[14] Z. Guo, P. Nilsson, Algorithm and implementation of the K-Best sphere decoding for MIMO detection, IEEE JSAC 24 (no 3) (Mar 2006) 491–503.
[15] K. Wong, C. Tsui, R.S. Cheng, W. Mow, A VLSI architecture of a K- best lattice decoding algorithm for MIMO channels, IEEE Int. Symp. Circuits Syst. 3 (May 2002) 273–276.
[16] K. Amiri, J.R. Cavallaro, C. Dick, R. Rao, A high throughput configurable SDR detector for multi-user MIMO wireless systems, Springer Journal of Signal Processing (2009).
[17] K. Amiri, C. Dick, R. Rao, J.R. Cavallaro, Flex-Sphere: An FPGA Configurable Sort-Free Sphere Detector for Multi-user MIMO Wireless Systems, Proc. of SDR Forum (Oct 2008).

[18] L. G. Barbero and J. S. Thompson, 'FPGA design considerations in the implementation of a fixed-throughput sphere decoder for MIMO systems,' Field Programmable Logic and Applications, 2006. FPL '06. International Conference on, Aug. 2006.

[19] C.P. Schnorr, M. Euchner, Lattice basis reduction: improved practical algorithms and solving subset sum problems, Math. Programming 66 (no 2) (Sep 1994) 181−191.

[20] K. Amiri, C. Dick, R. Rao, J.R. Cavallaro, Novel sort-free detector with modified real-valued decomposition (M-RVD) ordering in MIMO systems, Proc. of IEEE Globecom (Dec 2008).

[21] [Online]. Xilinx: http://www.xilinx.com/

[22] L.G. Barbero, J.S. Thompson, A fixed-complexity MIMO detector based on the complex sphere decoder, IEEE 7th Workshop on Signal Processing Advances in Wireless Communications, 2006. SPAWC '06 (Jul 2006).

[23] L. Zheng, D. Tse, Diversity and multiplexing: A fundamental tradeoff in multiple-antenna channels, IEEE Trans. Inform. Theory 49 (Oct 2003) 1073−1096.

[24] V. Tarokh, N. Seshadri, A.R. Calderbank, Space-time codes for high data rate wireless communication: Performance criterion and code construction, IEEE Trans. Inform. Theory 44 (1998) 44−765.

[25] A. Hottinen, M. Kuusela, K. Hugl, J. Zhang, B. Raghothaman, Industrial embrace of smart antennas and MIMO, IEEE Wireless Communications 13 (Aug 2006) 8−16.

[26] 'IEEE standard for local and metropolitan area networks part 16: Air interface for fixed and mobile broadband wireless access systems,' IEEE Std 802.16e−2005 and IEEE Std 802.16-2004/Cor 1-2005, 2006.

[27] D. Love, R. Heath, V.K.N. Lau, D. Gesbert, B.D. Rao, M. Andrews, An overview of limited feedback in wireless communication systems, IEEE JSAC 46 (Oct 2008) 1341−1365.

[28] [Online]. Available: http://www.xilinx.com/products/virtex4

[29] [Online]. Available: http://www.xilinx.com/products/virtex5

[30] M. Duarte, A. Sabharwal, C. Dick, R. Rao, A vector mapping scheme for efficient implementation of beamforming MIMO systems, MILCOM (2008).

[31] S.M. Alamouti, A simple transmit diversity technique for wireless communications, IEEE JSAC 16 (1998) 1451−1458.

[32] [Online]. Available: http://warp.rice.edu

[33] [Online]. Available: http://warp.rice.edu/trac

[34] [Online]. Available: http://warp.rice.edu/trac/wiki/WARPLab

CHAPTER 6

The DSP Hardware/Software Continuum

Mike Brogioli

Chapter Outline

Introduction

Hardware systems and system topologies used in digital signal processing (DSP) can vary greatly in design. Quite often, each system design and component within the system comes with its own programmability, power, and performance tradeoffs. What may be appropriate for one system designer's needs may often not be suitable for another system designer, for a varied number of reasons. This chapter details various components of DSP platform designs as they pertain to system configurability, system programmability, algorithmic demands, and system complexity. At one end of the spectrum, application specific integrated circuits (ASICs) are detailed as a high performance, low configurability solution. At the other end of the spectrum, general purpose software programmable embedded microprocessors are presented as a highly configurable solution. Various design points are discussed along the way such as reconfigurable FPGA based solutions and hardware acceleration. Subsequent sections detail tradeoffs for each manner of system design, and aim to provide the system designer with insights into when each solution is appropriate both for current system designs as well as those that may need to scale or migrate hardware platforms in the future.

DSP for Embedded and Real-Time Systems. DOI: 10.1016/B978-0-12-386535-9.00006-8

FPGA in embedded design

FPGA-based solutions for embedded systems often have the advantage of providing a more targeted solution to a given application than the more traditional one size fits most programmable DSPs and embedded processors. In the case of a system designer seeking to tailor the amount of hardware parallelism afforded to their applications, FPGAs provide an attractive alternative. By exploiting the available functional unit level parallelism that FPGAs can provide to signal processing system designers, an FPGA-based system component using modern FPGA platforms such as the Xilinx Virtex 6 can often provide 40–50× improvement in raw performance over its programmable signal processing core counterpart.

There are a number of emerging new use cases for FPGAs in the embedded processing domain. Video surveillance is one such application domain, incorporating real-time video analytics in the overall system design. In-car entertainment and rich media content is another, which strives to incorporate true multimedia experiences often associated with in-home entertainment with other data such as real-time traffic and weather. Much of these in-car applications exhibit large levels of parallelism that may outstrip programmable signal processors and are thus better suited to FPGAs and other highly parallel accelerator-like architectures.

In light of the growth in FPGA development tools, reusable logic components, and available third party designs, system designers are now free to incorporate FPGA- based solutions into a far broader set of designs. At the same time, given the reprogrammable nature of the FPGA, many system designers can retool or augment their designs with significantly less engineering overhead and recurring costs than their traditional discrete or custom logic counterparts.

While FPGAs do afford significantly higher levels of parallelism at the hardware block level versus their programmable microprocessor counterparts, there are a number of other factors that system designers should keep in mind. FPGAs tend to operate at significantly lower clock rates than their programmable processor counterparts, yet still tend to provide higher computational throughput per clock cycle due to the increase in parallel computational hardware. As an example, a given FPGA may operate in the hundreds of MHz range, but may be capable of performing tens of thousands of computations per clock cycle operating in the tens of watts power range. A comparable microprocessor may operate at the 1–2 GHz range, but will be much more limited in the amount of parallel computation that can be performed per clock cycle. Typical SIMD style architectures in this latter category may provide only four to eight computations per clock cycle, whereas an FPGA may provide a 50× improvement in raw computational throughput over its microprocessor counterpart. There are, however, other points that must be considered in using an FPGA within the system design.

The impressive 50× noted performance gains possible with FPGA based designs are predicated on whether or not the computational workload in the overall system is suitable for an FPGA based implementation. Typically, system designers must consider whether or not the workload is suitable in terms of the type of computation performed, whether or not the algorithm will require fixed point or floating point computation, and challenges that are associated with implementing FPGA based designs versus more traditional embedded software written in C/C++ like languages.

FPGA computational throughput and power

As has been mentioned previously, there are power consumption challenges with the usage of FPGAs in embedded designs. Typically in a signal processing workload, however, the high levels of computational parallelism and matrix like nature of computation can permit the FPGA to exploit the parallel nature of the computation to offset the lower clock rate versus their traditionally higher clock rated programmable microprocessor counterparts. In summary, while the FPGA hardware itself may have a significantly lower clock rate than the programmable microprocessor, suitable applications can exploit the vast increases in parallel FPGA hardware to provide significant computational throughput.

Algorithm suitability

When considering the use of an FPGA in an embedded system design, the type of algorithm to be implemented on the FPGA must be considered. FPGAs themselves tend to be suitable to what are traditionally referred to as embarrassingly parallel type problems — that is to say, those types of algorithms that perform parallel repeated tasks in a regular fashion, and may be easily decomposable into modular parts. Many algorithms are suitably parallel, such as radar applications, beam forming, certain types of image compression, and various signal processing kernels such as Fast Fourier Transforms and other matrix based computations.

Other algorithms and applications that are less predictable or require dynamic partitioning and load balancing are likely not suitable for FPGA implementation. In addition, algorithms with significant amounts of control plane processing typically implemented in control logic statements in higher level C/C++ like programming languages may also be unsuitable for FPGA implementation.

Fixed point versus floating point

Another issue that system designers must consider when utilizing FPGAs in their designs is whether or not key algorithms require fixed point or floating point calculations. Traditionally, software programmable microprocessors will support floating point calculations natively in hardware units within the processor pipeline, which affords significant performance improvement for floating point computation versus emulation on the host processor via a software runtime library. In addition, microprocessors will often have parallel hardware within the processor pipeline for vector or SIMD style floating point computation, such as Freescale's Altivec style SIMD extensions as well as other vendors.

While FPGAs can implement floating point computation, they are not particularly well suited to this type of computation. This is due to the large amount of logic that is required for the FPGA design to implement the floating point calculation. This large amount of logic serves to decrease the overall density of the FPGA design in terms of gates versus computational throughput, and in summary reduces the overall computational effectiveness of the FPGA based implementation in the case of primarily floating point based algorithms.

Implementation challenges

When deploying a given algorithm on an FPGA, there are a number of challenges that may arise in the programming of the device. The architectural abstraction of an FPGA is fairly low, compared with that of traditional high level programming languages. As such, programmer knowledge of the hardware design and device details is often required through such languages as Verilog or VHDL. Identifying the optimal mapping of application parallelism onto hardware is also not trivial, and the time consuming synthesis flow increases complexity in identifying performance optional aspects of the implementation. While it is true that high level synthesis tools can raise the level of hardware abstraction, parallelism extraction may be inherently limited by the programming model or may not offer the evaluation and selection of the best parallel extraction for system performance.

As FPGA technology has matured for embedded computing and signal processing, vendors have devised multiple ways to program and configure the FPGA functionality. The most common approach to specifying the FPGA design is by the use of a hardware description language as mentioned already, commonly Verilog or VHDL. These languages are used to describe the functionality and topology of the system. Follow-on tools are also often used in the design to further optimize the power and configuration of the design for peak performance.

System designers may also want to avail themselves of the modular and open source nature of many FPGA system components. As these types of devices have grown in complexity, designers have begun to produce blocks of hardware description language (HDL) code that can be reused in various products. These modular components allow system developers to reuse circuit designs and other system components from outside designs or third party sources. These blocks can provide very simple functionality, all the way up to fully functional codecs and microcontroller solutions. Programmable IP cores are often available directly from the FPGA vendors, as well as from third party suppliers or, in some cases, even as open source HDL code. Commercial IP is often fee based, but comes with expected documentation, verification tools, and support. These packages often include embedded development software kits, development boards, hard and/or soft processor configurations for programmable cores to be used within the design, and various other software tools and system profilers as necessary.

Given the large amounts of parallelism often available in FPGA based solutions, FPGA based computation is often suitable for high performance embedded solutions based around high rate data acquisition, digital signal processing, and software defined radio solutions. It is not uncommon for various solution providers to incorporate FPGA based components in their overall solution for the heavier lifting portions of the computational stack; this may also afford configurability and programmability that is desired by OEMs building their products upon the platform. The tradeoff here is usually system designer dependent; however, offloading significant computational portions into the FPGA can yield a higher computational throughput per watt for suitable parts of the overall application stack. Furthermore, by alleviating the need for development of these portions of the application stack in software, reduction in time to market and validation may be possible.

As various designers attempt to remain within budget while increasing system complexity, FPGA devices and development tools become increasingly important considerations in an overall platform design. The FPGA can facilitate multiple configurations and performance points within a single hardware design. These are often performance points that are difficult, if not impossible, to meet with software programmable processor solutions. While their recurring cost is typically higher than for other solutions, they are often very attractive solutions for low to medium sized projects that requires significant heavy lifting.

Application specific integrated circuits versus FPGA

There are multiple design decisions that must be considered when choosing between an FPGA or an application specific integrated circuit (ASIC) for the hardware design of a system. FPGAs are semiconductor devices that contain programmable logic components or logic blocks, and programmable interconnects. The logic blocks within the FPGA can be programmed with the functions of various logic gates such as logical AND/OR, as well as more complex digital functions such as decoding and multiplication. In most FPGAs, the logic block component of the system also includes memory elements for local data storage. These memory storage elements can be either simple flip flops or local RAM arrays within the device.

Modern FPGAs have a number of uses in embedded systems. At the prototyping stage, they can be used for ASIC prototyping. Due to the high cost of ASIC chips, FPGAs can be used to first verify the logic of the application by programming and uploading the HDL code into the FPGA. This permits faster testing time with reduced cost, and permits the ASIC to be manufactured only upon verification of the design. FPGAs similarly benefit applications that make use of large amounts of parallelism, as was detailed earlier in this chapter. In summary, FPGAs are very attractive for computationally intensive kernels such as FFTs, whereby the data flow graph of the computation maps more directly to the FPGAs compute resources than to those of a programmable embedded architecture. In implementing such kernels in FPGAs, the overhead of software executing on a programmable processor is eliminated, and the massively parallel hardware can be targeted more directly.

Advantages of ASICs over FPGAs

ASICs have a number of advantages over FPGAs, depending on the system designer's goals. ASICs, for instance, permit fully custom capability for the system designer as the device is manufactured to custom design specifications. Additionally, for very high volume designs, an ASIC implementation will have a significantly lower cost per unit. It is also likely that the ASIC will have a smaller form factor since it is manufactured to custom design specifications. ASICs will also benefit from higher potential clock speeds over their FPGA counterparts.

A corresponding FPGA implementation, on the other hand, will typically have a faster time to market as there is no need for layout of masks and manufacturing steps. FPGAs will also benefit from simpler design cycles over their ASIC counterparts, due to software development tools that handle placement, routing, and timing restrictions. FPGAs also benefit from being reprogrammable, in that a new bit stream can quickly be uploaded, during system development as well as when deployed in the field. This is one large advantage over the ASIC counterparts.

While FPGA devices at one time were selected for lower volume and capacity systems, modern FPGAs have competitive clock rates for use in high capacity systems. In addition, FPGAs have benefited from increased logic density in recent years, as well as other features such as integration of embedded processors, DSP blocks, and high speed serial input/output. As such, they may offer competitive solutions in the signal processing space depending on system requirements and flexibility.

Software programmable digital signal processing

Unlike the aforementioned ASIC and FPGA based solutions, modern DSPs are software programmable by the user using standard C-like programming languages. As such, they provide a greater amount of flexibility to the system developer since their function within the system is determined by the application layer software. Unlike their general purpose processor counterparts, or even embedded general purpose microprocessors, DSPs have a number of high performance, application domain specific features that require a certain level of expertise on the part of system programmers in order to fully exploit the performance of the processor. Examples of such features are advanced addressing modes such as bit reversed and modulo addressing, non-standard bit widths such as wide accumulators, saturating arithmetic operations and perhaps limited integer computational support, memory alignment constraints, and non-uniform and non-orthogonal instruction sets.

Each of these application domain specific, high performance features described previously may introduce programming hurdles for high performance DSP codes. In addition, they may contribute to barriers in software portability across vendor solutions, although increasingly vendors are providing software emulation libraries to tackle this hurdle. This may be a challenge in the traditional DSP market space as OEMs traditionally do not want to lock their software solution to a given vendor's silicon. As evidence, a number of vendors have recently begun providing software migration packages that allow native software based intrinsic functions for one vendor's architecture to execute on another vendor's architecture. Examples of this are both CEVA and Freescale, who have begun offering software solutions whereby Texas Instruments C6000 intrinsics functions can be emulated in software on the CEVA and Freescale DSPs. In doing this, vendors lower the barrier to entry on migrating legacy software solutions to their own architecture.

General purpose embedded cores

While the DSP solutions described previously offer a software programmable solution that is much more flexible than their FPGA and ASIC counterparts, as can be seen, the application developer must be aware of the underlying architecture and proprietary software constructs needed to target said architecture's performance. Additionally, once a high performance software solution is achieved on a given architecture, a clear software migration path to implementing that solution on differing architectures is not always apparent. General purpose embedded architectures tend to provide a more application generic solution for embedded computing, often incorporating some limited set of features to handle signal processing components of a given application. These embedded general purpose architectures are designed for wide ranges of applications, varying from consumer electronics to communications and automotive solutions. They usually employ standard 32-bit data paths, and are often scaled back versions of existing microprocessors. Examples of these are ARM, MIPS, and PowerPC architectures. It is not uncommon for these devices to be incorporated into larger dedicated systems or system on chip architectures. Many of these embedded processors include functionality for lighter weight signal processing needs, such as SIMD extensions within the instruction set that are targeted to multimedia or signal processing workload requirements. Examples of this are the ARM NEON general purpose SIMD extensions, which are targeted multimedia, signal processing, and gaming extensions. Quite often these SIMD extensions are supported by the build tools allowing for efficient vectorization of SIMD operations without the requirement of custom intrinsics and pragmas, as was the case for their DSP counterparts. These are often attractive solutions when the application requires modest amounts of signal or image processing, but also requires a general purpose embedded processor solution. Software developers can often aim to exploit the SIMD functionality afforded on these general purpose embedded architectures only when required by key computational bottlenecks in the application, while relying on the general purpose nature of the architecture for the remaining portion of the overall software application.

Putting it all together

DSP and embedded multiprocessor system on chip architectures and their related hardware constructs are a unique area of computer architecture as driven by the requirements placed on these systems, such as real-time deadline demands, low power consumption, and the multitasking requirements as well as often standardized components of the system. Rather than utilize commodity processors scaled down to fit a standard chip size, unique components are often used for the embedded and signal processing components to meet the strict requirements of each system. While some system designers do incorporate custom-like hardware acceleration blocks in their overall system topology, these are quite often in the form of microcode controlled accelerators rather than true non-configurable custom circuit designs. While there are still significant challenges in maintaining the intellectual property for vast libraries of hardware acceleration blocks, by utilizing a configurable hardware acceleration block, vendors can tailor the functionality as markets change, or open up the microcode driven control plane to customers as needed for custom functionality.

Architecture

The large majority of multiprocessor systems for embedded computing follow the standard design methodology of programmable CPU cores with associated memories connected via an interconnect framework. Unlike their general purpose or scientific computing counterparts, however, most of these processing solutions are highly heterogeneous in nature. Examples of these are the Texas Instruments OMAP series of processors for mobile handsets, and the Freescale MSC8156 series of processing platforms for wireless infrastructure.

General purpose architectures, such as those used in the general purpose and scientific computing, are typically not amenable to embedded and signal processing applications and platforms. This is due to the fact that the applications of interest to most embedded developers have high levels of performance that must be met in conjunction with the real-time and predictive nature. Traditional general purpose and scientific computing has striven for high performance and computational throughput, but often without regard for hard limits on how long such computation should require. These real-time demands of embedded systems are further complicated by strict power requirements and cost requirements placed on these devices.

Embedded systems hardware and platform designers can often exploit various characteristics of not only the workloads to be run, but also the system being developed to help meet these deadlines, such as strict power consumption budgets. Irregular memory systems may save power consumption by reducing RAM ports, and dedicated software controlled scratch pad memories may provide real-time predictability into a system design, as well as irregular interconnect networks amongst processing components. Even varying the types of processors used in the system topology can increase performance according to the strict criteria above. For example, a system design that requires large amounts of control processing code typically associated with user interfaces and operating systems, may reside on an ARM-like core or general purpose processing core within the system topology. Numerically intensive portions of the system software may be better suited to reside on a programmable signal processor architecture similar to a Texas Instruments DSP, or in some cases, may benefit from further parallelism afforded by a GPU style processor. As various components in the software application stack require differing levels of instruction, data, thread, and task level parallelism, so too the system designer should strive to map those computational requirements to the appropriate computational component within the system.

Application driven design

Unlike their general purpose computing counterparts, many of the applications with heterogeneous multiprocessing hardware solutions which are applied in the embedded space are not single algorithms but complex systems consisting of many algorithms. As such, the requirements of the computation to be performed at various points in the system can vary widely over both physical location within the system and over time amongst components within the system. As an example, algorithmic components within the application may vary widely in terms of types of operations or computation to be performed, memory system bandwidth requirements, memory access patterns that may relate to computational efficiency depending on processing component architecture, activity profiles, and total memory footprint.

Many multimedia codecs such as MPEG-2 and H.264 have components within them that have widely varying computational demands at runtime (Haskell, 1997). Major computational blocks within these systems can vary from having small data sets, requiring fairly regular systems of multiplication and addition operations, to other data dependent computation that operates on large data sets. Additionally, in large scale systems various components may be fairly standardized in terms of algorithmic requirements (such as a Fast Fourier Transform within the system), whereas for other algorithmic components it may be desirable to keep a proprietary implementation executing in software due to key algorithmic insights from specific vendors.

Bibliography

B.G. Haskell, A. Prui, A.N. Netravali, Digital Video: An Introduction to MPEG-2, Chapman & Hall, New York, 1997.

CHAPTER 7

Overview of DSP Algorithms

Robert Oshana

Chapter Outline

Applications of DSP

Why do we want to learn DSP? DSP can be used for many things. Some of the most popular are:

- Filtering — removing unwanted frequencies from a signal
- Spectrum analysis — determining the frequencies in a signal

DSP for Embedded and Real-Time Systems. DOI: 10.1016/B978-0-12-386535-9.00007-X

- Synthesis — generating complex signals such as speech
- System identification — identifying the properties of a system by numerical analysis
- Compression — reducing the memory/bandwidth it takes to store a signal, such as audio or video

Doing things in the digital world also has its advantages over the analog world such as:

- Programmability — the algorithm for filter specifications can be updated easily, without changing the hardware
- Stability — the system components are not subject to drifting by changes in temperature or aging of the parts
- Cost — as the complexity of the system increases, so does the cost of adding electronics to the system
- Functionality — with a DSP, new functionality can be added with leftover cycles in the processor
- Availability — new chips are being made at lower costs
- Analog limitations — DSPs can be programmed to perform tasks not possible with their analog counterparts

However, DSP does have its limitations including:

- Cost — For simple circuits, the cost of adding a DSP can be much more expensive.
- Speed — Digital techniques have a much higher delay than analog components.
- Precision — DSPs are limited to the number of bits available by the processor or the analog to digital (A/D) converter.
- Complexity — Designing a DSP system not only requires knowledge of analog electronics, but DSP theory and programming as well. There is much more room for things to go wrong with a DSP system.

Systems and signals

Before we start talking about DSP systems it is necessary to introduce a few concepts that are common to all systems. However, emphasis will be placed on how these concepts are used in DSP systems. The systems generally used in DSP applications are Linear Time-Invariant Systems. The property of linearity is important for a couple of reasons. The most important reason is that the system is not dependent on the order in which processes are applied. For example, it does not matter if we scale our input before or after we run it through a filter; the output is the same. As a result, a complex system can be divided into multiple systems. In addition, an eighth order system can be divided into four second order systems and still produce the same output. Time invariance is important because we need to know that our system will react in the same way to an input, and that the output is not dependent on when the input is applied. These two qualities make our system very predictable.

Signals and systems are usually graphed to show how the input and output relate to each other. There are two typical ways of viewing the data — in the time domain and in the frequency domain. The time domain is especially handy for applications such as control systems, where response times are important. The frequency domain is useful for viewing filter results to see which frequencies will pass and which will be attenuated.

We usually describe our systems in the time domain in terms of impulse response. An impulse is a stimulus of infinite magnitude and zero duration and its integral is equal to one. Since it is only a single instant of time, and spans all frequencies, the impulse makes a good test to show how a system will react to an input. The response to an impulse input, called the Impulse Response, can be used to describe the system. In the digital domain, an impulse is an input signal whose magnitude is 1 at time 0 and 0 everywhere else, as shown in Figure 7-1. The way a system reacts to an impulse input can also be considered the system's transfer function. The transfer function (or impulse response) can provide us with everything we need to determine how a system will react to an input in the time or the frequency domain.

A system response is usually plotted in the frequency domain to show how it will affect signals of different frequencies. When plotted in the frequency domain, there are two characteristics to be concerned about, magnitude and phase. Magnitude is the ratio of the output's strength compared to the strength of its input. For example, how much of a radio signal will pass through a filter designed to pass a specific frequency. The phase is how much a frequency of the signal will be changed by the filter, usually lagging or leading. While not important in all applications, sometimes the phase can be particularly important, such as in music or speech applications.

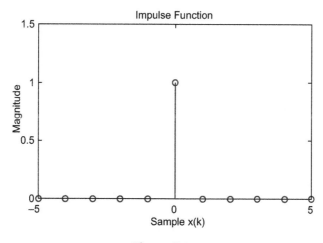

Figure 7-1:
Impulse response.

DSP systems

Analog signals are converted to digital signals through a process called sampling. Sampling is the process of taking an analog signal and converting it to discrete numbers. The sampling frequency (or sampling rate) is how many times per second the signal will be sampled. This is important because it restricts the highest frequency that can be present in a signal. Any signal greater than half the sampling rate will be folded into a lower frequency that is less than half the sampling rate. This is known as aliasing and will be discussed in more detail in the next section.

The inverse of the sampling frequency is known as the sampling period. The sampling period is the amount of time that elapses between the samples of the analog signal. This conversion is done by hardware that takes the best approximation of the voltage level and converts it to the nearest digital level that can be represented by the computer. The loss of information during the conversion is referred to as quantization.

Aliasing

An important concept to understand when working with digital systems is that of aliasing. Without an understanding of this point, numerous unexpected problems can arise when implementing your digital system. Aliasing is due to the fact that, mathematically, the following two equations are equal:

$$X(n) = \sin(2*pi*fo*n*ts) = \sin(2*pi*fo*n*ts) + 2 * pi * k$$

These equations can be re-written as:

$$X(n) = \sin (2*pi*fo*n*ts) = \sin (2*pi*\mathbf{fo} + \mathbf{k}*fs) * n * ts)$$

A visual example of aliasing is provided in Figure 7-2.

To avoid this ambiguity, we must restrict all frequencies in our signal to be in the range of 0 to $f_s/2$. The frequency $f_s/2$ is called the Nyquist Frequency (or Nyquist Rate). While this provides a limitation to the frequencies available in our system, there is no other choice but to do this. The way this is done is by building an analog (anti-aliasing) filter and placing it before our digital to analog (D/A) converter.

The basic DSP system

The basic DSP system, illustrated in Figure 7-3, consists of an analog to digital (A/D) converter, a digital signal processor (DSP) and a digital to analog (D/A) converter. Typically, DSP systems have analog filters before and after the converters to make the signals more pure. Let's discuss each component in detail.

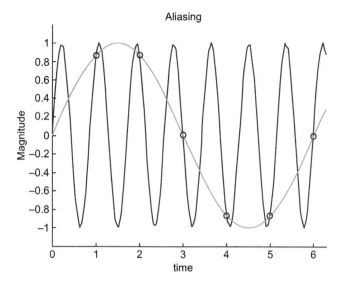

Figure 7-2:
Aliasing.

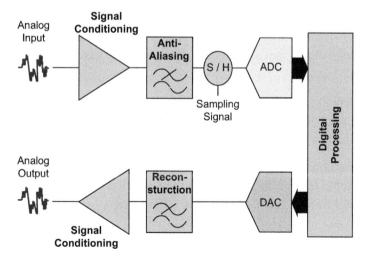

Figure 7-3:
The basic DSP system.

Since the signal can contain no frequencies above the Nyquist frequency, steps must be taken to ensure that such high frequencies are eliminated from the signal. This is done with an analog low pass filter with a cutoff rate set around the Nyquist frequency. This analog filter is known as the anti-aliasing filter. The signal is then used as input into an A/D converter where the signal is converted to a digital signal so that the DSP can handle it and process it. The DSP will perform the actions required of it, such as filtering, and then pass the new signal to the D/A. The D/A then converts the digital output back to an analog signal. This analog output usually contains high frequencies introduced by the D/A, so a low pass filter is needed to smooth the waveform back to its intended shape. This filter is known as a reconstruction filter.

Filters

Filtering is a fundamental process in the analog world as well as the digital world. Almost any filter that can be implemented with analog techniques can also be implemented using digital techniques. Digital filters are by far the most widely used application in a DSP system. Digital filters operate on different frequencies in different ways, allowing some to pass, while attenuating others. Filters can also change the phase of a signal, which can be important in applications such as digital video. The most popular types of filters are:

- Low Pass Filters — filters to block out high frequencies
- High Pass Filters — filters to block out low frequencies
- Band Pass Filters — filters which allow only a range of frequencies to pass
- Band Stop Filters — filters which only reject a range of frequencies
- Comb Filters — a filter designed to reject a specific frequency and all its harmonics
- All Pass Filters — filters that allow all frequencies to pass and which modify the phase of a signal

As you can see, deciding which type of filter is needed for an application is straightforward, depending on the needs of the application. Since the filters are so varied, there is usually no confusion as to which type of filter is required for the specific application.

However, some decisions are necessary when implementing the filter, and there are trade-offs between the different types of filters. The two main filters discussed here are Finite Impulse Response (FIR) and Infinite Impulse Response (IIR) filters.

FIR filters

An FIR filter is a filter with no feedback in its equation. This can be an advantage because it makes an FIR filter inherently stable. Another advantage of FIR filters is the fact that they can produce linear phases. So, if an application requires linear phases, the decision is simple, an FIR filter must be used. The main drawback of a digital FIR filter is the time that it takes to execute. Since the filter has no feedback, many more coefficients are needed in the system equation to meet the same requirements that would be needed in an IIR filter. For every extra coefficient, there is an extra multiply and extra memory requirements for the DSP. For a demanding system, the speed and memory requirements to implement an FIR system can make the system unfeasible.

IIR filters

To reduce the demand on the system while maintaining requirements, an IIR filter can be used. An IIR filter uses both inputs and past outputs in its equation, allowing it to operate much more efficiently. However, a drawback of having feedback is that linear phase is almost impossible to maintain. So, if phase doesn't matter, and the designer wants to reduce the number of taps (coefficients) in the system, an IIR filter is the filter of choice.

Frequency analysis

Almost every application in DSP deals with frequencies in some way. Therefore, a tool is needed that allows conversion from the time domain to the frequency domain, and vice versa. This tool builds on the Fourier Transform, an equation to calculate the frequencies in a signal, and is known as the Discrete Fourier Transform (DFT). The Discrete Fourier Transform takes a set of time domain samples as inputs, and returns frequencies in the range of negative ½ the sampling frequency to ½ the sampling frequency. This allows viewing of the spectral content of the signal. To transform the information from the frequency domain back to the time domain, the data is run through an algorithm called the Inverse Discrete Fourier Transform.

Convolution

In linear time-invariant systems, the system response is always predictable no matter when it occurs or at what frequency it occurs. Every response is the result of the current impulse plus the sum of past impulses. To get the output, one must multiply the current input by the first filter coefficient. In addition to that, the second coefficient is multiplied by the previous input, and the third coefficient is multiplied by the N-2 input. Basically, the system response is reversed, multiplied by the corresponding previous inputs, and the results are summed, as shown in the equation:

$$y(n) = \sum_{k=0}^{M-1} h(k)x(n-k)$$

This gives a weighted sum of the current input and all the past inputs. This process is known as convolution, and is the basis for most filter algorithms.

Correlation

Correlation is a technique used to determine the similarity between two signals or between a signal and itself. The process of correlating a signal with itself is called autocorrelation. Correlation, like convolution, is a sum of products. When correlating two signals, the higher the value of the result, the more similar the signals. A threshold value can be set to determine if the signal is actually a replica of itself.

This process is useful in several situations. Radar is one application where correlation is very useful. For example, a distance can be found by correlating a reference with the incoming signal point by point. When the threshold is met, the delay from the time the signal was sent to when we received it can be used to determine the distance the signal traveled.

This same technique can also be useful in finding a known signal in a noisy input signal. When the threshold is met, it can be assumed that the known signal is found.

Designing an FIR filter

The simplest design of an FIR filter is an averaging filter, where all the coefficients have the same value. However, this filter does not give a very desirable magnitude response. The trick to designing an FIR filter is getting the right coefficients. Today there are several good algorithms that can be used to find these coefficients, and several software design programs to assist in calculating them. Once the coefficients are obtained, it is a fairly simple matter to place them in an algorithm to implement the filter. Let's talk about some of the techniques used in selecting these coefficients.

Parks-McClellan algorithm

One of the best 'catch-all' algorithms used to determine the filter coefficients is the Parks-McClellan algorithm. Once the specifications are obtained (cut-off frequency, attenuation, band of filter), they can be supplied as parameters to the function, and the output of the function will be the coefficients for the filter. The program works by spreading out the error over the entire frequency response. So, an equal amount of minimized error will be present in the passband and stopband ripple. Also, the Parks-McClellan algorithm isn't limited to the types of filters discussed earlier (low-pass, high-pass). It can have as many bands as are desired, and the error in each band can be weighted. This facilitates building filters of arbitrary frequency response. To design the filter, first calculate the order of the filter with the following equations:

$$\hat{M} = \frac{-20 \log_{10} \sqrt{A \delta_1 \delta_2} - 13}{14.6 \Delta f}; \quad \Delta f = \frac{w_s - w_p}{2\pi}$$

where $\hat{M}$ is the order, w_s and w_p are the passband and stopband frequencies, and δ_1 and δ_2 are the ripple on the passband and stopband respectively.

δ_1 and δ_2 are calculated from the desired passband ripple and stopband attenuation with the following formulas:

$$\delta_1 = 10^{Ap/20} - 1 \text{ and } \delta_2 = 10^{-As/20}$$

Once these values are obtained, the results can be plugged into the MATLAB remez function to get the coefficients. For example, to obtain a filter that cuts off between 0.25 and 0.3 with a passband and stopband ripple of 0.2 and 50 DB respectively, the following specifications can be plugged into the MATLAB script to get the filter coefficients:

```
% design specifications
  wp=.23; ws=.27; ap=.025; as=40;
  %calculate deltas
  d1=10^(ap/20)-1; d2=10^(-as/20); df=ws-wp;
  % calculate M
  M=((((-10 * log10(d1*d2))-13) / (14.6 * df))+1);
  M=ceil(M);
  % plug numbers into remez function for low pass filter
  ht=remez(M-1, [0 wp ws 1], [1 1 0 0]);
```

ht will be a vector array containing the 35 (the value of M) coefficients. To get a graph of the frequency response, use the following MATLAB commands:

```
[h,w]=freqz(ht); % Get frequency response
w=w/pi;          % normalize frequency
m=abs(h);        % calculate magnitude
plot(w,m);       % plot the graph
```

The graph is shown in Figure 7-4.

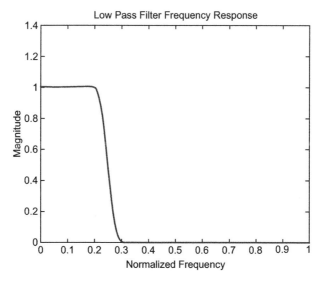

Figure 7-4:
Low pass filter frequency response.

Windowing

Another popular technique of the FIR filter is the ability to generate the frequency coefficients from an ideal impulse response. The time domain response of this ideal impulse response can then be used as coefficients for the filter. The problem with that approach is that the sharp transition of frequencies in the frequency domain will create a time domain response that is infinitely long. And when the filter is truncated, ringing will be created around the cut-off frequency of the frequency domain due to the discontinuities in the time domain. To reduce this problem, a technique called windowing is used.

Windowing consists of multiplying the time domain coefficients by an algorithm to smooth the 'edges' of the coefficients. The trade-off here is reducing the ringing but increasing the transition width. There are several windows discussed, each with a trade-off in transition width versus stopband attenuation.

The following are several types of popular windows:

*   Rectangular — sharpest transition, least attenuation in the stopband (21 dB)
*   Hanning — over 3× transition width of rectangular, but 30 dB attenuation
*   Hamming — winder transition, but 40 dB
*   Blackman — 6× transition of rectangular, but 74 dB
*   Kaiser — any (custom) window can be generated based on a stopband attenuation

When designing a filter using the windowing technique, the first step is to use response curves or trial and error and decide which window would be appropriate to use. Then, the desired number of filter coefficients is chosen. Once the length and type of window are determined, the window coefficients can be calculated. Then, the window coefficients are multiplied by the ideal filter response. Here is the code and frequency response for the same filter as before with a Blackman window (Figure 7-5):

```
%lowpass filter design using 67 coefficient hamming window
%design specifications
ws = .25; wp = .3;
N = 67;

wc = (wp - ws) / 2 + ws %calculate cutoff frequency
%build filter coefficients ranges
n = -33:1:33;
hd = sin(2 * n * pi * wc) ./ (pi * n); % ideal freq
hd(34) = 2 * pi * wc / pi; %zero ideal freq
hm = hamming(N); % calculate window coefficients
hf = hd .* hm'; %multiply window by ideal response
```

Adding feedback using IIR filters

Another class of digital filters is built by incorporating feedback into the equation. This class of filters is called Infinite Impulse Response (IIR). Adding feedback allows the equation to

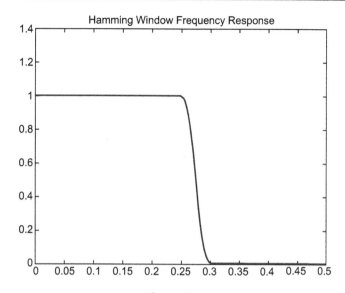

Figure 7-5:
Hamming window frequency response.

contain 5–10 times fewer coefficients than their FIR counterparts. However, it does mangle the phase and make designing and implementing the filter more complicated.

While filters will usually be designed by software, it is a good idea to know the techniques involved in designing the filter so the designer has some idea of what the software is trying to accomplish, and what methods it is using to meet these goals. There are two primary techniques involved in designing IIR filters. They are direct and indirect design. Direct design does all its work in the z-domain (digital domain), while indirect design works in the s-domain (analog domain) and converts the results to the z-domain. Most of the time IIR filters are designed using analog techniques. While it may seem a less efficient way of doing things, analog methods for designing filters have been around a lot longer than digital design methods, and these proven techniques can be applied to digital filters in the same way.

With indirect design, the designer relies on optimized analog design techniques to develop the filter. Once they have an optimized solution for their analog filter, the problem lies in converting the analog solution to the digital solution. Since the analog domain can contain an infinite number of frequencies, and the digital domain is limited to half the sampling rate, the two domains will not match up perfectly, and the frequencies must be mapped. There are two popular techniques used to accomplish this mapping. One is by wrapping the s-domain around the unit circle in the z-domain, and the other is done by compressing the s-domain into the unit circle.

There are several techniques which have been optimized for analog design over the years, most of which excel at one particular area or specification, such as passband ripple, transition, or phase. The most popular analog techniques and their useful characteristics are mentioned below.

- Butterworth — use for a flat passband ripple. Also, the magnitude response will not increase as frequency increases
- Chebychev — sharper transition than Butterworth, with the cost of more ripple in the passband
- Chebychev II — monotonic passband but adds ripples to the stopband
- Bessel — when phase is important in an IIR filter
- Eliptical — sharpest transition, but allows ripples in the stopband and passband

Once the filter's poles and zeros are determined, they must be converted to the z-domain for use by the digital filter. The most popular technique for doing this is the Bilinear Transform. The Bilinear Transform method does this by mapping (or compressing) all the frequencies into the unit circle. It does this in a non-linear manner, and to compensate for this 'warping of frequencies,' the frequencies must be 'pre-warped' before the filter is designed.

So, to develop a filter using the Bilinear Transform technique, the designer should follow the following specs.

1. Determine the filter critical frequencies and sampling rate.
2. Pre-warp the filters critical frequencies.
3. Design an analog filter using 'classic' techniques with these pre-warped frequencies.
4. Convert the filter to the z-domain using the Bilinear Transform.

Another technique used to transform poles and zeros from the s-domain to the z-domain is called the impulse invariance method. The impulse invariance method takes the s-domain only up to half the sampling rate and converts it to the z-domain. For this reason, it is limited to low pass and band pass filters only. This has the benefit of creating an impulse response that is a sampled version of the s-domain impulse response.

There are MATLAB functions to assist in designing filters. The functions to design the popular analog filters are BUTTER, CHEB1AP, CHEB2AP, and ELLIPAP. These functions will return the coefficients for the IIR filter, and there are two additional functions for converting the analog coefficients to the digital domain: BILINEAR and IMPINVAR. The function BILINEAR does the necessary 'pre-warping.'

Typically, when designing IIR filters by hand, only low pass filters would be used. They would then be converted to the appropriate specifications using complex formulas. However, when designing with a software package such as MATLAB, the user does not have to worry about this transformation.

Algorithm implementation — DSP architecture

Today's DSP architectures are made specifically to maximize throughput of DSP algorithms, such as a DSP filter. Some of the features of a DSP include:

- On-chip memory — internal memory allows the DSP fast access to algorithm data such as input values, coefficients and intermediate values.
- Special MAC instructions — for performing a multiply and accumulate, the crux of a digital filter, in one cycle.
- Separate program and data busses — this allows the DSP to fetch code without affecting the performance of the calculations.
- Multiple read busses — for fetching all the data to feed the MAC instruction in one cycle.
- Separate write busses — for writing the results of the MAC instruction.
- Parallel architecture — DSPs have multiple instruction units so that more than one instruction can be executed per cycle.
- Pipelined architecture — DSPs execute instructions in stages so more than one instruction can be executed at a time. For example, while one instruction is doing a multiply, another instruction can be fetching data with other resources on the DSP chip.
- Circular buffers — to make pointer addressing easier when cycling through coefficients and maintaining past inputs.
- Zero overhead looping — special hardware to take care of counters and branching in loops.
- Bit-reversed addressing — for calculating FFTs.

Number format

As discussed earlier, when converting an analog signal to digital format, the signal has to be truncated, due to the limited precision of a DSP. DSPs come in fixed- and floating-point format. When working with a floating-point format, this truncation is not usually much of a factor due to its good mix of precision and dynamic range. However, implementing hardware to deal with floating-point formats is harder and more expensive, so most DSPs on the market today are fixed-point format. When working with fixed-point format, a number of considerations have to be taken into account. For example, when two 16-bit numbers are multiplied, the result is a 32-bit number. Since we ultimately want to store the final result in 16-bit format, we need to handle this loss of data. Clearly, by just truncating the number, we would lose a significant portion of the number. To deal with this issue we work with a fractional format called Q format. For example, in Q15 (or 1.15) format the most significant digit is used to represent the sign and the rest of the digits represent the fractional part of the data. This allows for a dynamic range of between −1 and just less than 1. However, the results of a multiply will never be greater than 1. So, if the lower 16 bits of the result are dropped, a very insignificant portion of the results is lost. One nuance of the multiply is that there are two sign bits, so the result will have to be shifted to the left one bit to eliminate the redundant information. Most processors will take care of this, so the designer doesn't have to waste cycles when doing many multiplications in a row.

Overflow and saturation

Two other problems that can occur when using fixed-point arithmetic are overflow and saturation. However, DSPs help the programmer deal with these problems. One way a DSP does this is by providing guard bits in the accumulator. In a normal 16-bit processor, the accumulator may be 40 bits: 32 bits for the results (keep in mind a 16×16 bit multiplication can be up to 32 bits) and an extra 8 bits to guard against overflow (of multiple multiplies in a repeat block).

Even with the extra guard bits, multiplications can provide overflow situations, where the result contains more bits than the processor can hold. This situation is handled with a flag called an overflow bit. The processor will set this automatically when the results of a multiplication overflow the accumulator.

When an overflow occurs the results in the accumulator usually become invalid. So what can be done? Another feature of DSPs can be taken advantage of: saturation. When the saturate instruction on a DSP is executed, the processor sets the value in the accumulator to the largest positive or negative value the accumulator can handle. That way, instead of possibly flipping the result from a high positive number to a negative number, the result will be the highest positive number the processor can handle.

There is also a mode DSP processors have that will automatically saturate a result if the overflow flag gets set. This saves the code from having to check the flag and manually saturating the results.

Implementing an FIR filter

We will begin our discussion of implementing an algorithm on a DSP by examining the C code required to implement an FIR filter. The code is pretty straightforward and looks like this.

```
long temp;
int block_count;
int loop_count;

// loop through inputs
for (block_count=0;block_count<output_size;block_count++)
{
    temp=0;
    for (loop_count =0;loop_count<coeff_size;loop_count++)
    {
        temp += (((long)x[block_count+loop_count]*(long)a[loop_count]) << 1);
    }
    y[block_count]=(temp >> 16);
}
```

This code is a very simple sum of projects written in C with a few caveats. The caveats stem from the number format issues we discussed earlier. First of all, temp must be declared as long so that it is represented by 32 bits for temporary calculations. Also, the value of the MAC must be shifted to the left one bit because of the fact that multiplying two 1.15 numbers results in a 2.30 number, and we want our result in 1.15. Finally, the temp value is shifted 16 bits to the right before writing to output so that we take the most significant bits of our result.

As you can see, this is a very simple algorithm to write, and central to most DSP applications. The problem with implementing the code in C is that it is too slow. A general rule in designing embedded DSP applications is known as the 90/10 rule. It is used to determine what to write in C and what to write in assembly. It states that a DSP application generally spends 90% of its time in 10% of the code. This 10% of the code should be written in assembly, and in this case the code is the FIR filter code.

Here is an example of a filter written in TMS320c5500 assembler language for a DSP:

```
fir:
        AMOV  #184, T1         ; block_count=184
        AMOV  #x0, XAR3        ; init input pointer
        AMOV  #y, XAR4         ; init output pointer
                              ; do
oloop:  SUB        #1, T1      ; block_count-
        AMOV  #16, T0          ; loop_count=16
        AMOV  #a0, XAR2        ; init coefficient pointer
        MOV        #0, AC0           ; y[block_count]=0
                              ; do
loop:   SUB  #1,T0             ; loop_count-
        MPYM *AR2+, *AR3+, AC1 ; temp1=x[] * a[]
        nop
        ADD        AC1, AC0                    ; temp=temp1
        BCC        loop, T0 != #0  ; while (loop_count > 0)
        nop
        nop
        MOV  HI(AC0), *AR4+ ; y[block_count]=temp >> 16
        SUB  #15, AR3         ; adjust input pointer
        BCC        oloop, T1 != 0  ; while (block_count > 0)
        RET
```

This code does the same thing as the C code, except it is written in assembly. However, it does not take advantage of any of the DSP architecture. We will now start re-writing this code to take advantage of the DSP architecture.

Utilizing on-chip RAM

Typically, data such as filter coefficients are stored in ROM. However, when running an algorithm, the designer would not want to have to read the next coefficient value from ROM.

Therefore, it is a good practice to copy the coefficients from ROM into internal RAM for faster execution. The following code is an example of how to do so.

```
copy: AMOV #table, XAR2
      AMOV #a0, XAR3
      RPT #7
          MOV dbl(*ar2+), dbl(*ar3+)
      RET
```

Special MAC instruction

All DSPs are built to do a Multiply-Accumulate (MAC) in 1 instruction cycle. There are a lot of things going on in the MAC instruction. If you notice, there is a multiply, an add, an increment of the pointers, and a load of the values for the next MAC, all in one cycle. Therefore, it is efficient to take advantage of this useful instruction in the core loop. The new code will look like this:

```
MAC *AR2+, *AR3+, AC0 ; temp += x[] * a[]
```

Block filtering

Typically, an algorithm is not performed one cycle at a time. Usually a block of data is processed. This is known as block filtering. In the example, looping was used to apply the filter algorithm on a hundred inputs rather than just one, thus generating 100 outputs at a time. This technique allows us to use many of the optimizations we will now talk about.

Separate program and data busses

The 55x architecture has three read busses and two write busses, as shown in Figure 7-6. We will take advantage of all three read busses and both write busses in the filter by using what's called a coefficient data pointer and calculating two outputs at a time. Since the algorithm uses the same coefficients in every loop, one bus can be shared for the coefficient pointer and the other two busses can be used for the input pointer. This will also allow the use of the two output busses and two MAC units each inner loop, allowing the values to be calculated more than twice as fast. Here is the new code to optimize the MAC hardware unit and the busses:

```
AMOV #x0, XAR2            ; x[n]
AMOV #x0+1, XAR3          ; x[n+1]
AMOV #y, XAR4            ; y[n]
AMOV #a0, XCDP           ; a[n] coefficient pointer

MAC AR2+, CDP+, AC0      ; y[n]=x[n] * a[n]
 :: MAC *AR3+, CDP+, AC1 ; y[n+1]=x[n+1] * a[n]
MOV pair(hi(AC0)), dbl(*AR4+); move AC0 and AC1 into mem pointed to by AR4
```

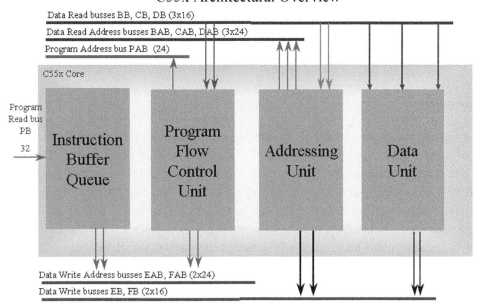

Figure 7-6:
Separate program and data busses.

Notice that a colon separates the two MAC instructions. This tells the processor to execute the instructions in parallel. By executing in parallel we take advantage of the fact that the processor has two MAC units in hardware, and the DSP is instructed to execute two MAC instructions in one cycle by using both hardware units.

Zero overhead looping

DSP processors have special hardware to take care of the overhead in looping. The designer need only set up a few registers and execute the RPT or RPTB instruction (for a block of instructions) and the processor will execute the loop the specified number of times. Here is the code taking advantage of zero overhead looping:

```
MOV #92, BRC0 ; calculating 2 coefficients at a time, block loop is 184/2
```

And here is the actual loop code:

```
RPTBlocal   endfir                          ; repeat to this label;, loop start
            MOV #0, AC1                     ; set outputs to zero
            MOV #0, AC0
            MOV #a0, XCDP                   ; reset coefficient pointer
            RPT          #15                ; inner loop
                MAC *AR2+, *CDP+, AC0
```

```
          :: MAC *AR3+, *CDP+, AC1

        SUB  15, AR2                    ; adjust input pointers
        SUB  15, AR3
   MOV  pair(hi(AC0)), dbl(*AR4+)       ; write y and y+1 output values
endfir:  nop
```

Circular buffers

Circular buffers are useful in DSP programming because most implementations include
a loop of some sort. In the filter example, all the coefficients are processed, and then the
coefficient pointer is reset when the loop is finished. Using circular buffering, the coefficient
pointer will automatically wrap around to the beginning when the end of the loop is
encountered. Therefore, the time that it takes to update the pointers is saved. Setting up
circular buffers usually involves writing to some registers to tell the DSP the buffer start
address, buffer size, and a bit to tell the DSP to use circular buffers. Here is the code to set up
a circular buffer:

```
; setup coefficient circular buffer
        AMOV #a0, XCDP ; coefficient data pointer
        MOV        #a0, BSAC        ; starting address of circular buffer
        MOV        #16, BKC         ; size of circular buffer
        MOV        #0, CDP          ; starting offset for circular buffer
        BSET CDPLC     ; set circular instead of linear
```

Another example where circular buffers are useful is when working with individual inputs
and only saving the last N inputs. A circular buffer can be written so that when the end of the
allocated input buffer is reached, the pointer automatically wraps around to the beginning of
the buffer. Writing to the correct memory is then ensured. This saves the time of having to
check for the end of the buffer and resetting the pointer if the end is reached.

System issues

After the filter code is set up, there are a few other things to take into consideration when
writing the code. First, how does the DSP get the block of data? Typically, the A/D and D/A
would be connected to serial ports built into the DSP. The serial ports will provide a common
interface to the DSP, and will also handle many timing considerations. This will save the DSP
a lot of cycles. Also, when the data comes in to the serial port, rather than having the DSP
handle the serial port with an interrupt, a Direct Memory Access (DMA) can be configured to
handle the data. A DMA is a peripheral designed for moving memory from one location to the
other without hindering the DSP. This way, the DSP can concentrate on executing the
algorithm and the DMA and serial port will worry about moving the data. The system block
diagram for this type of implementation is shown in Figure 7-7.

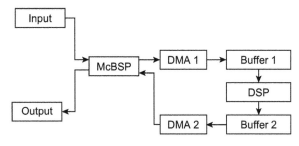

Figure 7-7:
Basic I/O for a DSP system.

Conclusion

As DSP systems become more advanced, more and more tools are being introduced to make developing with them easier. There are software packages that will design the digital system for you, and sometimes even generate the code for a specific processor.

Also, there are complete integrated environments that make programming a DSP very similar to programming an embedded system. These DSP systems can even be programmed in C. Today, modern C compilers are built to take advantage of the special features of a DSP architecture, to make the execution of these applications as efficient as possible. However, sometimes it is still necessary to hand optimize the code to make the application as efficient as possible. We will learn more about this in later chapters.

CHAPTER 8

High-level Design Tools for Complex DSP Applications

Yang Sun[1], Guohui Wang[1], Bei Yin[1], Joseph R. Cavallaro[1], Tai Ly[2]

[1]*Rice University, ECE Department, Houston, TX* [2]*National Instruments Corporation, Austin, TX*

Chapter Outline

High-level synthesis design methodology

High level synthesis (HLS) [1], also known as behavioral synthesis and algorithmic synthesis, is a design process in which a high level, functional description of a design is automatically compiled into a RTL implementation that meets certain user specified design constraints. The HLS design description is 'high level' compared to RTL in two aspects: design abstraction, and specification language:

I. **High level of abstraction**: HLS input is an untimed (or partially timed) dataflow or computation specification of the design. This is higher level than RTL because it does not describe a specific cycle by cycle behavior and allows HLS tools the freedom to decide what to do in each clock cycle.

II. **High level specification language**: HLS input is specified in languages like C, C++, System C, or even Matlab, and allows use of advanced language features like loops, arrays, structs, classes, pointers, inheritance, overloading, template, polymorphism, etc.

DSP for Embedded and Real-Time Systems. DOI: 10.1016/B978-0-12-386535-9.00008-1

This is higher level than (synthesizable subset of) RTL description languages and allows concise, reusable, and readable design descriptions.

The objective of HLS is to extract parallelism from the input description, and construct a micro architecture that is faster and cheaper than simply executing the input description as a program on a microprocessor. The micro architecture contains a pipelined datapath and a cycle-by-cycle description of how data is routed through this datapath. The output of HLS may include:

I.   **RTL implementation**: This includes the RTL netlist that contain the datapath, control logic, interfaces to I/O, host, and memories; as well as scripts, libraries, and synthesis timing constraints required to synthesize the RTL netlist using conventional logic synthesis flows.

II. **Analysis feedback**: This includes GUI and reports on performance bottlenecks, mapping of high level source code to RTL, hardware costs, etc, to help user understand and improve the micro architectures.

III. **Verification Artifacts**: This includes simulation test bench, linting checks, scripts, and library for code coverage, etc., to help user develop and debug high level language test suite and reuse these tests for RTL verification.

User specified constraints help HLS construct the desired micro architecture. These constraints include:

I.   **Target hardware**: This includes the platform, technology library, clock frequency, etc, that the design is intended for. HLS uses this information to estimate sub-cycle timing and cost of the datapath.

II. **Performance constraint**: This may be expressed in the form of input sampling rate, output production rate, input-to-output latency, loop initiation intervals, loop latency, etc. These constraints impose cycle level timing constraints on the micro architecture.

III. **Memory architecture**: This specifies how multi-dimensional arrays are mapped to memories and memory interfaces, allowing HLS to construct micro architectures containing mult-port, multi-bank, arbitrated, external and internal memories.

IV. **Interface constraint**: This includes the protocol, ports, and handshake/arbitration logic to create at each input, output, host, and external memory interface. HLS generates these interface ports and logic in the RTL netlist so it can be easily integrated with other hardware blocks.

V.   **Design hierarchy**: This partitions a design using hierarchy in the high level input description, allowing HLS to manage design complexity by divide-and-conquer.

HLS is not a substitute for a good RTL designer. For example, if the micro architecture is given, designing in RTL is easier and sufficient. HLS is designed for exploring different algorithms and architectures to find the best micro architecture under a variety of constraints.

The primary benefits of HLS derive from its support for high level of abstraction and high level specification language:

I.   Benefits of designing at a high level of abstraction:
   a)   Allows focus on designing core functionality, not implementation details. Easily explore different architectures.
   b)   Easily evaluate algorithmic changes.
   c)   Easily generate memory, IO, and host interfaces, as well as pipeline, stall, handshake and arbitration logic, etc.
   d)   Easily retarget for different hardware constraints or performance from the same input description.
II. Benefits of verifying at a high level of abstraction:
   a)   Easily debug and test functionality of input descriptions.
   b)   Fast and free simulation.
   c)   Test suite is reusable for RTL verification.
   d)   Code coverage and functional coverage is more meaningful and easier to achieve.
III. Benefits of high level specification language:
   a)   Legacy code can be reused for design and for verification.
   b)   Software development tools (e.g., Visual Studio) can be used.
   c)   Support advanced language features like polymorphism and templatized classes for concise, reusable input descriptions.

High-level design tools

To meet high performance and low power requirement of the VLSI digital signal processing system, traditional hardware design methods require designers to manually write the RTL codes which are too time-consuming for both designing and debugging.

Catapult C

Catapult C Synthesis [2] is an algorithmic synthesis tool that provides implementations from C++ working specifications. Catapult C Synthesis employs the industry standard ANSI C++ and System C to describe algorithms or functionality of the circuit. The output of Catapult C is a RTL netlist in either VHDL or Verilog HDL that can be synthesized to the gate level by using Precision RTL Synthesis, Design Compiler or similar RTL synthesis tools. Figure 8-1 shows the comparison between the traditional RTL design and Catapult C RTL design workflow.

With this approach, full hierarchical systems comprised of both control blocks and algorithmic units are implemented. By speeding time to RTL and automating the generation

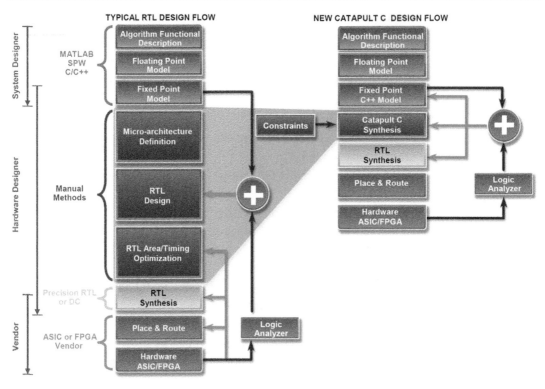

Figure 8-1:
Catapult C Synthesis RTL design flow versus the traditional RTL design flow.

of bug free RTL, the designer can significantly reduce the time to verified RTL. Catapult's workflow provides modeling, synthesizing, and verifying complex ASICs and FPGAs allows hardware designers to explore micro-architecture and interface options. The designer should carefully tune architectural constraint parameters in the interactive environment to generate different micro architectures which can meet the design specifications.

The designers should start the Catapult C HLS design process from describing and simulating an algorithm. The algorithm description is written in pure ANSI C++ or SystemC source, describing only the functionality and data flows. Hardware requirements are not considered in this stage. The next step is to determine the target technology and architecture constraints. The target technology defines the building blocks in a design which can be ASIC or FPGA. Then the clock frequency and hardware constraints should be set. Hardware requirements such as parallelism and interface protocols can be set in Catapult through constraints, which, in turn, guide the synthesis process. Catapult C determines the micro-architecture of the generated RTL code based on the technology and clock frequency chosen by the designer. As depicted in Figure 8-2, with different technology settings, the architectures are different for the generated RTL codes even though the same clock frequency is chosen.

```
int multaddadd (short A[4], short B[4])
{
   return (A[0]*B[0]) + (A[1]*B[1]) + \
          (A[2]*B[2]) + (A[3]*B[3]);
}
```

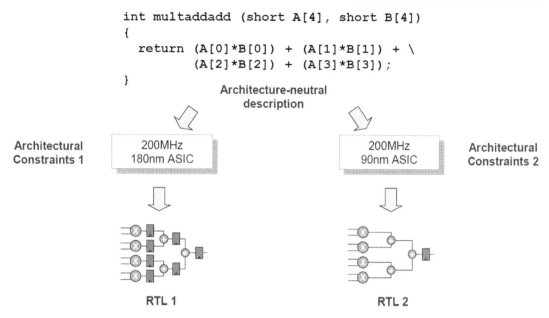

Figure 8-2:
Target optimized RTL code generation.

Once the target technology and clock frequency are specified, the designer is free to begin exploring the design space using built-in HLS process. The designer can look at a wide range of alternatives, explore the trade-off between area and performance, and finally generate the hardware implementation which meets the design goals. Optimization methods including interface streaming, loop unrolling, loop merging and loop pipelining, and so on, can be selected by the designer to create a wide range of micro-architectures from the sequential to fully parallel implementations. However, the tool is not able to directly digest the design specification and automatically generate the desired hardware architecture. Therefore, the designer should still keep the big picture of the hardware architecture in mind and guide the tools to generate the optimized architecture by tuning the optimization options step by step.

Once all the steps described above are finished, an RTL module is generated based on the specification and optimization methods. The designer can use RTL verification tools to verify the correctness of the design. If there is anything that does not meet the requirements of the specification, the designer should go back to the Catapult C to modify the design.

A DSP system such as wireless communication signal processing system usually contains very complicated processing blocks which are developed and verified independently. Catapult C provides integration with many third-party tools which allow the designers to synthesize, simulate and verify the blocks for the DSP system. Figure 8-3 briefly shows one of the typical Catapult C tool flow for DPS system design.

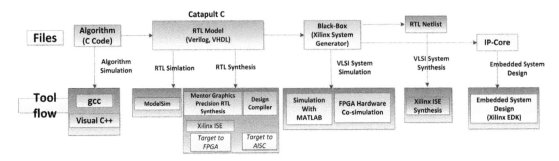

Figure 8-3:
Catapult C HLS design tool flow.

During the algorithm development and simulation stage, the designer can write C++ program for fast algorithm development. The C++ program can be compiled using Microsoft Visual C++ or GCC compiler and Catapult C IDE has provided the interface with these compilers. By using bit-accurate arithmetic library (for example, AC data types), the fixed-point algorithm simulation can be done by converting the floating-point C++ model. Afterwards, Catapult C generates the RTL model according to the architecture constraints provided by the designer. The designer can employ ModelSim to simulate and verify the functionality of the RTL model or directly synthesize these RTL model by Precision RTL (for FPGA), Xilinx ISE (for FPGA) or Design Compiler (for ASIC). The designer can also generate a bigger system with the RTL model generated by Catapult C and blocks generated by other tools such as Xilinx System Generator and Xilinx EDK and so on.

PICO

The PICO C-Synthesis [3, 4] creates application accelerators from un-timed C for complex processing hardware in video, audio, imaging, wireless, and encryption domains. Figure 8-4 shows the overall design flow for creating application accelerators using PICO. The user provides an algorithm in C along with functional test inputs and design constraints such as the target throughput, clock frequency, and technology library. The PICO system automatically generates the synthesizable RTL, customized test benches, and System C models at various levels of accuracy as well as synthesis and simulation scripts. PICO is based on an advanced parallelizing compiler that finds and exploits the parallelism at all levels in the C code. The quality of the generated RTL is competitive with manual design, and the RTL is guaranteed to be functionally equivalent to the algorithmic C input description. The generated RTL can then be taken through standard simulation, synthesis, place, and route tools and integrated into the SoC through automatically configured scripts.

Figure 8-5 shows the general structure of hardware generated by PICO from a high level C procedure. This architecture template is called a pipeline of processing arrays (PPA). Using

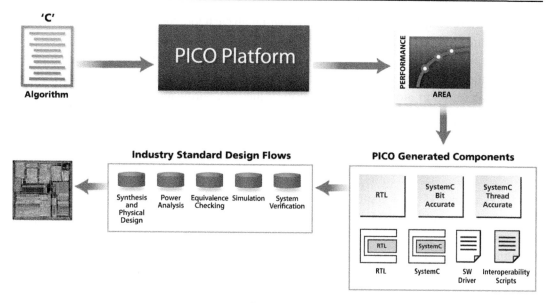

Figure 8-4:
System level design flow using PICO.

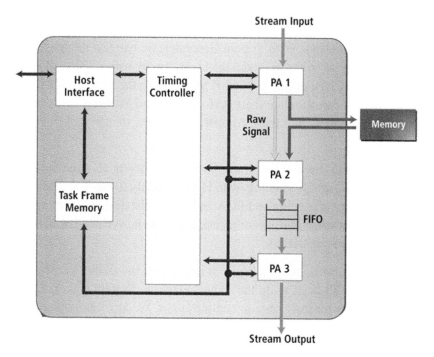

Figure 8-5:
The PPA architecture template.

this architecture template, the PICO compiler will map each loop in the top level C procedure to a hardware block or a processing array (PA). PAs communicate with each other via FIFOs, memories, or raw signals. A timing controller is used to schedule the pipeline, and to preserve the sequential semantics of the original C procedure. The host interface and the task frame memory were used to provide the integration of the PPA hardware into a system using memory mapped IO.

System Generator

System Generator [5] is a system-level DSP design tool from Xilinx. It uses the Simulink design environment for FPGA design, which is very suitable for hardware design. Designs are implemented by using modules from Xilinx blockset. Many modules are provided for Simulink, from common building blocks such as adders, multipliers, and registers to complex building blocks such as forward error correction blocks, FFTs, filters, and memories. These blocks make the design process much easy and deliver optimized results for the selected device. Furthermore, all of the FPGA implementation steps including synthesis and place and route are automatically performed to generate an FPGA programming file.

The design flow of using System Generator is described below. Usually, design starts from algorithm exploration. System Generator is very useful for algorithm exploration, design prototyping, and model analysis. With System Generator, we can fast model the problem and different algorithms. We not only can model the algorithm by using Xilinx blockset, but also can integrate the Simulink blocks and MATLAB M-code into the System Generator. After we model the algorithm, we can simulate their performance and estimate hardware cost in System Generator. With the help of these comparison results, we can make a fast and proper choice of the algorithm.

To speed up the simulation, System Generator provides hardware co-simulation. This makes it possible to incorporate a design running in an FPGA directly into a Simulink simulation. In this way, part of the design is running on FPGA and part of the design is running in Simulink. This is particularly useful when portions of a design need to be verified but need a lot of time when doing pure software simulation.

After we choose the algorithm, we can begin to implement the design part by part. In this level, we can use Xilinx blockset to do the implementation. System Generator provides various Xilinx blockset, from basic ones to complex ones. We also can implement the design in HDL and use an HDL wrapper to make it a component in System Generator. There are many parameters inside each module. We can refer Matlab variable in the parameters. When we want to change the parameter, we just need to assign a different value in Matlab. This makes the design very flexible and easy to maintain.

When we have designed all the modules, we can integrate them together into a whole system. We also can integrate a Matlab testbench to the system. By inputting Matlab variables to the design, we can design much complicated verification method in a short time. If there the design is all correct, we can generate the netlist to the ISE to the specific FPGA. Then you can synthesize the design in ISE and download to FPGA.

Case studies

In the following case studies, we will present three complex DSP accelerator designs using high-level design tools: 1) LDPC decoder accelerator design using PICO C; 2) Matrix multiplication accelerator design using Catapult C; and 3) QR decomposition accelerator design using System Generator.

LDPC decoder design example using PICO

Low-density parity-check (LDPC) codes [6] have received tremendous attention in the coding community because of their excellent error correction capability and near-capacity performance. Some randomly constructed LDPC codes, measured in Bit Error Rate (BER), come very close to the Shannon limit for the AWGN channel with iterative decoding and very long block sizes (on the order of 10^6 to 10^7). The remarkable error correction capabilities of LDPC codes have led to their recent adoption in many standards, such as IEEE 802.11n, IEEE 802.16e, and IEEE 802.15.3c.

As wireless standards are rapidly changing and different wireless standards employ different types of LDPC codes, it is very important to design a flexible and scalable LDPC decoder that can be tailored to different wireless applications. In this section, we will explore the design space of efficient implementations of LDPC decoders using the PICO high level synthesis methodology. Under the guidance of the designers, PICO can effectively exploit the parallelism of a given algorithm, and then create an area-time-power efficient hardware architecture for the algorithm. We will present a partial-parallel LDPC decoder implementation using PICO.

A binary LDPC code is a linear block code specified by a very sparse binary M by N parity

check matrix:

$$\mathrm{H}\cdot\mathbf{x}^T = 0,$$

where x is a codeword and H can be viewed as a bipartite graph where each column and row in H represents a variable node and a check node, respectively. Each element of the parity check matrix is either a zero or a one, where nonzero entries are typically placed at random to achieve good performance. During the encoding process, N-K redundant bits are added to the K information bits to create a codeword length of N bits. The code rate is the ratio of the information bits to the total bits in a codeword. LDPC codes are often represented by a bi-partite graph called a Tanner graph. There are two types of nodes in a Tanner graph, variable nodes and check nodes. A variable node corresponds to a coded bit or a column of the parity check matrix, and a check node corresponds to a parity check equation or a row of the parity check matrix. There is an edge between each pair of nodes if there is a one in the corresponding parity check matrix entry. The number of nonzero elements in each row or column of a parity check matrix is called the degree of that node. An LDPC code is regular or irregular based on the node degrees. If variable or check nodes have different degrees, then the LDPC code is called irregular, otherwise, it is called regular. Generally, irregular codes have better performance than regular codes. On the other hand, irregularity of the code will result in more complex hardware architecture.

Non-zero elements in **H** are typically placed at random positions to achieve good coding performance. However, this randomness is unfavorable for efficient VLSI implementation that calls for structured design. To address this issue, block-structured quasi-cyclic LDPC codes are recently proposed for several new communication standards such as IEEE 802.11n, IEEE 802.16e, and DVB-S2. As shown in Figure 8-6, the parity check matrix can be viewed as a 2-D array of square sub matrices. Each sub matrix is either a zero matrix or a cyclically shifted identity matrix I_x. Generally, the block-structured parity check matrix **H** consists of a j-by-k array of z-by-z cyclically shifted identity matrices with random shift values x $(0 = <x< = z)$.

A good tradeoff between design complexity and decoding throughput is partially parallel decoding by grouping a certain number of variable and check nodes into a cluster for parallel processing. Furthermore, the layered decoding algorithm [7] can be applied to improve the decoding convergence time by a factor of two and hence increases the throughput by two times.

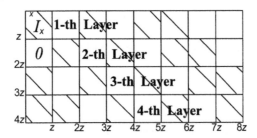

Figure 8-6:
A block structured parity check matrix with block rows (or layers) $j = 4$ and block columns $k = 8$, where the sub-matrix size is z-by-z.

In a block-structured parity-check matrix, which is a j by k array of z by z sub-matrices, each sub-matrix is either a zero or a shifted identity matrix with random shift value. In every layer, each column has at most one 1, which satisfies that there are no data dependencies between the variable node messages, so that the messages flow in tandem only between the adjacent layers. The block size z is variable corresponding to the code definition in the standards.

To simplify the hardware implementation, the scaled min-sum algorithm [8] is used. This algorithm is summarized as follows. Let Q_{mn} denote the variable node log likelihood ratio (LLR) message sent from variable node n to the check node m, R_{mn} denote the check node LLR message sent from the check node m to the variable node n, and APP_n denote the *a posteriori* probability ratio (APP) for variable node n, then:

$$Q_{mn} = APP_n - R_{mn}$$

$$R'_{mn} = s \times \prod_{j:j \neq n} sign(Q_{mj}) \times (\min_{j:j \neq n} |Q_{mj}|)$$

$$APP'_n = Q_{mn} + R_{mn}$$

where s is a scaling factor. The APP messages are initialized with the channel reliability values of the coded bits.

Hard decisions can be made after every horizontal layer based on the sign of APP_n. If all parity-check equations are satisfied or the pre-determined maximum number of iterations is reached, then the decoding algorithm stops. Otherwise, the algorithm repeats for the next horizontal layer.

To implement this algorithm in hardware, we use a block-serial decoding method [9]: data in each layer is processed block-column by block-column. The decoder first reads APP and R messages from memory, calculates Q, and then finds the minimum and the second minimum values for each row m over all column n. Then, the decoder computes the new R and APP values based on the two minimum values, and writes the new R and APP values back to memory. The algorithm is coded in an un-timed C code. A section of the C code is shown in Figure 8-7, which depicts the PPA architecture generated by the PICO C compiler. The parallelism of this architecture is at the level of the sub-matrix size z. Note that the 'pragma unroll' statement in the C code will be used by the PICO C compiler to determine the parallelism level. Multiple instances of the decoding cores are generated by the PICO C compiler to achieve a large decoder parallelism.

As a case study, a flexible LDPC decoder which fully supports the IEEE 802.16e standard was described in an un-timed C procedure, and then the PICO software was used to create synthesizable RTLs. The generated RTLs were synthesized using Synopsys Design Compiler, and placed & routed using Cadence SoC Encounter on a TSMC 65nm 0.9V 8-metal layer CMOS technology. Table 8-1 summarizes the main features of this decoder.

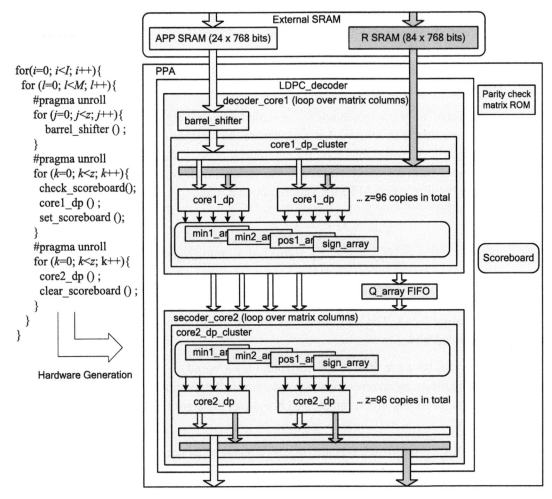

```
for(i=0; i<I; i++){
  for (l=0; l<M; l++){
    #pragma unroll
    for (j=0; j<z; j++){
      barrel_shifter () ;
    }
    #pragma unroll
    for (k=0; k<z; k++){
      check_scoreboard();
      core1_dp () ;
      set_scoreboard ();
    }
    #pragma unroll
    for (k=0; k<z; k++){
      core2_dp () ;
      clear_scoreboard () ;
    }
  }
}
```

Hardware Generation

Figure 8-7:
Pipelined LDPC decoder architecture generated by PICO.

Table 8-1: ASIC synthesis result.

Core area	1.2 mm^2
Clock frequency	400 MHz
Power consumption	180 mW
Maximum throughput	415 Mbps
Maximum latency	2.8 μs

Compared to the manual RTL designs [10, 11] which usually took 6 months to finish, the C based design using PICO technology only took 2 weeks to complete, and is able to achieve high performance in terms of area, power, and throughput. The area overhead is about 15% compared to the manual LDPC decoders [10, 11] that we have implemented before at Rice University.

Matrix multiplication design example using Catapult C

In this section, we will take matrix multiplication as an example and use Catapult C to explore the design space to achieve different design goals. Matrix multiplication is a very important computation block for many signal processing applications, such as MIMO detection in wireless communication, multimedia encoding/decoding, and so on.

The computational complexity for an N $\times$ N matrix multiplication is $O(N^3)$. Usually parallel architecture is employed to accelerate the matrix multiplication computations. According to different design requirements, different parallel architectures can be utilized and there is trade-off between the throughput performance and hardware cost. By using interactive GUI of Catapult C, we can explorer different the micro-architectures and generate the RTL models.

Assume A and B are two matrices, compute C = A $\times$ B. The problem is depicted in Figure 8-8, in which we can see the result matrix C has the same width as matrix B, while its height is equal to the width of matrix A. The C code for the matrix multiplication is listed below. To make the problem easy to describe, we assume A and B are both 4 $\times$ 4 matrices.

```
#define NUM 4
void Matrix_multiplication(int A[NUM][NUM], int B[NUM][NUM], int C[NUM][NUM])
{
  for(int i=0; i<NUM; i++)
  {
    for(int j=0; j<NUM; j++)
    {
      C[i][j]=0;
      for(int k=0; k<NUM; k++)
      {
        C[i][j]+=A[i][k] * B[k][j];
      }
    }
  }
}
```

Checking the code carefully, we notice that the main body of the function is a three-level nested for loop. Since the multiplication computations are operated on different sets of data, there are neither data dependencies nor data races. For instance the computations for each C_{ij} can be run in parallel as long as there are enough computation units. To compute C_{ij}, a dot product with a row of matrix A and a column of matrix B is performed, in which all the multiplications can be executed in parallel. There are many parameters we can explore in the design space to parallelize the matrix multiplication. If we check this function again from a hardware designer's perspective, we can regard this function as a description of a hardware model. The loops inside the function correspond to the pipeline structure in the hardware architecture. Since the main part of the function is a three-level nested for loop, we should focus more on the loop optimization to get high performance. In addition, we notice the input

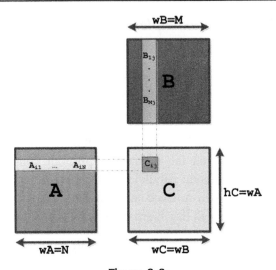

Figure 8-8:
Matrix multiplication.

and output interfaces of this function are two-dimension arrays which represent memory in the hardware.

First, we convert the code to fixed-point version, and add the necessary compiling pragma according to Catapult C specification. In this program, we define the data type of the elements in matrix A and B as int16 (16-bit integer), which has been defined in ac_int.h. For C, int34 is used to avoid overflow. We do a software simulation using the fixed-point C program and verify the results.

```
#define NUM 4
#include 'ac_int.h'
#pragma hls_design top
void Matrix_multiplication(int16 A[NUM][NUM], int16 B[NUM][NUM], int C34[NUM][NUM])
{
  OUTERLOOP: for(int i=0; i<NUM; i++)
  {
    INNERLOOP: for(int j=0; j<NUM; j++)
    {
      C[i][j]=0;
      RESULTLOOP: for(int k=0; k<NUM; k++)
      {
        C[i][j]+=A[i][k] * B[k][j];
      }
    }
  }
}
```

To begin the design with Catapult C, we set up the basic technology configurations. In this example, we select Xilinx Virtex-II Pro FPGA chip as the technology target, and set the

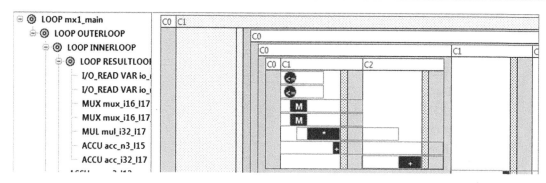

Figure 8-9:
Gantt chart for design with the default configuration.

design frequency to 100 MHz. For the interface setting, we use the default configurations, which include a clock and a synchronous reset signal. We also keep the default configurations for the architectural constraints. We set area as our design goal and try to minimize the area of the logic core. Figure 8-9 shows the generated Gantt chart of this design. From the Gantt chart, it is clear that Catapult C generates a serial implementation. In this serial implementation, the circuit reads two data from the memory and computes the multiplication product in the first clock cycle. Then the circuit performs an addition operation in the second clock cycle. The logic repeats the same operations until the computations are finished.

By pipelining the loops, we are able to hide some execution latency to increase the throughput. We can decide at which level the loop should be pipelined and we can change the setting for each loop separated. In Catapult C, it is still necessary for the designer to keep in mind the low-level details of the hardware, so that he can take advantage of loop pipelining by carefully applying the right loop optimization techniques to the right loops.

Next, we further optimize the implementation by using loop unrolling and loop merging techniques. By enabling loop unrolling, we increase the hardware parallelism to trade higher throughput. For example, we can unroll the INNERLOOP and RESULTLOOP. Since these loops are executed for 16 times, the generated hardware uses 16 multipliers after loop unrolling. The Gantt chart of the loop unrolling result is shown in Figure 8-10. As we have expected, sixteen multipliers are employed and running in parallel.

So far, we only changed the loop optimization settings. It is also possible to change the interface to meet the design specification and the design goal. In the previous design, matrices A, B, and C are all represented by two dimension arrays and they are mapped to memory inside the chip. For example, both A and B are mapped to 1×256 bits memory ($4 \times 4 \times 16$ bits). By default, Catapult C assumes all of the data should be written into the memory before performing all the computations. In the real world application, such as signal processing for wireless communications and multimedia processing, the data are streamed into the circuit

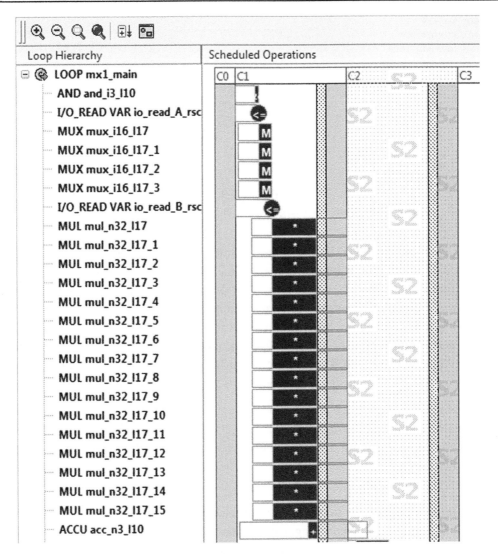

Figure 8-10:
Gantt chart for loop unrolling.

instead of the coming in blocks. Therefore, in the next step, we try to change the interface options to make the matrix multiplication block support streaming data I/O.

By setting the streaming options to 1, we are trying to extract one of the matrix's dimension and stream this dimension's data into the circuit. For example, to input matrix A (defined as: **int A[4][4]**), we input four successive vectors (A[0], A[1], A[2], A[3]) in order. By changing the streaming setting to 1, the interface for A and B now becomes 1×64 bits memory.

Table 8-2: Performance comparison for different optimizations.

	Latency Cycles	Total Area
Default	159	1314
Pipelining	65	1995
Loop unrolling	5	9235
Streaming with partial loop unrolling	17	2293

So far, we have explored several optimization techniques. Table 8-2 shows the performance comparison designs with different design parameters. Since we only care about the latency and area, only these two terms are listed. Table 8-2 shows the performance comparison.

From the table, we can see by using the pipelining and loop unrolling techniques, the latency cycles are reduced significantly from 159 to 5 while the cost of area increases from 1314 to 9235. By using the streaming I/O and partial loop unrolling, we found a sweet point between latency and area, which shows a small latency of 17 cycles and a relatively small area 2293.

QR decomposition design example using System Generator

In this section, we will design a 4x4 QR decomposition hardware accelerator by using the System Generator. The accelerator decomposes a 4x4 matrix A into two 4x4 matrixes Q and R, $A = QR$ where Q is a unitary matrix and R is an upper triangle matrix. Nowadays, QR decomposition is wildly used. For example if we want to solve a matrix equation $Ax = B$, we can decompose A into Q and R. Then the equation becomes $QRx = B$. Now we can move Q to the other side of the equation. The equation becomes $Rx = Q^*B$. Q^* is the conjugate transpose of Q, which is equal to Q^{-1}. Because now R is an upper triangle matrix, we can do back substitution to figure out x.

Many methods are proposed to perform QR decomposition. Here we will focus on Givens rotation. The idea of Givens rotation is to rotate the current vector with an angle, which makes part of the current vector become 0. It is shown below:

$$\begin{bmatrix} \cos\theta & \sin\theta \\ -\sin\theta & \cos\theta \end{bmatrix} \begin{bmatrix} a \\ b \end{bmatrix} = \begin{bmatrix} r \\ 0 \end{bmatrix},$$

and

$$r = \sqrt{a^2 + b^2}$$

$$\cos\theta = a/r,$$

$$\sin\theta = b/r$$

By repeatedly applying Givens rotation to matrix A, we can decompose it into Q and R.

The system architecture is shown in Figure 8-11. The light gray node is delay node. It delays the input data one cycle. The dark gray node is processing node. It has two modes: vectoring and rotating. In the vectoring mode, the horizontal output is the magnitude of the input vector and the vertical output is 0, and the angle of the vector is stored inside the node. In the rotating mode, the stored angle is used to rotate the input vector. The horizontal output is the rotated value of the input X, and the vertical output is the rotated value of the input Y. Only the first time the processing node receives the data of the matrix, the node will operate in the vectoring mode. In other time, the processing node works in the rotating mode. For example, when A11 and A21 arrive at the upper left processing node, the node operates in the vectoring mode. The angle between A11 and A21 is stored inside the node. When the next set of data A12 and A22 arrives at that processing node, the node works in the rotating mode to rotate the A12 and A22 by using the stored angle. And the node will continue working in the rotating mode until the next matrix. By connecting the delay nodes and the processing nodes in the following way, the system can decompose a 4x4 matrix. When inputting a data matrix A and an identity matrix I, the output will be matrixes R and Q.

After we choose the algorithm and architecture, we can begin to do the implementation by using the Xilinx Blockset and Xilinx Reference Blockset from the Simulink library. These libraries will appear in Simulink after we install the System Generator.

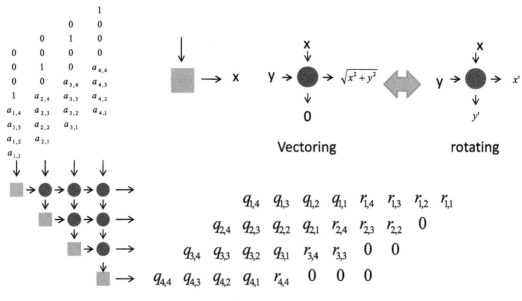

Figure 8-11:
System architecture.

First, we need to implement the processing node. As we know, processing node has two modes. For the vectoring mode, CORDIC ATAN is used. The module is in Xilinx Reference Blockset/Math. The module has two inputs: X and Y. They represent a vector. The module also has two outputs. MAG is the magnitude of the input vector, and ATAN is the angle of the vector. As stated in its document, the MAG needs to be compensated by 1/1.646760 after outputting. This is implemented by a CMULT at the magnitude output of the CORDIC ATAN, as shown in Figure 8-12. CMULT is in Xilinx Blockset/Math. For the rotating mode, CORDIC SINCOS is used. The module has three inputs. THETA is the angle used for rotating. X and Y represent a vector. According to the document, two inputs X and Y need to be compensated by 1/1.646760 before inputting to the module. CMULT is used to compensate both. Two outputs, COS and SIN, are the ROTATED_X and ROTATED_Y, respectively.

After we implement both vectoring and rotating modules, we can use them to implement the processing node. This is shown in Figure 8-13. The processing node has four inputs. X and Y represent the vector. They are connected to the vectoring and rotating modules. MODE controls if the processing node works in the vectoring mode or the rotating mode. OUT_EN is the enable signal for the output. Register is connected between the vectoring module and the rotating module. It keeps the angle calculated from the vectoring module. Register can be found from Xilinx Blockset/Memory. The processing node has four outputs. OUT_X and OUT_Y represent output vector. Muxes from Xilinx Blockset/Control Logic control which value sent to OUT_X and OUT_Y. In vectoring mode, the processing node will output the magnitude of the X and Y to OUT_X and 0 to OUT_Y. Zero is implemented by Constant from Xilinx Blockset/Basic Elements. In rotating mode, the processing node will output the ROTATED_X to the OUT_X and the ROTATED_Y to OUT_Y. MODE_TO_NEXT and OUT_EN_NEXT are the control signals sent to the next processing mode. They are just the delayed version the MODE and OUT_EN signals. The delay is 15. This is because vectoring or rotating needs 14 cycles and Mux needs 1 cycle.

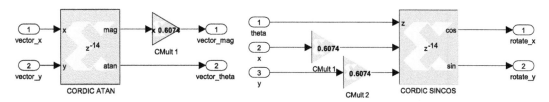

Figure 8-12:
Implementation of vectoring and rotating.

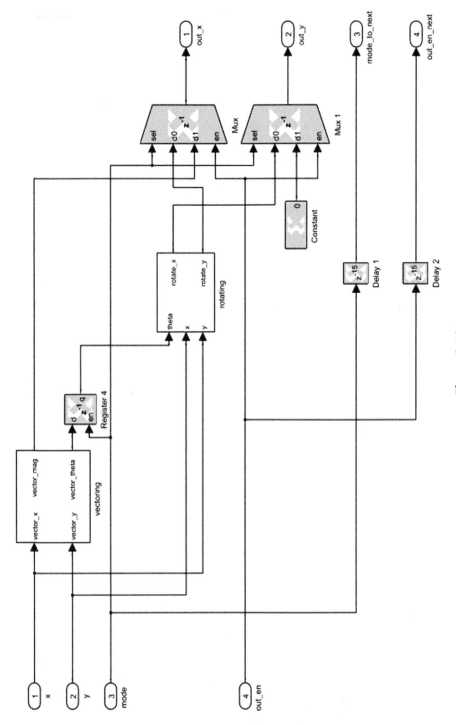

Figure 8-13:
Implementation of processing node.

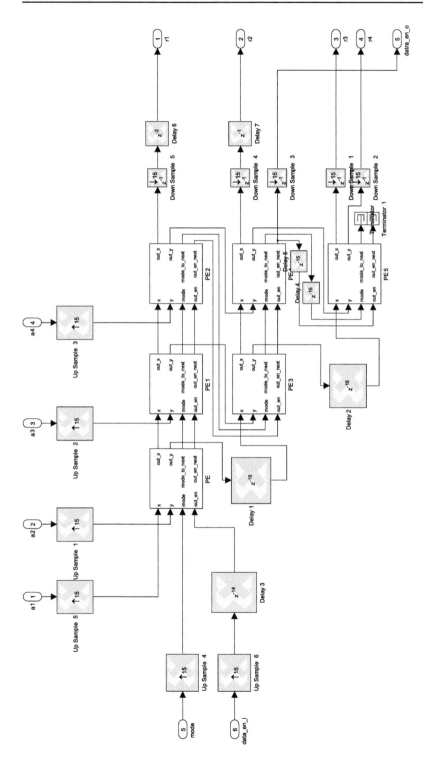

Figure 8-14:
Top level block diagram for the QR decomposition system.

Now we can connect processing node to each other to implement the system as shown in Figure 8-14. Delay node is implemented by Register with specified delay. In this implementation, the delay is set to 15, because each processing node will consume 15 cycles to output a data. The OUT_X of the processing node is connected to the next horizontal node. The OUT_Y of the processing node is connected to the next vertical node. The whole system has six inputs. A1, A2, A3, and A4 are input data. MODE is control signal and DATA_EN_I is input data enable signal. It also has five outputs. R1, R2, R3, and R4 are output data and DATA_EN_O represent when the output data is available. A few Up Sample and Down Sample are used in the system. This is because each processing node needs 15 cycles to output the data, but there will a new input data in each cycle. By using the Up Sample and Down Sample, the system creates a new clock domain, which is 15 times faster than the main clock.

Conclusion

As the demand for high performance DSP systems is rapidly increasing, the chip designer faces the challenge of implementing complex algorithms quickly and efficiently without compromising on power consumption. High level synthesis, which can automatically create efficient hardware from an untimed C algorithm, can provide the solution. With the high level design tools, the designers work at a higher level of abstraction, starting with an algorithmic description in a high-level language such as ANSI C/C++ and System C.

In this chapter, we have introduced the basic methodology of the high level design tools for DSP systems and summarized some important features of the high level synthesis design flow. We have introduced several high level synthesis tools for ASIC/FPGA implementation of the complex DSP systems. In the case studies, we gave three DSP design examples using high level design tools. The design created from high level design tools is comparable to a manual hand design in terms of area-power-timing but with a much faster design cycle.

Today, there are several very successful high level synthesis tools that provide effective solutions in building complex DSP systems. The high level design tools for DSP systems have great potentials to be widely used in the DSP designers' community.

References

[1] P. Coussy, A. Morawiec, High-Level Synthesis rom Algorithm to Digital Circuit, Springer Netherlands (2008).
[2] Catapult C Synthesis official website, http://www.mentor.com/esl/catapult.
[3] Synfora PICO Product, http://www.synfora.com.
[4] Xilinx System Generator Product, http://www.xilinx.com/tools/sysgen.htm.
[5] S. Aditya, V. Kathail, Algorithmic Synthesis Using PICO, Springer Netherlands (2008) 53–74.
[6] R. Gallager, Low-density parity-check codes, IEEE Transaction on Information Theory vol. 8 (Jan 1962) 21–28.

[7] D.E. Hocevar, A reduced complexity decoder architecture via layered decoding of LDPC codes, in: IEEE Workshop on Signal Processing Systems, SIPS, 2004, pp. 107–112.

[8] J. Chen, A. Dholakia, E. Eleftheriou, M. Fossorier, X. Hu, Reduced-complexity decoding of LDPC codes, IEEE Transactions on Communications vol. 53 (2005), pp. 1288–1299.

[9] Y. Sun, J.R. Cavallaro, A low-power 1-Gbps reconfigurable LDPC decoder design for multiple 4G wireless standards, in: IEEE International SOC Conference, SoCC, Sept. 2008, pp. 367–370.

[10] Y. Sun, M. Karkooti, J.R. Cavallaro, High throughput, parallel, scalable LDPC encoder/decoder architecture for OFDM systems, in: 2006 IEEE Dallas/CAS Workshop on Design, Applications, Integration and Software, Oct. 2006, pp. 39–42.

[11] Y. Sun, M. Karkooti, J.R. Cavallaro, VLSI decoder architecture for high throughput, variable block-size and multi-rate LDPC codes, in: IEEE International Symposium on Circuits and Systems, ISCAS, May 2007.

CHAPTER 9

Optimizing DSP Software — Benchmarking and Profiling DSP Systems

Michelle Fleischer

Chapter Outline

Introduction

A key part of optimizing DSP software is being able to properly profile and benchmark the DSP kernel and the DSP system. With a solid benchmark and profiling of the DSP kernel, both best-case and in-system performance can be assessed. Proper profiling and benchmarking can often be an art form. It is often the case that an algorithm is tested in nearly ideal conditions, and that performance is then used within a performance budget. Truly understanding the performance of an algorithm requires being able to model system effects along with understanding an algorithm's best-case performance. System effects can include changes such as a running operating system, executing out of memories with different latencies, cache overhead, and managing coherency with memory. All of these effects require a carefully crafted benchmark, which can model these behaviors in a standalone fashion. If modeled correctly, the standalone benchmark can very closely replicate a DSP kernel's execution, as it would behave in a running system. This chapter discusses how to perform this kind of benchmark.

The key ingredients to a proper benchmark include being able to build the DSP kernel as a single isolated module. Modularity helps to isolate a kernel for testing and measurement. A DSP kernel should always be separated from its test harness and from the runtime libraries used by a test harness. Modularity also ensures that when it is implemented in a system along with other kernels, it is still measurable and comparable to any standalone benchmarks. Also of key importance is to create a flexible test harness. This should be code that is able to exercise the kernel, possibly multiple times, provide a means of measuring the performance of the kernel, provide input test configuration, input test vectors, output test vectors, and often act as a means to self check an algorithm for correctness. Other test harness considerations should include being able to move code and data content to and from memory regions with differing latencies and cache policies, and to be able to adjust memory sizes and allocations such that the algorithm can more closely replicate how it would behave in a fully operational system.

Writing a test harness

The best method for getting optimized code is to break the work into modules with distinct and measureable inputs and outputs. In this way a DSP functional block or kernel can be worked on in isolation. This removes system impacts such as operating systems, interrupts, and other system-level interference from impacting performance measurements. By isolating a DSP kernel, its performance can also be isolated, measured, and re-measured to show improvements or degradation in the kernel's performance. Getting the best possible performance in a DSP kernel, in isolation, almost always results in getting the best performance from the algorithm once it is integrated back into the running system.

In order to properly test and measure the performance of a DSP kernel, it is typically 'wrapped' in a test harness. A test harness is simply a bit of standard C code (typically) which is used to test the DSP kernel. Writing a test harness is not difficult, but it requires some careful planning. A consistent methodology should also be used so that the harness itself is re-usable, extendable, and consistent with other DSP kernel test harnesses. Other things to consider in writing a test harness include profiling capability of the hardware or simulator, ease of substitution of new versions of a DSP kernel, the ability to automate the testing and profiling process, and the profiling or optimization goals. A test harness for a DSP kernel should be able to:

- Handle multiple input and output vectors in a manner that does not impact the kernel's functionality or performance
- Isolate a DSP kernel for testing and profiling purposes
- Verify results with a 'golden' model or other reference vectors
- Run under multiple architectures and tooling

Test harness inputs, outputs, and correctness checking

Test harnesses all require a method for input of test vectors outside of a real system. This is normally achieved either by reading the vectors from an external text or binary file, or by placing the vector data within the test harness's memory. In modern DSP systems, one of the more critical aspects of optimization is code placement in various memories. Latency and cache policy can vary significantly from external to internal memories and from architecture to architecture. So placement of input control structures and data, output data, and other results is critical in creating a test harness, which gives realistic performance numbers. The test case itself should be allowed some flexibility in its memory placement, so that performance tuning can be performed based on code and data location. The inclusion of data vectors can be carried out via several methods. Common approaches include:

- Placing the test vector data inline in a header or source file
  ```
  #include "my_input_vector.dat"
  #include "my_control_parameters.dat"
  ```
- Using a #include "my_vector.dat" inside a variable definition or for the variable definition
  ```
  UINT32 my_IQ_buffer[] = {#include "my_vector.dat"};
  ```
- Reading the data from an external file
  ```
  FILE input_file;
  Char my_input_filename[] = "./../../vectors/my_input_vector";
  if ((input_file = fopen(my_input_filename, "rb") == NULL)
  {
      return (FILE_OPEN_ERROR);
  }
  //Read the file
  .
  .
  .
  //Close the file
  fclose(input_file);
  ```

- Pulling the data using a development tool script

```
#Load the Input Buffer with the Input Test Vector and Go
## Input Test Vector Naming Convention --> IF2_Tx_<sbfn>_<user_id>_<rvindex>.lod
  set TVIn _$RVIndx.lod
  set TVIn _$UserId$TVIn
  set TVIn _$SubFrN$TVIn
  set TVIn IF2_Tx$TVIn
  set TVIn /$TVIn
  set TVIn ..//..//vector/in/TC$TC$TVIn
  set dummy_addr [evaluate #x dummy_sequence]
  set in_addr [evaluate #x IF2a_Tx_Data]
  puts "input_sequence ="
  puts $in_addr
  restore -b $TVIn m:$in_addr
  restore -b $TVIn m:$dummy_addr
  go 10000
```

These approaches each have some benefits and drawbacks of course. Placing vectors inline with the test bench can lead to difficulty changing the test vectors and can also lead to very large source files which may take longer for a compiler to parse. Also if multiple buffers of a large size need to be placed sequentially or randomly, inline data sizes can grow significantly. Using a #include statement allows for an external text file to be used, which tends to be easier to change for updates to test vectors, and also makes for cleaner, more readable code. However, it has the same drawbacks for compile time and space usage. In all of these cases, changes such as swapping to a different set of test vectors would generally require a rebuild, unless all test cases could fit into memory, but again for general debugging, this has drawbacks as well. This approach tends to limit the number of test cases that can be run, so keep this in mind when developing a test bench. Reading data from an external file is also a typical approach and this tends to be the most flexible one for updating test vectors, as it typically requires no rebuild of the test bench itself. Most Silicon vendors provide at least a basic <stdio> library which includes functions such as fopen, fread, fwrite, fseek, fclose, and printf. A drawback of this approach is that it adds runtime library code to the test harness and it may have an impact on the test execution's performance if care is not taken to exclude any cache or IO operations which may take time from the test harness's execution. Runtime effects can be mitigated by judicious use of cache control operations such as synchronizing or flushing around the runtime library calls.

In both cases care should be taken in the scope of the input data relative to the function being tested. This is particularly important when using global optimizations. It is often the case where compilers will constant propagate or even dead strip algorithmic code if it can find an optimal solution. This can lead to misleading results and incorrect behavior modeled from a single test vector rather than an entire system. (Example of a flag being const propagated, or an output becoming the answer). For example in the following code sequence:

```
my_config = TEST_FOR_120_FRAMES;
  if(my_config == TEST_FOR_60_FRAMES)
{
   RLSIP_Frame60();
} else if(my_config == TEST_FOR_120_FRAMES)
  {
   RLSIP_Frame120();
  } else
{
   RLSIP_Frame180();
}
```

In this example code the if-else ladders which would call RLSIP_Frame60() and RLSIP_Frame180() would be optimized away, and If this is the only place these functions are called the functions themselves could be dead stripped. Additionally, no comparisons for my_config would occur. This could change the code layout which can impact instruction cache performance and removes some comparison functions which changes how the code is generated by the compiler.

Isolating a DSP kernel

Isolating a DSP kernel is a straightforward process. The best approach here is always to keep any code that will be running on the target system in some form or another in its own source files and header files. This prevents a compiler from doing things that would not occur in the real system, such as moving functions inline with the test harness, constant propagation, and in some cases, algorithms or control code being stripped and replaced with results. This can all occur if the test kernel is within the same scope of the test harness and test input vectors. Also, if global optimizations are available, then these should either not be used, or should be used with caution. If they are helpful, then using a pragma to disable automatic inlining might be a good idea, but this may not prevent constant propagation, which would essentially remove code dependent on values perceived as constants by the compiler.

Protecting against aggressive build tools

Some build tools will optimize away significant portions of an algorithm or control code. This often occurs when using global optimizations or if the test vectors, input control structures, or even the output data is within the scope of the compiler to optimize away. For example, if testing for a case where control code has four options and only one of the four are used in the test vectors, an aggressive compiler may throw away the control paths for the other three options. Additionally, if the output data appears not to be used, some compilers can decide to skip the algorithm entirely. Care should be taken in specifying the scope and global compiler options. It is also often a good idea to inspect assembly listings to verify that code is in place for parts of the DSP kernel being tested and also for parts that may not be exercised by the current set of test vectors. If the results look too good to be true, in general they are.

Allowing flexibility for code placement

In modern DSP systems, object placement is often critical to the overall system's performance. Multiple levels of internal and external memories are available, and it is often necessary to be able to move test harness code, test vectors, output vectors, kernel code and kernel data to lower or higher latency memories. This is typically done at link time, and less often in a dynamic environment. Keep in mind questions such as:

- Where would input vectors be in your software design?
- Where do output vectors go in your design?
- Where is the DSP kernel going to be?
- What is the size of these objects and will they all fit into a low latency memory?
- Do you need to model the benchmark as it would be in your system or can some assumptions be made?
- Will there be multicore effects such as reduced bandwidth to some memories?
- What will the cache loading be when your algorithm starts?

Modeling of true system behaviors

Cache effects

Today with DSP kernels, it's really all about the caches. Level 1 caches tend to be zero wait state, and ideally a DSP kernel executes from cache for both instruction and data accesses. To run a fair test, a developer needs to know what the status of data should be in caches. Does the system support hardware coherency or does it require software coherency support? Is it a mix of these two? Is it data coming from a DMA copy? Is it located in an external high-speed interface? In these cases the data would not be in the caches at all when the kernel first starts. Is it appropriate to prefetch the data or instructions in the kernel or in the test harness? Does the test harness also occupy space in the caches prior to running the kernel? Is this a valid state for the hardware or should test harness code and data not being used by the kernel be flushed? Does a run with the cache not initiated, or cold, give a valid representation of the kernel benchmark or does a warmed cache run make more sense? Data and instruction alignment is also often used as a cache optimization. Other considerations should include the cache behavior for fetching, synchronization, flushing, and invalidation operations.

Memory latency effects

Memory latency has a large impact on a DSP kernel's performance. A developer must understand the latencies to various memories, but even more, the rules of thumb of any memory controllers with regard to access alignments and page switching.

System effects

RTOS overhead

An RTOS adds overhead. Even the thinnest, barest RTOS will be written to the lowest needed API functionality of a system. Additionally, it allows a DSP kernel to run in a much more realistic environment and it often makes porting a DSP kernel into the overall system less time consuming, as much of the work in integrating the algorithm with the RTOS is already completed. Considerations here include the RTOS's usage of system resources such as low latency memory and caches, and also any differences in the system configuration enforced by the RTOS. Other considerations include the overhead of the RTOS, as well as interrupts, which in a standalone test case are seldom enabled, but are in general always enabled under an RTOS, unless explicitly disabled.

Execution in a multicore/multidevice environment

It is quite rare to test a DSP kernel in a multicore environment. This is because this kind of testing is difficult to do, and unless a full system is ready, it does not give an indication of the true multicore system's effects on algorithm performance. These effects include the potential for longer memory stalls, lack of availability of peripherals and other system resources, interrupts which would not occur in a single core system, changes to system timing, and even to order of execution. Also, with multicore systems there is often a need to synchronize messaging and data exchange between cores.

Methods for measuring performance

There are many methods of measuring performance on a DSP system. These include time-based measurements using a real-time clock, hardware timer or RTOS tick counter, performance counters which run at the core clock rate or an integer division of the core clock rate, or at its most basic, using IO triggers and a logic analyzer or oscilloscope.

This latter method is often unnecessary for anything but a sealed or very simple DSP system.

Time-based measurement

Time-based measurements are the most frequently used methods of measurement in most modern DSP systems. Most RTOSs provide a timer service of some kind that can be co-opted for performance measurements. It is also often an RTOS event, which is used as a trigger point to enable and disable measurements. For example, an RTOS context switch to and from a task is often a simple way to take measurements and many RTOSs provide hooks in the context switch code of their kernels to easily create these measurement points.

Hardware timers

If a service is unavailable, or not precise enough, for use in benchmarking, then a hardware timer can also be used. This approach has the big drawback of using a hardware resource (the timer) for what may be a non-essential task. When using hardware timers it's a good idea to verify the input clock frequency, pre-scale settings, and general operation of the hardware timer. It sometimes makes sense to output the timing to a scope to verify and sanity check that the settings being used correspond to some real-world fixed timing and that no mistakes or omissions have occurred in the documentation for the device being used.

Performance counter-based measurement

Most DSP cores today offer 32- or even 64-bit performance counters, and often multiple counters are provided. These are often capable of measuring precise numbers of clock cycles since they run at the same clock rate as the core itself. Some hardware will give additional inputs to the counters to measure other system activities as well. These include cache misses, memory subsystem accesses, stall occurrences, speculation success rates, and other useful details on what is happening within the DSP kernel being tested. Some of these hardware blocks even allow for advanced triggering to enable and disable the profiling at specific locations in the code based on prior execution paths, data patterns, counted events or cycles, or other events.

Profiler-based measurement

Many DSP devices have built-in profiling hardware. This hardware is useful for point to point profiling, often showing a function or loop level of resolution. For more in depth resolution there are many simulators, which can show performance at an instruction-by-instruction resolution. This level of resolution is often very useful in fine tuning and optimizing a DSP kernel. It is often the case that these profiling capabilities are exposed by a development tool and are often one of the more complex aspects of the development tool. Using these features can sometimes be difficult. In these cases seeking assistance from the tool vendor is usually a good option to pursue.

Measuring the measurement

When taking very accurate measurements of a DSP kernel, or when running test vectors that do not create more than several hundred cycles, it is important to measure the latency of your measurement technique. This is often performed as a read of a counter and perhaps some small calculations for offsets of the count values. Latency to hardware counter registers and other memory mapped registers in an embedded system can vary quite a bit. Typically latency to these registers can be anywhere from 25 to 80 cycles! This means that if your DSP kernel executes in 400 cycles, the activity of reading registers may account for 20% of the total benchmark number! A typical approach to this would be to place a few assembly 'nop' instructions in a test function, and to measure the number of cycles it takes to benchmark that function. It will include the nop function and also the latency of reading the counters in the final number.

Excluding non-related events

When running a DSP kernel benchmark there can often be unrelated events that might impact benchmark results. These include system interrupts, runtime library functions, and other hostio-based interactions that might be used in the benchmarking process.

Interrupts

Interrupts, if at all possible, should be disabled during a benchmark. Many profilers do not distinguish between the context of being in a DSP kernel and being in an interrupt service routine. This means any interrupts which occur mid-benchmark will usually have the cycles associated with them buried in the benchmark results. If interrupts cannot be disabled, then care must be taken to exclude the interrupt prologue, epilogue, and body from the benchmark results. Also be aware that executing an ISR will have an impact on cache behavior, especially instruction cache behavior, as it is essentially a change of flow and often one that speculation will not predict. This may also impact branch speculation by evicting some of the DSP kernel's already present speculation entries from the hardware's speculation lookup tables. Interrupts do change the benchmark behavior, and in almost all cases, they will degrade benchmark performance. If performance increases in the presence of interrupts, it would be advisable to carefully check what is happening, as it may be indicative of cache thrashing or a software error.

Runtime library functions used in the benchmark

Runtime library code is a valid item to measure and to use within a benchmark. Care should be taken around 'special' functions that may interact with the hardware debug environment or the simulator environment. Functions that perform any kind of file IO or console IO, or accelerate memory clearing for simulation, should be used with caution. Contact the tools provider to find out which of these functions might change its behavior based on real-world execution versus execution in a debug environment.

Simulated measurement

Modern simulators for DSPs have become very complex, and are now able to simulate hardware timing behavior to a very high degree of accuracy and precision. That said, many vendors offer multiple simulator models with varying degrees of hardware modeling. These can be something very basic, such as an instruction set simulator (ISS), which performs a functional modeling only. Timing information in this model is not generated, nor is it available. These are often used to check functional correctness on simulations that may run too long on a more complex model. The more commonly used simulators for profiling and optimization purposes are labeled cycle accurate simulators (CAS) or performance accurate simulators (PAS). These can model behavior from the DSP core to the DSP subsystem to the entire device. Additionally they model caches, cache controllers, memory busses, and latencies to provide a very accurate accounting of everything that executes within a DSP kernel. Often these simulators have software hooks enabling a very large amount of detailed information to be gathered during profiling. Details on cache behaviors, core stalls, memory stalls, and an instruction-by-instruction account are all made available in these simulator models. This makes them an excellent choice for profiling and optimization work on standalone kernels or code which does not require external stimuli such as hardware interrupts or external ports or busses.

Figure 9-1 shows a sample profiling output for a DSP kernel. Each line of disassembled code which executed has core stalls, data bus stalls, and program bus stalls associated with them. Developers can use this information to better understand the details of how their software interacts with the hardware.

Simulator models also present a model for onboard profiling hardware as well. This is particularly useful when detailed views of the system may not be necessary, and also to do back-to-back comparisons of the simulator's results against actual hardware. By providing a model of profiling counters in the simulator code which reports from these counters will work identically on both the simulator and on actual hardware.

Hardware measurement

Measurements on hardware are often simpler than what is available in a simulator. Often the hardware provides one or more high-speed counters and several hardware timers which can

Line no. / A...	Disassembly	# exec...	cycles-total	execution stalls
63.	**DataOut[2*i] = round(YN);**	1	3	1
0xC0000072	mpy d8,d14,d15 & rnd d1,d1 & move.l #-$3fffbce8,r1	1	2	1
0xC00000DE	mac #-$199a,d10,d12 & rnd d5,d10 & mpy d8,d14,d11 & moves.4f d0:d1:d...	1	2	0
0xC0000114	mpy d13,d14,d0 & rnd d7,d4 & adr d0,d6 & move.l #-$3fffbce0,r3	1	1	0
0xC0000130	mpy d12,d14,d15 & move.4f (r2)+n3,d0:d1:d2:d3 & moves.4f d4:d5:d6:d7...	4	4	0
0xC00001BE	mpy d12,d14,d15 & rnd d1,d6 & mac #-$199a,d4,d7 & moves.4f d0:d1:d2:...	4	8	0
0xC00001FC	lpmarkb mac #-$199a,d6,d1 & add d9,d15,d2 & rnd d0,d6	4	4	0
0xC0000210	tfr #$28,d7 & moves.4f d4:d5:d6:d7,(r3)	1	1	0
64.	**DataOut[2*i+1] = round(YNP1);**	4	11	6
0xC0000208	add d1,d11,d0 & rnd d2,d5	4	4	2
0xC000020E	rnd d0,d7	4	7	4
65.	**}**			
66.				
67.	**for (i = 0; i < DataBlockSize; i++)**	40	470	234
0xC0000230	deceq d7	40	40	0
0xC0000232	jf $c0000224	40	430	234
68.	**printf("Output %d\n",DataOut[i]);**	40	282	200
0xC0000218	move.l #-$3fffbce8,r6	1	1	0
0xC000021E	moveu.l #$c0004164,d6	1	1	0
0xC0000224	move.w (r6)+,r0 & move.l d6,(sp-$4)	40	40	0
0xC0000228	jsrd $c0002cf0	40	40	0
0xC000022E	move.l r0,(sp-$8)	40	200	200
69.				
70.	**return(0);**	1	16	8
0xC0000238	sub d0,d0,d0 & suba #$8,sp	1	2	0
0xC000023C	pop r6 & pop.2l d6:d7	1	3	0
0xC0000242	rts	1	11	8

Figure 9-1:
Critical Code analysis view.

be used for profiling purposes. In some cases hardware trace can be configured to add these counters as tags to each trace message. When making measurements it is always a good idea to ask the vendor to provide any software setup code or detailed documentation on the profiling hardware. Some of the key questions about using hardware profiling counters are:

- How the counters count
  - What should the counter's initial value be?
  - Does it count up or down?
  - Does it interrupt on overflow or underflow?
  - Does the counter run at the core's clock rate or a division of it, or from an external oscillator?
  - Are there any pre-scale factors or other scaling factors that would affect the count rate?
  - What is the latency to read the counter values?
  - What events on the system can be counted?

Other considerations for hardware counters would be to check if they are accurately modeled in a simulator. This would allow use of these counters both with hardware and while hardware is inaccessible. It is often the case that some of the counters may be modeled while others are not.

Profiling results

Profiling results can be in the form of a database full of trace messages, or can be a simple reporting of a counter, scaled to some meaningful value relevant to the benchmark being run. Results will vary, and it is important to be able to identify quickly and easily where the majority of cycles are being spent in a DSP kernel. Some tools provide detailed information to identify these areas (see Figure 9-1). Also profiling results shows by inspection, core stalls, data accesses, cache misses, control code paths, and even branch speculation behaviors.

How to interpret results

Interpreting results should be done with some care. A close inspection of the assembly code is really the only method to truly ensure that the benchmark makes sense and does what was expected. The reason for this is that modern DSP compilers tend to obfuscate code, and they can strip code unexpectedly in a benchmark as being "dead" simply because the benchmark may not use a particular code path. By inspecting the code items such as how parallel the arithmetic operations are, algorithm efficiencies, and control code efficiencies can be determined. These items can be used to select areas to optimize by re-working the DSP kernel, and tend to be the valuable part of testing and benchmarking once the basic goal of functional verification is achieved.

How to use them to optimize code

   i.   Excessive movement in or out of memory
  ii.   Parallelism of code execution
 iii.   Cache behavior
  iv.   Control code efficiency
   v.   Algorithm efficiency

CHAPTER 10

Optimizing DSP Software — High-level Languages and Programming Models

Stephen Dew

Chapter Outline

Assembly language

In the early days of DSP, software was exclusively written in assembly language. This was due to the slow nature of DSPs and the need to make productive use of every cycle. Now, C and other languages based on C are in widespread use. These days, application programmers craft the bulk or all of their software in C and only optimize in assembly when absolutely necessary. The overhead generated by writing in C is outweighed by the simplicity of development in a high-level-language. However, assembly is still used in some cases where performance is absolutely critical. For critical DSP kernels, programmers may write functions in assembly or use vendor-supplied reference code, hand coded in assembly by experts on that processor platform.

Assembly languages are proprietary in nature and based on the particular instruction set of the DSP being programmed. In assembly, the programmer specifies the exact instruction

DSP for Embedded and Real-Time Systems. DOI: 10.1016/B978-0-12-386535-9.00010-X (and additionally any variants such as addressing mode, positive or negative accumulation, etc.) explicitly. The programmer must do the instruction selection, scheduling and register allocation (all performed by C compilers when writing in C).

Syntax varies from processor to processor. For example, on the ADI Blackfin DSP, the syntax is algebraic, as shown in Figure 10-1.

```
R0 = 1;
R1 = 2;
R3 = R1 + R0;
```

Figure 10-1:
Simple addition in assembly on ADI Blackfin DSP.

On the Freescale StarCore DSP, the actual instruction must be specified as shown in Figure 10-2.

```
tfr #1,d0
tfr #2,d1
iadd d0,d1
```

Figure 10-2:
Simple addition in assembly on Freescale StarCore DSP.

Advantages and disadvantages

The advantages to assembly are primarily performance and secondarily the ability to have complete control over the code. No compiler will match the performance of a good assembly programmer who knows the architecture well and has the time to develop good code in assembly.

The disadvantages are that highly skilled programmers who know the architecture are needed, assembly programming takes time (more than writing in C for example), and the code is not portable between platforms, and only sometimes portable to a newer platform from the same vendor.

C Programming language with intrinsics and pragmas

The C programming language, developed between 1969 and 1973 by Dennis Ritchie, is extensively used in embedded and DSP programming. C compilers have existed for DSPs for quite some time. DSPs were originally programmed in assembly language to extract the best performance possible, but with the advent of the modern compiler, the vast majority of DSP code is written in C. Only the most speed-critical routines may be hand optimized, if at all.

The compiler maps general-purpose high-level code to a target platform. Mapping must preserve the defined behavior of the high-level language definition. With DSPs, the target platform may provide functionality that is not directly mapped into the high-level language definition and application space may use algorithm concepts that are not handled by the high-level language definition. Therefore, custom language extensions are often necessary. In addition, understanding how the compiler generates code is important to writing code that will achieve desired results.

Standard C integral types

The ANSI C standard allows for some flexibility in data size for standard C types (char, short, int, long, etc.).

For example, both the TI and StarCore DSP cores can work with 8, 16, 32, 40, and 64-bit data. As noted in Table 10-1, the TI C64x+ has 64 32-bit general purpose registers whereas the SC3400 and SC3850 cores have sixteen 40-bit data registers and sixteen 32-bit address registers. Due to the difference in register size, there are some important differences in the way 40-bit types are handled, as shown in Table 10-1.

Table 10-1: Data type differences on different DSPs.

C Data Type	Size on StarCore-DSPs	Size on TI c64x/c64x+
char	8	8
unsigned char	8	8
short	16	16
unsigned short	16	16
int	32	32
unsigned int	32	32
long int (long)	32	40
unsigned long int (unsigned long)	32	40
long long	64*	64
unsigned long long	64*	64
float	32	32
double, long double	32	64
double, long double (with 64-bit types enabled)	64	
pointer	32	32

*means "if enabled"

Some implications from the table above:

- On TI 64x+ processors, the C type long is used to specify 40-bit data. On the hardware, 40-bit data types are represented across two registers on 64x+ (since registers are 32-bit). On StarCore, long is 32-bit (in a 40-bit register)
- 8- and 16-bit data types are supported in both architectures as one value or multiple packed values stored in a larger register (e.g., two 16-bit values in a 32- or 40-bit register)

FIR in C

The classic finite impulse response (FIR) filter can be written in C, as shown in Figure 10-3. As mentioned, DSP concepts such as fractional data and arithmetic saturation cannot be expressed in C.

```
short SimpleFir0(
    short *x,
    short *y)
{
    int i;
    short ret;

    ret = 0;
    for(i=0;i<16;i++)
        ret += x[i]*y[i];

    return(ret);
}
```

Figure 10-3:
C code for classic FIR filter.

Intrinsic functions

Intrinsic functions, or intrinsics for short, are a way to express operations not possible or convenient to express in C, or target-specific features. Intrinsics in combination with custom data types can allow the use of non-standard data sizes or types. They are used like function calls but the compiler will replace them with the intended instruction or sequence of instructions. There is no calling overhead (Table 10-2).

Some examples of features accessible via intrinsics are:

- Saturation
- Fractional types
- Disabling/enabling interrupts

For example, the FIR filter shown previously can be rewritten to use intrinsics and therefore to specify DSP operations natively (Figure 10-4). To do so, simply replace the multiply and add operations with the intrinsic L_mac (for long-multiply-accumulate). This will replace the two operations with one and add the saturation function to ensure that DSP arithmetic is handled properly.

Fractional types and saturation

Fractional types refer to data types which represent a fraction (values between -1 and almost $+1$). Fractional arithmetic and integer arithmetic have some differences. Fractional arithmetic usually has the option to saturate (clip to the minimum or maximum value),

Table 10-2: Intrinsics have no overhead.

Example Intrinsic (C)	Generated Assembly Code
d = L_add(a,b);	add d0,d1,d2

```
short SimpleFir1(
 short *x,
 short *y)
{
 int i;
 long acc;
 short ret;

 acc = 0;
 for(i=0;i<16;i++)
 // multiply, accumulate and saturate
 acc = L_mac(acc,x[i],y[i]);
 ret = acc>>16;

 return(ret);
}
```

Figure 10-4:
FIR filter re-written with intrinsic.

whereas integer behavior is typically wrap-around. In addition, fractional multiplication requires a left shift by 1 bit if the programmer wants to keep the values normalized between -1 and $+1$. Consequently, fractional arithmetic in the high-level language may generate different assembly instructions in some cases.

Fractional support is not native to (normal) C but may be available via intrinsic functions or directly available via custom types or the embedded C language. There are several ways to handle fractional data types and saturation. Saturation can be handled explicitly (i.e., data saturates only when programmer specifies, perhaps by invoking a saturate instruction) or implicitly (i.e., fractional instructions automatically saturate before they write the result back to the register).

For example, the StarCore cores support fractional and integer operations simultaneously in parallel. The type of arithmetic is specified by the instruction; there are instructions for fractional operations and integer operations. There are also move instructions to load and store the data for both data types. The fractional arithmetic instructions perform a left shift by 1 bit after a multiply operation and also saturate at the relevant point (either 32- or 40-bits depending on the configuration of the core). The code in Table 10-3 compares instruction sequences for a simple load of two data values, multiply and accumulate and store for fractional operations (on the right) and integer operations (on the left).

Floating point

Floating point types are typically emulated on fixed-point DSP platforms. On dedicated floating point DSPs, and on newer DSPs, there is native support for floating point operations along side integer and fractional operations. If floating point types are not native, they will be implemented via runtime library. There will be a performance hit in this case.

Table 10-3: Instruction sequences for different operations.

Example Integer Operations	Example Fractional Operations
short a,b; int c; c = a*b;	short a,b; int c; c = L_mpy(a,b); // use of intrinsic − fractional
Generated Code for Integer Operations	**Generated code for fractional operations**
move.w (r0)+,d0 move.w (r1)+,d1 impy d0,d1,d2 move.l d2,(r3)	move.f (r0)+,d0 move.f (r1)+,d1 mpy d0,d1,d2 move.l d2,(r3)

Custom types

Some platforms will use custom data types that will allow the programmer to access features specific to that processor. For example, on StarCore processors, 40-bit data types are specified by a custom defined type, Word40, which is a structure with 8-bit and 32-bit members. In the hardware, this is represented as one register. The compiler will effectively optimize away the structure when generating code.

Pragmas

Pragmas are used to express information which can be used by the compiler to help optimization or control code generation that cannot be expressed in the C language. Generally, they fall into three categories: function, statement, or variable.

Function pragmas

Function pragmas affect a particular function. Some examples include:

- Inlining of functions
- Indicating that the function is an interrupt service routine

```
int InterruptHandler(void) {
#pragma interrupt InterruptHandler
```

In the above example, the compiler would treat the function as an interrupt handler. This would trigger the generation of proper change-of-flow instructions to return from the interrupt routine which would typically differ from a function return.

Statement pragmas

Statement pragmas provide information about a particular statement only. These are typically used by the compiler to help optimization. If the loop minimum and maximum are known, for example, the compiler may be able to make more aggressive optimizations.

```
void correlation2 (short vec1[], short vec2[], int N, short *result)
{
 long int L_tmp = 0;
 int i;
 for (i = 0; i < N; i++)
#pragma loop_count (4,512,4)
 L_tmp = L_mac (L_tmp, vec1[i], vec2[i]);
 *result = round (L_tmp);
}
```

Figure 10-5:
Using a pragma to specify the loop count bounds.

Some examples include:

- Loop count
- Profiling information (if/then execution likelihood)

In the example in Figure 10-5, a pragma is used to specify the loop count bounds to the compiler. In this syntax, the parameters are minimum, maximum, and multiple respectively. If a non-zero minimum is specified, the compiler can avoid generation of costly zero-iteration checking code. The compiler can use the maximum and multiple parameters to know how many times to unroll the loop if possible.

Variable pragmas

Variable pragmas affect a particular variable only. Some examples are:

- Alignment (either specify that a pointer points to aligned memory or force a higher alignment than default)
- Pragma to place a variable in a certain area of memory (normally via the linker)

In the example in Figure 10-6, pragmas are used to tell the compiler that the incoming pointers vec1 and vec2 are aligned to a boundary of 4 bytes. This would not ordinarily be assumed since variables of type short are typically aligned to their size (2 bytes). This allows the compiler to potentially load two elements at once.

```
short Cor(short vec1[], short vec2[], int N)
{
#pragma align *vec1 4
#pragma align *vec2 4
 long int L_tmp = 0;
 long int L_tmp2 = 0;
 int i;
 for (i = 0; i < N; i += 2)
 {
 L_tmp = L_mac(L_tmp, vec1[i], vec2[i]);
 L_tmp2 = L_mac(L_tmp2, vec1[i+1], vec2[i+1]);
 }
 return round(L_tmp + L_tmp2);
}
```

Figure 10-6:
Using pragmas for data alignment.

Embedded C

Embedded C is a set of extensions for the programming language C to support embedded processors, enabling portable and efficient application programming for embedded systems.* Embedded C is a variant of the C language, specified in a technical report ratified by the ISO on February 2004. Its motivation is to allow better DSP programming, expose architectural features to the high level programmer and provide better portability between DSP targets. It supports DSP constructs such as:

- Fixed point data types
- Saturating data types
- Named address spaces
- Input/Output accesses

Although Embedded C has shown promise, it is only used by a few organizations currently and no major DSP development is taking place in Embedded C.

C++ for embedded systems

As embedded system developers move to higher level programming languages such as C++, they are presented with opportunities for higher levels of abstraction and productivity. At the same time, however, an embedded system's rigid requirements mean that developers must be careful in potentially impacting the runtime costs of these higher levels of abstraction that are not incurred with traditional lower level languages such as native assembly and C.

Some of the benefits of migration to a higher level object oriented language such as C++ are those of code reuse, syntactic similarity with C, type checking, memory allocation and so forth. Drawbacks of C++ can include larger code size when considering libraries, and performance and memory overhead of certain language features.

One benefit of C++ use for embedded development is the availability of static constants, which often replace the use of preprocessor macros. In many instances, the compiler may opt to fold these literals into their uses, depending on compile time constraints. Another beneficial feature of C++ is the use of namespaces, which can prevent name space collisions that were previously common within large scale C projects. Namespaces take care of this problem, and any names appearing in application code such as variables, functions, or enumerators are resolved to a given namespace including a globally unnamed namespace.

---

* (http://www.embedded-c.org/)

New and delete constructors are also available within C++ application development for handling of allocation and initialization of heap based in memory objects, and typically are not more expensive than typical calls to their malloc() counterparts within C. Additionally, many of the typical user errors commonly encountered with malloc() can be avoided.

Function inlining is functionality that can be used in C++, although many C compilers also offer manual inlining of functions as well. Inlining of procedures can afford runtime performance gains by removing the function call overhead and branching instructions to the callee procedure from a caller procedure. This feature does not come for free, however, in that excessive inlining of functions can cause non-trivial amounts of code size gain with associated performance losses due to caching behavior.

C++ also contains a number of features that are nice to have for large scale software applications, but come with runtime costs users may want to avoid in embedded application development. The first of these is run time type identification, or RTTI. Usage of this feature requires a runtime representation of the type hierarchy, and tracking of object types. There is the cost in terms of space as data structures are required at runtime for the book keeping logic. There is also the associated runtime cost of updating these data structures as well.

Exception handling is another feature that has associated overhead in embedded systems, although more and more C++ application developers are beginning to use this feature. Exception handling is an elegant way of handling abnormal executions at runtime, versus traditional return codes. At the same time, it is problematic for many embedded system developers as it demands the RTTI functionality described previously. In addition, "throw" clauses need to be forwarded up the call chain and require considerable run-time resource consumption and handling of objects within the current scope. Further information on the actual compile time structure and layout of runtime exception handling is beyond the scope of this section; however, due to the associated runtime overhead incurred in handling this feature, it is commonly disabled for runtime sensitive embedded systems applications.

Auto-vectorizing compiler technology

Many modern processors used in embedded computing contain Single Instruction Multiple Data (SIMD) instruction set architecture extensions. These powerful instructions allow multiple discrete data elements to reside within a given register, and be computed on in parallel across SIMD parallel ALUs within the processor's pipeline architecture. Programmers often take advantage of these powerful parallel instructions via native proprietary intrinsics, as described in (see the section on "Intrinsic functions" earlier in the chapter). It is often desirable to have the compiler automatically parallelize as much computation as possible, often across these SIMD style parallel ALUs. Auto-vectorizing compiler technology is one means by which this process can be automated, whereby the compiler will map parallel computation within loop bodies into vector arithmetic that executes on parallel ALUs. In addition, more advanced vectorizers may also strive to parallelize computation within single basic blocks that may reside outside of loop bodies. By relying on compiler technology, rather than manual programmer parallelization, shorter development times can be achieved by developers. In addition, by avoiding proprietary architecture specific intrinsics, source code can be more easily migrated across architectures.

Matlab, Labview and FFTW-like generator suites

Matlab is a matrix manipulation tool suite commonly used in signal processing applications. Matlab itself was first introduced in 1979 at Stanford University, as an interactive shell into FORTRAN routines, later being marketed under the name Matlab by MathWorks. It provides a powerful engineering environment and programming language for data analysis, modeling and visualization within the digital signal processing domain with support for most all modern operating systems. It also incorporates a number of built-in functions for performing linear algebra, polynomials, Fourier analysis, differential equations and so forth.

As a programming environment for signal processing applications, Matlab supports being used in interactive mode or as a programming language either interpreted or compiled. At the same time, it is possible to link in code from other languages such as compiled C code. The programming language itself supports standard programming constructs such as conditionals, looping, functions, and globals with corresponding debugger support.

In addition to a text-based programming environment, Matlab has a companion environment called Simulink which is useful for graphically-based modeling of systems. Use cases typically involve such things as graphical modeling of dynamic systems, using a GUI containing building blocks, and block diagrams. The models constructed are typically done in a hierarchical manner, allowing system composition and element reuse. This feature set is similar to National Instrument's LabView software, except that the building blocks used are Matlab functions.

Given the interpreted nature of Matlab, however, the runtime performance of various kernels is typically not that of their natively compiled code counterparts. To circumvent this, MathWorks offers Mex files, which allow scripts and routines within the Matlab environment to be output as C or C++ code which can be compiled natively on the target architecture. Additionally, Mex files can be used for integration of externally written C or C++ code to be integrated and called from within the Matlab environment. This offers two valuable use cases and tool chain flows for the DSP developer.

Matlab and native compiled code

DSP systems developers often have algorithms that are prototyped in the Matlab programming environment, due to its robust functionality and built-in components that facilitate rapid exploration. When the algorithm is complete, users have the option of exporting the code from Matlab to C or C++ code that can be compiled using a compiler for the host system's target processor or an embedded target. MCC is the Matlab C/C++ compiler and is used in generating the resulting executables from Matlab code. The MCC compiler allows users to create the following output from their Matlab code:

- C-language translation from Matlab code
- C++ language translation from Matlab code
- MEX file wrapper from Matlab code
- Creation of executables of dynamically linked libraries from Matlab code

Native code to Matlab and silicon emulation

The MEX file format also allows C and C++ developers to create kernels or code that can be incorporated into the Matlab environment as well, providing a two way street between native code compilation and the Matlab environment. This is a valuable resource in some cases in the signal processing domain. For instance, consider a target DSP which supports fractional arithmetic with saturation on 16-bit and 32-bit data types. These types of computations are not natively represented in Matlab, and may effect the resolution with which computation can be performed by a given kernels. Vendors may often supply emulation of their target DSP instruction set in MEX file format for integration into the Matlab environment. In doing this, Matlab programmers can see the bit-exact behavior of their Matlab algorithms when lowered onto the target DSP architecture implemented in C-like programming languages. This helps to close some of the gap in the algorithmic design process, possibly reducing the number of versions of a given kernel a team must have in bringing their algorithms to production.

CHAPTER 11

Optimizing DSP Software – Code Optimization

Stephen Dew

Chapter Outline

DSP for Embedded and Real-Time Systems. DOI: 10.1016/B978-0-12-386535-9.00011-1

Optimization process

Prior to beginning the optimization process, it's important to verify functional accuracy. In the case of standards-based code (e.g., voice or video coder), there may be reference vectors already available. If not, then at least some basic tests should be written to ensure that a baseline is obtained before optimization. This enables easy identification that an error has occurred during optimization — incorrect code changes done by the programmer or any overly aggressive optimization by a compiler. Once tests in place, optimization can begin. Figure 11-1 shows the basic optimization process.

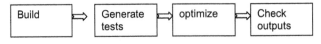

Figure 11-1:
Basic flow of optimization process.

Using the development tools

It's important to understand the features of the development tools as they will provide many useful, time-saving features. Modern compilers are increasingly better performing with DSP code and will reduce the development time required. Linkers, debuggers and other components of the tool chain will have useful features, but for the purpose of this chapter, we will focus only on the compiler.

Compiler optimization

From the compiler perspective, there are two basic ways of compiling an application: traditional compilation or global (cross-file) compilation. In traditional compilation, each source file is compiled separately and then the generated objects are linked together. In global optimization, each C file is preprocessed and passed to the optimizer in the same file. This enables greater optimizations (inter-procedural optimizations) to be made as the compiler has complete visibility of the program and doesn't have to make conservative assumptions about the external functions and references. Global optimization does have some drawbacks, however. Programs compiled this way will take longer to compile and are harder to debug (as the compiler has taken away function boundaries and moved variables). In the event of a compiler bug, it will be more difficult to isolate and workaround when built globally. Global or cross-file optimizations result in full visibility into all the functions, enabling much better optimizations for speed and size. The disadvantage is that since the optimizer can remove function boundaries and eliminate variables, the code becomes difficult to debug. Figure 11-2 shows the compilation flow for each.

Basic compiler configuration

Before building for the first time, some basic configuration will be necessary. Perhaps the development tools come with project stationary which has the basic options configured, but if not, these items should be checked (see Table 11-1 for examples):

Target architecture

Specifying the correct target architecture will allow the best code to be generated. For example, even though on Freescale devices, code built for the SC3400 core runs unmodified on the SC3850 core, greater performance and better code size will be achieved when recompiling for SC3850 as the compiler will be able to leverage new architectural features (e.g., new instructions, new pipeline schedule).

Endianness

Perhaps the vendor sells silicon with only one endianness, and perhaps the silicon can be configured. There will likely be a default option.

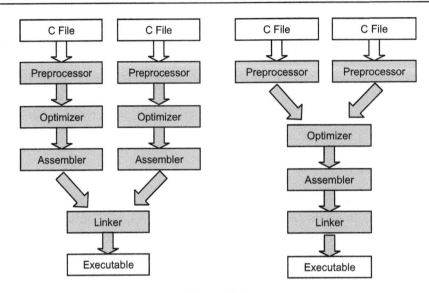

Figure 11-2:
Traditional versus global compilation.

Table 11-1: Basic configuration options on the Freescale StarCore CodeWarrior compiler.

Setting	Description
-arch	Specifies the target processor architecture (e.g., −arch sc3850)
-mb	Specifies big memory model (required for SC3850)
-be	Specifies big endian
-O0	Disables optimizations

Memory model

Different processors may have options for different memory model configurations.

Initial optimization level

It's best to disable optimizations initially.

Enabling optimizations

Optimizations may be disabled by default when no optimization level is specified and either new project stationary is created or code is built on the command line. Such code is designed for debugging only. With optimizations disabled, all variables are written and read back from the stack, enabling the programmer to modify the value of any variable via the debugger when stopped. The code is inefficient and should not be used in production code.

The levels of optimization available to the programmer will vary from vendor to vendor, but there are typically three levels (e.g., from zero to three), with three producing the most optimized code (Table 11-2). With optimizations turned off, debugging will be simpler but the code will obviously be much slower (and larger). As the level of optimization increases, more and more compiler features will be activated and compilation time will be larger.

Note that typically, optimization levels can be applied at the project, module, and function level by using pragmas, allowing different functions to be compiled at different levels of optimization.

Additional optimization configurations

In addition, there will typically be an option to build for size, which can be specified at any optimization level. On Freescale devices, in practice, two optimization levels are most often used: O3 (optimize fully for speed) and O3Os (optimize for size). In a typical application, critical code is optimized for speed and the bulk of the code may be optimized for size (Table 11-3).

Using the profiler

All DSP development environments have a profiler, which enables the programmer to analyze where cycles are spent. These are valuable tools and should be used to find the critical areas. The function profiler works in the IDE and also with the command line simulator. Table 11-4 below shows a sample function profile output which has been reformatted for clarity.

Analyzing compiled code

To correlate generated assembly with C source, use the line number of the C source code as shown in the comments that follow each assembly instruction. The line number is the first of

Table 11-2: Optimization levels on the Freescale StarCore CodeWarrior compiler.

Setting	Description
O0	Optimizations disabled. Outputs un-optimized assembly code.
O1	Performs target-independent high-level optimizations but no target-specific optimizations.
O2	Target independent and target-specific optimizations. Outputs non-linear assembly code.
O3	Target independent and target-specific optimizations, with global register allocation. Outputs non-linear assembly code. Recommended for speed-critical parts of application.

Table 11-3: Additional optimization settings for Freescale CodeWarrior compiler for StarCore DSPs.

Setting	Description
O3Os	Adds space optimizations for the indicated optimization level. Outputs assembly code which is small. Recommended to use in size-critical parts of application.
Og	Global (cross file) optimization.
-u0/-u2/-u4	Disables loop unrolling
-mod	Enables modulo buffer support. On most DSPs, the hardware will support modulo (circular) buffers. This enables support in the compiler for this feature.
-align	Forces alignment (code aligned to a multiple of 16-byte boundaries). Increases performance on change of flow to various jump points but increases code size slightly. 0 = disables alignment 1 = aligns hardware loops 2 = aligns hardware and software loops 3 = aligns all labels 4 = aligns all labels and function call return points

Table 11-4: Example function profiler results.

Module	Function	PC	No of Calls	Stack Size	Percentage	Total Cycles	Min Cycles	Max Cycles	Mean Cycles
fr_long_term_ asm	_Fr Gsm Calculation Of The LTP Parameter smaxCC	0x00005030	2080	16	13.98	2303712	1104	1113	1107
fr_structures	_Fr Gsm Short Term Synthesis Filtering	0x000059a0	2080	0	11.69	1927034	303	2764	926
fr_structures	Fr Gsm Short Term Analysis Filtering	0x00005a40	2080	0	10.12	1667640	260	2400	801

the two numbers in brackets. In the example in Figure 11-3, three lines of C source generated two VLES with three instructions in each.

Background — understanding the DSP architecture

Resources

Before writing code for an embedded processor, it's important to assess the architecture itself and understand the resources and capabilities available. Modern DSP architectures have many features to maximize throughput. Table 11-5 below shows some features that should be understood and questions the programmer should ask.

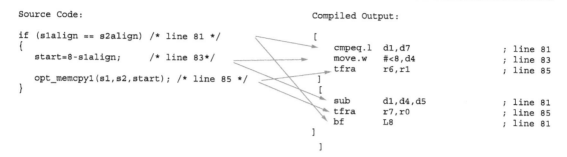

Figure 11-3:
Analyzing compiled code.

Table 11-5: DSP architectural features.

Instruction Set Architecture	Native Multiply or mpy followed by add? Is saturation implicit or explicit? Which data types are supported —8, 16, 32, 40? Fractional and/or floating point will be supported SIMD operations. Present? Does the compiler auto-vectorize? Use via intrinsic functions?
Register File	How many registers are there and what can they be used for? Implication: How many times can a loop be unrolled before performance is worsened due to register pressure?
Predication	How many predicates does the architecture support? Implication: more predicates means better control code performance
Memory system	What kind of memory is available and what are the speed tradeoffs between them? How many busses are there? How many reads/writes can be performed in parallel? Can bit-reversed addressing be performed? Is there support for circular buffers in hardware?
Other	Zero-overhead looping

To illustrate, features of some TI and Freescale DSP cores are presented. Table 11-6 below compares the SC3400, SC3850, Texas Instruments C64x and C64x+ core architectures from the perspective of available resources. The major difference is how the core architectures are partitioned. The TI C64x and C64x+ are divided into two identical blocks with four functional units each (each containing multiplication, arithmetic operations, add and load/store functionality and a register file). The SC3400 and SC3850 are divided into two different parts — the Data ALU (DALU) and Address Generation Unit (AGU). Both the DALU and AGU contain a separate register file. The DALU performs arithmetic operations such as multiply and accumulate whereas the AGU performs load and store operations. Six instructions can execute each clock cycle (four ALU and two AAU or one AAU and one BMU).

Basic C optimization techniques

This section contains basic C optimization techniques that will benefit code written for all DSP processors. The central ideas are to ensure the compiler is leveraging all features of the

Table 11-6: High-level architectural comparison of some DSP cores.

	Freescale SC3400	Freescale SC3850	TI C64x	TI C64x+
Available Parallelism	6 (plus hardware loop control)	6 (plus hardware loop control)	8	8
Multipliers (width)	4 (16-bit)	8 (16-bit) organized as 4x dual-mac	2 (32-bit)	2 (32-bit)
Native multiplies (no/cyc x width)	8x8-bit 4x16-bit	8x8 bit 4x16 bit	4x16-bit 8x8-bit	2x32-bit 4x16-bit 8x8-bit
Total Multiples per cycle (by data type)	Four 16x16-bit	Two 32x32-bit (effective), Four 16x32-bit, Eight 16x16-bit, Eight 8x8-bit	Four 16x16-bit Eight 8x8-bit	Two 32x32-bit Four 16x16-bit Eight 8x8-bit
Accumulators (width) (assuming load-stores in parallel)	4 (40-bit)*	4 (40-bit)*	4 (32-bit)*	4 (32-bit)*
Native adds (no/cyc x width)	4x40-bit 4x32-bit 8x16-bit	4x40-bit 4x32-bit 8x16-bit	4x32-bit 8x16-bit 16x8-bit	4x32-bit 8x16-bit 16x8-bit
Available Data bandwidth	2x64-bit	2x64-bit	2x64-bit	2x64-bit
Load/store units	2*	2*	2*	2*
Registers	16 Data (40-bit) 16 Address (32-bit) + 4 modulus, 4 offset, loop counter, stack pointer	16 Data (40-bit) 16 Address (32-bit) + 4 modulus, 4 offset, loop counter, stack pointer. Data registers can be used for temporary predicates	64 general (32-bit), divided into two banks of 32. Used for address, data and predicates	64 general (32-bit), divided into two banks of 32. Used for address, data and predicates

architecture and to communicate to the compiler additional information about the program which is not communicated in C.

Choosing the right data types

It's important to learn the sizes of the various types on the DSP before starting to write code. A compiler is required to support all the required types but there may be performance implications and reason to choose one type over another.

For example, a processor may not support a 32-bit multiplication. Use of a 32-bit type in a multiply will cause the compiler to generate a sequence of instructions. If 32-bit precision is not needed, it would be better to use 16-bit. Similarly, using a 64-bit type on a processor which does not natively support it will result in a similar construction of 64-bit arithmetic using 32-bit operations (Table 11-7).

Table 11-7: Illustration of assembly code for 32-bit multiplication.

C Multiplication of 32-Bit Values	Generated Assembly Sequence on an Older StarCore Processor. Highlighted Operations Show Assembly for the 32x32 Multiply.
```int GoofyBlockInt(int x) { return x*value; // value is a int }```	```[move.1 <_value,d4 adda #8,sp] [impysu d0,d4,d5 impyuu d0,d4,d2] imacus d0,d4,d5 aslw d5,d4 iadd d2,d4 move.1 d4,(sp-8) move.1 (sp-8),d0 suba #8,sp rts```

Use of intrinsics to leverage DSP features

Intrinsic functions, or intrinsics, for short, are a way to express operations not possible or convenient to express in C, or target-specific features. Intrinsics, in combination with custom data types, can allow the use of non-standard data sizes or types. An example is shown in Table 11-8. They can also be used to get to application-specific instructions (e.g., viterbi or video instructions) which cannot be automatically generated from ANSI C by the compiler. They are used like function calls but the compiler will replace them with the intended instruction or sequence of instructions. There is no calling overhead.

Some examples of features accessible via intrinsics are:

- Saturation
- Fractional types
- Disabling/enabling interrupts

For example, the FIR filter shown previously can be rewritten to use intrinsics and therefore to specify DSP operations natively. In this case, by simply replacing the multiply and add operations with the intrinsic L_mac (for long multiply-accumulate) which replaces two operations with one and adds the saturation function to ensure that DSP arithmetic is handled properly (See Figure 11-4).

Table 11-8: Example intrinsic.

Example Intrinsic (C)	Generated Assembly Code
d = L_add(a,b);	iadd d0,d1

```
short SimpleFir1( short *x, short *y)
{

    int i;

    long acc;

    short ret;


    acc = 0;

    for(i=0;i<16;i++)

        // multiply, accumulate and saturate

        acc = L_mac(acc,x[i],y[i]);

    ret = acc>>16;


    return(ret);

}
```

Figure 11-4:
Simple FIR filter with intrinsic.

Functions

Calling conventions

Each processor or platform will have different calling conventions. Some will be stack based, others register based or a combination of both. Typically, default calling conventions can be overridden though, which is useful. The calling convention should be changed for functions unsuited to the default, such as those with many arguments. In these cases, the calling conventions may be inefficient. Table 11-9 compares the default calling conventions of TI and StarCore processors.[*]

StarCore ABI Reference Manual 6 Dec 2010 revision, Freescale Semiconductor, Freescale. com

The advantages of changing a calling convention include the ability to pass more arguments in registers rather than on the stack. For example, on the StarCore DSPs, custom calling

---

[*] TMS320C6000 Optimizing Compiler v 7.0 User's Guide, spru187q, Feb 2010. www.ti.com

Table 11-9: Calling conventions of TI C6x and StarCore processors.

Processor	Calling Convention (TI 6x)	Calling Convention (StarCore DSP)
First Arguments	Up to the first ten arguments are placed in registers A4, B4, A6, B6, A8, B8, A10, B10, A12, and B12. If longs, long longs, doubles, or long doubles are passed, they are placed in register pairs A5:A4, B5:B4, A7:A6, and so on.	Up to the first two arguments are passed in registers (either d0 or r0, then second argument in d1 or r1 — scalar values in d registers, pointers in r registers). If first argument is long long or double (64-bits), then it's passed in register pair d0:d1.
Remaining arguments	Any remaining arguments are placed on the stack.	Any remaining arguments are placed on the stack.
Ellipsis	The last explicitly declared argument is passed on the stack, so that its stack address can act as a reference for accessing the undeclared arguments.	Pass the last fixed argument and all subsequent variable arguments on the stack. Such arguments of fewer than 4 bytes are located on the stack as if the argument had been promoted to 32 bits. The preceding rules apply to arguments before the last fixed argument.
Structures	Passed as the address of the structure. It is up to the called function to make a local copy.	Passed on the stack.
Return value	Integer, pointer, or float type: A4 register. Double, long double, long, or long long type: A5:A4 register pair. Structure: the caller allocates space for the structure and passes the address of the return space to the called function in A3.	Scalar: passed in d0. Pointer: passed in r0. Long long or double: returned in D0 and D1. Structure: receives in R2 the address of the returned structure or union. The caller allocates space for the returned object. A float value is returned in D0.

conventions can be specified for any function through an application configuration file and pragmas. It's a two step process:

- Custom calling conventions are defined by using the application configuration file (a file which is included in the compilation). See Figure 11-5.
- They are invoked via pragma when needed. The rest of the project continues to use the default calling convention. In Figure 11-6, the calling convention is invoked for function TestCallingConvention. The generated code is shown in Figure 11-7.

```
configuration
call_convention mycall (

  arg [1 : ( * $r9 , $d9),
 2 : ( * $r1 , $d1),
 3 : ( * $r2 , $d2),
 4 : ( * $r3 , $d3),
 5 : ( * $r4 , $d4),
      6 : ( * $r5 , $d5) ];     // argument list
  return $d0; // return value
  saved_reg   [
     $d6, $d7,          // callee must save and restore
       $d8,
       $d10, $d11,
       $d12, $d13,
      $d14, $d15,
     $r6, $r7,
       $r8,
       $r10, $r11,
       $r12, $r13,
       $r14, $r15,
       $n0, $n1,
      $m0, $m1,
       $n2, $n3,
       $m2, $m3
       ];
  deleted_reg [       // caller must save/restore
     $d0, $d1, $d2, $d3, $d4, $d5,
    $r0, $r1, $r2, $r3, $r4, $r5
             ];
  save = [ ];
)

  view default
  module "main" [
        opt_level = size
         function _GoofyBlockChar [
             opt_level = 03
             ]
     ]
      end view
      use view default
   end configuration
```

Figure 11-5:
Configuration of calling conventions.

```
char TestCallingConvention (int a, int b, int c, char d, short ve)
{
    return a+b+c+d+e;
}

#pragma call_conv TestCallingConvention mycall
```

Figure 11-6:
Invoking calling conventions.

The generated code shows the parameters passed in registers as specified:
```
;*******************************************************************************
;
;  Function Name:   _TestCallingConvention
;  Stack Frame Size:0 (0 from back end)
;  Calling Convention:14
;  Parameter:          a    passed in register d9
;  Parameter:          b    passed in register d1
;  Parameter:          c    passed in register d2
;  Parameter:          d    passed in register d3
;  Parameter:          e    passed in register d4
;
;  Returned Value:     returned in d0
;
;*******************************************************************************
      GLOBAL    _TestCallingConvention
      ALIGN     2
_TestCallingConvention   TYPE func OPT_SIZE
      SIZE _TestCallingConvention,F_TestCallingConvention_end-
_TestCallingConvention,2
;PRAGMA stack_effect _TestCallingConvention,0
   tfr      d9,d0      ;[30,1]
   add      d0,d1,d0   ;[33,1]
   add      d2,d0,d0   ;[33,1]
   add      d3,d0,d0   ;[33,1]
   add      d4,d0,d0   ;[33,1]
   rtsd                ;[33,1]
   sxt.b    d0,d0      ;[33,1]
```

Figure 11-7:
Generated code for function with modified calling conventions.

Pointers and memory access

Ensuring alignment

Most DSPs support loading of multiple data values across the busses as this is necessary to keep the arithmetic functional units busy. These moves are called multiple data moves (to not be confused with *packed* or *vector* moves). They move adjacent values in memory to different registers. In addition, many compiler optimizations require these multiple register moves because there is so much data to move to keep all the functional units busy. Figure 11-8 shows an example of a full bus usage.

Typically, however, a compiler aligns variables in memory to their access width. For example, an array of short (16-bit) data is aligned to 16 bits. However, to leverage multiple data moves, the data must be aligned to a higher alignment. For example, to load two 16-bit values at once, the data must be aligned to 32 bits.

For example, in order for the Freescale StarCore compiler to be able to use multiple register moves, the following must be true:

- The data must be aligned to the combined access width.
- The compiler must be informed of this alignment (for example, across a function boundary).

These requirements are met using the alignment pragma:

- Step 1: Align the data (Figure 11-9).

- Step 2: Indicate to the compiler that any pointers pointing to that data are aligned (Figure 11-10). This is especially important when pointers are passed into functions. In this case, place the pragma inside the function itself. In See Function Using Aligned Data and Resulting Code, the pragma indicates to the compiler that inputPtr points to an array that is aligned to 8 bytes (Figure 11-11).

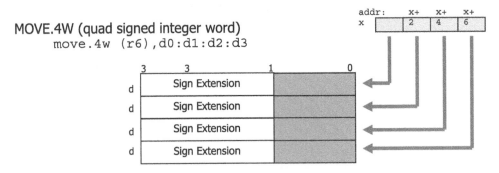

Figure 11-8:
Example of full bus usage with quad-word move (move.4w) on StarCore DSPs.

```
/* Aligning this vector to a boundary of 4 enables the Optimized
   function to generate the desired 2 ALU loop. Aligning this vector
   to 8 enables the compiler to use the move.4w instruction in some loops,
   resulting in even better optimization. */
Word16 gInAry [NO_INPUTS];
#pragma align gInAry 8
```

Figure 11-9:
Example of aligning an array via use of pragma

```
void DcOffsetRemovalOpt (Word16 * inputPtr) {
#pragma align * inputPtr 8

    int i;
    Word32 temp=0,temp2=0;

    /* Compute DC offset */
    for (i=0;i<NO_INPUTS/2;i++) {

        temp=L_add(inputPtr[2*i],temp);
        temp2=L_add(inputPtr[2*i+1],temp2);
    }

    temp=L_add(temp,temp2);

    /* divide by 32 */
    temp=L_shr(temp,5);

    /* Remove average */
    for (i=0;i<NO_INPUTS;i++) {

        inputPtr[i]=(L_sub(inputPtr[i],temp));
    }
}
```

Figure 11-10:
Example of function which tells the compiler the incoming pointer points to aligned data.

```
        LOOPSTART2
    [
        sub        d8,d0,d4     ;[212,1] 1%=1 [1]
        sub        d8,d1,d5     ;[212,1] 1%=1 [1]
        sub        d8,d2,d6     ;[212,1] 1%=1 [1]
        sub        d8,d3,d7     ;[212,1] 1%=1 [1]
        move.4w    d4:d5:d6:d7,(r0)+  ;[0,1] 2%=2 [0]
        move.4w    (r5)+,d0:d1:d2:d3  ;[212,1] 0%=0 [2]
    ]
        LOOPEND2
```

Figure 11-11:
Resulting assembly code after specifying greater alignment.

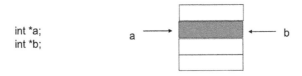

```
int *a;
int *b;
```

Figure 11-12:
Illustration of pointer aliasing.

Restrict and pointer aliasing

When pointers are used in the same piece of code, make sure that they cannot point to the same memory location (alias). When the compiler knows the pointers do not alias, it can put accesses to memory pointed to by those pointers in parallel, greatly improving performance. Otherwise, the compiler must assume that the pointers could alias. Communicate this to the compiler by one of two methods: using the restrict keyword or by informing the compiler that no pointers alias anywhere in the program.

The restrict keyword is a type qualifier that can be applied to pointers, references, and arrays. Its use represents a guarantee by the programmer that within the scope of the pointer declaration, the object pointed to can be accessed only by that pointer. Table 11-10 below shows an example loop before the restrict keyword is added to the parameters. A violation of this guarantee can produce undefined results. Table 11-11 shows the same example loop after adding restrict to the parameters.

Loops

Communicating loop count information

Pragmas can be used to communicate to the compiler information about loop bounds to help loop optimzation. If the loop minimum and maximum are known, for example, the compiler may be able to make more aggressive optimizations.

Table 11-10: Example loop before restrict added to parameters.

Example Loop with Restrict Qualifiers Added. Note: Pointers a and b must not Alias (Ensure Data is Located Separately).	Generated Assembly Code. Note: Now Accesses for a and b can be Issued in Parallel
```void foo (short * restrict a,short * restrict b,int N) int i; for (i=0;i<N;i++) { b[i]=shr(a[i],2); } return; }```	```move.w (r0)+,d4 asrr #<2,d4 doensh3 d2 FALIGN LOOPSTART3 [move.w d4,(r1)+ ; parallel move.w (r0)+,d4 ; accesses] asrr #<2,d4 LOOPEND3 move.w d4,(r1)```

Table 11-11: Example loop after restrict added to parameters.

Example Loop	Generated Assembly Code
```void foo (short * a,short * b,int N) { int i; for (i=0;i<N;i++) { b[i]=shr(a[i],2); } return; }```	```doen3 d4 FALIGN LOOPSTART3 move.w (r0)+,d4 asrr #<2,d4 move.w d4,(r1)+ LOOPEND3```

In Figure 11-13, a pragma is used to specify the loop count bounds to the compiler. In this syntax, the parameters are minimum, maximum, and multiple respectively. If a non-zero minimum is specified, the compiler can avoid generation of costly zero-iteration checking code. The compiler can use the maximum and multiple parameters to know how many times to unroll the loop if possible.

Note that nassert (on TI) or cw_assert (Freescale) can also be used to communicate the same information:

```
cw_assert(4<=N<512)
cw_assert(N%4==0)
```

Hardware loops

Hardware loops are mechanisms built into the DSP core which allow zero-overhead (in most cases) looping by keeping the loop body in a buffer or prefetching. Hardware loops are faster

```
void correlation2 (short vec1[], short vec2[], int N, short *result)
{

  long int L_tmp = 0;

  int i;

  for (i = 0; i < N; i++)

  #pragma loop_count (4,512,4)

    L_tmp = L_mac (L_tmp, vec1[i], vec2[i]);

  *result = round (L_tmp);

}
```

Note that nassert (on TI) or cw_assert (Freescale) can also be used to communicate the same information:

```
cw_assert(4<=N<512)

cw_assert(N%4==0)
```

Figure 11-13:
Example of communicating loop information to a correlation function.

than normal software loops (decrement counter and branch) because they have less change-of-flow overhead. Hardware loops typically use loop registers that start with a count equal to the number of iterations of the loop, decrease by 1 each iteration (step size of −1), and finish when the loop counter is zero.

Compilers most often automatically generate hardware loops from C even if the loop counter or loop structure is complex. However, there will be certain criteria under which the compiler will be able to generate a hardware loop (vary from compiler/architecture to others). In some cases, the loop structure will prohibit generation but if the programmer knows about this, the source can be modified so the compiler can generate the loop using hardware loop functionality. The compiler may have a feature to tell the programmer if a hardware loop was not generated (compiler feedback). Alternatively, the programmer should check the generated code to ensure hardware loops are being generated for critical code.

Hardware loop mapping (Figure 11-13). The StarCore architecture supports four hardware loops. Note the LOOPSTART and LOOPEND markings, which are assembler directives marking the start and end of the loop body, respectively, as shown in Figure 11-14.

Additional tips and tricks

The following are some additional tips and tricks to use for further code optimization.

```
          doensh3   #<5
            move.w   #<1,d0
          LOOPSTART3
      [   iadd      d2,d1
          iadd      d0,d4
          add       #<2,d0
          add       #<2,d2    ]
          LOOPEND3
```

Figure 11-14:
Example showing generated assembly code with hardware loop.

Memory contention

When data is placed in memory, be aware of how the data is accessed. Depending on the memory type, if two busses issue data transactions in a region/bank etc., they could conflict and cause a penalty. Data should be separated appropriately to avoid this contention. The scenarios that cause contention are device-dependent because memory bank configuration and interleaving differs from device to device.

Use of unaligned accesses

Some DSP processors, notably the TI C64x-based devices, support unaligned memory access. This is particularly useful for video applications. For example, a programmer might load four byte-values which are offset by one byte from the beginning of an area in memory. Typically there is a performance penalty for doing this.

Cache accesses

In the caches, place data that are used together next to each other in memory so that pre-fetching the caches is more likely to obtain the data before it is accessed. In addition, ensure that the loading of data for sequential iterations of the loop is in the same dimension as the cache prefetch.

Inline small functions

The compiler normally inlines small functions, but the programmer can force inlining of functions if for some reason it isn't happening (for example in the case that size optimization is activated). For small functions the save, restore, and parameter passing overhead can be significant relative to the number of cycles of the function itself. Therefore, inlining is beneficial. An example is shown in Figure 11-15. Also, inlining functions decreases the chance of an instruction cache miss because the function is sequential to the former caller

```
int foo () {
#pragma inline
...
}
```

Figure 11-15:
Example forcing a function to inline via pragma.

function and is likely to be prefetched. Note that inlining functions increases the size of the code. On the StarCore DSPs, pragma inline forces every call of the function to be inlined.

Use vendor DSP libraries

DSP vendors typically provide optimized library functions for common DSP routines like FFT, FIR, complex operations, etc. Normally, these are hand written in assembly as it still may be possible to improve performance over C. These can be invoked by the programmer using the published API without the need to write such routines, speeding time to market.

General loop transformations

The optimization techniques described in this section are general in nature. They are critical to taking advantage of modern multi-ALU DSPs. A modern compiler will perform many of these optimizations, perhaps simultaneously. In addition, they can be applied on all DSP platforms, at the C or assembly level. Therefore, throughout the section, examples are presented in general terms, in C and in assembly.

Loop unrolling

Background

Loop unrolling is a technique whereby a loop body is duplicated one or more times. The loop count is then reduced by the same factor to compensate. Loop unrolling can enable other optimizations such as:

- Multisampling
- Partial Summation
- Software Pipelining

Once a loop is unrolled, flexibility in coding is increased. For example, each copy of the original loop can be slightly changed. Different registers could be used in each copy. Moves

Table 11-12: Unrolling a loop by a factor of two.

Loop Prior to Unrolling	After Unrolling by Factor of 2
| ```
for (i=0;i<10;i++)
   operation();
``` | ```
for (i=0;i<5;i++) {
   operation();
   operation();
}
``` |

```
loopstart1
[ move.f (r0)+,d2  ; Load some data
  move.f (r7)+,d4   ; Load some reference
  mac d2,d4,d5      ; Do correlation
]
[ move.f (r0)+,d2  ; Load some data
  move.f (r7)+,d4   ; Load some reference
  mac d2,d4,d5      ; Do correlation
]
loopend1
```

Figure 11-16:
Example of unrolling a loop in assembly language.

can be done earlier and multiple register moves can be used. A simple example of a loop unrolled twice is shown in Table 11-12.

- Unrolling procedure
- Duplicate loop body N times
- Decrease loop count by factor of N

Implementation

Figure 11-16 is an example of a correlation inner loop which as been unrolled by a factor of two.

Multisamping

Background

Multisampling is a technique for maximizing the usage of multiple ALU execution units in parallel for the calculation of independent output values that have an overlap in input source data values. In a multisampling implementation, two or more output values are calculated in parallel by leveraging the commonality of input source data values in calculations. Unlike partial summation, multisampling is not susceptible to output value errors from intermediate calculation steps.

Multisampling can be applied to any signal processing calculation of the form:

$$y[n] = \sum_{m=0}^{M} x[n+m]h[n]$$

Where:

y[0] = x[0+0]h[0] + x[1+0]h[1] + x[2+0]h[2] + ... + x[M+0]h[M]

y[1] = x[0+1]h[0] + x[1+1]h[1] + ... + x[M-1+1]h[M-1] + x[M+1]h[M]

Thus, using C pseudo code, the inner loop for the output value calculation can be written as:

```
tmp1 = x[n];
for(m=0; m<M; m+=2)
{
tmp2 = x[n+m+1];
y[n] += tmp1*h[m];
y[n+1] += tmp2*h[m];
tmp1 = x[k+m+2];
y[n] += tmp2*h[m+1];
y[n+1] += tmp1*h[m+1];
}
tmp2 = x[n+m+1];
y[n+1] += tmp2*h[m];
```

Implementation procedure

- Unrolling the inner loop N times to allow for common data elements in the calculation of the N samples to be shared

The multisampled version works on N output samples at once. Transforming the kernel into a multisample version involves the following changes:

- Changing the outer loop counters to reflect the multisampling by N
- Use of N registers for accumulation of the output data
- Unrolling the inner loop N times to allow for common data elements in the calculation of the N samples to be shared
- Reducing the inner loop counter by a factor of N to reflect the unrolling by N

Implementation

Example implementation on a two-MAC DSP is shown in Figure 11-17.

```
[ clr d5 ; Clears d5 (accumulator)
clr d6 ; Clears d6 (accumulator)
move.f (r0)+,d2 ; Load data
move.f (r7)+,d4 ; Load some reference
]
move.f (r0)+,d3 ; Load data
InnerLoop:
loopstart1
[ mac d2,d4,d5 ; First output sample
mac d3,d4,d6 ; Second output sample
move.f (r0)+,d2 ; Load some data
move.f (r7)+,d4 ; Load some reference
]
[ mac d3,d4,d5 ; First output sample
mac d2,d4,d6 ; Second output sample
move.f (r0)+,d3 ; Load some data
move.f (r7)+,d4 ; Load some reference
]
loopend1
```

Figure 11-17:
Example of multisampling a loop in assembly language.

Partial summation

Background

Partial summation is an optimization technique whereby the computation for one output sum is divided into multiple smaller, or partial, sums. The partial sums are added together at the end of the algorithm. Partial summation allows more use of parallelism since some serial dependency is broken, allowing the operation to complete sooner.

Partial summation can be applied to any signal processing calculation of the form:

$$y[n] = \sum_{m=0}^{M} x[n+m]h[n]$$

Where:

y[0] = x[0+0]h[0] + x[1+0]h[1] + x[2+0]h[2] + ... + x[M+0]h[M]

To do a partial summation, each calculation is simply broken up into multiple sums. For example, for the first output sample, assuming M=3:

```
sum0 = x[0+0]h[0] + x[1+0]h[1]
sum1 = x[2+0]h[0] + x[3+0]h[1]
y[0] = sum0 + sum1
```

Note that the partial sums can be chosen as any part of the total calculation. In this example, the two sums are chosen to be the first + the second, and the third + the fourth calculations.

> Partial summation can cause saturation arithmetic errors. Saturation is not associative.

For example:

Saturate (a*b) + c may not equal saturate (a*b+c). Care must be taken to ensure that such differences don't affect the program.

Implementation procedure

The partial summed implementation works on N partial sums at once. Transforming the kernel involves the following changes:

- Use of N registers for accumulation of the N partial sums
- Unrolling the inner loop. The unrolling factor depends on the implementation, how values are reused, and how multiple register moves are used
- Changing the inner loop counter to reflect the unrolling

Implementation

Figure 11-18 shows the implementation on a 2-MAC StarCore DSP:

Software pipelining

Background

Software pipelining is an optimization whereby a sequence of instructions is transformed into a pipeline of several copies of that sequence. The sequences then work in parallel to leverage more of the available parallelism of the architecture. The sequence of instructions can be duplicated as many times as needed, substituting a different set of registers for each sequence. Those sequences of instructions can then be interwoven.

```
[ move.4f (r0)+,d0:d1:d2:d3 ; Load data - x[..]
move.4f (r7)+,d4:d5:d6:d7 ; Load reference - h[..]
]
InnerLoop:
loopstart1
[ mpy d0,d4,d8 ; x[0]*h[0]
mpy d2,d6,d9 ; x[2]*h[2]
]
[ mac d1,d5,d8 ; x[1]*h[1]
mac d3,d7,d9 ; x[3]*h[3]
move.f (r0)+,d0 ; load x[4]
]
add d8,d9,d9 ; y[0]
[ mpy d1,d4,d8 ; x[1]*h[0]
mpy d3,d6,d9 ; x[3]*h[1]
moves.f d9,(r1)+ ; store y[0]
]
[ mac d2,d5,d8 ; x[2]*h[2]
mac d0,d7,d9 ; x[4]*h[3]
move.f (r0)+,d1 ; load x[5]
]
add d8,d9,d9 ; y[1]

[ mpy d2,d4,d8 ; x[2]*h[0]
mpy d0,d6,d9 ; x[4]*h{1]
moves.f d9,(r1)+ ; store y[1]
]
[ mac d3,d5,d8 ; x[3]*h[2]
mac d1,d7,d9 ; x[5]*h[3]
move.f (r0)+,d2 ;load x[6]
]
add d8,d9,d9 ; y[2]
[ mpy d2,d4,d8 ; x[3]*h[0]
mpy d0,d6,d9 ; x[5]*h[1]
moves.f d9,(r1)+ ; store y[2]
]
[ mac d3,d5,d8 ; x[4]*h[2]
mac d1,d7,d9 ; x[6]*h[3]
move.f (r0)+,d3 load x[7]
]
add d8,d9,d9 ; y[3]
moves.f d9,(r1)+ ; store y[3]
    loopend1
```

Figure 11-18:

Example of partial summation of a loop in assembly language.

For a given sequence of dependent operations:

```
a=operation();
b=operation(a);
c=operation(b);
```

```
sub d0,d1,d2
impy d2,d2,d2
asr d2,d2
```

Figure 11-19:
Example sequence to be software-pipelined.

Software pipelining gives (where operations on the same line can be parallelized):

```
a0=operation();
b0=operation(a); a1=operation();
c0=operation(b); b1=operation(a1);
c1=operation(b1);
```

Implementation

A simple sequence of three dependent instructions can easily be software pipelined; for example see the sequence shown in Figure 11-19:

A software pipelining of three sequences is shown. The sequence of code in the beginning where the pipeline is filling up (when there are less than three instructions grouped) is the prologue. Similarly, the end sequence of code with less than three instructions grouped is the epilogue. The grouping of three instructions in parallel can be transformed into a loop kernel as shown in Figure 11-20.

> Software pipelining will increase the code size. Ensure that optimizations are worth the increase in size.

Example application of optimization techniques: cross correlation

A good example for a case study in optimization using the techniques above is the implementation of the cross-correlation algorithm. Cross correlation is a standard method of estimating the degree to which two series are correlated. It computes the best match for the supplied reference vector within the supplied input vector. The location of the highest cross correlation (offset) is returned. The Cross-Correlation algorithm is illustrated in Figure 11-21.

Setup

The cross-correlation function is set up in a project consisting of a test harness with input vectors and reference output vectors. Testing is done using two different set of inputs: input vector length 24, reference 4; input size 32, reference 6. For performance measurement, the profiler in the CodeWarrior 10 tool chain for StarCore is used. The minimum and maximum function execution times are measured (corresponding in this case to the shorter and longer vectors respectively). For the purposes of illustration, three different implementations of cross-correlation are presented:

```
sub d0,d1,d2 ; Prologue

[ impy d2,d2,d2 ; Prologue

sub d3,d4,d5

]

[ asr d2,d2 ; Can be transformed into loop

impy d5,d5,d5

sub d6,d7,d8

]

[ asr d5,d5 ; Epilogue

impy d8,d8,d8

]

asr d8,d8 ; Epilogue
```

Note: software pipelining will increase the code size. Ensure that optimizations are worth the increase in size.

Figure 11-20:
Example sequence after software pipelining.

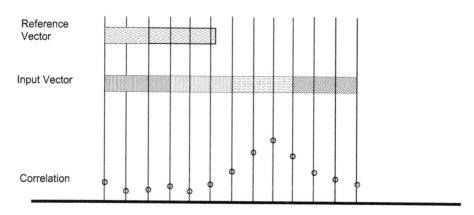

Figure 11-21:
Cross Correlation Algorithm.

Initial port

Step 1: Intrinsics for fractional operations
Step 2: Optimized by specifying alignment and using multisampling techniques
Step 3: Optimized in assembly

Original implementation

The original implementation is written in ANSI C. It consists of two nested loops: the outer loop computes each correlation value (i.e., match) and the inner loop computes each calculation that forms part of that correlation value. So, the outer loop steps through the input vector, while the inner loop steps through the reference vector. The C code and generated assembly is shown in Figure 11-22.

Performance analysis — Freescale StarCore SC3850 Core

Assumptions: zero wait state memory (all in cache). Core only benchmarked.

Test 1 (short vector)	734 cycles
Test 2 (long vector)	1258 cycles

Step 1: Use intrinsics for fractional operations and specify loop counts

In the first step, ensure that intrinsic functions are used to specify fractional operations to ensure the best code is generated. On SC3850, there's a multiply-accumulate instruction which does the left shift after the multiply and saturates after the addition operation. This combines many operations into one. Replacing the inner loop body with an L_mac intrinsic will ensure that the mac assembly instruction is generated. This is shown in Figure 11-23.

Performance analysis — Freescale StarCore SC3850 Core

Assumptions: zero wait state memory (all in cache). Core only benchmarked.

Test 1 (short vector)	441 cycles
Test 2 (long vector)	611 cycles

Step 2: Specify data alignment and modify for multisampling algorithm

In the final step, multisampling techniques are used to transform the cross-correlation algorithm. The cross-correlation code is modified so that two adjacent correlation samples are calculated simultaneously. This allows data reuse between samples and a reduction in data loaded from memory. In addition, aligning the vectors and using a factor of two for multisampling ensures that when data is loaded, the alignment stays a factor of two, which means multiple-register moves can be used (in this case, two values at once). In summary, the changes are:

C Implementation	Generated Assembly

`// Receives pointers to input and reference vectors`
`short CrossCor(short *iRefPtr, short *iInPtr)`
`{`
`    long acc;`
`    long max = 0;`
`    int offset = -1;`
`    int i, j;`
 	
`// For all values in the input vector`
`  for(i=0; i<(inSize-refSize+1); i++)`
`  {`
`    acc = 0;`
`// For all values in the reference vector`
`    for(j=0; j<refSize; j++)`
`    {`
`// Cross-correlation operation:`
`//Multiply integers Shift into fractional representation`
`//Add to accumulator`
`      acc += ((int)(iRefPtr[j] * iInPtr[j])) << 1;`
`    }`
`    iInPtr++;`
`    if(acc > max)`
`    {`
`// Save location (offset) of maximum correlation result`
`      max = acc;`
`      offset = i;`
`    }`
`  }`
`  return offset;`
`}`
	`3 cycle inner loop shown:`
	
	`    FALIGN`
	`    LOOPSTART3`
	`[`
	`    move.w   (r14)+,d4`
	`    move.w   (r4)+,d3`
	`]`
	`[`
	`    impy      d3,d4,d5`
	`    addl1a    r2,r3`
	`]`
	`    move.l   d5,r2`
	`    LOOPEND3`

Figure 11-22:
Initial ANSI C implementation.

C Implementation	Generated Assembly

long acc;
 	
long max = 0;
 	
int offset = -1;
 	
int i, j;
 	
for(i=0; i<inSize-refSize+1; i++) {
#pragma loop_count (24,32)
 	
    acc = 0;
  for(j=0; j <= refSize+1; j++) {
        #pragma loop_count (4,6)
 	
        acc =**L_mac** (acc, iRefPtr[j],
iInPtr[j]);
    }
    iInPtr++;
  if(acc > max) {
        max = acc;
        offset = i;
    }
  }
return offset;
	One Inner Loop Only shown:
	
	    skip1s ; note this was added due to
	pragma loop count. Now if zero, skips loop
	
	    FALIGN
	    LOOPSTART3
	DW17 TYPE debugsymbol
	    [
	    mac d0,d1,d2
	    move.f (r2)+,d1
	    move.f (r10)+,d0
	    ]
	    LOOPEND3

Figure 11-23:
Use Intrinsics for fractional operations and Specify Loop Counts.

- Multisampling — perform two correlation calculations each correlation per loop. Zero pad first multiplication of second correlation (then compute the last multiply outside the loop).
- Reuse of data — since two adjacent correlation computations use some of the same values, they can be reused, removing the need to refetch them from memory. In addition, the one value reused between iterations is saved in a temporary variable.

Multisampling techniques are shown in Figure 11-24.

Since InSize-refSize+1 correlations are needed and our vectors are even, there will be one remaining correlation to calculate outside the loop. The resultant code for this operation is shown in Figure 11-25.

Leveraging some assumptions made about the data set (even vectors), some more aggressive optimizations could be done at the C level. This optimization means some flexibility is given up for the sake of performance.

Performance analysis — Freescale StarCore SC3850 Core

Assumptions: zero wait state memory (all in cache). Core only benchmarked.

Test 1 (short vector)	227 cycles
Test 2 (long vector)	326 cycles

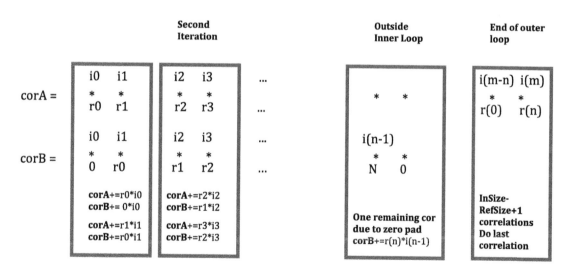

Reference: r(0)..r(n)
Input: i(0)..i(m)
Perform two correlations at once.
Zero pad first multiplication of second correlation
Perform two correlation calculations each correlation per loop
Reuse data each loop

Figure 11-24:
Diagram of multisampling technique.

C Implementation	Generated Assembly
	Both loop bodies shown:
`#pragma align *iRefPtr 4` `#pragma align *iInPtr 4`	`        skipls   PL001` `  ]`
`long accA, accB;` `long max = 0;`	`  FALIGN` `  LOOPSTART3`
`short s0,s1,s2,s3,s4;`	`  FALIGN` `  LOOPSTART2`

`int offset = -1;`	`  [`
	`        tstgt    d12`
	`        clr      d8`

`int i, j;`	`        clr      d4` `        clr      d5`

`for(i=0; i<inSize-refSize; i+=2) {`
`#pragma loop_count (4,40,2)`
`        accA = 0;`
`        accB = 0;`
	`        suba     r5,r5`
	`        move.1   d13,r8`
	`  ]`
	`  [`

`        s4 = 0;`	`        tfra     r1,r2`
	`        jf       L5`
	`  ]`

`    for (j=0; j<refSize; j+=2) {`
`#pragma loop_count (4,40,2)`
`        s0 = iInPtr[j];`
`        s1 = iInPtr[j+1];`
	`  [`
	`        tfra     r0,r3`
	`        addl1a   r4,r2`
	`  ]`
	`  [`

`        s2 = iRefPtr[j];`
`        s3 = iRefPtr[j+1];`
	`        move.2f  (r2)+,d0:d1`
	`        move.2f  (r3)+,d2:d3`
	`  ]`

`        accA = L_mac(accA, s2, s0);`
`        accB = L_mac(accB, s4, s0);`
	`  [`
	`        mac      d8,d0,d4`
	`        mac      d2,d0,d8`

`        accA = L_mac(accA, s3, s1);`
`        accB = L_mac(accB, s2, s1);`
	`        tfr      d3,d5`
	`        suba     #<1,r8`
	`        tfra,r5  r9`

`        s4 = s3;`
`    }`
`        s0 = iInPtr[j];`
`        accB = L_mac(accB, s4, s0);`
	`  ]`
	`        doensh3  r8`
	`  FALIGN`
	`  LOOPSTART3`
	`  [`

Figure 11-25:
Specify data alignment and modify for multisampling algorithm.

```
    if(accA > max) {                         mac      d3,d1,d8
            max = accA;                       mac      d2,d1,d4
            offset = i;                       move.2f  (r2)+,d0:d1      ; packed moves
        }                                     move.2f  (r3)+,d2:d3
    if(accB > max) {                      ]
            max = accB;               [
            offset = i+1;                     mac      d2,d0,d8
        }                                     mac      d5,d0,d4
                                              tfr      d3,d5
        iInPtr +=2;                       ]
    }
                                          LOOPEND3
    accA = 0;                         [
    accB = 0;                             mac      d2,d1,d4
    for(j=0; j<refSize; j+=2) {           mac      d3,d1,d8
#pragma loop_count (4,40,2)               ]
        accA = L_mac(accA, iRefPtr[j],    [
iInPtr[j]);                               cmpgt    d9,d8
        accB = L_mac(accB, iRefPtr[j+1],  tfra,r2  r1
iInPtr[j+1]);                             adda     r4,r5
    }                                     ]
    accA = L_add(accA, accB);             [
                                              tfrt     d8,d9
    if(accA > max) {                          tfrt     d11,d10
        max = accA;                           add11a   r5,r2
        offset = i;                           adda     #<2,r4
    }                                     ]
                                              move.f   (r2),d1
return offset;                                mac      d5,d1,d4
                                              cmpgt    d9,d4
                                          [
                                          IFT addnc.w  #<1,d11,d10
                                          IFA tfrt     d4,d9
                                          IFA add      #<2,d11
                                          ]
                                              LOOPEND2
```

Figure 11-26:
Assembly language optimization and the resultant cycle count.

Step 3: Assembly language optimization

Assembly language for DSP is still used in cases where performance is critical. In the example below, we will take the cross correlation from earlier and write an assembly function, integrate into a C framework and then optimize it, as shown in Figure 11-26.

Example Assembly Integration

```
;
; Function : CrossCor
;
; Prototype: extern short CrossCor(short *iRefPtr, short *iInPtr) ;
```

```
;
; Description : Cross correlates input data stream with smaller reference
; sample stream. Input arguments passed through function
; global variables. Return in d0 the offset from the
; beginning of the input data stream where the highest value
; was found.
;
; Inputs : inSize (global variable) - number of samples in input
; data stream.
; refSize (global variable) - number of samples in the
; reference stream.
; refPtr (param 0 - r0) - pointer to the reference
; sample stream. Reference samples are
; 16-bits.
; inDataPtr (param 0 - r1) - pointer to the input
; sample stream. Input samples are 16-bits.
;
; Outputs : d0 - Offset from inDataPtr where the max value can be
; found.
;
; Assumptions : Uses stack for temporarily holding cross correlation values.
;
;*****************************************************************************
align $10
global _CrossCor
_CrossCor: type func
; RefPtr passed in register r0
tfra r0,r9 ; save a copy
; InDataPtr passed in register r1
tfra r1,r2 ; save a copy
dosetup0 CrossCorTopLoop

move.w _inSize,d0 ; load the data size into d0
move.w _refSize,d1 ; load the reference data size into d1

[
sub d1,d0,d0 ; iterate the loop inSize-refSize+1 times
clr d11 ; cor index
clr d12 ; current index.
]
[
clr d10 ; cor max
add #1,d0 ; iterate the loop inSize-refSize+1 times
]
doen0 d0
loopstart0
CrossCorTopLoop:
[
tfra r9,r0 ; reset refPtr to start
doensh1 d1 ; do the inner loop refSize times
clr d2 ; d2 is the accumulator. clear it.
```

```
]
[
move.f (r1)+,d3 ; load data value before loop
move.f (r0)+,d4 ; load reference value before loop
]
CrossCorInnerLoop:
loopstart1
[
mac d3,d4,d2 ; ref[i]*data[i]
move.f (r1)+,d3 ; load data value
move.f (r0)+,d4 ; load reference value
]
loopend1

CrossCorInnerLoopEnd:
cmpgt d10,d2 ; if d2>d10, SR:T set
[
tfrt d2,d10 ; save max corr
tfrt d12,d11 ; save new max index
adda #2,r2,r2 ; increment InPtr start by 2 bytes
adda #2,r2,r1 ; increment InPtr start by 2 bytes
add #1,d12
]
loopend0

CrossCorReport:
tfr d11,d0 ; save max index
        global F_CrossCor_end
F_CrossCor_end
rts
```

Performance analysis — Freescale StarCore SC3850 Core

Assumptions: zero wait state memory (all in cache). Core only benchmarked.

Test 1 (short vector)	296 cycles
Test 2 (long vector)	424 cycles

The assembly language implementation shown is not fully optimized and as such doesn't beat the prior C implementation. However, it's a framework that can be the basis for further optimizations such as multisampling techniques.

CHAPTER 12

DSP Optimization — Memory Optimization

Robert Oshana

Chapter Outline

Introduction

Optimization metrics for compiled code are not always measured in resulting execution clock cycles on the target architecture. Consider modern cellular telephone or wireless device which may download executables over a wireless network connection or backhaul infrastructure. In such cases, it is often advantageous for the compiler to reduce the size of the compiled code which must be downloaded to the wireless device. By reducing the size of the code needed to be downloaded to the wireless device, savings are achieved in terms of bandwidth required for each wireless point of download.

Optimization metrics such as the memory system performance of compiled code are another metric which are often important to developers. These are metrics correlated to the dynamic runtime behavior of not only the compiled code on the target processor, but also the underlying memory system, caches, DRAM, and busses etc. By efficiently arranging the data within the application, or more specifically, the order in which data and corresponding data structures are accessed by the application dynamically at runtime, significant performance improvements can be gained at the memory system level. In addition, vectorizing compilers can also improve performance due to spatial locality of data when SIMD instruction sets are present and varying memory system alignment conditions are met.

The next section illustrates optimization techniques that may be used to improve application code size. The first techniques presented fall under the category of compiler 'flag mining,' which is the means by which different permutations of compile time options are used to achieve the desired results on generated code. In addition, the subject of lower level system details are presented such as application binary interfaces, and multiple encoding instruction set architectures as vehicles to further reduce code size in the presence of resource constrained systems.

Code size optimizations

Compiler flags and flag mining

In compiling a source code project for execution on a target architecture, it is often desirable for the resulting code size to be reduced as much as possible. Reasons for this pertain to the amount of space in memory the code will occupy at program runtime, as well as potentially reducing the amount of instruction cache needed by the device. In reducing the code size of a given executable, a number of factors can be tweaked during the compilation process to accommodate this.

Typically, users will first begin by configuring the compiler to build the program for size optimization, frequently using a compiler command line option such as −Os as is available in the GNU GCC compiler as of version 4.5. When building for code size, it is not uncommon for the compiler to disable other optimizations that frequently result in improvements in the runtime performance of the code. Examples of these might be loop optimizations such as loop unrolling, or software pipelining, which typically are performed in an attempt to increase the runtime performance of the code at the cost of increases in the compiled code size. This is due to the fact that the compiler will insert additional code into the optimized loops such as prolog and epilog code in the case of software pipelining, or additional copies of the loop body in the case of loop unrolling.

In the event that users do not want to disable all optimization or build exclusively at optimization level −O0 with code size optimization enabled, users may also want to disable functionality such as function inlining via either a compiler command line option or compiler pragma, depending on the build tools system and functionality supported. It is often the case that at higher levels of program optimization, specifically when optimizing for program runtime performance, compilers will attempt to inline copies of a function, whereby the body of the function code is inlined into the calling procedure, rather than the calling procedure being required to make a call into a callee procedure, resulting in a change of program flow and obvious system side effects. By specifying at either the command line option, or via a customer compiler pragma, the user can prevent the tools from inadvertently inlining various functions which would result in the increase in the overall code size of the compiled application.

When a development team is building code for a production release, or a user case scenario when debugging information is no longer needed in the executable, it may also be beneficial to strip out debugging information and symbol table information. In doing this, significant reductions in object file and executable file sizes can be achieved. Furthermore, in stripping out all label information, some level of IP protection may be afforded to the user in that consumers of the executable will have a difficult time reverse engineering the various functions being called within the program.

Target ISA for size and performance tradeoffs

Various target architectures in the embedded and DSP spaces may afford additional degrees of freedom when trying to reduce code size of the input application. Quite often it is advantages for the system developer to take into consideration not only the algorithmic complexity and software architecture of their code, but also the types of arithmetic required and how well those types of arithmetic and system requirements map to the underlying target architecture. For example, an application that requires heavy use of 32-bit arithmetic may run functionally correctly on an architecture that is primarily tuned for 16-bit arithmetic; however, an architecture tuned for 32-bit arithmetic can provide a number of improvements in terms of performance, code size, and perhaps power consumption.

Variable length instruction encoding is one particular technology that a given target architecture may support, which can be effectively exploited by the build tools to reduce overall code size. In variable length instruction coding schemes, certain instructions within the target processors ISA may have what is referred to as 'premium encodings,' whereby those instructions most commonly used can be represented in a reduced binary footprint. One example of this might be a 32-bit embedded Power Architecture device, whereby frequently used instructions such as integer addition are also represented with a premium 16-bit encoding. When the source application is compiled for size optimization, the build tools will attempt to map as many instructions as possible to their premium encoding counterpart in an attempt to reduce the overall footprint of the resulting executable.

Freescale Semiconductor supports this feature in the Power Architecture cores for embedded computing, as well as their StarCore line of DSPs. Other embedded processor designs such as those by ARM Limited and Texas Instruments' DSPs have also employed variable encoding formats for premium instructions, in an effort to curb the resulting executable's code size footprint. In the case of Freescale's Power Architecture, Freescale states that both standard 32-bit code and 16-bit premium encoded code can be mixed interchangeably within the executable within a flash page size access basis. Other architecture may opt to specify the encoding within some format of prefix bits, allowing an even finer level of code intermingling.

It should be mentioned than often the reduced footprint premium encoding of instructions in a variable length encoding architecture often comes at the cost of reduced functionality. This is due to the reduction in the number of bits that are afforded in encoding the instruction, often reduced from 32 bits to 16 bits. An example of a non-premium encoding instruction versus a premium encoding instruction might be an integer arithmetic ADD instruction. On a non-premium encoded variant of the instruction, the source and destination operations of the ADD instruction may be any of the 32 general purpose integer registers within the target architecture's register file. In the case of a premium encoded instruction, whereby only 16 bits of encoding space are afforded, the premium encoded ADD instruction may only be permitted to use R0-R7 as source and destination register, in an effort to reduce the number of bits used in the source and register destination encodings. Although it may not readily be apparent to the application programmer, this can result in subtle, albeit minor, performance degradations. These are often due to additional copy instructions that may be required to move source and destination operations around to adjacent instructions in the assembly schedule due to restrictions placed on the premium-encoded variants.

As evidence of the benefits and potential drawbacks of using variable length encoding instruction set architectures as a vehicle for code size reduction, benchmarking of typical embedded codes when targeting Power Architecture devices has shown that VLE (variable length encoding) enabled code has been shown to be approximately 30% smaller in code footprint size than standard Power Architecture code while only exhibiting a 5% reduction in code performance. Resulting minor degradations in code performance are typical, due to limitations in functionality when using a reduced instruction encoding format of an instruction.

Floating point arithmetic and arithmetic emulation may be another somewhat obfuscated source of code size explosion. Consider the case in which the user's source code contains loops of intensive floating point arithmetic when targeting an architecture lacking native floating point functionality in hardware. In order to functionally support the floating point arithmetic, the build tools will often need to substitute in code to perform floating point arithmetic emulation at program runtime. This typically entails trapping to a floating point emulation library that provides the required functionality, such as floating point division, using the existing non-floating point instructions natively supported on the target architecture.

As one might predict, it is not uncommon for a given floating point emulation routine to require hundreds of target processor clock cycles to emulate the floating point operation, which execute over tens if not hundreds of floating point emulation instructions. In addition to the obvious performance overhead incurred versus code targeting a processor with native floating point support in hardware, significant code size increases will occur due to the inclusion of floating point emulation libraries or inlined floating point emulation code. By correctly matching the types of arithmetic contained in the source application with the underlying native hardware support of the target architecture, reductions in the overall resulting executable size can be achieved with some effort.

Tuning the ABI for code size

In software engineering, the application binary interface, or ABI, is the low level software interface between a given program and the operating system, system libraries, and even inter-module communication within the program itself. The ABI itself is a specification for how a given system represents items such as: data types, data sizes, alignment of data elements and structures, calling conventions and related modes of operations. In addition, a given ABI may specify the binary format of object files and program libraries. The calling convention and alignment may be areas of interest to those wishing to reduce the overall code size of their application by using a custom calling convention within their particular application.

A given target processor and related ABI will often specify a calling convention to be used between functions within the application, as well as calls to the underlying operating system, runtime libraries, and so forth. It is often desirable for a vendor to specify a default calling convention that affords a reasonable level of performance for the general use case when making calls between caller and callee procedures within an application. At the same time, such a default calling convention may also attempt to make reasonable reductions in the code size generated at both the caller and callee procedures for maintaining machine level state consistency between both the caller and callee. Oftentimes, however, this is not ideal for an application developer who demands either tight restrictions on code size, or in other cases, high levels of compiled code performance in key system kernels of hot paths within the call graph.

Consider for example the function in Figure 12-1 below, which passes a large number of 16-bit integer values from the caller to the callee procedure.

Looking at this example, it can be seen that the caller procedure computes a number of 16-bit values that must be passed as input parameters to the callee procedure. The callee procedure shown in Figure 12-2 will then use these input values to compute some result that is passed back to the caller procedure to use in subsequent computation.

Let's also assume that we are dealing with a somewhat trivialized ABI that is succinct for this illustrative example. The ABI assumes a 32-bit general purpose embedded architecture that has a 32-bit general purpose register file. The default calling convention for this ABI states that the first two char, short, or integer, values that are passed to a callee procedure get passed in general purpose register R00 and R01, with subsequent parameters being passed from the

```
void caller_procedure(void)
{
      short tap_00, tap_01, tap_02, tap_03,
            tap_04, tap_05, tap_06, tap_07;
            long callee_result;


      // some computation occurs, setting up taps
      callee_result = callee_procedure(tap_00, tap_01,
                                       tap_02, tap_03,
                                       tap_04, tap_05,
                                       tap_06, tap_07);


      // subsequent computation occurs based on results
}
```

Figure 12-1:
Caller procedure in C.

caller to the callee via the stack. This might be typical for a processor targeting a mobile embedded device sensitive to both performance and code size. The resulting assembly might look something like what is shown in Figures 12-3 and 12-4.

We can see here that the default ABI has been used as the vehicle for communications between both the caller and callee procedures named caller_procedure() and callee_procedure() respectively. Looking at the assembly code generated for the caller_procedure, we can see that the local variables computed within caller_procedure, namely tap_00 through tap_07, are loaded from memory within the local procedure, and copied onto the stack for passing to the callee routine, namely callee_procedure. Because the default calling convention as specified by this example ABI states that the first two char, short, or integral type parameters may be passed via registers from the caller to the callee

```
long callee_procedure(short tap_00, short tap_01,
                      short tap_02, short tap_03,
                      short tap_04, short tap_05,
                      short tap_06, short tap_07)
{
      long result;
      // do computation.
      return result;
}
```

Figure 12-2:
Callee Procedure in C.

```
;*****************************************************************
;* NOTE:  Using default ABI, R00 and R01 can be used to pass
;*        parameters from caller to callee, all other parameters
;*        must be passed via the stack.
;*
;* SP+TAP_00 contains tap_00
;* SP+TAP_01 contains tap_01
;* SP+TAP_02 contains tap_02
;* SP+TAP_03 contains tap_03
;* SP+TAP_04 contains tap_04
;* SP+TAP_05 contains tap_05
;* SP+TAP_06 contains tap_06
;* SP+TAP_07 contains tap_07
;*
;*****************************************************************
__caller_procedure:
        ;* some computation setting tap_00 .. tap_07 in local memory
        ;* and various bookkeeping.

        ;* all parameters that can not be passed via default ABI
        ;* configuration must be pushed onto the stack.
        ;*
        LOAD    R00,(SP+TAP_03);
        PUSH    R00;                        ;* SP+=4
        LOAD    R00,(SP+TAP_04);
        PUSH    R00                         ;* SP+=4
        LOAD    R00,(SP+TAP_05);
        PUSH    R05                         ;* SP+=4
        LOAD    R00,(SP+TAP_06);
        PUSH    R00                         ;* SP+=4
        LOAD    R00,(SP+TAP_07);
        PUSH    R00                         ;* SP+=4
```

Figure 12-3:
Caller procedure in assembly.

procedures, the compiler has taken the liberty of passing tap_00 and tap_01 using target processor registers R00 and R01 respectively.

It is interesting to note that fewer instructions are required for setting up parameters to be passed via registers than those to be passed via the stack. Additionally, it can be

```
;* all parameters to pass to callee_procedure via stack are
;* loaded from callee_procedure's memory and put on stack,
;* we can low load tap_00 and tap_01 into register and pass those
;* first two parameters via register
;*
LOAD    R00,(SP+TAP_00);
LOAD    R01,(SP+TAP_01);
CALL    __callee_procedure;
NOP;
NOP;

;* store the value of result computed by callee_procedure
;* into our local space.
;*
STORE   R00,(SP+RESULT);

;* subsequent computation using result returned in R00
;*
RTS;
NOP;
NOP;
__end_caller_procedure:
```

Figure 12-3:
Continued

seen from the callee procedure that significantly more instructions must be inserted into the callee procedure by the compiler to restore parameters passed via copy on the stack from the caller function into registers for local computation within the callee routine. While this affords a very nice abstraction of computation between the caller and the callee routines, clearly if the user wishes to reduce the code size of their resulting executable, one might consider alternative means of communication between caller and callee routines.

This is where custom calling conventions may be used with a given ABI to further improve performance, or in the case of this example to further reduce the code size and increase performance as well. Suppose that now the user has altered the calling convention within the ABI to use for these two procedures. Let's call this new calling convention as specified by the user 'user_calling_convention.' The user has now stated that rather than pass only the first two parameters from the caller function to the caller in registers, with subsequent parameters being passed via the stack, that the user_calling_convention may pass up to

```
;*****************************************************************************
;* R00 contains tap_00
;* R01 contains tap_01;
;* tap_02 through tap_07 have been passed via the stack, as seen
;* previously being setup in caller_procedure via the push operations.
;* Upon entry, callee_procedure must transfer all of the input parameters
;* passed via the stack into registers for local computation.  This
;* requires additional instructions both on the caller side (to put on
;* the stack) as well as the callee size (to restore from the stack).
;*
;* NOTE:  INSERT PROS AND CONS
;*
;*
;*
;*****************************************************************************
__callee_procedure:

        ;* ADJUST STACK POINTER TO NOW POINT TO CALLEE'S STACK FRAME
        ;* SO WE CAN ACCESS DATA PASSED VIA THE STACK IN ABI COMPLIANCE
        ;*
        POP     R07;            ;* tap_07 into R07, SP-=4
        POP     R06;            ;* tap_06 into R06, SP-=4
        POP     R05;            ;* tap_05 into R05, SP-=4
        POP     R04;            ;* tap_04 into R04, SP-=4
        POP     R03;            ;* tap_03 into R03, SP-=4
        POP     R02;            ;* tap_02 into R02, SP-=4

        ;* perform local computation on input parameters now stored
        ;* in registers R00-R07, storing result into
        ;* SP+RESULT_OFFSET
        ;*
        ;* move result into register R00
        ;*
        STORE R00,(SP+RESULT_OFFSET)
        RTS;
        NOP;
        NOP;
__end_callee_procedure:
```

Figure 12-4:
Callee procedure in assembly.

eight parameters from the caller to the callee function via registers, namely R00–R07. In doing this, the tools will need to account for additional registers being used for parameter passing, and the bookkeeping required on both sides of the caller/callee world, however for this user's example code which passes large numbers of parameters from caller to callee, a benefit can be gained. Figure 12-5 illustrates what assembly code the user could expect to be generated using this user_calling_convention customization as specified by the developer.

Referring to the figure above, it can be seen that by using the user_calling_convention, the resulting assembly generated by the compiler looks quite different from that of the default calling convention. By permitting the build tools to pass additional parameters between caller and callee functions using registers, a drastic reduction in the number of instructions generated for each procedure is evident. Specifically, the callee_procedure can be shown to require far fewer moves to the stack before the call to the callee_procedure (Figure 12-6). This is due to the fact that additional hardware registers are now afforded to the calling convention, whereby values from the caller's memory space may simply be loaded into registers before making the call rather than loading into registers and then copying onto the stack (and possibly adjusting the stack pointer explicitly).

Similarly, referring to the callee_procedure, it can be seen that a number of instructions have been removed from the previous example's generated assembly. Once again, this is due to the fact that parameters are now being passed from the caller to the callee function via the register file, rather than pushing onto and pulling off of the stack. As such, the callee does not need the additional instruction overhead to copy local copies from the stack into registers for local computation. In this particular example, not only is it likely that performance improvements will be seen due to fewer instructions required to execute dynamically at runtime, but code size has also been reduced due to the number of instructions statically reduced in the executable.

While this example has shown how custom calling conventions can be used as part of an embedded system's larger ABI to reduce code size, and tailor memory optimization, there are a number of other concepts that may also play into this. Subjects such as spill code insertion by the compiler, the compiler's ability to compute stack frame sizes to utilize standard MOVE instructions to/from the stack frame rather than PUSH/POP style instructions, and also SIMD style move operations to the stack whereby increased instruction density is obtained further increasing performance and reducing code size overhead are left as further reading and considered beyond the scope of this example.

Caveat emptor: compiler optimization orthogonal to code size!

When compiling code for a production release, developers often want to exploit as much compile time optimization of their source code as possible in order to achieve the best performance

```
;************************************************************
;* NOTE:  Using default ABI, R00 and R01 can be used to pass
;*        parameters from caller to callee, all other parameters
;*        must be passed via the stack.
;* NOTE:  NEEDS UPDATING AS OF 11/11/2011 AWAITING EDITOR COMMENTS
;*
;* SP+TAP_00 contains tap_00
;* SP+TAP_01 contains tap_01
;* SP+TAP_02 contains tap_02
;* SP+TAP_03 contains tap_03
;* SP+TAP_04 contains tap_04
;* SP+TAP_05 contains tap_05
;* SP+TAP_06 contains tap_06
;* SP+TAP_07 contains tap_07
;*
;*****************************************************************************
__caller_procedure:
        ;* some computation setting tap_00 .. tap_07 in local memory
        ;* and various bookkeeping.

        ;* all parameters that can not be passed via default ABI
        ;* configuration must be pushed onto the stack.
        ;*
        LOAD    R00,(SP+TAP_03);
        PUSH    R00;                          ;* SP+=4
        LOAD    R00,(SP+TAP_04);
        PUSH    R00                           ;* SP+=4
        LOAD    R00,(SP+TAP_05);
        PUSH    R05                           ;* SP+=4
        LOAD    R00,(SP+TAP_06);
        PUSH    R00                           ;* SP+=4
        LOAD    R00,(SP+TAP_07);
```

Figure 12-5:
Sample caller assembly code using user_calling_convention.

possible. While building projects with —Os as an option will tune the code for optimal code size, it may also restrict the amount of optimization that is performed by the compiler due to such optimizations resulting in increased code size. As such, a user may want to keep an eye out for errant optimizations performed typically around loop nests and selectively disable them on a one

```
PUSH    R00                              ;* SP+=4


        ;* all parameters to pass to callee_procedure via stack are

        ;* loaded from callee_procedure's memory and put on stack.

        ;* we can low load tap_00 and tap_01 into register and pass those

        ;* first two parameters via register

        ;*

LOAD    R00,(SP+TAP_00);

LOAD    R01,(SP+TAP_01);

CALL    __callee_procedure;

NOP;

NOP;


        ;* store the value of result computed by callee_procedure

        ;* into our local space.

        ;*

STORE   R00,(SP+RESULT);


        ;* subsequent computation using result returned in R00

        ;*

RTS;

NOP;

NOP;

__end_caller_procedure:
```

Figure 12-5:
Continued

```
;*************************************************************************
;* NOTE:  NEEDS UPDATING AS OF 11/11/2011 AWAITING EDITOR COMMENTS.
;* R00 contains tap_00
;* R01 contains tap_01;
;* tap_02 through tap_07 have been passed via the stack, as seen
;* previously being setup in caller_procedure via the push operations.
;* Upon entry, callee_procedure must transfer all of the input parameters
;* passed via the stack into registers for local computation.  This
;* requires additional instructions both on the caller side (to put on
;* the stack) as well as the callee side (to restore from the stack).
;*
;* NOTE:  INSERT PROS AND CONS
;*
;*
;*
;*************************************************************************
__callee_procedure:

        ;* ADJUST STACK POINTER TO NOW POINT TO CALLEE'S STACK FRAME
        ;* SO WE CAN ACCESS DATA PASSED VIA STACK IN ABI COMPLIANCE
        ;*
        POP    R07;          ;* tap_07 into R07, SP-=4
        POP    R06;          ;* tap_06 into R06, SP-=4
        POP    R05;          ;* tap_05 into R05, SP-=4
        POP    R04;          ;* tap_04 into R04, SP-=4
        POP    R03;          ;* tap_03 into R03, SP-=4
        POP    R02;          ;* tap_02 into R02, SP-=4

        ;* perform local computation on input parameters now stored
        ;* in registers R00-R07, storing result into
        ;* SP+RESULT_OFFSET
        ;*
        ;* move result into register R00
        ;*
        STORE R00,(SP+RESULT_OFFSET)
        RTS;
        NOP;
        NOP;
__end_callee_procedure:
```

Figure 12-6:
Sample callee assembly code using user_calling_convention.

by one use case rather than disable them for an entire project build. Most compilers support a list of pragmas that can be inserted to control compile time behavior. Examples of such pragmas can be found with the documentation for your target processor's build tools.

Software pipelining is one optimization that can result in increased code size due to additional instructions that are inserted before and after the loop body of the transformed loop. When the compiler or assembly programmer software pipelines a loop, overlapping iterations of a given loop nest are scheduled concurrently with associated 'set up' and 'tear down' code inserted before and after the loop body. These additional instructions inserted in the set up and tear down, or prolog and epilog as they are often referred to in the compiler community, can result in increased instruction counts and code sizes. Typically a compiler will offer a pragma such as '#pragma noswp' to disable software pipelining for a given loop nest, or given loops within a source code file. Users may want to utilize such a pragma on a loop by loop basis to reduce increases in code size associated for select loops that may not be performance critical or on the dominant runtime paths of the application.

Loop unrolling is another fundamental compiler loop optimization that often increases the performance of loop nests at runtime. By unrolling a loop so that multiple iterations of the loop reside in the loop body, additional instruction level parallelism is exposed for the compiler to schedule on the target processor; in addition fewer branches with branch delay slots must be executed to cover the entire iteration space of the loop nest, potentially increasing the performance of the loop as well. Because multiple iterations of the loop are cloned and inserted into the loop body by the compiler, however, the body of the loop nests typically grows as a multiple of the unroll factor. Users wishing to maintain a modest code size may wish to selectively disable loop unrolling for certain loops within their code production, at the cost of compiled code runtime performance. By selecting those loop nests that may not be on the performance critical path of the application, savings in code size can be achieved without impacting performance along the dominant runtime path of the application. Typically compilers will support pragmas to control loop unrolling related behavior, such as the minimum number of iterations a loop will exist for various unroll factors to pass to the compiler. Examples of disabling loop unrolling via a pragma are often of the form '#pragma nounroll.' Please refer to your local compiler's documentation for correct syntax on this and related functionality.

Procedure inlining is another optimization that aims to improve the performance of compiled code at the cost of compiled code size. When procedures are inlined, the callee procedure that is the target of a caller procedures callee invocation site is physically inlined into the body of the caller procedure. Consider the example in Figure 12-7.

Instead of making a call to callee_procedure() every time caller_procedure() is invoked, the compiler may opt to directly substitute the body of callee_procedure into the body of caller_procedure to avoid the overhead associated with the function call. In doing this, the statement a + b will be substituted into the body of caller_procedure in the hopes of improving runtime performance by eliminating the function call overhead, and hopefully proving better instruction cache performance. If this inlining is performed for all call sites of callee_procedure within the application, however, one can see how multiple inlinings quickly can lead to an explosion in the sizes of the application especially for examples where callee_procedure contained more than a simple addition statement. As such, users

```
int caller_procedure(void)
{
    int result, a, b;

    // intermediate computation
    result = callee_procedure();
    return result;
}

int callee_procedure(void)
{
    return a + b;
}
```

Figure 12-7:
Procedure inlining.

may wish to manually disable function inlining for their entire application or for selective procedures via a compiler provided pragma. Typical pragmas are of the form '#pragma noinline' and will prevent the tools from inlining the procedure marked at compilation time.

Memory layout optimization

In order to obtain sufficient levels of performance, application developers and software systems architects must not only select the appropriate algorithms to use in their applications, but also the means by which those applications are implemented. Quite often this also crosses the line into data structure design, layout, and memory partitioning for optimal system performance. It is true that senior developers often have insight into both algorithms and their complexity, as well a toolbox of tips and tricks for memory optimization and data structure optimization. At the same time, the scope of most embedded software engineering projects prohibits manual code and data hand optimization due to time, resource, and cost constraints. As such, developers must often rely on the tools as much as possible to optimize the general use cases, only resorting to hand level tuning and analysis to tweak performance on those performance critical bottlenecks after the initial round of development. This last round of optimization often entails using various system profiling metrics to determine performance critical bottlenecks, and then optimizing these portions of the application by hand using proprietary intrinsics or assembly code, and in some cases rewriting performance critical kernel algorithms and/or related data structures. This section details design decisions that may prove useful for embedded system developers concerned with those topics mentioned above.

Overview of memory optimization

Memory optimizations of various types are often beneficial to the runtime performance and even power consumption of a given embedded application. As was mentioned previously, these optimizations can often be performed to varying degrees by the application build tools such as compilers, assemblers, linkers, profilers, and so forth. Alternately, it is often valuable for developers to go into the application and either manually tune the performance, or design in consideration of memory system optimization a priori for either given performance targets, or so as to design the software architecture to be amenable to automated tools optimization at subsequent phases in the development cycle.

In tuning a given application, quite often the baseline or 'out of box' version of the application will be developed. Once functionality is brought online, the development team or engineers may select to profile the application for bottlenecks that require further optimization. Often these are known without profiling, if certain kernels within the application must execute within a given number of clock cycles as determined by a spreadsheet or pen and paper exercise during system definition. Once these key kernels are isolated or key data structures are isolated, optimization typically begins by those experts with knowledge of software optimization techniques, compiler optimizations, the hardware target, and perhaps details of the hardware target instruction set.

Focusing optimization efforts

Amdahl's law plays an interesting role in the optimization of full application stacks, yet is not always appreciated by the software system developer. If only 10% of the dynamic runtime of a given application can benefit from SIMD or instruction level parallel optimizations versus the 90% of dynamic runtime that must be executed sequentially, then inordinate amounts of effort on parallelizing the 10% portion of the code will still only result in modest performance improvements. Conversely, if 90% of the total applications dynamic runtime is spent in code regions exhibiting large amounts of instruction level parallelism and data level parallelism, it may be worthwhile to focus engineering effort on parallelizing these regions to obtain improved dynamic runtime performance.

In determining those portions of the code which dominate the dynamic application runtime, and may be the best candidate for either hand optimization or hand adjustment for applicability to automated tools optimization, application developers typically use a software profiler in conjunction with either the silicon target or software based system simulation. Intel's VTUNE is one such example of a profiling framework; alternately the GNU GCC compiler and GPROF are open source solutions that provide dynamic runtime information. Many silicon vendors such as Freescale Semiconductor and Texas Instruments also offer their own proprietary solutions for use with their respective silicon targets, allowing for either traces collected on software based simulation platforms, or alternately larger application level traces that can be collected on the native silicon target.

Vectorization and the dynamic code-compute ratio

Vectorization of loops is an optimization whereby computation performed across multiple loop iterations can be combined into single vector instructions, effectively increasing the instruction to compute ratio within the applications dynamic runtime behavior. Consider the example in Figure 12-8.

In the first loop nest, we can see that each iteration of the loop contains a single 16-bit by 16-bit multiply instruction whose result is written to the a[] array as output. One multiplication instruction is performed for each iteration of the loop, resulting in 16 16-bit multiplies. The second loop, however, shows pseudo-code for how the compiler or application developer might vectorize the loop when targeting and architecture that supports a 4-way SIMD multiply instruction over 16-bit integer elements. In this case, the compiler has vectorized multiple iterations of the loop together into the multiply instruction, as denoted by the array [start_range:end_range] syntax denoted in the second loop nest. Note that the loop counter is incremented by the vectorized length for each iteration of the loop now. Clearly only four iterations over the loop are now needed to compute the resulting a[] output array, as each iteration of the loop now contains a single vector multiply instruction that computes four elements of the output vector in parallel.

```
short a[16], b[16], c[16];
for(iter=0; iter<16; ++iter)
{
    // results in single 16-bit MPY instruction
    // generated in assembly listing
    //
    a[iter] = b[iter] * c[iter]
}


short a[16], b[16], c[16];
for(iter=0; iter<16 iter+=4)
{
    // with high level compiler vectorization,
    // results in 4-WAY parallel 4x16-BIT multiply
    // vector instruction, effectively performing the
    // computation of four iterations of the loop in
    // a single atomic SIMD4 instruction.
    //
    a[iter:iter+4] = b[iter:iter+4] * c[iter:iter+4];
}
```

Figure 12-8:
Vectorization of loops.

There are many benefits to vectorizing code in this manner, either by hand in the event that the application developer uses intrinsics that are proprietary with respect to the target architecture, or in the event that the compiler is able to vectorize the code. One such benefit is the increase in performance, as the code now exploits dedicated SIMD hardware, often providing a multiplication in improvement over the vectorized loop on the order of the underlying SIMD vector hardware. Other benefits are the reduction in code size, as loops are no longer unrolled resulting in explosions in the code size, but rather more dense instructions of vector format are used rather than atomic scalar instructions. This may have secondary benefits in reducing the number of instruction fetch transactions that go out to memory as well. Lastly, the overall ratio of dynamically issued instructions to computation performed within the application is also increased.

There are a number of challenges to the development tools as well as the application developers when trying to vectorize code at the loop level. One such challenge is the code shape of loop nests that are candidate for vectorization. Typically, build tools need to understand the loop iteration space of a loop, so using constant loop bounds rather than runtime computed values may be beneficial depending on the advancement of the underlying compiler's vectorization technology. Secondly, the types of computation performed within the loop nest must be amenable to vectorization. For example, in the example above simple 16-bit integer multiplication is performed for a target architecture supporting a supposed 16-bit 4-way SIMD multiply instruction. If the underlying target architecture only supports 8-bit SIMD multiplication, it may be advantageous to avoid 16-bit multiplication wherever possible if vectorization is desired.

Loop dependence analysis is another concern when vectorizing or parallelizing loop nests, in the compiler must be able to prove the safety of loop transformations. Loop dependence analysis is the means by which the compiler or dependence analyzer determines whether statements within a loop body form a dependence with respect to array accesses and data modifications, various data reduction patterns, simplification of loop independent portions of the code and management of various conditional execution statements within the loop body.

As an example, consider the fragment of C-language code in Figure 12-9.

For the loop next above, the compiler's data dependence analyzer will attempt to find all dependences between the statements reading the array b[] and writing the array a[]. The challenge for the data dependence analyzer is to find all possible dependences between the statements that write to array a[] and read from array b[]. To ensure safety, the data dependence analyzer must ensure that it can explicitly prove safety, or in other words, any dependence that cannot be proven false must be assumed to be true to ensure safety!

```
for(iter_a=0; iter<LOOP_BOUND_A; ++iter_b)
  for(iter_b=0; iter_b<LOOP_BOUND_B; ++iter_b)
    a[iter_a+4-iter_b] =
        b[2*iter_a-iter_b] + iter_a*iter_b;
```

Figure 12-9:
Loop dependence analysis.

The data dependence analysis shows independence between references by solving that no two instances of statements to array a[] and array b[] access or modify the same spot in array a[]. In the event that a possible dependence is found, loop dependence analysis will make an attempt to characterize the dependences as some types of optimizations over loop nests may still be possible and profitable. It may also be possible to further transform the loop nests so as to remove the dependence.

In summary, writing loop nests so that a minimum of data dependences exist between array references will benefit vectorization and other loop transforms as much as possible. While the compiler technology used in analyzing data dependences and auto-vectorizing serial code for vector hardware stems from the supercomputing community, improperly written code with troublesome data dependences and loop structure may still thwart the vectorization efforts of the most advanced tool sets. At a high level, simply writing code which is easiest for humans to understand is usually the easiest for the vectorizer to understand; in addition the vectorizer and data dependence analyzers can easily recognize what the programmer intended. In other words, highly hand tuned code with a priori knowledge of the underlying target architecture is not the best candidate for automated vectorization at the tools level.

There are a number of things that application developers may want to keep an eye out for when developing code with the intent of auto-vectorization by the build tools.

Pointer aliasing in C

One challenge for vectorizers and data dependence analysis is the user of pointers in the C language. When data is passed to a function via pointers as parameters, it is often difficult or impossible for the data dependence analyzer and subsequent vectorizer to guarantee that the memory regions pointed to by the various pointers do not overlap in the interaction spaces of the loops in which they are computed. As the C standard has evolved over time, support for the 'restrict' keyword has been added as can be seen in the example in Figure 12-10.

By placing the restrict keyword qualifier on the pointers passed to the procedure, this ensures to the compiler that the data accessed by a given pointer with the restrict keyword does not alias

```
void restrict_compute(restrict int *a, restrict int
*b, restrict int *c)
{
    for(int i=0; i<LIMIT; ++i)
        a[i] = b[i] * c[i];
}
```

Figure 12-10:
Use of the 'restrict' keyword.

with anything else the function may modify using another pointer. Note that this only applies to the function at hand, not the global scope of the application itself. This permits the data dependence analyzer to recognize that arrays are not aliased or modified by references with other side effects, and allows more aggressive optimization of the loop nest including vectorization amongst other optimizations.

Data structures, arrays of data structures, and adding it all up!

Appropriate selection of data structures, before the design of kernels, which compute over them, can have significant impact when dealing with high performance embedded DSP codes. This is often especially true for target processors that support SIMD instruction sets and optimizing compiler technology as was detailed previously in this chapter. As an illustrative example, this section details the various tradeoffs between using an array-of-structure elements versus a structure-of-array elements for commonly used data structures. As an example data structure, we'll consider a set of six dimensional points that are stored within a given data structure as either an array-of-structures or a structure-of-arrays as detailed in Figure 12-11.

The array of structures, as depicted on the left hand side of the table in Figure 12-11, details a structure which has six fields of floating point type, each of which might be the three coordinates of the ends of a line in three dimensional space. The structures are allocated as an array of size elements. The structure of arrays, which is represented on the right hand side above, creates a single structure which contains six arrays of floating point data types, each of

```
/* array of structures*/          /* structure of arrays */
struct {                          struct {
  float x_00;                       float x_00[SIZE];
  float y_00;                       float y_00[SIZE];
  float z_00;                       float z_00[SIZE];
  float x_01;                       float x_01[SIZE];
  float y_01;                       float y_01[SIZE];
  float z_01;                       float z_01[SIZE];
} list[SIZE];                     } list;
```

Figure 12-11:
Example data structure.

which is of size elements. It should be noted that each of the data structures above are functionally equivalent, but have varying system side effects in regard to memory system performance and optimization.

Looking at the array of structures example above, it can be seen that for a given loop nest that is known to access all of the elements of a given struct element before moving onto the next element in the list, good locality of data will be exhibited. This will be due to the fact that as cache lines of data are fetched from memory into the data caches, adjacent elements within the data structure will be fetched contiguously from memory and exhibit good local reuse.

The downside when using the array of structures data structure, however, is that each individual memory reference in a loop that touches all of the field elements of the data structure do not exhibit unit memory stride. For example, consider the illustrative loop in Figure 12-12.

Each of the field accesses in the loop above access different fields within an instance of the structure, and do not exhibit unit stride memory access patterns which would be conducive to compiler level auto-vectorization. In addition, any loop that traverses the list of structures and accesses only one or few fields within a given structure instance will exhibit rather poor spatial locality of data within the cases, due to fetching cache lines from memory that contain data elements which will not be reference within the loop nest.

We can contrast this rather bleak use case depicted above by migrating the array-of-structures format to the structure-of-arrays format, as depicted in the loop nest in Figure 12-13.

By employing the structure-of-arrays data structure, each field access within the loop nest exhibits unit stride memory references across loop iterations. This is much more conducive to auto-vectorization by the build tools in most cases. In addition, we still see good locality of data across the multiple array streams within the loop nest. It should also be noted that in contrast to the previous scenario, even if only one field is accessed by a given loop nest,

```
for(i=0 i<SIZE; ++i)
{
        local_struct[i].x_00 = 0.00;
        local_struct[i].y_00 = 0.00;
        local_struct[i].z_00 = 0.00;
        local_struct[i].x_01 = 0.00;
        local_struct[i].y_01 = 0.00;
        local_struct[i].z_01 = 0.00;
}
```

Figure 12-12:
Data structure that does not exhibit unit memory stride.

```
for(i=0 i<SIZE; ++i)
{
        local_struct.x_00[i] = 0.00;
        local_struct.y_00[i] = 0.00;
        local_struct.z_00[i] = 0.00;
        local_struct.x_01[i] = 0.00;
        local_struct.y_01[i] = 0.00;
        local_struct.z_01[i] = 0.00;
}
```

Figure 12-13:
Structure of arrays format.

locality within the cache is achieved due to subsequent elements within the array being prefetched for a given cache line load.

While the examples presented previously detail the importance of selecting the data structure that best suites the application developer's needs, it is assumed that the developer or system architect will study the overall application hot spots in driving the selection of appropriate data structures for memory system performance. The result may not be a clear case of black and white, however, and a solution that employs multiple data structure formats may be advised. In these cases, developers may wish to use a hybrid type approach that mixes and matches between structure-of-array and array-of-structure formats. Furthermore, for legacy code bases which are tightly coupled to their internal data structures for various reasons beyond the scope of this chapter, it may be worthwhile to runtime convert between the various formats as needed. While the computation required to convert from one format to another is non-trivial, there may be use cases where the conversion overhead is dramatically offset by the computational and memory system performance enhancements achieved once the conversion is performed.

Loop optimizations for memory performance

In addition to structuring loops for targetability by auto-vectorizing compiler technology, and tailoring data structures over which loops compute, there are loop transformations themselves which may benefit the memory system performance of an application as well. This section details various loop transformation that can be performed either manually by the application developer, or automatically by the development tools to improve system performance.

Data alignment's rippling effects

The alignment of data within the memory system of an embedded target can have rippling effects on the performance of the code, as well as the development tools' ability to optimize certain use cases. On many embedded systems, the underlying memory system does not support unaligned memory accesses or such accesses are supported with a certain performance penalty. If the user does not take caution in aligning data properly within the memory system layout, performance can be lost. In summary, data alignment details the manner in which data is accessed within the computer's memory system. When a processor reads or writes to memory, it will often do this at the resolution of the computer's word size which might be 4 bytes on a 32-bit system. Data alignment is the process of putting data elements at offsets that are some multiple of the computers word size so that various fields may be accessed efficiently. As such, it may be necessary for users to put padding into their data structures or for the tools to automatically pad data structures according to the underlying ABI and data type conventions when aligning data for a given processor target.

Alignment can play an impact on compiler and loop optimizations such as vectorization. For instance, if the compiler is attempting to vectorize computation occurring over multiple arrays within a given loop body, it will need to know if the data elements are aligned so as to make efficient use of packed SIMD move instructions, and to also know whether certain iterations of the loop nest must be peeled off that execute over non-aligned data elements. If the compiler cannot determine whether or not the data elements are aligned, it may opt to not vectorize the loop at all thereby leaving the loop body sequential in schedule. Clearly this is not the desired result for the best performing executable. Alternately, the compiler may decide to generate multiple versions of the loop nest with runtime test to determine at loop execution time whether or not the data elements are aligned. In this case the benefits of a vectorized loop version are obtained; however, the cost of a dynamic test at runtime is incurred and the size of the executable will increase due to multiple versions of the loop nest being inserted by the compiler.

Users can often do multiple things to ensure that their data is aligned, for instance padding elements within their data structures and ensuring that various data fields lie on the appropriate word boundaries. Many compilers also support sets of pragmas to denote that a given element is aligned. Alternately, users can put various asserts within their code to compute at runtime whether or not the data fields are aligned on a given boundary before a particular version of a loop executes.

Selecting data types for big payoffs

It is important that application developers also select the appropriate data types for their performance critical kernels in addition to the aforementioned strategies of optimization. When the minimal acceptable data type is selected for computation, it may have a number of secondary effects that can be beneficial to the performance of the kernels. Consider, as an example, a performance critical kernel that can be implemented in either 32-bit integral computation or 16-bit integral computation due to the application programmer's knowledge of the data range. If the application developer selects 16-bit computation using one of the built-in C/C++ language data types such as 'short int,' then the following benefits may be gained at system runtime.

By selecting 16-bit over 32-bit data elements, more data elements can fit into a single data cache line. This allows fewer cache line fetches per unit of computation, and should help alleviate the compute to memory bottleneck when fetching data elements. In addition, if the target architecture supports SIMD style computation, it is highly likely that a given ALU within the processor can support multiple 16-bit computations in parallel versus their 32-bit counterparts. For example, many commercially available DSP architectures support packed 16-bit SIMD operations per ALU, effectively doubling the computational throughput when using 16-bit data elements versus 32-bit data elements. Given the packed nature of the data elements, whereby additional data elements are packed per cache line or can be placed in user managed scratchpad memory, coupled with the increased computational efficiency, it may also be possible to improve the power efficiency of the system due to the reduced number of data memory fetches required to fill cache lines.

CHAPTER 13

Software Optimization for Power Consumption

Andrew Temple

Chapter Outline

DSP for Embedded and Real-Time Systems. DOI: 10.1016/B978-0-12-386535-9.00013-5

Introduction

One of the most important considerations in the product lifecycle of a DSP project is to understand and optimize the power consumption of the device. Power consumption is highly visible for handheld devices which require battery power to be able to guarantee certain minimum usage / idle times between recharging. The other main DSP applications — medical equipment, test, measurement, media, and wireless base station, are very sensitive to power as well, due to the need to manage heat dissipation of increasingly powerful processors, power supply cost, and energy consumption cost [1], so the fact is that power consumption cannot be overlooked.

The responsibility of setting and keeping power requirements often falls on the shoulders of hardware designers, but the software programmer has the ability to provide a large contribution to power optimization. Often, the impact that the software engineer has to influence the power consumption of a device is overlooked or underestimated, as Oshana notes in the *Introduction to Power [1]*.

The goal of this work is to discuss how software can be used to optimize power consumption, starting with the basics of what power consumption consists of, how to properly measure power consumption, and then moving on to techniques for minimizing power consumption in software at the algorithmic level, the hardware level, and with regard to data flow. This will include demonstrations of the various techniques and explanations using Freescale StarCore DSPs of both how and why certain methods are effective at reducing power, so the reader can take and apply this work to their application right away.

Understanding power consumption

In general, when power consumption is discussed, the four main factors discussed for a device are the application, the frequency, the voltage and the process technology, so we need to understand why exactly it is that these factors are so important.

The application is highly important, so much so that the power profile for two handheld devices could differ to the point of making power optimization strategies the complete opposite. While we will be explaining more about power optimization strategy later on, the basic idea is clear enough to introduce in this section.

Take for example a portable media player versus a cellular phone. The portable media player needs to be able to run at 100% usage for a long period of time to display video (full length movies), audio, etc. We will discuss this later, but the general power consumption profile for this sort of device would have to focus on algorithmic and data flow power optimization more than on efficient usage of low power modes.

Compare this to the cellular phone, which spends most of its time in an idle state, and during call time, the user is only talking a relatively small percentage of the time. For this small percentage of time, the processor may be heavily loaded performing encode/decode of voice and transmit/receive data. For the remainder of the call time, the phone is not so heavily tasked, performing procedures such as sending heartbeat packets to the cellular network and providing 'comfort noise' to the user to let the user know the phone is still connected during silence. For this sort of a profile, power optimization would be focused first around maximizing processor sleep states to save as much power as possible, and then on data flow / algorithmic approaches.

In the case of process technology, the current cutting edge DSPs are based on 45 nm technology, a decrease in size from their predecessor, the 65 nm technology. What this smaller process technology provides is a smaller transistor. Smaller transistors consume less power and produce less heat, so are clearly advantageous compared to their predecessors.

Smaller process technology also generally enables higher clock frequencies, which is clearly a plus, providing more processing capability; but higher frequency, along with higher voltage, comes at the cost of higher power draw. Voltage is the most obvious of these; as we learned in physics (and EE101), power is the product of voltage times current. So if a device requires a large voltage supply, power consumption increase is a fact of life.

While staying on our subject of $P = V \times I$, the frequency is also directly part of this equation because current is a direct result of the clock rate. Another thing we learned in physics and EE101: when voltage is applied across a capacitor, current will flow from the voltage source to the capacitor until the capacitor has reached an equivalent potential. While this is an over-simplification, we can imagine that the clock network in a DSP consumes power in such a fashion. Thus at every clock edge, when the potential changes, current flows through the device until it reaches the next steady state. The faster the clock is switching, the more current is flowing; therefore a faster clocking implies more power consumed by the DSP. Depending on the device, the clock circuit is responsible for consuming between 50% and 90% of dynamic device power, so controlling clocks is a theme that will be covered very heavily here.

Static versus dynamic power consumption

Total power consumption consists of two types of power: dynamic and static (also known as static leakage) consumption, so total device power is calculated as:

$$P_{total} = P_{Dynamic} + P_{Static}$$

As we have just discussed, clock transitions are a large portion of the dynamic consumption, but what is this 'dynamic consumption'? Basically, in software we have control over dynamic consumption, but we do not have control over static consumption.

Static power consumption

Leakage consumption is the power that a device consumes independent of any activity or task the DSP is running, because even in a steady state, there is a low 'leakage' current path (via transistor tunneling current, reverse diode leakage, etc.) from the device's V_{in} to ground. The only factors that affect the leakage consumption are: supply voltage, temperature, and process.

We have already discussed voltage and process in the introduction. In terms of temperature, it is fairly intuitive to understand why heat increases leakage current. Heat increases the mobility of electron carriers, which will lead to an increase in electron flow, causing greater static power consumption. As the focus of this chapter is software, this will be the end of static power consumption theory. (For a quick read on temperature and carrier mobility, refer to Wikipedia's basic electron mobility article [2].)

Dynamic power consumption

The dynamic consumption of the DSP includes the power consumed by the device actively using the cores, core subsystems, peripherals such as DMA, I/O (radio, Ethernet, PCIe, CMOS camera), memories, PLLs, and clocks. At a low level, this can be translated as dynamic power – the power consumed by switching transistors, which are charging and discharging capacitances.

Dynamic power increases as we use more elements of the system, more cores, more arithmetic units, more memories, higher clock rates, or anything that could possibly increase the amount of transistors switching, or the speed at which they are switching. The dynamic consumption is independent of temperature, but still depends on voltage supply levels.

Maximum, average, worst case, and typical power

When measuring power, or determining power usage for a system, there are four main types of power that need to be considered: maximum, average, worst, and typical power consumption.

Maximum and average power are general terms, used to describe the power measurement itself more than the effect of software or other variables on a device's power consumption.

Simply stated, maximum power is the highest instantaneous power reading measured over a period of time. This sort of measurement is useful to show the amount of decoupling capacitance required by a device to maintain a decent level of signal integrity (required for reliable operation).

Average power is intuitive at this point: technically the amount of energy consumed in a time period, divided by that time (power readings averaged over time). Engineers do this by calculating the average current consumed over time and use that to find power. Average power readings are what we are focusing on optimizing, as this is the determining factor for how much power a battery or power supply must be able to provide for a DSP to perform an application over time, and this is also used to understand the heat profile of the device.

Both worst case and typical power numbers are based on average power measurement. Worst case power, or the worst case power profile, describes the amount of average power a device will consume at 100% usage over a given period time. 100% usage infers that the processer is utilizing the maximum number of available processing units (data and address generation blocks in the core, accelerators, bit masking, etc.), memories, and peripherals simultaneously. This may be simulated by putting the cores in an infinite loop performing 6 or more instructions per cycle (depending on the available processing units in the core) while having multiple DMA channels continuously reading and writing from memory, and peripherals constantly sending and receiving data. Worst case power numbers are used by the system architect or board designer in order to provide adequate power supply to guarantee functionality under all worst case conditions.

In a real system, a device will rarely if ever draw the worst case power, as applications are not using all the processing elements, memory, and I/O for long periods of time, if at all. In general, a device provides many different I/O peripherals, though only a portion of them are needed, and the device cores may only need to perform heavy computation for small portions of time, accessing just a portion of memory. Typical power consumption then may be based on the assumed 'general use case' example application that may use anywhere from 50 to 70% of the processors available hardware components at a time. This is a major aspect of software applications that we are going to be taking advantage of in order to optimize power consumption.

In this section we have explained the differences of static versus dynamic power, maximum versus average power, process effect on power, and core and processing power effect on power. Now that the basics of what makes power consumption are covered, we will discuss power consumption measurement before going into detail about power optimization techniques.

Measuring power consumption

Now that background, theory, and vocabulary have been covered, we will move on to taking power measurements. We will discuss the types of measurements used to get different types of power readings (such as reading static versus dynamic power), and use these methods in order to test optimization methods used later in the text.

Measuring power is hardware dependent: some DSPs provide internal measurement capabilities. DSP manufacturers also may provide 'power calculators' which give some power information. There are a number of power supply controller ICs which provide different forms of power measurement capabilities; some power supply controllers called VRMs ('Voltage Regulator Modules') have these capabilities internal to them to be read over peripheral interfaces. And finally, the old fashioned method of connecting an ammeter in series to the DSP's power supply.

Measuring power using an ammeter

The 'old fashioned' method to measure power is via the use of an external power supply connected in series to the positive terminal of an ammeter, which connects via the negative connector to the DSP device power input, as shown in Figure 13-1.

Note that there are three different setups shown in Figure 13-1, which are all for a single DSP. This is due to the fact that DSP power input is isolated, generally between cores (possibly multiple supplies), peripherals, and memories. This is done by design in hardware as different components of a device have different voltage requirements, and this is useful to use to isolate (and eventually optimize) the power profile of individual components.

In order to properly measure power consumption, the power to each component must be properly isolated, which in some cases may require board modification, specific jumper settings, etc. The ideal situation is to be able to connect the external supply/ammeter combo as close as possible to the DSP's power input pins.

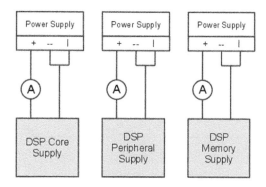

Figure 13-1:
Measuring power via ammeters.

Alternatively, one may measure the voltage drop across a (shunt) resister which is in series with the power supply and a DSP's power pins. By measuring the voltage drop across the resistor, current is found simply by calculating $I = V/R$.

Measuring power using a Hall Sensor type IC

In order to simplify efficient power measurement, many DSP vendors are building boards that use a Hall-Effect based sensor. When Hall sensors are placed on a board in the current path to the device's power supply, it generates a voltage equivalent to the current times some coefficient with an offset. In the case of Freescale's MSC8144 DSP Application Development System board, an Allegro ACS0704 Hall Sensor is provided on the board which enables such measurement. With this board, the user can simply place a scope to the board, and view the voltage signal over time, and use this to calculate average power using Allegro's current to voltage graph, shown in Figure 13-2.

Using Figure 13-2, we can calculate input current to a device based on measuring potential across V_{out} as:

$$I = (V_{out} - 2.5)^* 10A$$

Voltage regulator module power supply ICs

Finally, some voltage regulator module (VRM) power supply controller ICs, which are used to split a large input voltage into a number of smaller ones to supply individual sources at varying potentials, measure current/power consumption and store the values in registers to be read by the user. Measuring current via the VRM requires no equipment, but this sometimes comes at the cost of accuracy and real time measurement. For example, the PowerOne ZM7100 series VRM (also used on the MSC8144ADS) provides current readings for each

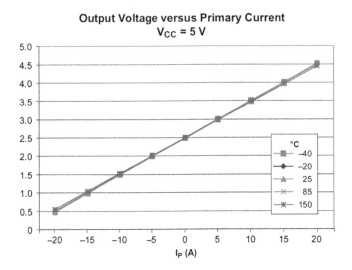

Figure 13-2:

Hall Effect IC voltage to current graph (www.allegromicro.com/en/Products/Part.../0704/0704-015.pdf).

supply, but the current readings are updated once every 0.5 to 1 seconds, and the reading accuracy is on the order of ~20%, so instantaneous reading for maximum power is not possible, and fine tuning and optimization may not be possible using such devices.

In addition to deciding a specific method for measuring power in general, different methods exist to measure dynamic power versus static leakage consumption. The static leakage consumption data is useful in order to have a floor for our low power expectations, and to understand how much power the actual application is pulling versus what the device will pull in idle. We can then subtract that from the total power consumption we measure in order to determine the dynamic consumption the DSP is pulling, and work to minimize that.

Static power measurement

Leakage consumption on the DSP can usually be measured while the device is placed in a low power mode, assuming that the mode shuts down clocks to all of the DSP core subsystems and peripherals. In the case that the clocks are not shut down in low power mode, the PLLs should be bypassed, and then the input clock should be shut down, thus shutting down all clocks and eliminating clock and PLL power consumption from the static leakage measurement.

Additionally, static leakage should be measured at varying temperatures since leakage varies based on temperature. Creating a set of static measurements based on temperature (and voltage) provides valuable reference points for determining how much dynamic power an application is actually consuming at these temperature/voltage points.

Dynamic power measurement

The power measurements should separate the contribution of each major module in the device to give the engineer information about what effect a specific configuration will have on a system's power consumption. As noted above, dynamic power is found simply by measuring the total power (at a given temperature) and then subtracting the leakage consumption for that given temperature using the initial static measurements from above.

Initial dynamic measurement tests include running sleep state tests, debug state tests, and a NOP test. Sleep state and debug state tests will give the user insight into the cost of enabling certain clocks in the system. A NOP test, as in a loop of NOP commands, will provide a baseline dynamic reading for your core's consumption when mainly using the fetch unit of the device, but no arithmetic units, address generation, bit mask, memory management, etc.

When comparing specific software power optimization techniques, we compare the 'before' and 'after' power consumption numbers of each technique in order to determine the effect of that technique.

Profiling your application's power consumption

Before optimizing an application for power, the programmer should get a baseline power reading of the section of code being optimized. In order to do this, the programmer needs to generate a sample power test which acts as a snapshot of the code segment being tested.

This power test case generation can be done by profiling code performance using a high end profiler to gain some base understanding of the % of processing elements and memory used. We can demonstrate this in Freescale's CodeWarrior for StarCore IDE, by creating a new example project using the CodeWarrior stationary with the profiler enabled, then compiling, and running the project. The application will run from start to finish, at which point the user may select a profiler view and get any number of statistics.

Using relevant data such as the % of ALU's used, AGUs used, code hot-spots, and knowledge of memories being accessed, we can get a general idea of where our code will spend the most time (and consume the most power). We can use this to generate a basic performance test which runs in an infinite loop, enabling us to profile the average 'typical' power of an important code segment.

In the standard Freescale CodeWarrior example project, there are 2 main functions: func1 and func2. Profiling the example code, we can see from Figure 13-3 that the vast majority of cycles are consumed by the func1 routine. This routine is located in M2 memory and is reading data from cacheable M3 memory (meaning possibly causing write back accesses to L2 and L1 cache). By using the profiler (as shown in Figure 13-4), information regarding the % ALU, % AGU can be extracted. We can effectively simulate this by turning the code into an

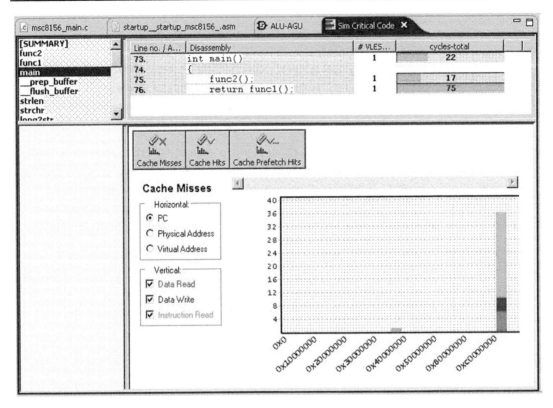

Figure 13-3:
Profiling for hot spots.

Function_Name	Num_VLSE	DALU_Parallelism	AGU_Parallelism	4 ALU / 2 AGU	4 ALU / 1 AGU	4 ALU / 0 AGU	3 ALU / 2 AGU
func2	7	0.14	0.86	0.00% / 0 VLES	0.00% / 0 VLES	0.00% / 0 VLES	0.00% / 0 VLES
func1	3958	0.20	0.80	0.00% / 0 VLES	0.00% / 0 VLES	0.00% / 0 VLES	0.00% / 0 VLES
main	7	0.00	1.00	0.00% / 0 VLES	0.00% / 0 VLES	0.00% / 0 VLES	0.00% / 0 VLES
_____ b.ff__	5	0.40	1.60	0.00% / 0 VLES	0.00% / 0 VLES	0.00% / 0 VLES	0.00% / 0 VLES

SUMMARY:

DALU Parallelism (0-4)	DALU Counter	AGU Parallelism (0-2)	AGU Counter
0.20	785	0.80	3173

% / No. VLES	4 ALU	3 ALU	2 ALU	1 ALU	0 ALU	Total AGU
2 AGU	0.00% / 0 VLES	0.00% / 0 VLES	0.00% / 0 VLES	0.00% / 0 VLES	0.00% / 0 VLES	0.00% / 0 VLES
1 AGU	0.00% / 0 VLES	0.00% / 0 VLES	0.00% / 0 VLES	0.00% / 0 VLES	80.17% / 3173 VLES	80.17% / 3173 VLES
0 AGU	0.00% / 0 VLES	0.00% / 0 VLES	0.00% / 0 VLES	19.83% / 785 VLES	0.00% / 0 VLES	19.83% / 785 VLES
Total ALU	0.00% / 0 VLES	0.00% / 0 VLES	0.00% / 0 VLES	19.83% / 785 VLES	80.17% / 3173 VLES	100.00% / 3958 VLES

Figure 13-4:
Core component (% ALU, % AGU) utilization.

infinite loop, adjusting the I/O, and compiling at the same optimization level, and verifying that we see the same performance breakdown. Another option would be to write a sample test in assembly code to force certain ALU/AGU usage models to match our profile, though this is not as precise and makes testing of individual optimizations more difficult.

We can then set a break point, re-run our application, and confirm that the device usage profile is in line with our original code. If not, we can adjust compiler optimization level or our code until it matches the original application.

This method is quick and effective for measuring power consumption for various loads. By having the infinite loop, testing is much easier as we are simply comparing steady state current readings of optimized and non-optimized code in hopes of getting lower numbers. We can use this to measure numerous metrics such as average power over time, average power per instruction, average power per cycle, and energy (power × time) in joules for some time t. For measuring specific algorithms and power saving techniques, we will form small routines using similar methods and then optimize the power savings over time.

This section has explained a few different methods for measuring static power and dynamic power, and how to profile power for an application. It also covered the availability of power calculators from DSP manufacturers, which sometimes may quicken the power estimation process. Using these tools will enable effectively measuring and confirming the knowledge shared in the next section of this text, which covers the software techniques for optimizing power consumption.

Minimizing power consumption

There are three main types of power optimization covered in this text: hardware supported features, data path optimization, and algorithmic optimization. Algorithmic optimization refers to making changes in code to affect how the DSP's cores process data, such as how instructions or loops are handled, whereas hardware optimization, as discussed here, focuses more on how to optimize clock control and power features provided in hardware. Data flow optimization focuses on working to minimize the power cost of utilizing different memories, busses, and peripherals where data can be stored or transmitted by taking advantage of relevant features and concepts.

Hardware support

Low power modes

DSP applications normally work on tasks in packets, frames, or chunks. For example, in a media player, frames of video data may be coming in at 60 frames per second to be decoded, while the actual decoding work may take the processor orders of magnitude less than $1/60^{th}$ of a second, giving us a chance to utilize sleep modes, shut down peripherals, and organize memory, all to reduce power consumption and maximize efficiency.

We must also keep in mind that power consumption profile varies based on application. For instance, two differing hand held devices: an mp3 player and a cellular phone, will have two very different power profiles.

The cellular phone spends most of its time in an idle state, and when in a call, is still not working at full capacity during the entire call duration as speech will commonly contain pauses which are long in terms of the DSP processor's clock cycles.

For both of these power profiles, software enabled low power modes (modes/features/ controls) are used to save power, and the question for the programmer is how to use them efficiently. The most common modes available consist of power gating, clock gating, voltage scaling, and clock scaling [3].

Power gating uses a current switch to cut off a circuit from its power supply rails during standby mode, to eliminate static leakage when the circuit is not in use. Using power gating leads to a loss of state and data for a circuit, meaning that using this requires storing necessary context/state data to active memory. As DSPs are moving more and more towards being full SoC solutions with many peripherals, some peripherals may be unnecessary for certain applications. Power gating may be available to completely shut off such unused peripherals in a system, and the power savings attained from power gating depends on the specific peripheral on the specific device in question.

It is important to note that, in some cases, documentation will refer to powering down a peripheral via clock gating, which is different from power gating. It may be possible to gate a peripheral by connecting the power supply of a certain block to ground, depending on device requirements and interdependency on a power supply line. This is possible in software in certain situations, such as when board/system level power is controlled by an on-board IC (such as the PowerOne IC), which can be programmed and updated via an I2C bus interface. As an example, the MSC8156 DSP has this option for the MAPLE accelerator and a portion of M3 memory.

Clock gating, as the name implies, shuts down clocks to a circuit or portion of a clock tree in a device. As dynamic power is consumed during state change triggered by clock toggling (as we discussed in the introductory portion of this chapter), clock gating enables the programmer to cut dynamic power through the use of a single (or a few) instructions. Clocking of a DSP is generally separated into trees stemming from a main clock PLL, into various clock domains as required by design for core, memories, and peripherals, and DSPs generally enable levels of clock gating in order to customize a power savings solution.

Freescale's MSC815x low power modes

Freescale DSPs provide various levels of clock gating in the core subsystem and peripheral areas. Gating clocks to a core may be done in the form of STOP and WAIT instructions. STOP mode gates clocks to the DSP core and the entire core subsystem (L1 and L2 Caches, M2 memory, memory management, debug and profile unit) aside from internal logic used for waking from STOP state.

In order to safely enter STOP mode, as one may imagine, care must be taken to ensure accesses to memory and cache are all complete, and no fetches/prefetches are underway.

The recommended process is:

1. Terminate any open L2 prefetch activity
2. Stop all internal and external accesses to M2/L2 memory
3. Close the subsystem slave port window (peripheral access path to M2 memory) by writing to the core subsystem slave port general configuration register
4. Verify slave port is closed by reading the register, and also testing access to the slave port (at this point, any access to the core's slave port will generate an interrupt)
5. Ensure STOP ACK bit is asserted in General Status Register to show subsystem is in stop state
6. Enter Stop mode

STOP state can be exited by initiating an interrupt. There are other ways to exit from STOP state, including a reset or debug assertion from external signals.

The WAIT state gates clocks to the core and some of the core subsystem aside from the interrupt controller, debug and profile unit, timer, and M2 memory, which enables faster entering and exiting from WAIT state, but at the cost of greater power consumption. To enter WAIT state, the programmer may simply use the WAIT instruction for a core. Exiting WAIT, like STOP, may also be done via an interrupt.

A particularly nice feature of these low power states on the Freescale DSPs is that both STOP and WAIT mode can be exited via either an enabled or disabled interrupt. Wake up via an enabled interrupt follows standard interrupt handling procedure: the core takes the interrupt, does a full context switch, and then the program counter jumps to the interrupt service routine before returning to the instruction following the segment of code that executed WAIT (or STOP) instruction. This requires a comparatively large cycle overhead, which is where disabled interrupt waking becomes quite convenient. When using a disabled interrupt to exit from either WAIT or STOP state, the interrupt signals the core using an interrupt priority that is not 'enabled' in terms of the core's global interrupt priority level (IPL), and when the core wakes, it resumes execution where it left off without executing a context switch or any ISR.

An example using a disabled interrupt for waking the MSC8156 is provided at the end of this section.

Clock gating to peripherals is also enabled, where the user may gate specific peripherals individually as needed. This is available for the MSC8156's serial interface, Ethernet controller (QE), DSP accelerators (MAPLE), and DDR. As with STOP mode, when gating clocks to any of these interfaces, the programmer must ensure that all accesses are completed beforehand. Then, via the System Clock Control register, clocks to each of these peripherals may be gated. In order to come out of the clock gated modes, a Power on Reset is required, so this is not something that can be done and undone on the fly in a function, but rather a setting that is decided at system configuration time.

Additionally, partial clock gating is possible on the High Speed Serial Interface components (SERDES, OCN DMA, SRIO, RMU, PCI Express) and DDR so that they may be temporarily put in a 'doze state' in order to save power, but still maintain the functionality of providing an acknowledge to accesses (in order to prevent internal or external bus lockup when accessed by external logic).

Texas Instruments C6000 low power modes

Another popular DSP family on the market is the C6000 series DSP from Texas Instruments (TI). TI DSPs in the C6000 family provide a few levels of clock gating, depending on the generation of C6000. For example, the previous generation C67x floating point DSP has low power modes called 'power down modes.' These modes include PD1, PD2, PD3, and 'peripheral power down,' each of which gates clocking to various components in the silicon.

For example, PD1 mode gates clocks to the C67x CPU (processor core, data registers, control registers, and everything else within the core aside from the interrupt controller). The C67x can wake up from PD1 via an interrupt into the core. Entering, power down mode PD1 (or PD2 / PD3) for the C67x, is done via a register write (to CSR). The cost of entering PD1 state is ~9 clock cycles plus the cost of accessing the CSR register. As this power down state only affects the core (and not cache memories), it is not comparable to the Freescale's STOP or WAIT state.

The two deeper levels of power down, PD2 and PD3, effectively gate clocks to the entire device (all blocks which use an internal clock: internal peripherals, the CPU, cache, etc.). The only way to wake up from PD2 and PD3 clock gating is via a reset, so PD2 and PD3 would not be very convenient or efficient to use mid-application.

The newer Keystone TI DSP family (C66x), which combine floating point and fixed point architectures from previous C6000 devices, retain the PD1, PD2, and PD3 states in the CSR register.

The C66xx provides the ability to gate a subset of the peripherals independently by clock domain, similar to the Freescale DSPs.

Clock and voltage control

Some devices have the ability to scale voltage or clock, which may help optimize the power scheme of a device/application. Voltage scaling, as the name implies, is the process of lowering or raising the power. In the section on measuring current, VRMs were introduced as one method. The main purpose of a VRM (Voltage Regulator Module) is to control the power/ voltage supply to a device. Using a VRM, voltage scaling may be done through monitoring and updating voltage ID (VID) parameters.

In general, as voltage is lowered, frequency / processor speed is sacrificed, so generally voltage would be lowered when demand of a DSP core or a certain peripheral is reduced.

The TI C6000 devices provide a flavor of voltage scaling called SmartReflex®. SmartReflex® enables automatic voltage scaling through a pin interface which provides VID to a Voltage Regulator Module (VRM). As the pin interface is internally managed, the software engineer does not have much effect over this, so we will not cover any programming examples for this.

Clock control is available in many DSPs, such as the MSC8144 from Freescale, which allows changing the values of various PLLs in runtime. In the case of the MSC8144, updating the internal PLLs requires relocking the PLLs, where some clocks in the system may be stopped, and this must be followed by a soft reset (reset of the internal cores). Because of this inherent latency, clock scaling is not very feasible during normal heavy operation, but may be considered if a DSP's requirements over a long period of time are reduced (such as during times of low call volume during the night for DSPs on a wireless base station).

When considering clock scaling, we must keep the following in mind: During normal operation, running at a lower clock allows for lower dynamic power consumption, assuming clock and power gating are never used. In practice, running a processor at a higher frequency allows for more 'free' cycles, which, as previously noted, can be used to hold the device in a low power / sleep mode — thus offsetting the benefits of such clock scaling.

Additionally, for the case of the MSC8144, updating the clock for custom cases is time intensive, and for many other DSPs, not an option at all — meaning clock frequency has to be decided at device reset/power on time, so the general rule of thumb is to enable enough clock cycles with some additional headroom for the real time application being run, and to utilize other power optimization techniques. Determining the amount of headroom varies from processor to processor and application to application — at which point it makes sense to profile your application in order to understand the amount of cycles required for a packet/ frame, and the core utilization during this time period.

Once this is understood, measuring the power consumption for such a profile can be done, as demonstrated earlier in this chapter in the Profiling Power section. Measure the average power consumption at your main frequency options. (In MSC8144 and MSC815x, this could be 800 MHz and 1 GHz), and then average in idle power over the headroom slots in order to get a head to head comparison of the best case power consumption.

Considerations and usage examples of low power modes

Here we will summarize the main considerations for low power mode usage, and then close with a coding example demonstrating low power mode usage in a real time multimedia application.

Consider available block functionality when in low power mode:

*   When in low power modes, we have to remember that certain peripherals will not be available to external peripherals, and peripheral busses may also be affected. As noted earlier in this section, devices may take care of this, but this is not always the case. If power gating a block, special care must be taken regarding shared external busses, clocks, and pins.
*   Additionally, memory states and validity of data must be considered. We will cover this when discussing cache and DDR in the next section.

Consider the overhead of entering and exiting low power modes:

*   When entering and exiting low power modes, in addition to overall power savings, the programmer must ensure the cycle overhead of actually entering and exiting the low power mode does not break real time constraints.
*   Cycle overhead may also be affected by the potential difference in initiating a low power mode by register access as opposed to by direct core instructions.

Low power example

To demonstrate low power usage, we will refer to the Motion JPEG (MJPEG) application as shown in Figure 13-5. As a quick intro: the MJPEG demo is a real time Smart DSP OS demo intended to be run on an MSC8144 or MSC8156 development board.

With the MJPEG demo, raw image frames are sent from a PC to the DSP over Ethernet. Each Ethernet packet contains 1 block of an image frame. A full raw QVGA image uses ~396 blocks plus a header. The DSP encodes the image in real time (adjustable from 1 to 30+ frames per second), and sends the encoded Motion JPEG video back over Ethernet to be played on a demo GUI in the PC. The flow and a screenshot of this GUI are shown in the following figure.

(a)

(b)

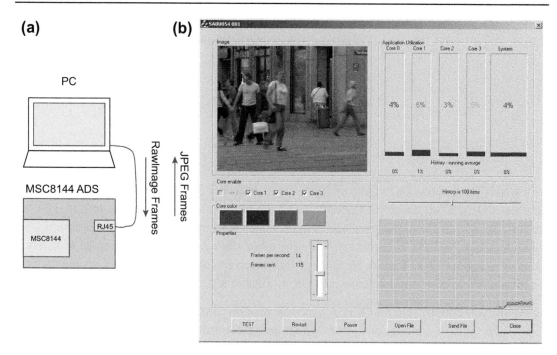

Figure 13-5:
SmartDSP OS motion JPEG demo.

The GUI will display not only the encoded JPEG image, but also the core utilization (as a percentage of the maximum core cycles available).

For this application, we need to understand how many cycles encoding a frame of JPEG consumes. Using this we can determine the maximum frame rate we can use and, in parallel, also determine the maximum down time we have for low power mode usage. If we are close to the maximum core utilization for the real-time application, then using low power modes may not make sense (may break real-time constraints).

As noted in previous chapters, we could simply profile the application to see how many cycles are actually spent per image frame, but this is already handled in the MJPEG demo's code using the core cycle counters in the OCE (On Chip Emulator). The OCE is the hardware block on the MSC81xx series DSPs that the profiler utilizes to get core cycle counts for use in code profiling.

The MJPEG code in this case counts the number of cycles a core spends doing actual work (handling an incoming Ethernet interrupt, dequeueing data, encoding a block of data into JPEG format, enqueueing/sending data back over Ethernet).

The # of core cycles required to process a single block encode of data (and supporting background data movement) is measured to be on the order of 13,000 cycles. For a full JPEG image (~396 image blocks and Ethernet packets), this is approximately 5 million cycles. So 1

JPEG frame a second would work out to be 0.5% of a core's potential processing power considering a 1 GHz core that is handling all Ethernet I/O, interrupt context switches, etc.

$$Cycles_{Block\ Mgmt\ \&\ Encode} = 13,000$$

$$Cycles_{JPEG\ Frame} = Cycles_{Block\ Mgmt\ \&\ Encode} \times 396 = 5,148,000$$

$$Core\ Utilization_{30FPS}(\%) = 30\frac{100 \times OCE\ Count}{1,000,000,000} = 15.4\%$$

As the MSC81xx series DSPs have up to six cores, and only one core would have to manage Ethernet I/O, in a full multicore system, utilization per core drops to a range of 3 to 7%. A master core acts as the manager of the system, managing both Ethernet I/O, intercore communication, and JPEG encoding, while the other slave cores are programmed to solely focus on encoding JPEG frames. Because of this intercore communication and management, the drop in cycle consumption from one core to four or six is not linear.

Based on cycle counts from the OCE, we can run a single core, which is put in a sleep state for 85% of the time, or a multicore system which uses sleep state up to 95% of the time.

This application also uses only a portion of the SoC peripherals (Ethernet, JTAG, a single DDR, and M3 memory). So we can save power by gating the full HSSI System (Serial Rapid IO, PCI Express), the MAPLE Accelerator, and the second DDR controller. Additionally, for our GUI demo, we are only showing four cores, so we can gate cores 4 and 5 without affecting this demo as well. Based on the above, and what we have discussed in this section, here is the plan we want to follow:

At application start up

- Clock Gate the unused MAPLE Accelerator Block
  - MAPLE power pins share a power supply with core voltage. If the power supply to MAPLE was not shared, we could completely gate power. Due to shared pins on the development board, the most effective choice we have is to gate the MAPLE clock.
  - MAPLE automatically goes into a doze state, which gates part of the clocks to the block, when it is not in use. Because of this, power savings from entirely gating MAPLE may not be massive.

- Clock gate the unused HSSI (High Speed Serial Interface)
  - We could also put MAPLE into a doze state, but this gates only part of the clocks. Since we will not be using any portion of these peripherals, complete clock gating is more power efficient.

- Clock gate the unused second DDR controller
    - When using VTB, SmartDSP OS places buffer space for VTB in the second DDR memory, so we need to be sure that this is not needed.

During application runtime

At runtime, QE (Ethernet Controller), DDR, and class CLASS, and cores 1-4 will be active. Things we must consider for these components include:

- The Ethernet Controller cannot be shut down or put into a low power state, as this is the block that receives new packets (JPEG blocks) to encode. Interrupts from the Ethernet Controller can be used to wake our master core from low power mode.
- Active core low power modes:
    - WAIT mode enables core power savings, while allowing the core to be woken up in just a few cycles by using a disabled interrupt to signal exit from WAIT.
    - STOP mode enables greater core savings by shutting down more of the subsystem than WAIT (including M2), but requires slightly more time to wake due to more hardware being re-enabled. If data is coming in at high rates, and the wake time is too long, we could get an overflow condition, where packets are lost. This is unlikely here due to the required data rate of the application.

The first DDR contains sections of program code and data, including parts of the Ethernet handling code. (This can be quickly checked and verified by looking at the program's .map file.) Because the Ethernet controller will be waking the master core from WAIT state, and the first thing the core will need to do out of this state is to run the Ethernet handler, we will not put DDR0 to sleep.

We can use the main background routine for the application to apply these changes without interfering with the RTOS. This code segment is shown below with power down related code in bold:

```
static void appBackground(void)
{
  os_hwi_handle hwi_num;
  if (osGetCoreID() == 0)
  {
    *((unsigned int*)0xfff28014) = 0xF3FCFFFB;//HSSI CR1
    *((unsigned int*)0xfff28018) = 0x0000001F;//HSSI CR2
    *((unsigned int*)0xfff28034) = 0x20000E0E; //GCR5
    *((unsigned int*)0xfff24000) = 0x00001500; //SCCR
  }
  osMessageQueueHwiGet(CORE0_TO_OTHERS_MESSAGE, &hwi_num);
  while(1)
  {
    osHwiSwiftDisable();
```

```
  osHwiEnable(OS_HWI_PRIORITY10);
  stop();//wait();
  osHwiEnable(OS_HWI_PRIORITY4);
  osHwiSwiftEnable();
  osHwiPendingClear(hwi_num);
  MessageHandler(CORE0_TO_OTHERS_MESSAGE);
  }
}
```

Note that the clock gating must be done by only one core as these registers are system level and access is shared by all cores.

This code example demonstrates how a programmer using the SmartDSP OS can make use of the interrupt APIs in order to recover from STOP or wait state without actually requiring a context switch. In the MJPEG player, as noted above, raw image blocks are received via Ethernet (with interrupts), and then shared via shared queues (with interrupts). The master core will have to use context switching to read new Ethernet frames here, but slave cores only need to wake up and go to the MessageHandler function.

We take advantage of this fact by enabling only higher priority interrupts before going to sleep:

```
osHwiSwiftDisable();
osHwiEnable(OS_HWI_PRIORITY10);
```

Then when a slave core is asleep, if a new queue message arrives on an interrput, the core will be woken up (on context switch), and standard interrupt priority levels will be restored. The core will then go and manage the new message without context switch overhead by calling the MessageHandler() function.

In order to verify our power savings, we will take a baseline power reading before optimizing across the relevant power supplies, and then measure the incremental power savings of each step.

The MSC8156ADS board has power for cores, accelerators, HSSI, and M3 memory connected to the same power supply, simplifying data collection. Since these supplies and DDR are the only blocks we are optimizing, we shall measure improvement based on these supplies alone.

Figure 13-6 provides a visual on the relative power consumed by the relevant power supplies (1V: Core, M3, HSSI, MAPLE Accelerators, and DDR) across the power down steps used above. Note that actual power numbers are not provided to avoid any potential non-disclosure issues.

The first two bars provide reference points — indicating the power consumption for these supplies using a standard FIR filter in a loop and the power consumption when the cores are held in debug state (not performing any instructions, but not in a low power mode). With our steps we can see that there was nearly a 50% reduction in power consumption across the

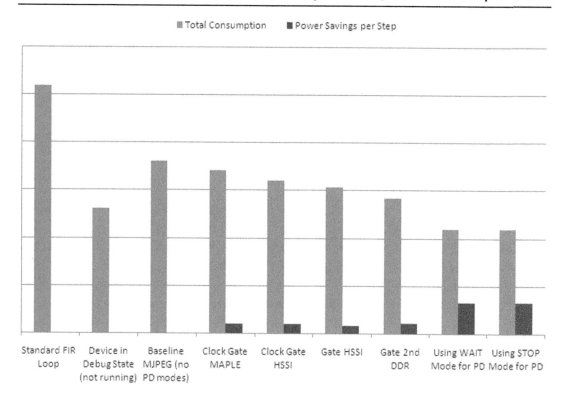

Figure 13-6:
Power consumption savings in PD modes.

relevant supplies for the Motion JPEG demo with the steps laid out above, with each step providing ~5% reduction in power with the exception of the STOP and WAIT power modes, which are closer to 15–20% savings.

One thing to keep in mind is that, while the MJPEG demo is the perfect example to demonstrate low power modes, it is not highly core intensive, so as we progress through different optimization techniques, we will be using other examples as appropriate.

Optimizing data flow

Reducing power consumption for memory accesses

Due to clocks having to be activated not only in the core components, but also in busses and memory cells, memory related functionality can be quite power hungry, but luckily, memory access and data paths can also be optimized to reduce power. This section will cover methods to optimize power consumption with regard to memory accesses to DDR and SRAM memories by utilizing knowledge of the hardware design of these memory types. Then we will cover ways to take advantage of other specific memory setups at the SoC level. Common practice is to optimize memory in order to maximize the locality of critical or heavily used data and code by placing as much in cache as possible. Cache misses incur not only core stall penalties, but also power penalties as more bus activity is needed, and higher level memories (internal device SRAM, or external device DDR) are activated and consume power. As a rule, accesses to higher level memory such as DDR are not as common as internal memory accesses, so high level memory accesses are easier to plan, and thus optimize.

DDR overview

The highest level of memory we will discuss here is external DDR memory. To optimize DDR accesses in software, first we need to understand the hardware that the memory consists of. DDR SDRAM, as the DDR (dual data rate) name implies, takes advantage of both edges of the DDR clock source in order to send data, thus doubling the effective data rate at which data reads and writes may occur. DDR provides a number of different types of features which may affect total power utilization, such as EDC (error detection), ECC (error correction), different types of bursting, programmable data refresh rates, programmable memory configuration allowing physical bank interleaving, page management across multiple chip selects, and DDR specific sleep modes.

Key DDR vocabulary to be discussed includes:

- **Chip Select** (also known as **Physical Bank**) — selects a set of memory chips (specified as a 'rank') connected to the memory controller for accesses.
- **Rank** - specifies a set of chips on a DIMM to be accessed at once. A **Double Rank** DIMM, for example, would have two sets of chips — differentiated by chip select. When accessed together, each rank allows for a data access width of 64 bits (or 72 with ECC).
- **Rows** are address bits enabling access to a set of data, known as a '**page**' — so row and page may be used interchangeably.
- **Logical banks**, like row bits, enable access to a certain segment of memory. By standard practice, the row bits are the MSB address bits of DDR, followed by the bits to select a logical bank, finally followed by column bits.
- **Column** bits are the bits used to select and access a specific address for reading or writing

On a typical DSP, the DSPs' DDR SDRAM controller is connected to either a discrete memory chips, or a DIMM (Dual Inline Memory Module), which contains multiple memory components (chips). Each discrete component/chip contains multiple logical banks, rows, and columns which provide access for reads and writes to memory. The basic idea of a discrete DDR3 memory chip's layout is shown in Figure 13-7.

Standard DDR3 discrete chips are commonly made up of 8 logical banks, which provide addressability as shown above. These banks are essentially tables of rows and columns.

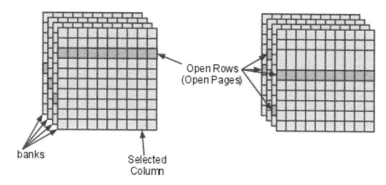

Figure 13-7:
Basic drawing of a discrete DDR3 memory chip's rows/columns.

The action to select a row effectively opens that row (page) for the logical bank being addressed. So different rows can be simultaneously open in different logical banks, as illustrated by the active or open rows highlighted in the picture. A column selection gives access to a portion of the row in the appropriate bank.

When considering sets of memory chips, the concept of chip selects is added to the equation. Using a chip selects, also known as 'PHYSICAL banks,' enables the controller to access a certain set of memory modules (up to 1 GB for the MSC8156, 2 GB for MSC8157 DSPs from Freescale for example) at a time. Once a chip select is enabled, access to the selected memory modules with that chip select is activated, using page selection (rows), banks, and columns. The connection of two chip selects is shown in Figure 13-8.

In Figure 13-8 we see at the bottom we have our DSP device which is intended to access DDR memory. There are a total of 16 chips connected to 2 chip selects: chip select 0 on the left, and 1 on the right. The 16 discrete chips are paired such that a pair of chips share ALL the same signals (address, bank, data, etc.), except for the chip select pin. (*Interesting note: This is basically how a dual rank DDR is organized, except each 'pair of chips' exists within a single chip.*) There are 64 data bits. So for a single chip select, when we access DDR and write 64 contiguous bits of data to DDR memory space in our application, the DDR controller is doing the following:

1. Selecting chip select based on your address (0 for example)
2. Opening the same page (row) for each bank on all 8 chips using the DDR address bits during the Row Access phase
    *   New rows are opened via the **ACTIVE command**, which copies data from the row to a 'row buffer' for fast access
    *   Rows that were already opened do not require an active command and can skip this step
3. During the next phase, the DDR controller will select the same column on all 8 chips. This is the column access phase

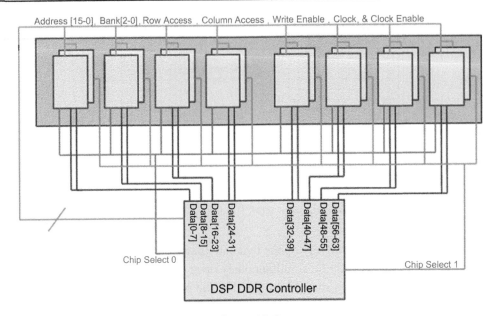

Address [15-0] , Bank[2-0] , Row Access , Column Access , Write Enable , Clock, & Clock Enable

Data[0-7]
Data[8-15]
Data[16-23]
Data[24-31]
Data[32-39]
Data[40-47]
Data[48-55]
Data[56-63]

Chip Select 0

Chip Select 1

DSP DDR Controller

Figure 13-8:
Simplified view of a DDR controller to memory connection: 2 chip selects.

4. Finally, the DDR controller will write the 64 bytes to the now open row buffers for each of the 8 separate DDR chips which each input 8 bits.

As there is a command to open rows, there is also one to close rows, called **PRECHARGE**, which tells the DDR modules to store the data from the row buffers back to the actual DDR memory in the chip, thus freeing up the row buffer. So when switching from one row to the next in a single DDR bank, we have to PRECHARGE the open row to close it, and then ACTIVATE the row we wish to start accessing.

A side effect of an ACTIVATE command is that the memory is automatically read and written, thus REFRESHing it. If a row in DDR is PRECHARGED, then it must be periodically refreshed (read/re-written with the same data) to keep data valid. DDR controllers have an autorefresh mechanism that does this for the programmer.

DDR data flow optimization for power

Now that the basics of DDR accesses have been covered, we can cover how DDR accesses can be optimized for minimal power consumption. As is often the case, optimizing for minimal power consumption is beneficial for performance as well.

The components of DDR power consumption are explained in [4]. DDR consumes power in all states, even when the CKE (clock enable – enabling the DDR to perform any operations) is disabled, though this is minimal. One technique to minimize DDR power consumption is made available by some DDR controllers which have a power saving mode that de-asserts the CKE pin — greatly reducing power. The Freescale DSP devices, including the MSC8156, call this mode Dynamic Power Management Mode, which can be enabled via the DDR_SDRAM_CFG[DYN_PWR] register. This feature will de-assert CKE when no memory refreshes or accesses are scheduled. If the DDR memory has self-refresh capabilities, then this power saving mode can be prolonged as refreshes are not required from the DDR controller.

This power saving mode does impact performance some, as enabling CKE when a new access is scheduled adds a latency delay.

Micron's DDR power calculator can be used to estimate power consumption for DDR. If we choose 1GB x 8 DDR chips with -125 speed grade, and we can see estimates for the main power consuming actions on DDR. Power consumption for non-idle operations is additive, so total power is the idle power plus non-idle operations.

- Idle with no rows open and CKE low is shown as: 4.3 mW (IDD2p)
- Idle with no rows open and CKE high is shown as: 24.6 mW (IDD2n)
- Idle with rows open and no CKE low is shown as: 9.9 mW (IDD3p)
- Idle with rows open and CKE high is shown as: 57.3 mW (IDD3n)
- ACTIVATE and PRECHARGE is shown as consuming 231.9 mW
- REFRESH is shown as 3.9 mW
- WRITE is shown as 46.8 mW
- READ is shown as 70.9 mW

We can see that using the Dynamic Power Management mode saves up to 32 mW of power, which is quite substantial in the context of DDR usage.

Also, it is clear that the software engineer must do whatever is possible to minimize contributions to power from the main power contributors: ACTIVATE, PRECHARGE, READ, and WRITE operations.

The power consumption from row activation/precharge is expected as DDR needs to consume a considerable amount of power in decoding the actual ACTIVATE instruction and address followed by transferring data from the memory array into the row buffer. Likewise, the PRECHARGE command also consumes a significant amount of power in writing data back to the memory array from row buffers.

Optimizing power by timing

One can minimize the maximum 'average power' consumed by ACTIVATE commands over time by altering the timing between row activate commands, t_{RC} (a setting the programmer can set at start up for the DDR controller). By extending the time required between DDR row activates, the maximum power spike of activates is spread, so the amount of power pulled by the DDR in a given period of time is lessened, though the total power for a certain number of accesses will remain the same. The important thing to note here is that this can help with limiting the maximum (worst case) power seen by the device, which can be helpful when having to work within the confines of a certain hardware limitation (power supply, limited decoupling capacitance to DDR supplies on the board, etc.).

Optimizing with interleaving

Now that we understand our main enemy in power consumption on DDR is the activate/precharge commands (for both power and performance), we can devise plans to minimize the need for such commands. There are a number of things to look at here, the first being address interleaving, which will reduce ACTIVATE/PRECHARGE command pairs via interleaving chip selects (physical banks) and additionally by interleaving logical banks.

In setting up the address space for the DDR controller, the row bits and chip select and bank select bits may be swapped to enable DDR interleaving, whereby changing the higher order address enables the DDR controller to stay on the same page while changing chip selects (physical banks) and then changing logical banks before changing rows. The software programmer can enable this in the MSC8156 DSP by enabling the BA_INTLV_CTL bits of the DDR_SDRAM_CFG register. One interleaving by physical and logical bank is enabled, the core-to-DDR bit addressing appears as shown in Figure 13-9 below.

Row x Col		MSB	Address from Core Initiator																											LSB
Row x Col		31	30	29	28	27	26	25	24	23	22	21	20	19	18	17	16	15	14	13	12	11	10	9	8	7	6	5	4	3	2–0
15 x 10 x3	MRAS	14	13	12	11	10	9	8	7	6	5	4	3	2	1	0	CS														
	MBA																SEL														
	MCAS																														
14 x 10 x3	MRAS		13	12	11	10	9	8	7	6	5	4	3	2	1	0	CS														
	MBA																SEL	2	1	0											
	MCAS																				9	8	7	6	5	4	3	2	1	0	
14 x 10 x2	MRAS			13	12	11	10	9	8	7	6	5	4	3	2	1	0	CS													
	MBA																	SEL	1	0											
	MCAS																				9	8	7	6	5	4	3	2	1	0	
13 x 10 x3	MRAS			12	11	10	9	8	7	6	5	4	3	2	1	0	CS														
	MBA																SEL	2	1	0											
	MCAS																				9	8	7	6	5	4	3	2	1	0	
13 x 10 x2	MRAS				12	11	10	9	8	7	6	5	4	3	2	1	0	CS													
	MBA																	SEL	1	0											
	MCAS																				9	8	7	6	5	4	3	2	1	0	

Figure 13-9:
64-bit DDR memory with chip select and logical bank interleaving.

By interleaving this way, once the 12 bits of column (and LSB) address space are used, then we will move to the next logical bank to start accessing (without necessarily requiring a PRECHARGE/ACTIVATE). And 15 bits of address space are available using different chip selects if there are multiple chip selects available on the specific board's memory layout.

Optimizing memory software data organization

We also need to consider the layout of our memory structures within DDR. In the case of using large ping-pong buffers for example, the buffers may be organized so that each buffer is in its own logical bank. This way, if DDR is not interleaved, we still can avoid unnecessary ACTIVATE/PRECHARGE pairs in the case that a pair of buffers is larger than a single row (page).

Optimizing general DDR configuration

There are other features available to the programmer, which can positively or negatively affect power, including 'open/closed' page mode. Closed page mode is a feature available in some controllers which will perform an auto-precharge on a row after each read or write access. This of course unnecessarily increases the power consumption in DDR as a programmer may need to access the same row 10 times; for example, closed page mode would yield at least 9 unneeded PRECHARGE / ACTIVATE command pairs. In the example DDR layout discussed above, this could consume an extra $2087.1 \, mW$ ($231.9 \, mW \times 9$).

As you may expect, this has an equally negative effect on performance due to the stall incurred during memory PRECHARGE and ACTIVATE.

Optimizing DDR burst accesses

DDR technology has become more restrictive with each generation: DDR2 allows 4 beat burst and 8 beat bursts, whereas DDR3 only allows 8. This means that DDR3 will treat all burst lengths as 8 beat (bursts of 8 accesses long). So for the 8 byte (64 bit) wide DDR accesses we have been discussing here, accesses are expected to be 8 beats of 8 bytes, or 64 bytes long.

If accesses are not 64 bytes wide, there will be stalls due to the hardware design. This means that if the DDR memory is accessed for only reading (or writing 32 bytes of data at a time, DDR will only be running at 50% efficiency, as the hardware will still perform reads/writes for the full 8 beat burst, though only 32 bytes will be used. Because DDR3 operates this way, the same amount of power is consumed whether doing 32 byte or 64 byte long bursts to our memory here. So for the same amount of data, if doing 4 beat (32 byte) bursts, the DDR3 would consume approximately twice the power.

The recommendation here then is to fill all accesses to DDR to be full 8 beat bursts in order to maximize power efficiency. To do this, the programmer must be sure to **pack data in the DDR so that accesses to the DDR are in at least 64 byte wide chunks**. The concept of data packing can be used to reduce the amount of used memory as well. For example, packing 8 single bit variables into a single character reduces memory footprint and increases the amount of usable data the core or cache can read in with a single burst.

In addition to data packing, **accesses need to be 8 byte aligned** (or aligned to the burst length). If an access is not aligned to the burst length, for example, if on the MSC8156, an 8 byte access starts with a 4 byte offset, both the first and second access will effectively become 4-beat bursts, reducing bandwidth utilization to 50% (instead of aligning to the 64 byte boundary and reading data in with 1 single burst).

SRAM and cache data flow optimization for power

Another optimization related to the usage of off chip DDR is avoidance: avoiding using external off chip memory, and maximizing accesses to internal on-chip memory saves the additive power draw that occurs when activating not only internal device busses and clocks, but also off chip busses, memories arrays, etc.

High speed memory close to the DSP processor core is typically SRAM memory, whether it functions in the form of cache or as a local on-chip memory. SRAM differs from SDRAM in a number of ways (such as no ACTIVATE/PRECHARGE, and no concept of REFRESH), but some of the principles of saving power still apply, such as pipelining accesses to memory via data packing and memory alignment.

The general rule for SRAM access optimization is that accesses should be optimized for higher performance. The fewer clock cycles the device spends doing a memory operation = less time that memory, busses, and core are all activated for said memory operation.

SRAM (all memory) and code size

As programmers, we can effect this in both program and data organization. Programs may be optimized for minimal code size (by a compiler, or by hand), in order to consume a minimal amount of space. Smaller programs require less memory to be activated to read the program. This applies not only to SRAM, but also DDR and any type of memory — the less memory that has to be accessed implies a lower amount of power drawn.

Aside from optimizing code using the compiler tools, other techniques such as instruction packing, which are available in architectures like the SC3850, enable fitting maximum code into a minimum set of space. The VLES (Variable Length Execution Set) instruction architecture allows the program to pack multiple instructions of varying sizes into a single execution set. As execution sets are not required to be 128 bit aligned, instructions can be packed tightly, and the SC3850 prefetch, fetch, and instruction dispatch hardware will handle reading the instructions and identifying start and end of each instruction set (via instruction prefix encodings prepended in machine code by the StarCore assembler tools).

Additionally, size can be saved in code by creating functions for common tasks. If tasks are similar, consider use the same function with parameters passed that determine the variation to run instead of duplicating the code in software multiple times.

Be sure to make use of combined functions where available in the hardware. In the Freescale StarCore architecture, using a Multiply Accumulate (MAC) instruction, which takes 1 pipelined cycle, saves space and performance in addition to power over using separate multiple and add instructions.

Some hardware provides code compression at compile time and decompression on the fly, so this may be an option depending on the hardware the user is dealing with. The problem with this strategy is related to the size of compression blocks. If data is compressed into small blocks, then not as much compression optimization is possible, but this is still desirable over the alternative. During decompression, if code contains many branches or jumps, the processor will end up wasting bandwidth, cycles, and power decompressing larger blocks that are hardly used.

The problem with the general strategy of minimizing code size is the inherent conflict between optimizing for performance and space. Optimizing for performance generally does not always yield the smallest program, so determining ideal code size vs. cycle performance in order to minimize power consumption requires some balancing and profiling. The general advice here is to use what tricks are available to minimize code size without hurting the performance of a program that meets real time requirements. The 80/20 rule of applying performance optimization to the 20% of code that performs 80% of the work, while optimizing the remaining 80% of code for size is a good practice to follow.

SRAM power consumption and parallelization

It is also advisable to optimize data accesses in order to reduce the cycles in which SRAM is activated, pipelining accesses to memory, and organizing data so that it may be accessed consecutively. In systems like the MSC8156, the core / L1 caches connect to the M2 memory via a 128-bit wide bus. If data is organized properly, this means that 128 bit data accesses from M2 SRAM could be performed in one clock cycle each, which would obviously be beneficial when compared to doing 16 independent 8 bit accesses to M2 in terms of performance and power consumption.

An example showing how one may use move instructions to write 128 bits of data back to memory in a single instruction set (VLES) is provided below:

```
[
MOVERH.4F d0:d1:d2:d3,(r4)+n0
MOVERL.4F d4:d5:d6:d7,(r5)+n0
]
```

We can parallelize memory accesses in a single instruction (as with the above where both of the moves are performed in parallel), and even if the accesses are to separate memories or memory banks, the single cycle access still consumes less than the power of doing two independent instructions in two cycles.

Another note: as with DDR, SRAM accesses need to be aligned to the bus width in order to make full use of the bus.

Data transitions and power consumption

SRAM power consumption may also be affected by the TYPE of data used in an application. Power consumption is affected by the number of data transitions (from 0's to 1's) in memory as well. This power effect trickles down to the DSP core processing elements as well, as found by Kojima, et al. [5]. Processing mathematical instructions using constants consumes less power at the core than with dynamic variables.

In many devices, because pre-charging memory to reference voltage is common practice in SRAM memories, power consumption is also proportional to the number of zeros as the memory is pre-charged to a high state.

Using this knowledge, it goes without saying that re-use of constants where possible, and avoiding zero-ing out memory unnecessarily will, in general, save the programmer some power.

Cache utilization and SoC memory layout

Cache usage can be thought of in the opposite manner to DDR usage when designing a program. An interesting detail about cache is: both dynamic and static power increase with increasing cache sizes; however, the increase in dynamic power is small. The increase in static power is significant, and becomes increasingly relevant for smaller feature sizes [6]. As the software programmer, we have no impact on the actual cache size available on a device, but when it is provided, based on the above, it is our duty to use as much of it as possible!!!

For SoC level memory configuration and layout, optimizing the most heavily used routines and placing them in the closest cache to the core processors will offer not only the best performance, but also better power consumption.

Explanation of Locality

The reason the above is true is thanks to the way caches work. There are a number of different cache architectures, but they all take advantage of the principle of locality. The principle of locality basically states that if one memory address is accessed, the probability of an address nearby being accessed soon is relatively high. Based on this, when a cache miss occurs (when the core tries to access memory that has not been brought into the cache), the cache will read the requested data in from higher level memory one line at a time. This means that if the core tries to read a 1 byte character from cache, and the data is not in the cache, then there is a miss at this address. When the cache goes to higher level memory (whether it be on-chip memory or external DDR, etc.), it will not read in an 8 bit character, but rather a full cache line. If our cache uses cache sizes of 256 bytes, then a miss will read in our 1 byte character, along with 255 more bytes that happen to be on the same line in memory.

This is very effective in reducing power if used in the right way. If we are reading an array of characters aligned to the cache line size, once we get a miss on the first element, although we pay a penalty in power and performance for cache to read in the first line of data, the remaining 255 bytes of this array will be in cache. When handling image or video samples, a single frame would typically be stored this way, in a large array of data. When performing compression or decompression on the frame, the entire frame will be accessed in a short period of time, thus it is spatially and temporally local.

In the case of the MSC8156, there are two levels of cache for each of the 6 DSP processor cores: L1 cache (which consists of 32 KB of instruction and 32 KB of data cache), and a 512 KB L2 memory which can be configured as L2 cache, or M2 memory. At the SoC level, there is a 1 MB memory shared by all cores called M3. L1 cache runs at the core processor speed (1 GHz), L2 cache effectively manages data at the same speed (double the bus width, half the frequency), and M3 runs at up to 400 MHz. The easiest way to make use of the memory heirarchy is to enable L2 as cache and make use of data locality. As discussed above, this works when data stored with high locality. Another option is to DMA data into L2 memory (configured in non-cache mode). We will discuss DMA in a later section.

When we have a large chunk of data stored in M3 or in DDR, the MSC8156 can draw this data in through the caches simultaneously. L1 and L2 caches are linked, so a miss from L1 will pull 256 bytes of data in from L2, and a miss from L2 will pull data in at 64 bytes at a time (64B line size) from the requested higher level memory (M3 or DDR). Using L2 cache has two advantages over doing directly to M3 or DDR. First, it is running at effectively the same speed as L1 (though there is a slight stall latency here, it is negligible), and second, in addition to being local and fast, it can be up to 16 times larger than L1 cache, allowing us to keep much more data in local memory than just L1 alone would.

Explanation of Set-Associativity

All caches in the MSC8156 are 8 way set-associative. This means that the caches are split into 8 different sections ('ways'). Each section is used to access higher level memory, meaning that a single address in M3 could be stored in one of 8 different sections (ways) of L2 cache for example. The easiest way to think of this is that the section (way) of cache can be overlaid onto the higher level memory x times. So, if L2 is set up as all cache, the following equation calculates how many times each set of L2 memory is overlaid onto M3:

$$\# \text{ of overlays } O = \frac{M3 \; size}{(L2 \; size/8 \; ways)}$$

$$= \frac{1 \; MB}{(512 \; KB/8)} = 16{,}384 \; overlays$$

In the MSC8156, a single way of L2 cache is 64 KB in size, so addresses are from 0x0000_0000 to 0x0001_0000 hexidecimal. If we consider each way of cache individually, we can explain how a single way of L2 is mapped to M3 memory. M3 addresses start at 0xC000_0000. So M3 addresses 0xC000_0000, 0xC001_0000, 0xC002_0000, 0xC003_0000, 0xC004_0000, etc. (up to 16K times) all map to the same line of a way of cache. So if way #1 of L2 cache has valid data for M3's 0xC000_0000, and the core processor wants to next access 0xC001_0000, what is going to happen?

If the cache has only 1 way set associativity, then the line of cache containing 0xC000_0000 will have to be flushed back to cache and re-used in order to cache 0xC001_0000. In an 8 way set associative cache, however, we can take advantage of the other 7 × 64 KB sections 'ways' of cache. So we can potentially have 0xC000_0000 stored in way #1, and the other 7 ways of cache have their first line of cache as empty. In this case, we can store our new memory access to 0xC001_0000 in way #2.

So, what happens when there is an access to 0xC000_0040? (0x40 == 64B). The answer here is that we have to look at the 2nd cache line in each way of L2 to see if it is empty, as we were only considering the 1st line of cache in our example above. so here we now have 8 more potential places to store a line of data (or program).

Figure 13-10 shows a 4 way set associative cache connecting to M3. In this figure, we can see that every line of M3 maps to 4 possible lines of the cache (one for each way). So line 0xC000_0040 maps to the 2nd line (second 'set') of each way in the cache. So when the core wants to read 0xC000_0040, but the first way has 0xC000_0100 in it, the cache can load the cores request into any of the other three ways if their 2nd line is empty (invalid).

The reason for discussing set associativity of caches is that it does have some effect on power consumption (as one might imagine). The goal for maximizing power consumption (and

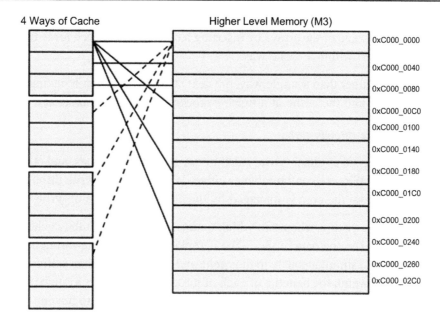

Figure 13-10:
Set Associativity by Cache Line: 4 way set associative cache.

performance) when using cache is to maximize the hit rate in order to minimize accesses to external busses and hardware caused by misses. Set-associativity is normally already determined by hardware, but in the case where the programmer can change set associativity, set-associative caches maintain a higher hit-rate than directly mapped caches, and thus draw lower power.

Memory layout for cache

While having an 8 way set associative architecture is statistically beneficial in improving hit ratio and power consumption, the software programmer may also directly improve hit ratio in the cache, and thus lower power by avoiding conflicts in cache. Conflicts in cache occur when the core needs data that will replace cache lines with currently valid data that will be needed again.

We can organize memory in order to avoid these conflicts in a few different ways. For memory segments we need simultaneously, it is important to pay attention to the size of ways in the cache. In our 8 way L2 cache, each way is 64 KB. As we discussed before, we can simultaneously load 8 cache lines with the same lower 16 bits of address (0x0000_xxxx).

Another example is if we are working with 9 arrays with 64 KB of data simultaneously. If we organize each array contiguously, data will be constantly thrashed, as all arrays share the same 64 KB offset. If the same indices of each array are being accessed simultaneously, we can offset the start of some of the arrays by inserting a buffer, so that each array does not map to the same offset (set) within a cache way.

When data sizes are larger than a single way, the next step is to consider reducing the amount of data that is pulled into the cache at a time — processing smaller chunks at a time.

Write back versus write through caches

Some caches are designed as either 'write back' or 'write through' caches, and others, such as the MSC815x series DSPs are configurable as either. Write back and write through buffering differs in how data from the core is managed by the cache in the case of writes.

Write back is a cache writing scheme in which data is written only to the cache. The main memory is updated when the data in the cache is replaced. In the write-through cache write scheme, data is written simultaneously to the cache and to memory. When setting up cache in software, we have to weigh the benefits of each of these. In a multicore system, coherency is of some concern, but so is performance, and power. Coherency refers to how up-to-date data in main memory is compared to the caches. The greatest level of multicore coherency between internal core caches and system level memory is attained by using write-through caching, as every write to cache will immediately be written back to system memory, keeping it up to date. There are a number of down sides to write-through caching including:

- Core stalls during writes to higher level memory
- Increased bus traffic on the system busses (higher chance for contention and system level stalls)
- Increased power consumption as the higher level memories and busses are activated for every single memory write

The write-back cache scheme on the other hand, will avoid all of the above disadvantages at the cost of system level coherency. For optimal power consumption, a common approach is to use the cache in write-back mode, and strategically flush cache lines/segments when the system needs to be updated with new data.

Cache coherency functions

In addition to write-back and write-through schemes, specific cache commands should also be considered. Commands include:

- Invalidation sweep: invalidating a line of data by clearing valid and dirty bits (effectively just re-labeling a line of cache as 'empty')
- Synchronization sweep: writing any new data back to cache and removing the dirty label
- Flush sweep: writing any new data back to cache and invalidating the line
- Fetch: fetch data into the cache

Generally these operations can be performed either by cache line or a segment of the cache, or as a global operation. When it is possible to predict that a large chunk of data will be needed in the cache in the near future, performing cache sweep functions on larger segments will make better use of the full bus bandwidths and lead to fewer stalls by the core. Memory accesses all require some initial memory access set up time, but after set up, bursts will flow at full bandwidth. Because of this, making use of large prefetches will save power when compared to reading in the same amount of data line by line. Still, using large prefetches should be done strategically so as to avoid having data thrashed before the core actually gets to use it.

When using any of these instructions, we have to be careful about the effect it has on the rest of the cache. For instance, performing a fetch from higher level memory into cache may require replacing contents currently in the cache. This could result in thrashing data in the cache and invalidating cache in order to make space for the data being fetched.

Compiler cache optimizations

In order to assist with the above, compilers may be used to optimizing cache power consumption by re-organizing memory or memory accesses for us. Two main techniques available are array merging and loop interchanging, explained below courtesy of [1].

Array merging organizes memory so that arrays accessed simultaneously will be at different offsets (different 'sets') from the start of a way. Consider the following two array declarations:

```
int array1[ array_size ];
int array2[ array_size ];
```

The compiler can merge these two arrays as shown below:

```
struct merged_arrays
{
  int array1;
  int array2;
} new_array[ array_ size ]
```

In order to re-order the way that high level memory is read into cache, reading in smaller chunks to reduce the chance of thrashing loop interchanging can be used. Consider the code below:

```
for (i=0; i<100; i=i+1)
  for (j=0; j<200; j=j+1)
    for (k=0; k<10000; k=k+1)
      z[ k ][ j ] = 10 * z[ k ][ j ];
```

By **interchanging** the second and third nested loops, the compiler can produce the following code, decreasing the likelihood of unnecessary thrashing during the innermost loop.

```
for (i=0; i<100; i=i+1)
  for (k=0; k<10000; k=k+1)
    for (j=0; j<200; j=j+1)
      z[ k ][ j ] = 10 * z[ k ][ j ];
```

Peripheral/communication utilization

When considering reading and writing of data, of course, we cannot just think about memory access: we need to pull data in and off from the device as well. As such, for the final portion of data path optimization we will look at how to minimize power consumption in commonly used DSP (I/O) peripherals.

Things to consider include the peripheral's burst size, speed grade, transfer width, and general communication modes. Main standard forms of peripheral communication for DSPs include DMA (direct memory access), SRIO (Serial Rapid I/O), Ethernet, PCI Express, and RF antenna interfaces. I2C and UART are also commonly used, though mostly for initialization and debug purposes.

The fact that communication interfaces usually require their own PLLs / clocks increases the individual power consumption impact. The higher clocked peripherals that we need to consider as the main power consumers are the DMA, SRIO, Ethernet, and PCI Express. Clock gating and peripheral low power modes for these peripherals have already been discussed in the Low Power Modes section of this paper, so this section will talk about how to optimize actual usage.

Although each protocol is different for the I/O peripherals and the internal DMA, they all share the fact that they are used to read/write data. As such, one basic goal is to maximize the throughput while the peripheral is active in order to maximize efficiency and the time the peripheral / device can be in a low power state, thus minimizing the active clock times.

The most basic way to do this is to increase transfer / burst size. For DMA, the programmer has control over burst size and transfer size in addition to start / end address (and can follow alignment and memory accessing rules we discussed earlier in the subsections on Data Path Optimization). Using the DMA, the programmer can not only decide the alignment, but also the transfer 'shape,' for lack of a better word. What this means is that, using the DMA, the programmer can transfer blocks in the form of 2-dimensional, 3-dimensional, and 4-dimensional data chunks, thus transferring data types specific to specific applications on the alignment chosen by the programmer without spending cycles transferring unnecessary data. Figure 13-11 demonstrates the data structure of a 3-dimensional DMA.

The user programs the start address for data, the length for the first dimension, the offset for the second dimension, and the number of transfers, followed by the offset for the 3rd dimension and the number of transfers. At the end of the all transfers, the programmer may

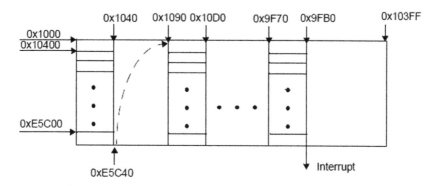

Figure 13-11:
Three-dimensional DMA data format.

also program the DMA to interrupt the core to signal data transfer completion. Having the DMA intelligently moving the data in the format and organization needed by the user's application helps optimize data flow and core processing by avoiding the need for the core to re-organize data or alter algorithms that are optimized for data in a specific format. This also simplifies the maintaining of certain alignments as the programmer can decide where each dimension of the data structure starts.

Other high speed peripherals generally will also use a DMA, whether it be the system DMA, or the peripheral's own private DMA for data passing. In the case of the MSC8156, the SRIO, PCI Express, and Ethernet controllers all have their own DMAs separate from the system DMA for data transfers. The basics still apply here: we want data transfers to be long (long bursts), we want bus accesses to be aligned, and additionally, one more thing we want is optimal access to the system bus! We will discuss system bus optimization later in this section.

DMA of data versus CPU

While on the topic of DMA, we need to consider whether the core should move data from internal core memory or if a DMA should be utilized in order to save power. As the DMA hardware is optimized solely for the purpose of moving data, it will move data while consuming less power than the core (which is generally running at much higher frequencies than the DMA). As the core runs at such a higher frequency, and is not intended solely for data movement, etc., the core utilizes more dynamic power while incurring heavy stall penalties when accessing external memory.

As some external memory access and stalls are incurred for writing to peripheral registers when setting up the DMA, there is a point where data accesses are too small or infrequent to justify DMA usage. In general, when moving larger chunks of data, or data in a predictable manner, for the purpose of power consumption (and core performance), DMA should be utilized for maximum power savings and application efficiency.

For transactions and I/O that are not large enough to justify DMA, we can consider caching as this assists with stalls and requires virtually no core intervention. Generally speaking, using the cache is much simpler than DMA, so this is the generally accepted solution for unpredictable data I/O, while DMA should be used for larger memory transfers. Due to the overhead for programming the DMA, and the unique properties of data per application, the trade-off between power savings, performance, and program complexity from DMA to cache has to be done on a case by case basis. Peripherals with their own DMA generally require the programmer to use that DMA for peripheral interaction, which is a good habit to force the programmer into, as we have just discussed.

Coprocessors

Just as the DMA peripheral is optimized for data movement and can do so more efficiently with less power consumption than the high frequency DSP core, so are other peripherals acting as coprocessors able to perform special functions more efficiently than the DSP core. In the case of the MSC8156, the MAPLE baseband coprocessor includes hardware for Fast Fourier Transforms, Discrete Fourier Transforms, and Turbo Viterbi. When a chain of transforms can be offloaded onto the MAPLE, depending on the cost of transferring data and transform sizes, the system is able to save power and cycles by offloading the core and having the MAPLE coprocessor do this work as the MAPLE is running at a much lower frequency than the core, has fewer processing elements aimed at a single function, and also has automatic lower power modes that are used when the MAPLE is not processing transforms, etc.

System bus configuration

System bus stalls due to lack of priority on the bus can cause a peripheral to actively wait unnecessarily for extra cycles when not set up properly. These extra active wait cycles mean more wasted power. Generally DSPs will have system bus configuration registers which allow the programmer to configure the priority and arbitration per bus initiator port. In the case of the MSC8156, the system bus (called the CLASS) has 11 initiator ports and 8 target ports (memory and register space). When the programmer understands the I/O needs of the application, it is possible to set up priority appropriately for the initiators that need the extra bandwidth on the bus so they can access memory and register space with minimal stalls.

There is not much of a trick to this; simply set priority based on I/O usage. Some devices such as the MSC815x series DSPs provide bus profiling tools which enable the programmer to count the number of accesses per initiator port to each target. This allows the programmer to see where congestion and bottlenecks are occurring in order to appropriately configure and tune the bus. Profiling tools also allow the programmer to see how many 'priority upgrades' are needed per port. This means that the programmer can temporarily assign test priorities to each port, and if some ports are constantly requiring priority upgrades, the programmer can decide to set the starting priority of these ports one level up and re-profile.

Peripheral speed grades and bus width

Like the case with the system bus access, the peripheral's external interface should be set up according to the actual needs of the system. The catch-22 of I/O peripherals is that some peripherals require being powered on all the time (so minimal to no use of low power modes is available). If a communication port such as SRIO is dedicated to receiving incoming blocks of data to process, when no data is coming in, clocks and low power modes for the SRIO port are not an option. As such, there is a balancing game to be played here.

In testing software and power consumption, we found on the MSC8156 that running 4 lanes of SRIO at 3.125 GHz with 40% utilization (~4 Gbps of data) consumes a comparable amount, or even less power as running 4 lanes of SRIO at 2.5 GHz with 50% utilization (the same data throughput). So the user needs to test various cases or make use of the device manufacturer's power calculator in order to make an informed decision. In a case like this, peripherals which have an auto-idle feature should make use of the higher speed bus in order to maximize sleep times.

SRIO, PCI Express, Ethernet over SGMII, and some antenna interfaces make use of the same serial I/O hardware, so similar care should be taken here. All could be required to be held in an active mode as a form of 'device wake' or signaling to the DSP core, meaning they may be restricted from going into sleep modes. In the case of antenna signaling, this is especially detrimental as the active antenna's RF interface has to constantly consume power to emit signal. If possible, it is ideal to use an alternative method for waking the DSP core in order to enable idle and sleep modes on the antenna.

Peripheral to core communication

When considering device waking, and general peripheral to core I/O, we have to consider how the peripheral interacts with the core processors. How does the core know that data is available? How often is the core notified of data available? How does the core know when to send data over the peripheral? There are three main methods for managing this: polling, time based processing, and interrupt processing.

Polling is by far the least efficient method of core-peripheral interaction as it has the core constantly awake and burning through high frequency clock cycles (consuming active current) just to see if data is ready. The only positive of using this method arise when the programmer is not concerned about power consumption. In this case, polling enables the core to avoid context switches that occur during interrupt processing, thus saving some cycles in order to access data faster. Generally this is only used for testing maximum peripheral bandwidth as opposed to being used in a real application.

Time based processing works on the assumption that data will always be available at a certain interval. For example, if a DSP is processing a GSM voice codec (AMR, EFR, HR, etc.). The core will know that samples will be arriving every 20 ms, so the core can go look for the new audio samples on this time basis and not poll. This process allows the core to sleep, and use a timer for wake functionality, followed by performing data processing. The downside of this is the complexity and inflexibility of this model: setup and synchronization requires a lot of effort on the programmer's side, and the same effect can be achieved using simple interrupt processing.

Interrupt processing

The final core to peripheral communication mechanism is also the most commonly used one as it allows the benefits of the time based processing without the complicated software architecture. We also briefly discussed using interrupt processing in the Low Power Modes section as a method for waking the core from sleep states: when new data samples and packets come in for processing, the core is interrupted by the peripheral (and can be woken if in sleep state) to start processing new data. The peripheral can also be used to interrupt the core when the peripheral is ready to transfer new data, so that the core does not have to constantly poll a heavily loaded peripheral to see when data is ready to send.

Power consumption results for polling versus interrupt processing have already been shown in Figure 13-6 when comparing the Baseline MJPEG versus the Using WAIT for PD and Using STOP for PD modes. When not using WAIT and STOP modes, the application would constantly check for new buffers without taking advantage of massive idle times in the application.

Algorithmic optimization

Of the three main areas of power optimization discussed here, algorithmic optimization requires the most work for a given amount of power savings. Algorithmic optimization includes: optimization at the core application level, code structuring, data structuring (in some cases, this could be considered as data path optimization), data manipulation, and optimizing instruction selection.

Compiler optimization levels

In the data path section, we briefly discussed that the compiler can be used to optimize code for minimal size. The compiler may also be used to optimize code for maximum performance (utilizing the maximum number of processing units per cycle and minimizing the amount of time code is running). This is also discussed in [7], where the TI C6000 DSP is used to test whether optimizing for performance will reduce power consumption. As expected, the results show that increasing the number of processing units will increase the power consumed per cycle, but the total power to perform a function over time will reduce as the number of cycles to perform the function is reduced. The question of when to optimize for performance versus code size generally still fits with the 80/20 rule (80% of cycle time is spent in 20% of the code), so as mentioned in the data path section, the general rule here is to optimize the cycle hungry (20%) portion of code for performance, while focusing on minimizing code size for the rest. Fine tuning this is the job of the programmer and will require power measurement (as discussed in section 2 "Measuring Power Consumption"). The rest of this section will cover specific algorithmic optimizations, some of which may be performed by the performance optimizer in the compiler.

Instruction packing

Instruction packing was already listed in the data path optimization section above, but may also be listed as an algorithmic optimization as it involves not only how memory is accessed, but also how code is organized. Refer to the section, *SRAM and cache data flow optimization for power* for details on instruction packing.

Loop unrolling

We briefly discussed using altering loops in code in order to optimize cache utilization. Another method for optimizing both performance and power in DSP processors is via loop-unrolling. This method effectively partially unravels a loop, as shown in the code snippets below:

Regular loop:

```
for (i=0; i<100; i=i+1)
  for (k=0; k<10000; k=k+1)
    a[ i ]= 10 * b[ k ];
```

Loop unrolled by 4x:

```
for (i=0; i<100; i=i+4)
  for (k=0; k<10000; k=k+4)
    {
      a[ i ]= 10 * b[ k ];
      a[ i+1 ]= 10 * b[ k+1 ];
      a[ i+2 ]= 10 * b[ k+2 ];
      a[ i+3 ]= 10 * b[ k+3 ];
    }
```

Unrolling code in this manner enables the compiler to make use of 4 MACs (Multiply-Accumulates) in each loop iteration instead of just one, thus increasing processing parallelization and code efficiency (more processing per cycle means more idle cycles available for sleep and low power modes). In the above case, we increase the parallelization of the loop by four times, so we perform the same amount of MACs in ¼ the cycle time, thus the effective active clock time needed for this code is reduced by 4x. Measuring the power savings using the MSC8156, we find that the above example optimization (saving 25% cycle time by utilizing 4 MACs per cycle instead of one enables the core a ~48% total power savings over the time this routine is executed).

Completely unrolling loops is not advisable as it is counterproductive to code size minimization efforts we discussed in the data path section, which would lead to extra memory accesses and possibility of increased cache miss penalties.

Software pipelining

Another technique common to both DSP performance optimization and DSP power optimization is software pipelining. Software pipelining is a technique where the programmer splits up a set of interdependent instructions that would normally have to be performed one at a time so that the DSP core can begin processing multiple instructions in each cycle. Rather than explaining in words, the easiest way to follow this technique is to see an example.

Say we have the following code segment:

Regular loop:

```
for (i=0; i<100; i=i+1)
{
  a[ i ]= 10 * b[ i ];
  b[ i ]= 10 * c[ i ];
  c[ i ]= 10 * d[ i ];
}
```

Right now, although we have 3 instructions occurring per loop, the compiler will see that the first instruction depends on the second instruction, and thus could not be pipelined with the second, nor can the second instruction be pipelined with the third due to interdependence: a[i] cannot be set to b[i] as b[i] is simultaneously being set to c[i], and so on. So right now the DSP processor has to execute the above loop 100 times with each iteration performing 3 individual instructions per cycle (not very efficient), for a total of 300 cycles (best case) performed by MACs in the core of the loop. With software pipelining, we can optimize this in the following way.

First we see where we can parallelize the above code by unrolling the loop to some extent:

Unrolled loop:

```
a[ i ]= 10 * b[ i ];
b[ i ]= 10 * c[ i ];
```

```
c[ i ]= 10 * d[ i ];
 a[i+1]= 10 * b[i+1];
 b[i+1]= 10 * c[i+1];
 c[i+1]= 10 * d[i+1];
   a[i+2]= 10 * b[i+2];
   b[i+2]= 10 * c[i+2];
   c[i+2]= 10 * d[i+2];
     a[i+3]= 10 * b[i+3];
     b[i+3]= 10 * c[i+3];
     c[i+3]= 10 * d[i+3];
```

Using the above, we can see that certain instructions are not interdependent. The first assignment of array 'a' relies on the original array 'b', meaning we can potentially assign a entirely before doing any other instructions. If we do this, this means that array 'b' would be entirely free of dependence and could be completely assigned to the original array 'c'. We can abstract this for 'c' as well.

We can use this idea to break the code apart and add parallelism by placing instructions together that can run in parallel when doing some assignment in advance.

First, we have to perform our first instruction (no parallelism):

```
a[ i ]= 10 * b[ i ];
```

Then we can have two instructions performed in one cycle:

```
b[ i ]= 10 * c[ i ];
a[i+1]= 10 * b[i+1];
```

Here we see that the first and second lines do not depend on each other, so there is no problem with running the above in parallel as one execution set.

Finally, we reach the point where three instructions in our loop are all being performed in one cycle:

```
c[ i ]= 10 * d[ i ];
b[i+1]= 10 * c[i+1];
a[i+2]= 10 * b[i+2];
```

Now we see how to parallelize the loop and pipeline; the final software pipelined will first have some 'setup,' also known as loading the pipeline. This consists of the first sets of instructions we performed above. After this we have our pipelined loop:

```
//pipeline loading — first stage
a[ i ]= 10 * b[ i ];
//pipeline loading — second stage
b[ i ]= 10 * c[ i ];
a[i+1]= 10 * b[i+1];
//pipelined loop
for (i=0; i<100-2; i=i+1)
```

```
{
  c[ i ]= 10 * d[ i ];
  b[i+1]= 10 * c[i+1];
  a[i+2]= 10 * b[i+2];
}
//after this, we still have 2 more partial loops:
c[i+1]= 10 * d[i+1];
b[i+2]= 10 * c[i+2];
//final partial iteration
c[i+2]= 10 * d[i+2];
```

By pipelining the loop, we enabled the compiler to reduce the number of cycles for MACs from 300 to:

- 3 MACs that can be performed in 2 cycles for pipeline loading
- 100 cycles (3 MACs each) in the core of our loop
- 3 MACs that can be performed in 2 cycles for pipeline loading

for a total of 104 cycles or roughly $\frac{1}{3}$ of the execution time, thus reducing the amount of time the core clocks must be active by 3x for the same functionality! Similar to the loop unrolling case, the pipelining case has enabled us to save substantially: ~43% total power over the time this routine is executed.

Eliminating recursion

An interesting technique suggested by Wolf [8] is that we want to eliminate recursive procedure calls in order to reduce function call overhead.

Recursive procedure calls require the functions' general context, etc. to be pushed onto the stack with each call. So in the classic case of the factorial example (n!), this can be calculated using recursion with a function as follows:

$fn!(0) = 1$ For n==0

$(n) = fn!(n-1)$ For n > 0

If this recursive factorial function is called with n = 100, there would be ~100 function calls entailing 100 branch to subroutines (which are change of flow routines which affect the program counter and software stack). Each change of flow instruction takes longer to execute because not only is the core pipeline disrupted during execution, but every branch adds at least a return address to the call stack. Additionally, in the case that multiple variables are being passed, this also must be pushed onto the stack.

This means that this recursive subroutine will require 100 times the amount of individual writes to memory and related stalls (as writes/reads to memory will not be pipelined), and also 100 additional pipeline stalls due to change of flows.

We can optimize this by moving to a simple loop:

```
int res=1;
for(int i=0; i<n; i++)
{
  res*=i;
}
```

This function requires no actual writes to the stack/physical memory as there are not any function calls/jumps. As this function only involves a single multiply, it qualifies as a 'short loop' on the MSC814x and MSC815x devices, whereby the loop is entirely handled in hardware. Thanks to this feature, there are no change of flow penalties, and no loop overhead either, so this effectively acts like a completely unrolled loop of multiplies (minus the memory cost).

Compared to the recursive routine, using the loop for *100 factorial* saves approximately:

- 100 change of flows (pipeline cycle penalties)
- 100+ pushes to the stack (100x memory accesses)

For the above example, avoiding recursion savings can be estimated as follows.

The loop method's change of flow savings from avoiding pipeline disruptions depend on the pipeline length and if there is any branch prediction available in the core hardware. In the case of the MSC8156's 12 stage pipeline, refilling it would potentially be a 12 cycle penalty. As branch target prediction is available on the MSC8156, this may reduce some of this penalty significantly, but not completely. We can multiply the estimated stall penalty by the factorial (iteration), which will indicate the additional active core clock cycles and thus active power consumption from the DSP core due to recursion.

The cost of recursion causing 100+ individual stack accesses is great, as even in internal device memory there is potentially an initial delay penalty for initial access. As these stack accesses will not be pipelined, initial memory delay is multiplied by the number of recursive calls. If we assume stack is stored low latency internal memory running at the core speed, initial latency still could be seen on the order of anywhere from 8 to 20 cycles. A 10 cycle latency for initial access would not be a problem if subsequent accesses were pipelined, meaning 100 reads have a total core stall time of 10 cycles, but in the case of recursion, we have non-pipelined accesses and thus 10×100 stalls, or 1000 additional core cycles of active clock consuming power.

In the above example, removing recursion and moving to loops reduces the total energy (power over time) consumed by the processor to complete the factorial function to less than half.

Reducing accuracy

Robert Oshana brings up an interesting point in his article [9] noting that often programmers will over-calculate mathematical functions, using too much accuracy (too much precision), which can lead to more complicated programs requiring the use of more functional units and more cycles.

In the case that 16-bit integers could be used, as the signal processing application is able to tolerate more noise, but 32-bit integers are used instead, this could cause additional cycles for just a basic multiply. A 16-bit by 16-bit multiply can be completed in 1 cycle on most DSP architectures, but a 32-bit by 32-bit may require more. Such is the case for the SC3400 DSP core; this requires two cycles instead of one, so the programmer is doubling the cycle time for the operation needlessly (inefficient processing and additional clock cycles where the core is consuming active dynamic power).

Low-power code sequences and data patterns

Another suggestion from Oshana's article is to look at the specific instructions used for an operation or algorithm. The programmer may be able to perform exactly the same function with different commands while saving power, though the analysis and work to do this is very time consuming and detail oriented.

Different instructions activate different functional units, and thus different power requirements. To accurately use this, it requires the programmer to profile equivalent instructions to understand the power tradeoffs.

Obvious examples could be using a MAC when only the multiply functionality is needed. Less obvious comparisons, such as the power consumption between using a subtraction to clear a register versus the actual clear instruction, require the programmer to profile power consumption for each instruction, as we may not know internally how the hardware goes about clearing a register.

Summary and closing remarks

In order to provide the reader with tools to optimize software for power, over thirty different optimization techniques in the areas of low power modes, current and voltage controls, memory optimization, data path optimization, and algorithmic strategies have been discussed. A summary of those techniques is provided in Table 13-1.

In the process of explaining how to optimize software to minimize power consumption we have also covered some basic details about the DSP devices from Freescale and Texas Instruments, and provided background into how low power modes work, how DDR and caches work, and how the compiler works in order to assist the programmer.

Table 13-1: Summary of power optimization techniques for DSP.

Category	Technique	Impact
Hardware Support	Power Gating: Via a VRM or processor supported interface, switch off current to specific logic or peripherals of the device	HIGH
Hardware Support	Clock Gating: Often provided as device low power modes maximize the amount of clocks that can be shut down for an application	HIGH
Hardware Support	Voltage and Clock scaling: where available, reduce frequency and voltage	Processor dependent
Hardware Support	Peripheral Low Power modes: gating power/clock to peripherals	Medium-High
Data Flow	DDR Optimizing Timing: Increasing timing between ACTIVATE commands	Low
Data Flow	DDR Interleaving: used to reduce PRECHARGE/ACTIVATE combinations	HIGH
Data Flow	DDR optimization of software organization: organizing buffers to fit into logical banks to avoid PRECHARGE/ ACTIVATE commands	Medium
Data Flow	DDR General Configuration: avoid using modes such as open/closed page mode, which would force a PRECHARGE/ ACTIVATE after each write	HIGH
Data Flow	DDR Burst Accesses: organize memory to make full use of the DDR burst size. This includes alignment and data packing	Medium
Data Flow	Code Size: Optimize code and data for minimal size via compiler tools	Application dependent
Data Flow	Code Size: code packing	Medium
Data Flow	Code Size: creating functions for common tasks	Application dependent
Data Flow	Code Size: Utilized combined function instructions (multiple instructions in one, which save size and cycles)	Processor dependent
Data Flow	Code Size: Use tools for compression on the fly	Processor dependent
Data Flow	Parallelize and pipeline accesses to memory	Medium
Data Flow	Use constants and avoid zeroing out memory	Processor dependent
Data Flow	Cache: Layout memory to take advantage of cache set associativity	Application dependent
Data Flow	Cache: Use write-back model when available and feasible for application	Application dependent
Data Flow	Cache: use prefetching to bring data in ahead of time and avoid miss penalties and extra dead clock cycles	Application dependent

Continued

Table 13-1: Summary of power optimization techniques for DSP.—cont'd

Category	Technique	Impact
Data Flow	Cache: array merging	Application dependent
Data Flow	Cache: interchanging	Application dependent
Data Flow	Take advantage of DMA for memory movement	Medium
Data Flow	Coprocessors: Use to perform functions instead of core	Medium
Data Flow	System Bus Configuration: configure bus to minimize stalls and bottlenecks	Application dependent
Data Flow	Peripheral speed grades and bus width: optimize per usage needs.	Application dependent
Data Flow	Peripheral to core flow: use interrupt processing when possible	HIGH
Algorithmic	Compiler Optimization levels: use compiler optimization tools to optimize for performance to minimize cycle time in critical areas, and optimize for code size elsewhere	Medium
Algorithmic	Instruction Packing: maximize code to functionality efficiency	Medium
Algorithmic	Loop Unrolling: maximizes parallelism, minimizes active clock time	HIGH
Algorithmic	Software Pipelining: another method to maximize parallelism and minimize active clock time	HIGH
Algorithmic	Eliminating Recursion: save cycle time from function call overhead	HIGH
Algorithmic	Reducing accuracy: saving cycles by reducing calculations	Application dependent
Algorithmic	Low power code sequences: using equivalent functions via a lower power set of instructions	Processor dependent

Numbers and references were provided for many of these optimization techniques as they apply to the hardware tested in this work, but the reader must understand that, as every application is different, and will work differently on other hardware, power consumption must be profiled as discussed in the section on measuring power consumption.

References

[1] R. Oshana, DSP Software Development Techniques for Embedded and Real-Time Systems, Newnes, 2005.
[2] Electron Mobility, http://en.wikipedia.org/wiki/Electron_mobility.

[3] P. Yeung, E. Marschner, Power aware verification of ARM-based designs.http://www.eetimes.com/design/embedded/4210422/Power-Aware-Verification-of-ARM-Based-Designs.

[4] Micron Technology, Inc. "Technical Note TN-41-01: Calculating Memory System Power for DDR3," Revision B. August 2007.

[5] Kojima et al., "Power analysis of a programmable DSP for architecture/program optimization," Proc. IEEE Int. Sym. On Low Power Electronics, pp. 26—27, Oct. 9-11, 1995.

[6] Dhireesha Kudithipudi, "Caches for Multimedia Workloads: Power and Energy Tradeoffs," IEEE Transactions on Multimedia, Vol. 10, No. 6, October 2008.

[7] Performance and power consumption trade-offs for a VLIW DSP. Signals, Circuits and Systems, 9-10 July 2009, pp. 1—4.

[8] W. Wolf, Basics of programming embedded processors. Part 8, http://www.eetimes.com/design/embedded/4007176/The-basics-of-programming-embedded-processors-Part-8.

[9] R. Oshana, Software programming techniques for embedded DSP software.http://www.dsp-fpga.com/articles/id/?2548.

CHAPTER 14

DSP Operating Systems

Michael Kardonik

Chapter Outline

DSP for Embedded and Real-Time Systems. DOI: 10.1016/B978-0-12-386535-9.00014-7

Introduction

Digital signal processors (DSPs) were first introduced commercially in the early 1980s with the first widely sold TIs TMS322010. The first DSPs had a very limited set of external interfaces (TDM, host interface), and high level language compilers were not available at all, or were not producing very efficient code. Most of the applications were focusing solely on 'data crunching' and did not contain much control code or multi-threaded execution paths. This is why, in the early days, most of the DSPs were utilizing the 'bare-metal' model where all the resources belonged to and were managed by the application itself, only rarely using very basic, typically home-grown OS code. This situation gradually changed during the early 1990s, mostly due to the emergence of the 2G wireless technology. Wireless infrastructure projects, more complicated peripherals, and networking stacks changed the basic requirements for OS support. More recently, the introduction of multicore heterogeneous and homogeneous platforms also affected basic requirements. As a result, we see that the importance of OSes has grown over the years. Firm evidence for this is the fact that two of the leaders in the DSP market (Texas Instruments and Freescale) nowadays provide their own, proprietary OS as part of their offering and this situation is markedly different from the 1980s and 1990s. It means that those companies recognize the importance of the OS offering and invest resources in this area. This chapter should provide a good introduction for DSP engineers and also contains some in-depth discussion on multicore and scheduling.

DSP OS fundamentals

What is an OS and what is an embedded OS? Generally speaking, OS responsibilities are twofold:

- Resource management: It includes computation resources sharing and management (multitasking and synchronization), I/O resources allocation, memory allocation, etc.

- Abstraction layer: provides a way to achieve application portability from one hardware platform to another

From this point of view, there is no principal difference between an embedded OS and a desktop OS — the difference is only in the set of peripherals/resources managed and in the typical applications supported. So, what is the most typical quality of an embedded system? We would suggest that an embedded system is focused on a limited amount of applications/ usages. It allows the design of a system using scarce resources.

Real-time constraints

DSP is all about real time and this is why real-time considerations are such an important factor in the design of an OS for DSP. Real-time constraints for a job mean that it has to produce results before some predefined deadline.

Real-time constraints of the job may be described using a usefulness of results function. This function is a relation between time and usefulness of the job's result as shown in Figure 14-1.

In Figure 14-1, job A has hard real time constraints because its usefulness function breaks sharply at deadline and is equal to zero (it may even be negative in some cases). The term real-time OS is somewhat confusing: real-time constraints always exist on a system, at application level, and an OS may only support the system to satisfy those constrains. So, it is

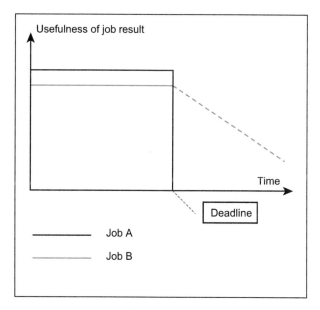

Figure 14-1:
Usefulness of results in a real-time system.

important to be aware of the fact that using real-time OS does not guarantee that the system will be real time.

If we think about real-time constraints in the broader context of 'real-time job must produce results before some deadline,' we notice that it implies that this job must be able to get all the resources it needs. In OS terms it means that resource allocation cannot fail for real-time jobs. In many cases, it means that all the resources for real-time jobs must be allocated in advance, even before the job is ready to be executed. If we talk about jobs that are cyclical or sporadic, those resources may never be released back to the system.

Processes, threads and interrupts

Surprisingly, there is some confusion around terminology in that area. Different OSes differ in what terminology they use to describe those entities. In this chapter we will use the following definitions of OS terms.

Process

Entity that consists of complete state of a program and at least one thread of execution. Classic processes are hardly ever used in DSP RTOSes. There is a tradeoff that exists between protection that OS guarantees and performance that one can achieve with such protection. For example, having user and supervisor modes would help tremendously to protect the kernel from user-level applications but at the same time might affect performance if each call to OS were to go through a system call interface. Another example is task-switch time that would grow significantly if the entire memory management unit (MMU) context had to be switched.

Thread

Schedulable entity within process. In some cases, it is beneficial to have several threads of execution within the same process. They share the same memory so it is possible to efficiently share data between them; there is no high cost of process switches (no need to switch memory context). In most DSP RTOSes, only one process will typically exist, meaning that the entire memory space is shared between all the threads. To make things more complicated most RTOSes call those threads tasks.

Task

Task is a type of thread that is scheduled by RTOS scheduler. It is different from hardware interrupt and software interrupts by having its own context where task state can be saved and restored. Typically, it has the following states:

- Run — task is currently executing on processor
- Ready — task is not executing as higher priority thread is running

- Blocked — task is waiting for a resource or I/O
- Suspended — task is temporally removed from scheduler ready queues

Interrupt

Interrupts can be seen as a special type of thread that is executed in reaction to hardware events. It is scheduled by OS and is executed in the context of OS kernel.

Interrupt latency

Interrupt latency measures the maximum response time of the system for a particular interrupt. In other words, how long it takes at maximum for specific interrupt service routine to be called in response to an interrupt. This parameter is important for real-time systems because they respond adequately fast to external events. Figure 14-2 describes interrupt latency as a function of contributing factors.

Interrupt latency is a sum of several elements:

- Hardware latency — time from event itself to first instruction executed from interrupt vector if interrupts were not disabled. It is typically the smallest part of interrupt latency.

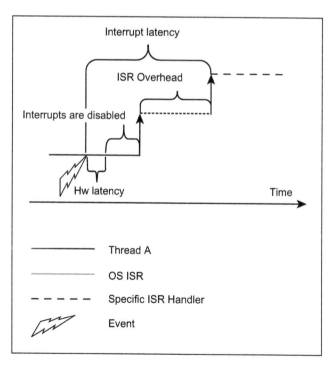

Figure 14-2:
Interrupt latency as a function of contributing factors.

- Interrupt disabled time — many OSes kernels use global interrupt disable instructions in order to guard local accesses to shared variables.
- Higher or equal priority interrupt is active — if higher or equal interrupt is active, lower priority interrupt cannot run and it adds to interrupt latency for this particular priority.
- Interrupt disable for particular priority — in some cases it is important that interrupts of particular priority and lower are disabled. For example, the system may have a specific thread (which is not interrupt service routine (ISR)) that is hard real time and it is important that none of the lower priority interrupts can interfere with it. In such a case, the user would set current ISR priority to an adequate level. Some OSes will have unified priority schemes for threads and interrupts that will do this automatically (OSE) and some will allow manual configuration (SmartDSP OS).
- Interrupt overhead — several actions that OS has to perform before it can call specific ISR. For example, updating TCB for preempted tasks like saving context and statuses, determining interrupt handler pointer, etc.

How do we determine the interrupt latency of the system? There is no simple answer to this question. Typically, any vendor of commercial RTOS has interrupt latency numbers available. However, those numbers do not take into account the users' application part. This is an important point to understand when designing real-time systems: the interrupt latency may depend on the entire system design and not only on OS. For example, if the user is implementing some high priority ISR, it will affect latency of all lower priority interrupts.

For complex systems that are not life critical, many times the interrupt latency is determined by thorough tests rather than analytically.

It is important to mention that high interrupt latency does not directly mean high interrupt overhead: it can also be explained by high interrupt disable time etc.

Now, one can ask the question: How may interrupt latency be controlled? Clearly, if we are talking about commercial OS, it would not be possible generally to change its source code so interrupt overhead could not be changed (it is possible sometimes to perform tricks like locking in the cache, etc. but it is beyond our interest here). However, users can still control other factors:

- Implementing 'short' ISR, possibly by breaking it into the ISR part and the lower priority software interrupt part
- Minimizing usage of global interrupts disable instructions
- Careful design taking into account desired priorities for all ISRs. It can help to keep interrupt latency of high priority tasks low
- Minimize number of hardware interrupts in the system. It sounds surprising but many hardware interrupts can be avoided. This can help for overall performance and latency. Generally speaking, interrupts should be used for sporadic events or for cyclic events that

are used to schedule the system. For example, imagine an interrupt that arrives at the system as result of an Ethernet frame being transmitted. We know that this interrupt is expected to arrive once the frame's transmission is completed within some period of time so this is not a sporadic event. In some cases this event can be handled in the next transmit operation instead of ISR.

It is important to understand the difference between thread scheduling and ISR scheduling:

- ISR does not have its own context and thus is always 'run to completion' execution. It can be preempted by a higher priority interrupt but the order is always preserved (see more details on this when software interrupts are discussed below).
- Hardware interrupts are caused by events that are external to the system. This is different from both threads and software interrupts

The opposite of interrupts is polling. The idea with polling is to poll interrupt events and call an interrupt handler. One advantage of polling is that there is no interrupt overhead involved typically.

Software interrupt

Another important threading technique that is typically used in RTOSes is 'software interrupts' (also known as softirq and tasklets in Linux). The term 'software-interrupt' is coined for two reasons:

- Software interrupts do not have context and thus are executed in 'run-to-completion' mode which is similar to hardware interrupts.
- They may be implemented using 'software interrupt' or 'system call' processor command. This command actually triggers hardware interrupt.

The idea behind the software interrupt is simple:

- Classic threads have relatively high task-switch time because they have its context.
- Hardware interrupts affect interrupt latency of other interrupts of the same and lower priority.

The concept is implemented in many OSes. For example, In SmartDSP OS there are three contexts where SWI can be activated:

- SWI of the same, lower or higher priority
- HWI
- Task (In SmartDSP OS as in many other RTOSes, task is a supervisor-level thread of execution)

Let's start with triggering of SWI from lower priority SWI. As SmartDSP OS is preemptable, priority based OS, it is required that at each instance of time highest priority thread should be executed (clearly, except in situations where scheduling is disabled in some way). Thus, higher priority thread must be activated immediately and it should preempt lower priority SWI. This may be accomplished by a simple function call. It is made possible because SWI does not have context and as such it will always be re-scheduled in the order of preemption so the semantics is the same as for function calls — they all reside on the same stack.

If SWI is activated from a task, the OS cannot just call an SWI function; it has to switch to interrupt stack and save current task state in its TCB (task control block). Instead, it executes a special instruction called 'sc' in the StarCore case or system call that generates special core interrupt. This interrupt's handler identifies which SWI was actually activated and will call the appropriate SWI handler.

If SWI is activated from HWI then the only immediate action that the scheduler takes is the indication to itself that SWI was activated. Once all HWI will be serviced, a scheduler will check if any SWI were activated and will call their respective handlers.

One important limitation of SWI is that it cannot block (meaning that it cannot be in the blocked state) so it is impossible to use any functionality that may block. Consider the following scenario: in SWI you want to wait for I/O and call read() function. This function will block until input actually arrives at the system. SWI does not have its context so OS cannot suspend it and start running another SWI or task. This inability to block should be not confused with ability to preempt to another SWI and HWI: when SWI is preempted by another SWI, its state resides on stack (as we explain below) and thus it will continue execution when higher SWI is finished and not because of some external I/O.

In Figure 14-3 we see how SWI A is preempted by SWI B on the same stack and then by SWI C. After SWI A completes, execution must return to B and then to C.

Some OSes like SYS/BIOS support only one process and multiple threads, some of them (OSEck); in particular, micro-kernel ones support multiple processes.

Multicore considerations

Multicore systems have now become commonplace and all DSP OSes must support them in some way. There are several questions that arise with multicore systems that should be answered during software development:

- How to map an application on different cores?
- How will those applications share resources (memory, DMA channels, I/O, cache)?
- How will those applications communicate?
- And most importantly: How do you achieve all of this and not make the software development process significantly more complex than in a single-core environment?

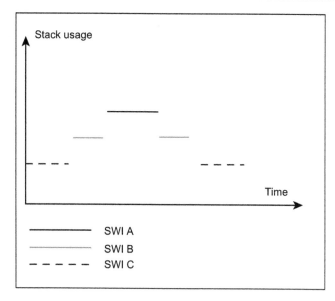

Figure 14-3:
SWI and stack.

This challenge is yet to be solved generically but RTOS can help resolve it in several ways, which we discuss in this chapter.

When we talk about multi-core systems we identify two major types:

- Homogeneous systems — contain only one type of processor, in our case DSP core. In some SoC's this line is not very clear, For example, some SoC's will contain very sophisticated co-processors (like MSC8156's Maple) that may even contain specialized cores that run specialized OS. We still consider MSC8156 and similar to be homogeneous SoC's as those coprocessors have a predefined, 'slave' role in the system and end-users cannot program them typically.
- Heterogeneous systems — the SoC that contains more than one type of core, typically DSP and general purpose cores, like ARM or PowerPC. When choosing how your RTOS will fit your application space in the case of heterogeneous systems one should consider the following topics specifically:
  - Inter-core communication between those different types of cores. The practical problem that sometimes arises here is the fact that two different OSes will likely run on those sub-systems such that protocols available on one OS maybe not be available on another and so forth. We will discuss the general topic of multicore communication later in this chapter
  - Booting of the system. How are sub-systems booted? Independently, or does one boot another?
  - Cache coherency and how it is supported by all OSes involved

In homogeneous systems where all the cores are of the same type and have shared memory, OS can be asymmetric multiprocessing (AMP) or asymmetric multiprocessing (SMP_). There is some confusion sometimes about AMP vs. SMP that we believe relates to attributing SMP or AMP characteristics to different layers of hardware and software. For example, OS can be SMP, which means that only one instance exists in multicore SoC but the application may execute core affine tasks thus making application level software asymmetric. It is also possible to see perfectly symmetric applications that are executed on rather asymmetric OS. In this chapter we adopt a narrow definition of the SMP term which is identical cores and only one instance of OS. In SMP systems user's process may migrate (implicitly or explicitly) from core to core while in AMP OS a process is always bound to the same core. We can state that majority of commercial DSP RTOSes in multicore environment support the AMP model rather than SMP.

AMP OS instances that are executed on different cores may have shared memory and may even cooperate on resource sharing. Shared memory is required for implementing inter-core communication protocols, and effective resource sharing. One example of an AMP system that has a layer that manages resource sharing is Freescale's SmartDSP OS implementation for MSC8156 processor. In Figure 14-4 we see an MSC8156 block diagram and how exactly SmartDSP OS is mapped to it. As we can see in this figure, the default mapping for SmartDSP OS is the following:

- Local memory is at M2 and it is mapped using MMU to have the same addresses across all OS instances. This allows having one only image that can be loaded to each core and

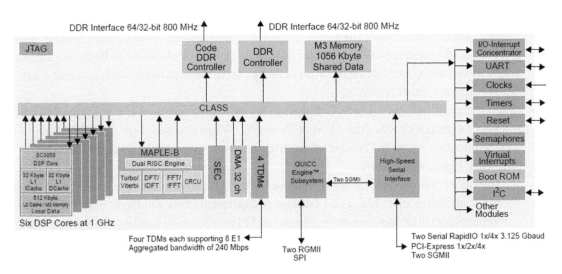

Figure 14-4:
MSC8156 block diagram. *(from [27])*

thus code may be shared across all instances. Consider the following situation: there is a global variable that is local for each instance and has an address 0xABC. This address is physically different for each partition's memory but identical from the program perspective thus allowing the possession of the same code image. Other approaches are possible as well, for example indirect access to such a memory but, depending on architecture, it may affect performance. Sharing code is important in systems where cache is shared between cores and will result in better cache utilization and a smaller footprint of the application. For example, the entire application may fit into some kind of internal memory (e.g., M3 on MSC8156) as a result of code sharing. As a side note, debuggers should take into account the possibility of code sharing as it may affect implementation of software breakpoints.

- Shared data is at M3 memory. All shared data is explicitly defined as such; it is different from the SMP environment where any global data is shared by default.
- Code is shared and resides at M3 or DDR depending on its performance impact.

This mapping is a typical AMP configuration with different OS instances on every core. However, SmartDSP OS is not a 'pure' AMP OS because all instances are sharing resources in a cooperative manner. For example, DMA channels will be assigned to each OS instance statically, at boot or compile time, such that each instance will use only those DMA channels, as shown in Figure 14-5.

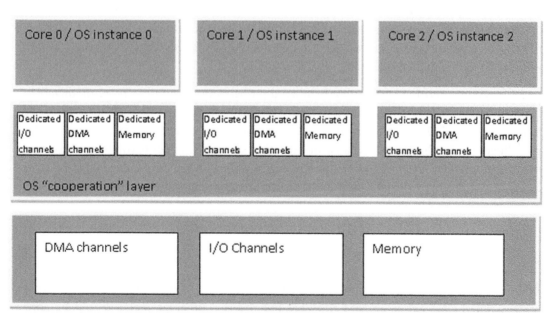

Figure 14-5:
Cooperative AMP style sharing.

As [1] states "… Classic AMP is the oldest multicore programming approach. A separate OS is installed on each core and is responsible for handling resources on that core only. This significantly simplifies the programming approach but makes it extremely difficult to manage shared resources and I/O. The developer is responsible for ensuring that different cores do not access the same shared resource as well as be able to communicate with each other." Because of those problems that the above quote mentions, 'Pure' AMP is rarely utilized in more complex systems. Instead, hybrid approaches are used. Enea OSEck and SmartDSP OS, as an example, will use 'multicore aware' device drivers [HoO] in the OS cooperation layer that we mention above. Let's define some more terms and try to understand how those techniques may be utilized:

- Resource partitioning is about the assigning of a particular resource to a particular partition. Partitioning may be done with hardware protection or just logically. In the case of DSP systems, the partition is typically a single core. The term partition is widely used in the Hypervisor world and means a set of resources for a virtual, logically independent computer that may execute its own instance of operating system.
- Virtualization of peripheral is creating a new, 'virtualized' instance of resource. When we talk about partitioning and virtualization, it is important to realize that an OS will have a direct access to resources while in case of virtualization, OS will work with a 'non-real,' virtualized resource.

To exemplify the difference between partitioning and virtualization we may look at an example of memory partitioning and virtualization. When memory is partitioned, each core can have an access only to its own region of it. In the case of virtualized memory, each core may have access to even more memory that is physically available (using swapping), which is impossible with simple partitioning.

In DSP systems, mostly, partitioning is utilized probably because it has more predictable behavior, which is very important for real-time applications and also because there is no need to execute several OSes concurrently on the same core. However, we do expect that in future this situation may change, not because of executing several OSes per partition but because in many-core environments it may be impossible to scale resources easily to the number of cores in hardware; resource virtualization may resolve this problem satisfactorily even with some overhead.

Peripherals sharing

Let's look at how resource partitioning can be implemented using as an example Ethernet network interface. Freescale's MSC8156 has two external Ethernet interfaces and six cores; a similar configuration is available for TI's C66x multicore DSP devices [25]. The Ethernet interface is responsible for sending Ethernet frames to the network and receiving frames into the system. A multicore system should be able to share the same interface so that each core is able to send and receive data. One way for doing this is to receive all the frames to the same core (master core) and then to redistribute them to other cores according to some predefined criteria (e.g., MAC address). This approach is essentially virtualization of the interface rather than partitioning and it has some serious drawbacks: the system becomes asymmetric and as such may require different software executing on cores; it will show less predictable behavior as cores will depend on each other and so on. Thus, such an approach in real-time applications should be avoided and partitioning should be preferred if possible.

Multicore devices mentioned above offer special hardware support that allow different cores to share peripherals. For example, Ti's KeyStone architecture [24] that C66x is based upon allows queues that connect packet processing interfaces with cores. From the same source: "The packet accelerator (PA) is one of the main components of the network coprocessor (NETCP) peripheral. The PA works together with the security accelerator (SA) and the gigabit Ethernet switch subsystem to form a network processing solution [as shown in Figure 14-6]. The purpose of PA in the NETCP is to perform packet processing operations such as packet header classification, checksum generation, and multi-queue routing."

Note that having classification and distribution is beneficial in the single core case as well, as it allows delivering a packet precisely to a specific thread instead of having some multiplexing thread in the middle.

OSes like Enea, SmartDSP OS, or SYS/BIOS support these hardware mechanisms with proper drivers and stacks so that user applications can take advantage of this model.

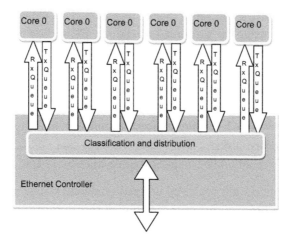

Figure 14-6:
The PA, the security accelerator (SA), and the gigabit Ethernet switch subsystem form a network processing solution.

Synchronization primitives

One of the reasons why multicore brings new complexity to software development is the fact that system resources are shared not only between different tasks running on the same core but also among tasks that are executed on different cores.

There are two problems that we look at here:

1. Cores are using shared structures in memory (shared data) and thus when they access it, race condition can exist that should be resolved.
2. When it is impossible to partition or virtualize a resource, two or more instances must share it and thus ensure exclusive access.

Both problems are of a similar nature and have similar solutions. There are different ways to control access to shared resources/structures in a multicore environment.

Spinlocks is a standard way to control access to a shared resource but not the only one. Another way to control access is to build a scheduler so that cores do not access the shared resources at the same time. In such case no explicit lock is required, which is always a preferable situation. It could be done more easily when offline scheduling methods are used (to be discussed later). However, it is impossible to use such a technique when two cores are executing in AMP mode and the tasks are asynchronous. In these cases, spinlock can be a solution for that problem. A typical spinlock implementation utilizes some special indivisible (atomic) operation such as test and set a key, and operates on special shared memory. Some architecture may provide special hardware semaphores support that are typically used when the only shared memory does not support atomic operations. When cores contend for a resource using spinlock they call a special OS function, e.g., get_spinlock(). This routine tries to get a lock on a resource and if the resource is not available, it continues to spin in a loop. This means that one of the cores waits till another finishes using the resource or handling some shared data and does nothing useful. It is not a desirable situation so developers should make sure that spinlocks are used very carefully and used for very short periods. It is not possible to use spinlocks in a single-core environment as this may cause dead-lock condition. It is easy to understand why: while one thread takes spinlock, it is possible that another, higher priority thread takes control and also competes for the same resource by trying to acquire spinlock; clearly, it never gets it as it continues to spin and the first process never runs. The way that this situation is resolved is to disable the interrupts or use any other means possible to ensure exclusivity.

In AMP systems, semaphores are typically utilized on each OS instance and spinlocks are used for synchronization between the cores. Although spinlocks and semaphores are similar in effect, they are significantly different in operation. The primary difference between a semaphore and a spinlock is that only semaphores support a 'pend' operation. In a pend operation, a thread surrenders control — in this case until the semaphore is cleared — and the RTOS transfers control to the highest priority thread that is ready to run. It is not practicable to use semaphores for synchronization purposes in a lightweight AMP environment, because there is no means for a 'post' operation on core A to make a pending thread on core B ready to run except in respect of expensive multicore interrupts.

Care should be taken in application design to ensure that spinlocks are used to block access to a resource for absolutely the minimum possible amount of time. Preferably such operations will only be performed with interrupts disabled. If a thread loses control when it holds the key to a spinlock, it could lead to deadlock. To use a spinlock to control access to a shared memory structure, simply include a 'lock' field as one of the elements in the data structure.

In an AMP environment, there is the additional synchronization concept of a barrier. A barrier may be shared by two or more cores, as shown in Figure 14-7. Barriers are used to synchronize activities across cores.

Figure 14-7 illustrates this concept; four cores must each reach a pre-defined state prior to any of the four cores continuing execution. As each of the four cores reaches its pre-defined state, it issues a 'Barrier_Wait' command, and the RTOS sits and waits until the last of the cores issues a Barrier_Wait, at which point all cores continue execution simultaneously. A barrier is a valuable tool to use during initialization. It ensures that all cores are fully initialized before any one of them commences active operation. Additionally, a barrier may be used to synchronize parallel computations across multiple cores.

Memory management

Memory allocation

Essentially any OS supports some form of memory allocation. RTOSes will typically have two forms of memory allocation: memory allocator for variably sized blocks (similarly to

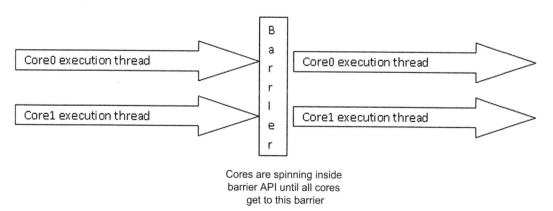

Cores are spinning inside
barrier API until all cores
get to this barrier

Figure 14-7:
Barrier.

malloc()) and fixed size buffers. The former is important for real-time applications, especially for DSP applications that always deal with 'crunching' chunks of data in memory, as it allows fast, bounded, and predictable memory allocation. This is possible with fixed size buffers because it enables the creation of a very simple and efficient way to allocate and release memory (essentially, any FIFO or LIFO will do it) and it does not introduce any fragmentation issues as variable size allocation does.

Virtual memory and memory protection

Memory protection OS features are completely dependent on the specific support of MMU. As most DPS cores do not have MMU that supports memory protection, DSP RTOSes do not typically support memory protection. The same is true for Virtual memory. However, this situation is currently changing for two reasons:

- More complicated applications require better protection between different tasks/processes. Otherwise, debugging becomes a very challenging task.
- Multicore introduction (which may in addition be regarded as a more complex system) forces designers to find ways to protect each core's memory from other core.

Another aspect of memory protection is protection from masters other than core. This is becoming more important as the number of masters on SoC is growing.

Different types of memory

In modern DSP SoC, typically several different types of memory exist. This may include internal memory, cache (several layers), and external memory.

All the types of memory play different roles in application and thus require different handling. For example, in DSP applications, it is very typical to explicitly hot-swap data into internal memory for processing. In hard real time application it is important to lock cache lines (both for data and program) to ensure predictable processing time for particular tasks.

Networking

Inter-processor communication

Many DSP applications consist of a host processor that communicates heavily with DSP processors. Also, it is not rare to see systems that include tens or more DSP and host processors. This is why inter-processor communication is a very important feature for DSP RTOSes. The very basic inter-processor communication involves the ability to send signals from processor to processor. This is typically achieved by using direct IRQ lines or by designated signaling of interconnecting busses (PCI, SRIO, etc.). More sophisticated and feature rich protocols exist that support reliable synchronous and asynchronous data transfer between the processors. Inter processor communication facilities become even more important with multicore inception. The IPC concept is surely not a new one. Let's not forget that multichip systems, specifically DSP systems, have been very common for a long time. Processors in those systems were connected by different types of backplanes; proprietary HW, Ethernet, and PCI are the common examples. There are many protocols that have been implemented and utilized over the years.

Enea's LINX has fairly rich functionality and mature implementations both on Linux and Enea's OSes. LINX on Linux is open source today and is released under BSD and GPL licenses. In reference [20] it is stated that "Enea LINX provides a solution for inter process communication for the growing class of heterogeneous systems using a mixture of operating systems, CPUs, microcontrollers, DSPs and media interconnects such as shared memory, RapidIO, Gigabit Ethernet or network stacks. Architectures like this pose obvious problems; endpoints on one CPU typically use the IPC mechanism native to that particular platform and they are seldom usable on platforms running other OSes. For distributed IPC, other methods, such as TCP/IP, must be used but that come with rather high overhead and TCP/IP stacks may not be available on small systems like DSPs. Enea LINX solves the problem since it can be used as the sole IPC mechanism for local and remote communication in the entire heterogeneous distributed system."

TIPC (Inter Process Communication Protocol) is another example of protocol and implementation that has been around for long time and also was pushed to the Linux kernel. It was initially designed by Ericsson and was adopted by VxWorks later. This is how reference [21] describes the motivation behind this protocol: "There are no standard protocols available today that fully satisfy the special needs of application programs working within highly available, dynamic cluster environments. Clusters may grow or shrink by orders of magnitude, having member nodes crashing and restarting, having routers failing and replaced, having functionality moved around due to load balancing considerations, etc. All this must be handled without significant disturbances of the service(s) offered by the cluster."

Generally, there are several aspects that have to be taken into account when deciding which IPC to adopt:

- Functionality: which includes features such as addressing mechanisms, blocking versus non-blocking API, supported transport layers and operating systems
- Interoperability/Portability
- Performance

As usual, there is a tradeoff between performance characteristics and functionality. For example, implementing generic connection-oriented, reliable protocol in general may be problematic from both the performance and the complexity perspective. This is one of the main reasons why standard existing protocols like TCP are not always adequate for IPC purposes.

Along with high level protocols DSP RTOSes export much lower level of API. SYS/BIOS, as an example, provide today two packages: SYS/LINK and IPC.

The IPC library provides OS level API, some of them (e.g., GateMP [23], Heap*MP) being based on single core versions. Below are some of the modules that are supported:

- MessageQ Module
- ListMP Module
- Heap*MP Modules
- GateMP Module
- Notify Module
- SharedRegion Module

Let's look at HeapBufMP which is part of Heap MP Modules for details. The idea here is to enable different cores to manage (allocate and release) blocks of memory of the same size. It is very useful for the following scenario: one of the cores produces some data and writes it to memory, and then it sends this data to another core (possibly using ListMP module), which consumes it and has to release the buffer.

First we need to create a pool:

```
heapBufMP_Params_init(&heap_params);
heap_params.name = "multicore_heap_1";
heap_params.align = 8;
heap_params.numBlocks = 100;
heap_params.blockSize = 512;
heap_params.regionId = 0; /* use default region */
heap_params.gate = NULL; /* use system gate */
heap = HeapBufMP_create(&heap_params);
```

On other processors we have to open this pool using its unique name (it uses the NameServer module to achieve that):

```
HeapBufMP_open("multicore_heap_1", &heap);
After we have initialized a heap we can allocate and release memory from this pool:
HeapBufMP_alloc(heap, 512, 8);
```

The implementation uses gate (GateMP) that was specified in HeapBufMP_create() call.

SYS/LINK (or just SysLink) is trying to go beyond providing IPC primitives and also supports the ability to load new components at run time (for co-processors), power management for the DSP side, and booting services.

- Loader: There may be multiple implementations of the Loader interface within a single Processor Manager. For example, COFF, ELF, dynamic loader, custom types of loaders may be written and plugged in.
- Power Manager: The Power Manager implementation can be a separate module that is plugged into the Processor Manager. This allows the Processor Manager code to remain

generic, and the Power Manager may be written and maintained either by a separate team, or by the customer.

- Processor: The implementation of this interface provides all other functionality for the slave processor, including setup and initialization of the Processor module, management of slave processor MMU (if available), functions to write to and read from slave memory, etc.

MCAPI (Multicore Communication Application Programming Interface) is a standard that is proposed by MCA (the Multicore Association http://multicore-association.org/). Its approach to standardization is different from TIPC, LYNX, or SysLink: instead of providing a standard for protocol or providing a de-facto standard by providing implementation, MCAPI chooses to define API only. Such an approach has both advantages and disadvantages: by definition it does not depend on underlying transport but it also does not enforce interoperability for different implementations (but it enforces portability). This is how MCAPI* states its main goals: "The primary goals were source code portability and the ability to scale communication performance and memory footprint with respect to the increasing number of cores in future generations of multicore chips."

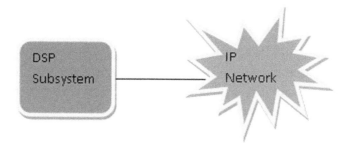

Figure 14-8:
DSP connected to network.

Internetworking

TCP/IP stack support is essential, especially in cases where the DSP sub-system is directly accessible by 'external-world' and not 'hides' behind Host processor (Figure 14-8). When DSP is encapsulated with the Host processor, the former typically implements the TCP/IP stack and DSP will only deal with specific DSP functions (Figure 14-9).

We can see that over the years DSP OSes started to include TCP IP stack as part of the standard offering. The functionality of those stacks may lag after that of more advanced operating systems like Windows or BSD and will focus primarily on protocols that are related more to data exchange and less to control. For example, it will typically have no support for routing protocols.

---

*http://www.multicore-association.org/workgroup/mcapi.php

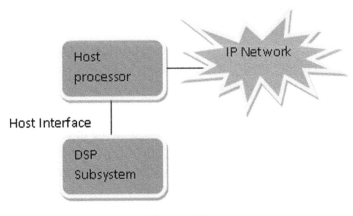

Figure 14-9:
DSP connected to host.

For example, SmartDSP OS provides TCP/IP stack that has proprietary API (not standard BSD socket) which allows very efficient, zero copy, zero context switch interfacing with a stack for UDP. SYS/BIOS does not include TCP/IP stack in its standard installation. Instead, it is possible to download NDK — Network Developer Kit, which contains portable TCP/IP stack that it is possible to run on top of SYS/BIOS [19].

Lately, IPv6 and IPSec protocols have become essential to support wireless access applications in particular. If we look at IPSec implementations for SmartDSP OS and SYS/BIOS platforms, we will find that both are utilizing hardware accelerators (SEC on Freescale MSC8156 and KeyStone for TMS320C66x) for IPsec and do not handle a typical control path of IPSec — IKE.

Scheduling

Designer should consider the following tasks when developing real-time system:

- Identifying inputs and outputs of the system
- Identifying control path
- Identifying algorithms that will be utilized
- Creating a proper software architecture for the application
- Selecting hardware components
- Mapping software into hardware and OS

These are not independent tasks as they affect each other and will be executed in parallel. In this article we intend to focus on how an application can utilize effectively different features that DSP RTOSes implement; specifically, we will research scheduling and threading mechanisms.

During our years in DSP systems development, we have found that there exists a gap between developments in real-time theory and engineering practice: in many cases, the development of DSP systems would be based on *ad hoc* utilization of threading and scheduling.

There are several possible reasons for this situation:

1. Lack of proper education in real-time theory for engineers who develop DSP systems.
2. It is possible that the current state of OS development or real-time scheduling theory does not provide answers that are suitable for real-life use-cases.

As stated in reference [2] "Although scheduling is an old topic, it has certainly not played out. A real-time scheduler provides some assurances of timely performance given certain component properties, such as a component's invocation period or task deadlines. ... Unfortunately, most methods are not compositional. Even if a method can provide assurances individually to each component in a pair, there is no systematic way to provide assurances for the aggregate of the two, except in trivial cases. Our goal here is to show how to apply some of the basic real-time theory results when using DSP."

We will need to describe a set of definitions that can be used across different RTOSs and will be based primarily on definitions found in academic real-time literature. We need to do this because different RTOSs are utilizing inconsistent terminology. This chapter is not an article on real-time theory so we will try focusing on very basic and useful ideas that can help practicing engineers to build efficient DSP systems. There are several classic books on real-time subjects, reference ([8] being an example) that readers can refer to for a more in-depth discussion.

Reference model

- Job: The smallest schedulable unit. Many times, when we have a general discussion about application, it is convenient to have a term that is not tight to any specific scheduling type. A job may be implemented in different ways: it may be an interrupt, software interrupt, process, function call in background loop, thread, etc.
- Task: Application level unit of work. For example, encoding a video frame is a task that the system should accomplish and it may consist of several (possibly interdependent) jobs, e.g., motion estimation and so on. Sometimes, a task may not be broken into several jobs. There are some common reasons to break tasks into jobs, including:
  - Finer decomposition of tasks into jobs helps to manage complexity.
  - In DSP systems, different jobs within the same task may be executed on different processors: a control job may be handled by a general purpose processor and DSP jobs on special purpose processors or co-processors.

Each job is characterized by the following parameters:

Timing constraints

Release time (r): instant of time when a job becomes available for execution. For example, in Media Gateway's PSTN interface we can say that the encoding job's release time is when data becomes available on the TDM interface.

Relative deadline (d) is an instance of time, relative to release time, when the job must be completed. In our example of PSTN interface, a typical requirement is that the encoding task must be finished before the next data portion is available.

Timing characteristics

Execution time: time it takes to accomplish a job. Typically we will consider only maximum execution time for a job because the system should be schedulable in the worst case.

Precedence graph

Precedence graph shows dependencies of jobs on the system. Vertexes — jobs and edges are dependencies.

Preemptable vs. non-preemptable scheduling

If a job is preemptable it means that it can be interrupted at any time and resumed later by a scheduler. We have to differentiate between preemptability on the application level and preemptability that is supported by a scheduler. For example, even if a particular thread is preemptable, a user may decide that in his specific system this thread should not be preempted because it has very tight real-time characteristics or it must have exclusive access to some resources.

Blocking vs. non-blocking jobs

Some types of jobs (ISR, software interrupts) cannot block (put into 'sleep' state to be revoked later) as they do not possess their own context. Others (threads, for example) can block. This distinction is orthogonal with preemptability. One can ask why not have all jobs having context so that each one could potentially block? The short answer is that it can make context switch time much higher for ISR, as an example, and thus having both types of jobs provides more flexibility for designers.

Cooperative scheduling

Cooperative scheduling means that a job can surrender control and activate another job. This is the opposite of preemtable scheduling where each job can be preempted without its cooperation. In its purest form, the scheduler, as an OS object, does not make any decisions on

Table 14-1: Threading characteristics.

Property/Thread	Task	SWI	HWI
Preemptable	Yes	Yes	Yes
Blocking	Yes	No	No
Cooperative	Yes	Yes (limited)	No

which job is executed next and all those decisions are made exclusively by jobs. One can see such a scheduling as an application level scheduler implementation.

Both techniques may co-exist in the same OS as well; for instance, threads may be preemptable and still be able to yield control for execution. SmartDSP OS provides such a hybrid implementation, as an example. Table 14-1 shows those characteristics for SmartDSP OS.

Types of scheduling

There exist two main types of scheduling that are different as regards how decisions are made [15]:

- Offline Scheduling (sometimes referred to as static or clock-driven scheduling). In this case all scheduling decisions are made offline based on full a priori knowledge of jobs' real-time parameters.
- Online Scheduling makes scheduling decisions in response to events that happen at run-time.

Multicore considerations on scheduling

Scheduling real-time jobs especially when they have dependencies is not an easy task. The situation becomes more complex when we allow tasks to dynamically migrate cores in run-time. Clearly, it can bring better resources (cores) utilization but at the same time it may result in a much more complex and unpredictable system. As Liu stated in 2000 in reference [8]: "...most hard real time systems built and in use today and in the near future are static [no dynamic job migration from core to core]."

Intensive research in that area continues [16] but even today Liu's statement holds. This means that most DSP systems today are built so that tasks do not migrate dynamically and we can continue to focus on single core scheduling techniques if jobs executing on different cores are independent. This is also why virtually all DSP OSes focus on scheduling of single core and providing the capability to partition the system in such a way that users work with independent single cores.

Offline scheduling and its possible implementation

In hard real-time systems new jobs must produce results before their deadline. In the offline type of scheduling all decisions are made off-line, using some tool or just pen and paper and based on the assumption that users know everything about the system at design time.

The idea with offline scheduling is to create an execution list for all jobs that is used in run time to decide when to run each job. This kind of approach allows producing systems that are very effective, compared to on-line scheduling, for a given set of jobs:

- Avoiding explicit synchronization (like semaphores) because the designer takes care of those issues at design time. It can provide significant performance advantages as no semaphores may mean no context switches and also simplify the job's code.
- The scheduling code overhead is very low. The decisions made offline and next job calculation does not exist. Also, there is no need to make any acceptance tests that dynamic cases may include.
- Very predictable system which simplifies testing, debugging: it may be enough to feed in the worst case data (if execution time depends on data, as an example) to see that the system performs correctly. For priority driven systems, this would not be sufficient as preemptive scheduling and resource sharing result in more complex scenarios.

Despite all these good features, offline scheduling is not explicitly supported by most available RTOSes (SmartDSP OS, SYS/BIOS, Enea OSEck). Surely, users can utilize these platforms to implement static scheduling (and we will see how) but there are no special offline tools that help crafting those schedulers (as in [9]). There are several possible reasons for this:

- It requires plenty of design time to create an offline scheduler
- It may require significant redesign in case of even minor application modifications, system updates, etc.
- It is impossible to create a generic tool that will create an offline schedules
- Many systems are dynamic in nature so it is very challenging to create a static model without compromising the system's performance. (But this is sometimes required in hard real-time systems.)

So, how is it possible to implement an offline scheduler on RTOS like SmartDSP OS? First of all, the system should be characterized using a terminology that we established earlier (job, task, relative deadline, release time, execution time, and precedence graph). For offline schedules we will typically find a hyperperiod (least common multiple of the periods of all the tasks) and all the release times. The reason is simple: we have to know all the information about the system *a priori* and this is only feasible if the system is periodic. Many DSP systems are periodic by nature: media gateway with PSTN interface, real-time media systems like set-top boxes, radar DSP sub-systems and others, and this is why we will focus here on tasks that are periodic. Table 14-2 summarizes job parameters for a sample application.

Table 14-2: Job parameters.

Task.Job/Parameter	Release Time	Relative Deadline	Maximum Execution Time
A.1	0	5	4
A.2	5	10	8
A.3	8	5	3
B.1	2	8	8
B.2	10	8	7

Task A has a period of 40 milliseconds and Task B has a period of 20 milliseconds and the same phase, thus the hyperperiod is 40 milliseconds and we need to build a schedule for 40 milliseconds.

When we look at a precedence graph like that in Figure 14-10, we can see that job A.3 depends on A.1. Typically this means that A.3 is consuming the results that A.1 produces. It means that A.3 cannot start before A.1 finishes so its effective release time is different and is equal to A.1 release time + A.1 execution time.

Another important factor that we must take into account a situation where there are some resources that are shared between different jobs. For example, if A.1 and A.2 are utilizing the same coprocessor, they cannot run in the same time on different cores without proper synchronization and such synchronization may cause higher execution time for each job. So, a designer should identify all shared resources at this stage and ascertain how it affects execution times and consider schedules that will prevent such sharing. One parameter that

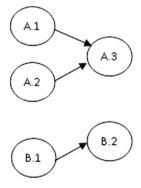

Figure 14-10:
Precedence graph.

relates to resource sharing is preemptability of a job, meaning that if it is possible to preempt job at some point, start another one and come back to it later.

Once we have found all that information and characterized the system we have to start a building schedule. Unfortunately, there is no generic algorithm to build such a schedule, especially when dealing with multicore systems, and dependencies between jobs. This is why often these schedules are designed by hand. Figure 14-11 shows a potential schedule for a two cores example while Figure 14-12 shows a similar schedule for a one core example. We will omit here how exactly those schedules can be designed and refer readers to books on real-time systems; instead, we want to focus on how they can be actually implemented using RTOSes.

An offline scheduler in its basic form (clock based) may be implemented as an endless loop where each job is implemented as a 'run to completion' procedure and the scheduler calls each function at a predefined time and order. This way of building a scheduler is sometimes called 'cyclic executive' [3].

For example:

```
while(1)
{
start_timer();
job_a_1();
wait_for_time(5);
job_a_2();
wait_for_time(8);
///etc.
}
```

In this case, at the beginning of the period the A.1 job is executed and then a scheduler waits for the next job A.2 release time and so forth. A trivial implementation for wait_for_time() function is polling time-base.

```
wait_for_time (int next_time)
{
while(read_time() < next_time)
;
}
```

The read_time() function could be implemented in different ways; the straightforward implementation involves reading a value of the hardware timer (the software timer cannot be used as it must be implemented as an interrupt). In our example with the voice media gateway, read_time() should read values of a timer that is connected to TDM interface: one that paces the entire system or even the status of this interrupt directly; it still can be seen as a clock driven scheduler. Furthermore, in many cases, if the system is designed to be paced by some internal

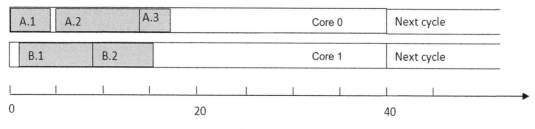

Figure 14-11:
Two cores example.

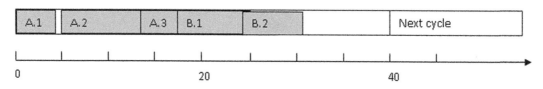

Figure 14-12:
One core example.

periodic event, e.g., TDM frame, or video synchronization signal, we still can use the same techniques.

SmartDSP OS allows to choose a source for hardware timers as a parameter of osHWTimerCreate() but does not allow reading of timer value directly so the users would have to access timer registers directly.

Another question is: in which context should this while() loop be executed. One possibility is to create a high priority task and assign it execution of this loop, and another possibility is to utilize an idle or background task. This task is executed when no other task is ready to run so if no tasks were defined in the system this task will always run: this is the behavior we want to achieve for the offline scheduler. Both SYS/BIOS and SmartDSP OS have an ability to modify background loop. In SmartDSP OS

os_status osActivate(os_background_task_function background_task);

accepts as it argument a pointer to the background task that OS will execute. In SYS/BIOS, there is more sophistication around the background task and it is possible to set functions that the idle task will execute or it is possible to substitute the background task in the same way that SmartDSP OS does. The most basic scheduling technique is calling jobs' subroutines in a background loop but there is some.

There are several things we could improve in the program we wrote earlier. What happens if we decide to change a predefined schedule table to another one at run time? To do that we would prefer having scheduler parameters to act as parameters to this cyclic scheduler function. So, the situation will look as follows:

```
cyclic_schedule(sched_params* params)
{
Int i;
start_timer();
for(i = 0;params[i].job != null; i++)
{
wait_for_time(params[i]->release_time)
params[i]->job();
}
}
```

Some possible enhancements

There are some cases where it is important to take into account the possibility that a job will be executing for a time longer then its maximum execution time. This situation is possible because of some error in a job's code or we might intentionally decide to take into account the rare case where some of the deadline is broken but keep our maximum execution time to a minimum. The question is how a scheduler can react to this situation. Sometimes it is possible to set some checkpoint in a job and once it is clear that the job is sliding beyond its maximum execution time to some predefined red-zone, the job will break execution in an orderly fashion. To implement it the job will have to read the timer from time to time and make sure it can finish execution before its maximum execution time. If we look at our example using media gateway, it is possible that for some worst case input data, we cannot process all the channels in the allowed time; what we can do in such a case is to send data from a previous cycle. What we essentially accomplish in this way is making sure that jobs will never exceed more than maximum execution time. What if it is not feasible to set such checkpoints in the code? In that case we may try to use RTOS services to signal a job to break.

One way to do this is to set up a timer for a time instance before a specific job must finish. It can be done in SYS/BIOS by utilizing the following sequence:

```
Timer_Params timer_params;
Timer_Params_init(&timer_params);
/* Default is periodic and we need here one shot */
timer_params.runMode = Timer_RunMode_ONESHOT;
/* Set time it expires */
Timer_params.period = 1000;
timer_handler = Timer_create(Timer_ANY, timer_handler, &timer_params, &eb);
/* Here we call a function that represent a task or activate specific OS thread */
Timer_reconfig(timer_handler, timer_handler, &timer_params, &eb);
timer_handler() Function will be called when timer expires.
```

If jobs are implemented as a function call then the straightforward way to accomplish a desired behavior is using the POSIX-style signal; we could define a signal to handle a 'run out of time' case per job and then use setjmp/longjmp functionality to get back to a main loop. However, neither SYSBIOS nor SmartDSP OS support such functionality as part of the run time library that is utilized.

Let's go back now to our scheduler and notice that although we perform a direct function call in the code, it should be possible to implement those jobs as independent threads. There are two available thread types for this purpose in SmartDSP OS and SYS/BIOS: tasks and software interrupts. We want to be able to start a job and halt execution of the thread externally and asynchronously with the thread itself. Software interrupt can be easily triggered but is not suitable as it cannot be halted — it runs always to completion. So, we have only one option to try: task. One can trigger a task execution by sending a mailbox message to a task that is waiting for it. In the case of SYSBIOS, it may look like this:

```
for (;;) {
//Wait here at the beginning of each cycle
Mailbox_pend(joba_trigger_mailbox, &some_parameters, BIOS_WAIT_FOREVER);
process_data(some_parameters);
}
```

How can we 'kill' a task from the outside context (externally) and let this task call some clean-up function? Neither SYSBIOS nor SmartDSP OS support such functionality in a straightforward way. However, it is still possible to accomplish this using existing API. In SmartDSP OS it is possible to suspend a task using osTaskSuspend() in timer ISR and then call osTaskDelete() function that removes a task from the scheduler but there is no way to define a hook function that will be called before the task is deleted in this task context and we need such a function in order to be able to release all the resources associated with a task and possibly also to perform some other actions to gracefully break a job. A workaround for this problem is to export all the task's resources to global space so that we can clean up the resources after osTaskDelete(). In the SYS/BIOS case, it is possible to dynamically delete a task from the system with a predefined hook. In timer's ISR we can trigger a high priority task (Task_delete() cannot be called from SWI or HWI). The task must be dynamically created in advance.

Another practical and important consideration is scheduling aperiodic jobs in this model. Those jobs do not fit directly the cyclic executive paradigm as they do not have a pre-defined period. One possible solution within the cyclic executive paradigm is to preserve some time for polling for aperiodic jobs when executing the cyclic executive. So this is what we will have:

```
cyclic_schedule(sched_params* params)
{
Int i;
```

```
start_timer();
for(i = 0;params[i].job != null; i++)
{
wait_for_time(params[i]->release_time)
params[i]->job();
}
poll_for_aperiodic_jobs(); // execute aperiodic jobs here
}
```

Here we practically convert aperiodic tasks to periodic. How can we implement event polling in SmartDSP OS, as an example? Let's assume that we want to handle network egress traffic in our cyclic scheduler. Typically it will be done using hardware interrupts that are automatically initialized by OS's stack but in our case we cannot use interrupts as it may change the timing for already scheduled tasks. So, we have the following alternatives:

- Disable interrupts when hard real-time tasks are executing and then re-enable when the system is ready to handle aperiodic jobs
- Disable interrupts and the explicitly call required functions

When we look at network driver, we also have to understand how exactly received packets are handled. Consider the situation where a handler will try to process all the frames that were received previously. If this is the implementation then we cannot guarantee maximum execution time (or more precisely the maximum execution time will be equal to the maximum number of frames that it is physically possible to receive, multiplied by the processing time of each frame — this could be a rather unrealistic number).

In SDOS, users may call the following function

```
osBioChannelCtrl(bio_rx, CPRI_ETHERNET_CMD_RX_POLL, NULL) != OS_SUCCESS);
```

This function will handle *one* Ethernet frame at a time: this is easy to see when we look at the low level driver that handles this call. Thus, it should be possible to handle Ethernet ingress traffic in the way we described earlier.

We notice that it is not always possible to 'convert' aperiodic tasks to periodic: they may have release times and deadlines that will not make this scheduling feasible. Clearly, because we are considering here off-line scheduling with a-priori known job parameters, introducing aperiodic jobs may create problem as it does not fit this model; thus more sophisticated schemes that include both approaches have been devised (e.g., [11]).

To conclude, in this chapter we have discussed how it is possible to build offline schedules and implement them using DSP RTOSes. We have found that basic implementations are possible in SYS/BIOS and SmartDSP OS. As regards more advanced features, we have found that some aspects of the functionality's implementation could be challenging (e.g., halting and restarting tasks) and that some of the functionality depends very much on specific implementation.

Online scheduling (priority based scheduling)

In contrast to the offline scheduler, the online scheduler makes decisions at run-time based on system events. Each job is assigned a priority according to some predefined algorithm and the decisions will be made on those priorities. The algorithms may set the priority once only (fixed-priority) or change it depending on system's state (dynamic priority).

Online scheduling has two roles:

* Accepting a new job into the system, verifying that jobs are schedulable
* Run-time decisions on which job will be executed next

In this article we will focus on the latter role as the former (schedulability test) has not much to do with the operating system and depends entirely on the algorithm and job set. As we have stated before, in this is not a chapter on real-time scheduling algorithms and our primary goal is to understand how this theoretical work translates into OS features that can be utilized by developers. We will start first with static priority algorithms.

Static priority scheduling

In this section we will learn two algorithms that are used determine the priorities for each task. Those priorities are called "static" as they do not change during system's execution. All DSP RTOSes support this kind of scheduling natively.

Rate monotonic scheduling

Rate monotonic method assumes the following (from reference [4]):

* (A1) The requests for all tasks for which hard deadlines exist are periodic, with constant intervals between requests.
* (A2) Deadlines consist of run-ability constraints only, i.e., each task must be completed before the next request for it occurs.
* (A3) The tasks are independent in that requests for a certain task do not depend on the initiation or the completion of requests for other tasks.
* (A4) Run-time for each task is constant for that task and does not vary with time. Run-time here refers to the time which is taken by a processor to execute the task without interruption.
* (A5) Any nonperiodic tasks in the system are special; they are initialization or failure-recovery routines; they displace periodic tasks while they themselves are being run, and do not themselves have hard, critical deadlines.

Because of these constraints, rate monotonic scheduling in its pure form is rarely used and we believe that the most important usable result of rate monotonic theory is the estimation of upper bound of core utilization.

From [15] "Rate-monotonic scheduling principles translate the invocation period into priorities. Priorities may also be based on semantic information about the application, reflecting the criticality with which the scheduler must deal with some event, for example." So, the idea is to assign priorities to each task according to their period: the shorter the period, the higher the priority. For example, if PRI_0 is highest and PRI_256 is lowest and the period of task A is 10, task B is 20 and task C is 15, the algorithm may assign PRI_10 to A, PRI_15 to C and PRI_20 to B. In reality the assumptions above rarely hold in pure form. For example, (A3) will not hold if two jobs have to use a shared resource using semaphore. In that case a priority inversion may happen that could cause jobs missing their deadlines even if otherwise the system is schedulable under RMS. We will discuss the priority inversion situation in more depth later in this paper.

Reference [4] states a very important result for the utilization bound of the system. If m is the number of tasks and U is the utilization of the system then under RM scheduling:

$$U = m\left(2^{\frac{1}{m}} - 1\right)$$

Table 14-3 summarizes the utilization bound for different task numbers.

This means that any jobs set has a feasible schedule under RM if its utilization does not exceed 0.69 (ln2). This is the worst case scenario the situation is better for random jobs sets where the RM scheduler can achieve up to 88% utilization [13]; for harmonic periods it is 100%. Those results are very important specifically for DSP applications as many of them

Table 14-3: Rate monotonic analysis.

Number of Tasks	Utilization Bound
1	1.0
2	.82
3	.78
4	.76
5	.74
6	.73
7	.73
8	.72
9	.72
infinity	.69

feature harmonic periods for jobs. More resent research by Bini at al. proposed a schedulability test called Hyperbolic Bound that provides better results compared to [14].

Deadline monotonic schedule

Deadline monotonic algorithm is similar to rate monotonic except that it assigns priorities reversely to relative deadline instead of period. In that sense, it is weakening one of rate monotonic algorithm's constrains (A2) so that RM is a degenerative form of DM when the period is equal (or proportional) to relative deadline. As [12] states "Deadline-monotonic priority assignment is an optimal static priority scheme (see theorem 2.4 in [5]). The implication of this is that if any static priority scheduling algorithm can schedule a process set where process deadlines are unequal to their periods, an algorithm using deadline-monotonic priority ordering for processes will also schedule that process set."

Dynamic priority scheduling

Earliest deadline first

EDF algorithm assigns highest priority to a job that has the closest absolute deadline.

Let's look at the example described in Table 14-4, and the resulting schedule in Figure 14-13. Tasks A.1 and A.2 are released at 0. At this point a scheduler should make a decision as to which one will run first. EDF does this by calculating which job has the closest deadline. In our case it is A.1 as its deadline is 7 and that of A.2 is 10. At time 7 A.1 finishes and A.2 takes control. At 10 A.3 is released and preempts A.2 as its deadline is 15 and A.3's deadline is 25. After A.3 finishes at 15; A.2 takes control and completes at 22.

EDF is an optimal scheduling algorithm for independent, preemptable jobs, which means that if there is any algorithm that can produce a feasible schedule EDF will produce a schedule too. This seems like a very appealing algorithm to use, however we can say with a high level of probability that the RM algorithm is used much more frequently in practice. So, what are the reasons for this? There are several reasons and we will start with how this algorithm may be implemented using SmartDSP OS or SYS/BIOS.

As we saw before, RM or DM's implementations are trivial as all DSP RTOSes support static priority scheduling and the above algorithms only show how the priorities are defined.

Table 14-4: Job characteristics.

Job\Parameter	Release Time	Relative Deadline	Maximum Execution Time
A.1	0	7	7
A.2	0	25	10
A.3	10	5	5

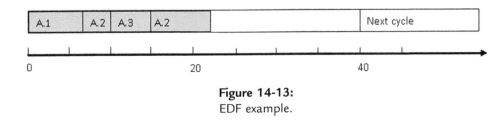

Figure 14-13:
EDF example.

Things are different for EDF. OS has to be able to change the priorities of a periodic job when it is released.

With SmartDSP, OS users can call the osTaskPrioritySet() function to adjust the priority. One important feature that the EDF algorithm requires is keeping FIFO for jobs in the same priority: for example, if task A is executing and task B is released so that at the release point the absolute deadlines are the same for both tasks. Task B should be put into the job queue and executed after A finishes. When we look at SmartDSP OS sources, we can see that in fact the following is what happens:

Function osTaskPrioritySet() will eventually call taskReadyAdd() for a task that its priority is set and it will call list_tail_add() putting task to the end of the list. So, from this perspective SmartDSP OS provides adequate implementation.

Another problem is that sometimes all priorities have to be remapped. Consider the following example [6]: "...consider the case in which two deadlines da and db are mapped into two adjacent priority levels and a new periodic instance is released with an absolute deadline dc, such that da <dc <db. In this situation, there is not a priority level that can be selected to map dc, even when the number of active tasks is less than the number of priority levels in the kernel. This problem can only be solved by remapping da and db into two new priority levels which are not consecutive. Notice that, in the worst case, all current deadlines may need to be remapped, increasing the cost of the operation." This is especially true for SmartDSP OS as it has only 32 priorities and such remapping may happen relatively often.

With SYS/BIOS users can use the following function:

UInt Task_setPri(Task_Handle handle, Int newpri);

It will also add task to the end of the priority queue which is expected by EDF. Number of different priorities levels for SYS/BIOS is 32 for some platforms and 16 for others so in that sense it may require even more remapping than SmartDSP OS.

Another possible approach to take is to implement your own scheduler that will bypass scheduler of OS. For SYS/BIOS one possibility is to use negative priorities (-1) for all the tasks except the one that we want to execute. So, scheduler implementation could use EDF to determine the next task to run and then set its task's priority to a non-negative number.

The scheduler also can intercept task switches if required, using a hook. This mechanism may be required to be used when jobs use semaphores, as an example.

With SmartDSP OS it is possible to use osTaskActivate()/osTaskSuspend() pair to control a schedule from outside; however, there is no mechanism to control the context switch so a more complex scenario with EDF may not work.

To conclude the RTOS implementation topic, it may be challenging to implement EDF on top of existing RTOSes schedulers because they support a different paradigm that does not directly fit EDF.

This is not the only reason EDF did not find its way into many real-time DSP designs. Another problem with EDF is that it may produce less predictable results under high load. As [8] states, "The timing behavior of a system scheduled according to fixed-priority algorithm is more predictable than that of system scheduled according to dynamic-priority algorithm. When task have fixed priorities, overruns of jobs in task can never affect higher-priority tasks. It is possible to predict which tasks will miss their deadlines during an overload."

Offline versus online scheduling

The difference between two types of scheduling is not as obvious as it seems at first glance. For both types of scheduling, users must perform offline research, at least related to schedulability of the jobs. As [10] states, "... we can conclude that the terms 'offline' and 'online' scheduling cannot be seen as disjointed in general. Real-time scheduling requires online guarantees, which require assumptions about online behavior at design time. At runtime, both offline and online execute according to some (explicitly or implicitly) defined rules, which guarantee feasibility. The question offline versus online is thus less black and white, but more about how much of the decision process is online. Thus, both offline and online are based on a substantial online part. The question is then where to set the tradeoff between determinism — all decisions online — and flexibility — some decisions online."

Priority inversion

As we have mentioned before, one of the effects that prevent simple deployment of scheduling techniques is priority inversion. Priority inversion is defined as a situation where a lower priority job is executed in time when a higher priority job should run instead. According to this definition calls that disable scheduler (e.g., osTaskSchedulerLock() in SmartDSP OS or SYS/BIOS' Task_disable()), disabling interrupts, utilizing semaphores or any other similar actions may cause priority inversion. Let's look at a couple of examples of priority inversion starting with Figure 14-14.

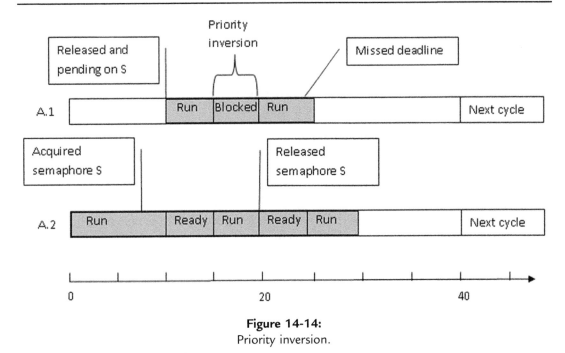

Figure 14-14:
Priority inversion.

In the above diagram A.1 job has higher priority than A.2. A.1 release time is 10 and A.2 is 0. Relative deadline and maximum execution time for A.1 are 15. Relative deadline for A.2 is 35 and maximum execution time is 20.

At time 0 job A.2 releases and begins executing.
At time 7 job A.2 acquires semaphore S.
At time 10 job A.1 releases and scheduler puts it to execution as one having higher priority.
At time 15 job A.1 tries to acquire semaphore S and as a result blocks and A.2 resumes.
At time 20 job A.2 releases semaphore S and A.1 resumes as it has higher priority.
At time 25 job A.1 misses its deadline and terminates (we assume here that the job terminates itself if it misses the deadline). Job A.2 resumes.
At time 30 job A.2 completes.

In this example, job A.1 misses its deadline as result of priority inversion that was in turn caused by resource contention. In such case the priority inversion was bound as A.2 kept the resource and continued execution and would eventually release the resource. It is important to mention that the same behavior would arise if an implementation used scheduler locking of any kind.

In Figure 14-15, the A.1 job is hard real time and it has highest priority. A.1 release time is 10, maximum execution time is 15 and relative deadline is 15. A.2 is a sporadic, non real-time job

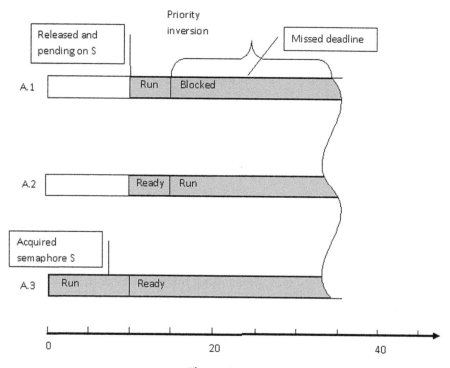

Figure 14-15:
Unbounded priority inversion.

and it is released at 10 as well and is executed for 25. A.3 is a non real-time job too and it is released at 0.

The problem here is that a medium priority sporadic job is executing while a higher priority job is blocked by a low priority job holding mutually exclusive resources.

Until this point we have described the problem by providing some basic examples for it. For the first type of priority inversion that we have described (bounded case) there is no 'magic' algorithm to apply and the solutions may include:

- Shorten length of shared resource usage.
- Use a synchronization method that is the weakest possible. For example, if threads that contend for resources are implemented as tasks then use semaphores or disable the scheduler etc.
- Avoid resource sharing. In many cases it is possible to avoid resource sharing by proper usage of hardware and partitioning.
- Deploy virtualization techniques so that resources are virtualized thru some central entity and thus can be used in parallel.

Table 14-5: Multithreading disable API.

	HWI	SWI	Task
SYS/BIOS	Hwi_disable(); Hwi_enable();	Swi_disable(); Swi_restore(key);	Task_disable(); Task_restore(key);
SmartDSP OS	osHwiDisable(); osHwiEnable();	osSwiDisable() osSwiEnable()	osTaskSchedulerLock() osTaskSchedulerUnLock()

Clearly, all these methods will help for an uncontrolled case as well, but for such a case there are some additional methods that can be deployed.

A very straightforward way is to use scheduler disabling methods as shown in Table 14-5.

Disabling scheduler

The simplest way to resolve uncontrolled priority inversion is by disabling scheduling of the appropriate threading layer before task begins using shared resource. It is easy to see why it resolves any uncontrolled priority inversion: no other thread can start executing while the resource is used (that can be weakened to some extent. For example, if we know that HWI are short enough we can only lock tasks and SWI). This approach will work adequately when the time period for which the resource is utilized is relatively short; with longer periods of time another problem arises: bounded priority inversion relative to *all* threads in the system. So, we basically traded a possibility for uncontrolled priority inversion to a generally higher possibility of controlled priority inversion.

Priority inheritance

Another possible solution that is proposed by [7] is uses of the priority inheritance algorithm. The idea behind this algorithm is simple: when job A holds resource and higher priority job B tries to use it as well, A inherits B's priority so it is able to complete its work with the resource. This approach has advantages over the previous one as tasks that have higher priority and do not contend over resources will preempt, while in the previous, simplistic approach this will not happen thus introducing additional priority inversion. Priority inheritance is often supported by OSes, for example, SYS/BIOS supports the priority inheritance algorithm with GateMutexPri module [18]: "To guard against priority inversion, GateMutexPri implements priority inheritance: when task High tries to acquire a gate that is owned by task Low, task Low's priority will be temporarily raised to that of High, as long as High is waiting on the gate. Task High will 'donate' its priority to task Low. When multiple tasks wait on the gate, the gate owner will receive the highest priority of any of the tasks waiting on the gate."

SmartDSP OS does not support priority inheritance directly. Some changes in API will be required if someone decides to implement this algorithm on top of existing API. For example, current API does not allow the getting of information on a pending task.

Priority ceiling

The inheritance algorithm does not resolve all possible issues (specifically when we consider several resources that need to be protected) such as deadlocks and may produce some timing anomalies [17] so a more sophisticated algorithm was introduced — the priority ceiling algorithm [7]. No DSP RTOS we looked at implemented this later algorithm.

Tools support for DSP OSes

Several types of tools exist today:

- Log visualization: tools visualizing an OS's log.
- Context support in IDE: browsing thread's variables.
- Configuration: visual configuration of OS supported modules, drivers, etc.

Let's discuss all the above tools in more detail.

When you debug a race condition it is beneficial to see if some thread is preempted; when you try to find issues related to priority inversion, it is important to understand the timeline of different events that happen in the system in run-time. Hardware interrupts, scheduling events, context switches, and user-defined events are logged and then visualized (see Figure 14-16). It is also very useful and easy with this kind of tool to measure intervals between different events. The implementation of such a tool has two parts: logging that happens on the target and tool itself that can get this data from the target and visualize. Implementation on the target should be very effective and not intrusive as it should not affect normal execution of the program.

SYS/BIOS's tool that supports above functionality is System Analyzer [22]. Users can also add functionality on the target using UIA (the Unified Instrumentation Architecture) concepts. For example, to measure time between two calls using System Analyzer users can call:

```
Log_write1(UIABenchmark_start, (xdc_IArg)"user_thread_1");
```

And then:

```
Log_write1(UIABenchmark_stop, (xdc_IArg)"user_thread_1");
```

SmartDSP OS has a similar tool (see screenshot in Figure 14-17).

Another important functionality that becomes important today is timeline correlation between different instances of OSes in an AMP system or even different OSes in heterogeneous systems (e.g., Linux and SYS/BIOS).

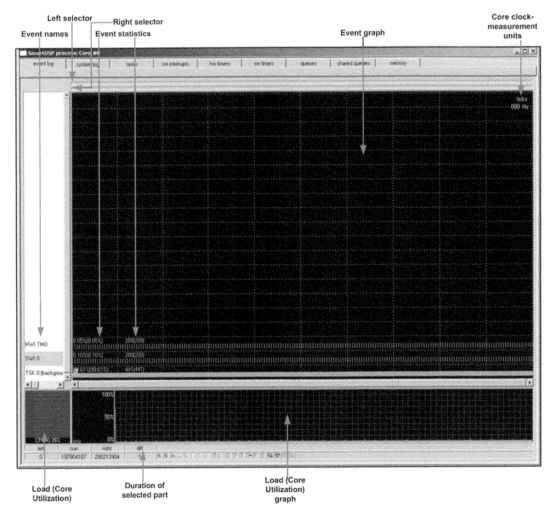

Left selector
Right selector
Event names
Event statistics
Event graph
Core clock-
measurement
units

Load (Core
Utilization)
Duration of
selected part
Load (Core
Utilization)
graph

Figure 14-16:
Log viewer.

Sometimes you need to debug an OS with a source level debugger. With DSPs you will probably use a JTAG connection and some IDE. When RTOS is executing on target the IDE has to be aware of its existence so that it can show information correctly. For example, show variables that are local to specific context, and show program counters for different threads of execution.

Another aspect that is handled by visual tools sometimes is initial configuration. Such a tool can help to include a particular module for OS or to help to configure hardware.

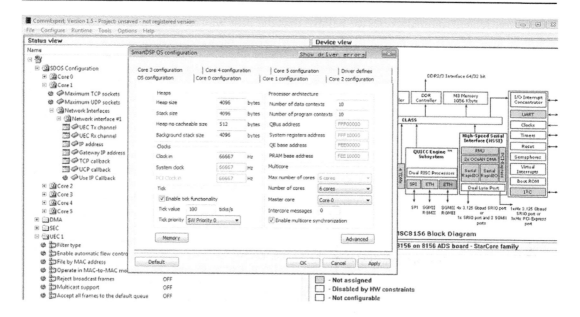

Figure 14-17:
Graphic configuration tool.

Figure 14-17 shows a snapshot of tool that accompanies SmartDSP OS — CommExpert. This tool allows configuration of the OS and underlying hardware using graphical user interface.

Conclusions

Today's DSP RTOSs have to support complex SoCs that may include several DSP cores, numerous peripherals that interact with external devices and networks, memory sub-systems that may have a few cache levels, bus or switch fabric, different types of memory with different accesses types, DMAs, or DSP co-processors. OSes have to abstract and encapsulate this complexity by proper partitioning layers of standard API to handle common functionality like DMA or Ethernet. Multicore platforms are still relatively new in the DSP world but we can already see how DSP RTOSs approach it: hardware resource partitioning, early multiplexing of network traffic, cooperative AMP, and well defined IPC are some of the most important features related to multicore.

On another front, scheduling of real-time jobs in such a complex environment still remains challenging and we can see that not all of those theoretical concepts that we have described made their way into RTOS's code. As for the resource contention topics that we discussed, clearly, there are no mechanisms in existence today that can completely resolve resource contention-related anomalies, so we believe that the best way to resolve this is to avoid resource sharing, with proper partitioning or virtualization-like mechanisms.

References

Refereed journal articles

[1] L.J. Karam, I. AlKamal, A. Gatherer, G.A. Frantz, D.V. Anderson, B.L. Evans, Trends in Multicore DSP Platforms, IEEE Signal Processing Magazine, Special Issue on Signal Processing on Platforms with Multiple Cores 26 (6) (Nov 2009) 38–49.

[2] Edward A. Lee, What is ahead of embedded software? Computer 33 (9) (Sep 2000) 18–26.

[3] C. Douglass Locke, Software architecture for hard real time applications: Cyclic Executives vs. Fixed Priority Executives, Real-time Systems, 4 (1) 37–53

[4] C.L. Liu, JAMESW. Layland, Scheduling Algorithms for Multiprogramming in a Hard Real-Time Environment, Journal of the ACM (JACM) 20 (1) (Jan 1973).

[5] J.Y.T. Leung, J. Whitehead, On the Complexity of Fixed-Priority Scheduling of Periodic, Real-Time Tasks, Performance Evaluation 2 (4) (1982) 237–250. December 1982.

[6] Giorgio C. Buttazzo, Rate Monotonic vs. EDF: Judgment Day. Real-time Systems, 29 (1) 5–26.

[7] Lui Sha, Ragunathan Rajkumar, John P. Lehoczky. Priority Inheritance Protocols: An Approach to Real-Time Synchronization, IEEE Transactions on Computers 39 (9) (Sep 1990) 1175–1185.

Books and book chapters

[8] Real-Time Systems, Prentice Hall, Publication Date: April 23, 2000. ISBN-13: 978-0130996510

Refereed conference papers

[9] Gerhard Fohler, Damir Isović, Tomas Lennvall, and Roger Vuolle. SALSART — AWeb Based Cooperative Environment for Offline Real-time Schedule Design, Proc. Parallel, Distributed and Network-based Processing, 2002, 10th Euromicro Workshop on Issue, 2002, pp. 63–70

[10] Gerhard Fohler, University of Kaiserslautern, How Different are Offline and Online Scheduling? *Proceedings of the 2nd International Real-Time Scheduling Open Problems Seminar*, July 5, 2011, pp. 5–6, Beto, Portugal

[11] Michal Young, Lih-Chyun Shu. Hybrid Online/Offline Scheduling for Hard Real-Time Systems, Proc. 2nd International Symposium on Real-Time and Media Systems. (1996) 231–240.

[12] C. Audsley, A. Burns, M.F. Richardson, A.J. Wellings, Hard Real-Time Scheduling: the Deadline-Monotonic Approach, Proc. IEEE Workshop on Real-Time Operating Systems and Software (1991) 133–137.

[13] John Lehoczky, Lui Sha, Ye Ding, The Rate Monotonic Scheduling Algorithm: Exact Characterization and Average Case Behavior, Proc. Real Time Systems Symposium (5–7 Dec 1989) 166–171. Santa Monica, CA, USA.

[14] E. Bini, G. Buttazzo, A hyperbolic bound for the rate monotonic algorithm, Proc. Real-Time Systems, 13th Euromicro Conference, 13 Jun 2001 — 15 Jun 2001, Delft, Netherlands, pp. 59–66.

Technical reports

[15] Richard M. Karp, On-Line Algorithm Versus Off-Line Algorithm: How Much is it Worth to Know the Future?, International Computer Science Institute, Technical report TR-92-044.

[16] Robert I. Davis, Alan Burns, A Survey of Hard Real-Time Scheduling Algorithms and Schedulability Analysis Techniques for Multiprocessor Systems (2011).http://www.cs.york.ac.uk/ftpdir/reports/2009/YCS/443/YCS-2009-443.pdf. Retrieved at 11.11.

[17] Victor Yodaiken, Against priority inheritance, FSMLABS Technical Report (September 23, 2004).

Product documentation

[18] SYS BIOS API Documentation, Version bios_6_32_00_28

[19] TMS320C6000 Network Developer's Kit (NDK) Software, TI Document identifier spru523g

[20] LINX Protocols, Document Version 21. http://linx.sourceforge.net/linxdoc/doc/book-linx-protocols.pdf. Retrieved 29.11.2011

[21] TIPC: Transparent Inter Process Communication Protocol http://tipc.sourceforge.net/doc/draft-spec-tipc-07.txt. Retrieved 29.11.2011

[22] System Analyzer User's Guide, TI Literature Number: SPRUH43B. July 2011

[23] SYS/BIOS Inter-Processor Communication (IPC) and I/O User's Guide, Literature Number: SPRUGO6C

[24] KeyStone Architecture Multicore Navigator, User Guide, TI Literature Number SPRUGR9D

[25] TMS320C66x DSP generation of devices, TI Literature Number SPRT580A

[26] SysLink User Guide, http://processors.wiki.ti.com/index.php/SysLink_UserGuide. Retrieved 29.11.2011

[27] MSC8156 Data Sheet, Freescale Semiconductor. Document Number: MSC8156, Rev. 4, 10/2011. http://www.freescale.com/webapp/sps/site/prod_summary.jsp?code=MSC8156. Retrieved 11.30.2011

CHAPTER 15

Managing the DSP Software Development Effort

Robert Oshana

Chapter Outline

Introduction

Software development using DSPs is subject to many of the same constraints and development challenges which other types of software development face. These include a shrinking time to market, tedious and repetitive algorithm integration cycles, time intensive debug cycles for real-time applications, and integration of multiple differentiated tasks running on a single DSP, as well as other real-time processing demands. Up to 80% of the

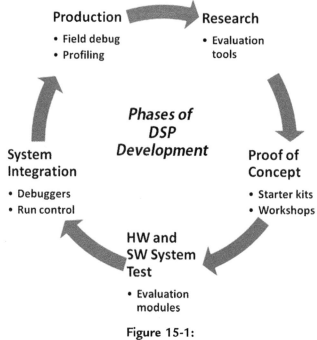

Figure 15-1:
Phases of DSP development.

development effort is involved in analysis, design, implementation, and integration of the software components of a DSP system.

Early DSP development relied on low level assembly language to implement the most efficient algorithms. This worked reasonably well for small systems. However, as DSP systems grow in size and complexity, assembly language implementation of these systems has become impractical. Too much effort and complexity is involved to develop large DSP systems within reasonable cost and schedule constraints. The migration has been towards higher level languages like C to provide the maintainability, portability, and productivity needed to meet cost and schedule constraints. Other real-time development tools are emerging to allow even faster development of complex DSP systems.

DSP development environments can be partitioned into host tooling and target content. Host tooling consists of tools to allow the developer to perform application development tasks such as program build, program debug, data visualization, and other analysis capabilities. Target content refers to the integration of software running on the DSP itself, including the real-time operating system (if needed), and the DSP algorithms that perform the various tasks and functions. There is a communication mechanism between the host and the target for data communication and testing.

DSP development consists of several phases as shown in Figure 15-1. During each phase there exists tooling to help the DSP developer quickly proceed to the next stage.

Challenges in DSP application development

The implementation of software for embedded digital signal processing applications is a very complex process. This complexity is a result of increasing functionality in embedded applications; intense time-to- market pressures; and very stringent cost, power, and speed constraints.

The primary goal for most DSP application algorithms focuses on the minimization of code size as well as the minimization of the memory required for the buffers that implement the main communication channels in the input dataflow. These are important problems because programmable DSPs have very limited amounts of on-chip memory, and the speed, power, and cost penalties for using off-chip memory are prohibitively high for many cost-sensitive embedded applications. Complicating the problem further, memory demands of applications are increasing at a significantly higher rate than the rate of increase in on-chip memory capacity offered by improved integrated circuit technology.

To help cope with such complexity, DSP system designers have increasingly been employing high-level, graphical design environments in which system specification is based on hierarchical dataflow graphs. Integrated development environments (IDE) are also being used in the program management and code build and debug phases of a project to manage increased complexity.

The main goal of this chapter is to explain the DSP application development flow and review the tools and techniques available to help the embedded DSP developer analyze, build, integrate, and test complex DSP applications.

Historically, digital signal processors have been programmed manually using assembly language. This is a tedious and error-prone process. A more efficient approach is to generate code automatically using available tooling. However, the auto-generated code must be efficient. DSPs have scarce amounts of on-chip memory and its use must be managed carefully. Using off-chip memory is inefficient due to increased cost, increased power requirements, and decreased speed penalty. All these drawbacks have a significant impact on real-time applications. Therefore, effective DSP-based code generation tools must specify the program in an imperative language such as C or C++ and use a good optimizing compiler.

The DSP design process

The DSP design process in many ways is similar to the standard software and system development process. However, the DSP development process also has some unique challenges that must be understood in order to develop efficient, high performance applications.

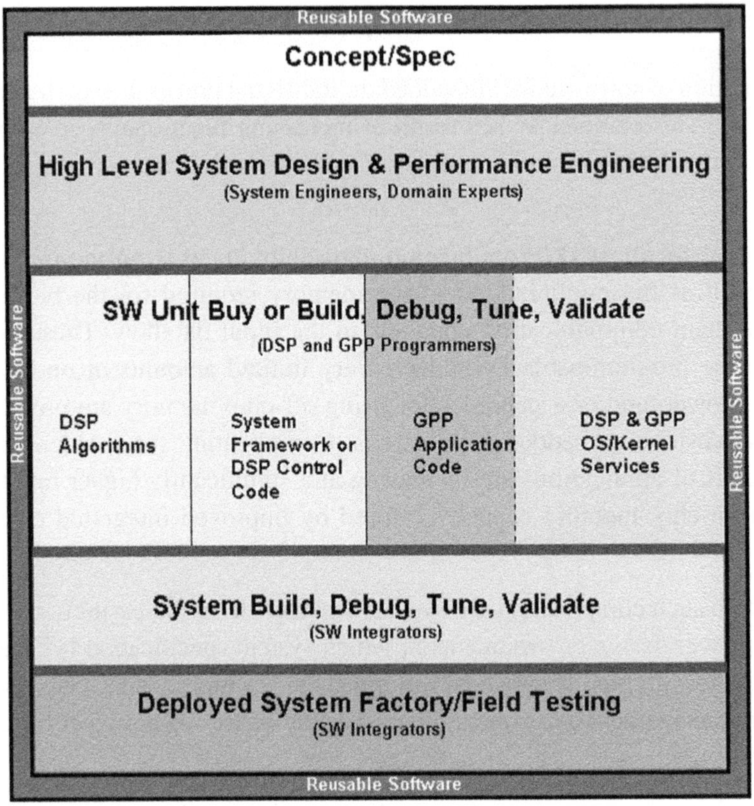

Figure 15-2:
A block diagram of the general system design flow.

A high level model of the DSP system design process is shown in Figure 15-2. As shown in this figure, some of the steps in the DSP design process are similar to those found in the conventional system development process.

Concept and specification phase

The development of any signal processing system begins with the establishment of requirements for the system. In this step, the designer attempts to state the attributes of the system in terms of a set of requirements. These requirements should define the characteristics of the system expected by external observers such as users or other systems. This definition becomes the guide for all other decisions made during the design cycle.

DSP systems can have many types of the requirements that are common in other electronic systems. These requirements may include power, size, weight, bandwidth, and signal quality. However, DSP systems often have requirements that are unique to digital signal processing. Such requirements can include sample rate (which is related to bandwidth), data precision (which is related to signal quality), and real-time constraints (which are related to general system performance and functionality).

The designer devises a set of specifications for the system that describes the sort of system that will satisfy the requirements. These specifications are a very abstract design for the system. The process of creating specifications that satisfy the requirements is called requirements allocation.

A specification process for DSP systems

During the specification of the DSP system, using a real-time specification process such as Sommerville's six-step real-time design process is recommended[1]. The six steps in this process include:

1. Identify the stimuli to be processed by the system and the required responses to these stimuli.
2. For each stimulus and response, identify the timing constraints. These timing constraints must be quantifiable.
3. Aggregate the stimulus and response processing into concurrent software processes. A process may be associated with each class of stimulus and response.
4. Design algorithms to process each class of stimulus and response. These must meet the given timing requirements. For DSP systems, these algorithms consist mainly of signal processing algorithms.
5. Design a scheduling system which will ensure that processes are started in time to meet their deadlines. The scheduling system is usually based on a pre-emptive multitasking model, using a rate monotonic or deadline monotonic algorithm.
6. Integrate using a real-time operating system (especially if the application is complex enough).

Once the stimuli and responses have been identified, a software function can be created by performing a mapping of stimuli to responses for all possible input sequences. This is important for real time systems. Any unmapped stimuli sequences can cause behavioral problems in the system.

Algorithm development and validation

During the concept and specification phase, a majority of the developer's time is spent doing algorithm development. During this phase, the designer focuses on exploring approaches to solve the problems defined by the specifications at an abstract level. During this phase, the algorithm developer is usually not concerned with the details of how the algorithm will be implemented. Moreover, the developer focuses on defining a computational process which can satisfy the

---

[1] *Software Engineering version 9* by Ian Sommerville, chapter 16.

system specifications. The partitioning decisions of where the algorithms will be hosted (DSP, GPP, Hardware acceleration such as ASIC or FPGA) are not the number one concern at this time.

A majority of DSP applications require sophisticated control functions as well as complex signal processing functions. These control functions manage the decision-making and control flow for the entire application (for example, managing the various functional operations of a cell phone, as well as adjusting certain algorithm parameters based on user input). During this algorithm development phase, the designer must be able to specify as well as experiment with both the control behavior and the signal processing behavior of the application.

In many DSP systems, algorithm development first begins using floating-point arithmetic. At this point, there is no analysis or consideration of fixed-point effects resulting from running the application on a fixed point processor. Not that this analysis is not important. It is very critical to the overall success of the application and is considered shortly. But the main goal is to get an algorithm stream working, providing the assurance that the system can indeed work! When it comes to actually developing a productizable system, a less expensive fixed point processor may be the choice. During this time transition, the fixed point effects must be considered. In most cases it will be advantageous to implement the productizable system using simpler, smaller numeric formats with lower dynamic range to reduce system complexity and cost. This can only be found on fixed point DSPs.

For many kinds of applications, it is essential that system designers have the ability to evaluate candidate algorithms running in real-time, before committing to a specific design and hardware/software implementation. This is often necessary where subjective tests of algorithm quality are to be performed. For example, in evaluating speech compression algorithms for use in digital cellular telephones, real-time, two-way communications may be necessary.

DSP algorithm standards and guidelines

DSPs are often programmed like 'traditional' embedded microprocessors. They are programmed in a mix of C and assembly language, they directly access hardware peripherals, and for performance reasons, almost always have little or no standard operating system support. Thus, like traditional microprocessors, there is very little use of commercial off-the-shelf (COTS) software components for DSPs. However, unlike general-purpose embedded microprocessors, DSPs are designed to run sophisticated signal processing algorithms and heuristics. For example, they may be used to detect important data in the presence of noise, to or for speech recognition in a noisy automobile traveling at 65 miles per hour. Such algorithms are often the result of many years of research and development. However, because of the lack of consistent standards, it is not possible to use an algorithm in more than one system without significant reengineering. This can cause significant time to market issues for DSP developers. Appendix F provides more detail on DSP algorithm development standards and guidelines.

High level system design and performance engineering

High level system design refers to the overall partitioning, selection, and organization of the hardware and software components in a DSP system. The algorithms developed during the specification phase are used as the primary inputs to this partitioning and selection phase. Other factors are considered as well:

- Performance requirements
- Size, weight, and power constraints
- Production costs
- Non-recurring engineering (engineering resources required)
- Time to market constraints
- Reliability

This phase is critical, as the designer must make trade-offs among these often conflicting demands. The goal is to select a set of hardware and software elements to create an overall system architecture that is well-matched to the demands of the application.

Modern DSP system development provides the engineer with a variety of choices to implement a given system. These include:

- Custom software
- Optimized DSP software libraries
- Custom hardware
- Standard hardware components

The designer must make the necessary system trade-offs in order to optimize the design to meet the system performance requirements that are most important (performance, power, memory, cost, manufacturability, etc.).

Performance engineering

Software performance engineering (SPE) aims to build predictable performance into systems by specifying and analyzing quantitative behavior from the very beginning of a system, through to its deployment and evolution. DSP designers must consider performance requirements, the design, and the environment in which the system will run. Analysis may be based on various kinds of modeling and design tools. SPE is a set of techniques for:

- Gathering data
- Constructing a system performance model
- Evaluating the performance model
- Managing risk of uncertainty
- Evaluating alternatives
- Verifying the models and results

SPE requires the DSP developer to analyze the complete DSP system using the following information[2]:

- *Workload*; worst case scenarios
- *Performance objectives*; quantitative criteria for evaluating performance (CPU, memory, I/O)
- *Software characteristics*; processing steps for various performance scenarios (ADD, simulation)
- *Execution environment*; platform on which the proposed system will execute, partitioning decisions
- *Resource requirements*; estimate of the amount of service required for key components of the system
- *Processing overhead*; benchmarking, simulation, prototyping for key scenarios

Software development

Most DSP systems are developed using combinations of hardware components and software components. The proportion varies depending on the application (a system that requires fast upgradability or changeability will use more software; a system whose deploys mature algorithms and requires high performance may use hardware). Most systems based on programmable DSP processors are often software-intensive.

Software for DSP systems comes in many flavors. Aside from the signal processing software, there are many other software components required for programmable DSP solutions:

- Control software
- Operating system software
- Peripheral driver software
- Device driver software
- Other board support and chip support software
- Interrupt routine software

There is a growing trend towards the use of reusable or 'off the shelf' software components. This includes the use of reusable signal processing software components, application frameworks, operating systems and kernels, device drivers, and chip support software. DSP developers should take advantage of these reusable components whenever possible. The topic of reusable DSP software components is that of a later chapter.

System build, integration, and test

As DSP system complexity continues to grow, system integration becomes paramount. System integration can be defined as the progressive linking and testing of system

---

[2] See *Performance Solutions* as an excellent reference to this technology

components to merge their functional and technical characteristics into a comprehensive interoperable system. System integration is becoming more common in DSP systems can contain numerous complex hardware and software subsystems. It is common for these subsystems to be highly interdependent. System integration usually takes place throughout the design process. System integration may first be done using simulations prior to the actual fabrication of hardware.

System integration proceeds in stages as subsystem designs are developed and refined. Initially, much of the system integration may be performed on DSP simulators interfacing with simulations of other hardware and software components. The next level of system integration may be performed using a DSP evaluation board (this allows the software to be integraed with device drivers, board support packages, kernel, etc.). A final level of system integration can begin once the remaining hardware and software components are available.

Factory and field test

Factory and field test includes the remote analysis and debugging of DSP systems in the field or in final factory test. This phase of the lifecycle requires sophisticated tools which allow the field test engineer to quickly and accurately diagnose problems in the field and report those problems back to the product engineers to debug in a local lab.

Design challenges for DSP systems

What defines a DSP system are signal processing algorithms used in the application. These algorithms represent the numeric recipe for the arithmetic to be performed. However, the implementation decisions for these algorithms are the responsibility of the DSP engineer. The challenge for the DSP engineer is to understand the algorithms well enough to make intelligent implementation decisions what endure the computational accuracy of the algorithm while achieving 'full technology entitlement' for the programmable DSP in order to achieve the highest performance possible.

Many computationally intensive DSP systems must achieve very rigorous performance goals. These systems operate on lengthy segments of real-world signals that must be processed in real-time. These are hard real time systems that must always meet these performance goals, even under worst case system conditions. This is orders of magnitude more difficult than with a soft real time system where the deadlines can be missed occasionally.

DSPs are designed to perform certain classes of arithmetic operations such as addition and multiplication very quickly. The DSP engineer must be aware of these advantages and be able to use the numeric formats and type of arithmetic wisely to have a significant influence on the overall behavior and performance of the DSP system. One important choice is the selection of fixed-point or floating-point arithmetic. Floating-point arithmetic provides much greater dynamic range than does fixed-point arithmetic. Floating point also reduces the probability of overflow and the need for the programmer to worry about scaling. This alone can significantly simplify algorithm and software design, implementation, and test.

The drawback to floating-point processors (or floating point libraries) is that they are slower and more expensive than fixed-point. DSP engineers must perform the required analysis to understand the dynamic ranges needed throughput the application. It is highly probable that a complex DSP system will require different levels of dynamic range and precision at different points in the algorithm stream. The challenge is to perform the right amount of analysis in order to use the required numeric representations to achieve the performance required from the application.

In order to properly test a DSP system, a set of realistic test data is required. This test data may represent calls coming into a base station or data from another type of sensor that represents realistic scenarios. These realistic test signals are needed to verify the numeric performance of the system as well as the real-time constraints of the system. Some DSP applications must be tested for long periods of time in order to verify that there are no accumulator overflow conditions or other 'corner cases' that may degrade or break the system.

High level design tools for DSP

System-level design for DSP systems requires both high-level modeling to define the system concept and low level modeling to specify behavior details. The DSP system designer must develop complete end-to-end simulations and integrate various components such as analog and mixed signal, DSP, and control logic. Once the designer has modeled the system, the model must be executed and tested to verify performance to specifications. These models are also used to perform design trade-offs, what-if analysis, and system parameter tuning to optimize performance of the system.

DSP modeling tools exist to aid the designer in the rapid development of DSP-based systems and models. Hierarchical block-diagram design and simulation tools are available to model systems and simulate a wide range of DSP components. Application libraries are available for use by the designer that provide many of the common blocks found in DSP and digital communications systems. These libraries allow the designer to build complete end-to-end systems. Tools are also available to model control logic for event driven systems.

DSP toolboxes

DSP toolboxes are collections of signal processing functions that provide a customizable framework for both analog and digital signal processing. DSP toolboxes have graphical user interfaces that allow interactive analysis of the system. The advantage of these toolbox algorithms is that they are robust, reliable, and efficient. Many of these toolboxes allow the DSP developer to modify the supplied algorithm to more closely map to the existing architecture as well as the ability for the developer to add custom algorithms to the toolbox.

DSP system design tools also provide the capability for rapid design, graphical simulation, and prototyping of DSP systems. Blocks are selected from available libraries and interconnected in various configurations using point and click operations. Signal source blocks are used to test models. Simulations can be visualized interactively passed on for further processing. ANSI standard C code is generated directly from the model. Signal processing blocks, including FFT, DFT; window functions; decimation/interpolation; linear prediction; and multi-rate signal processing are available for rapid system design and prototyping.

Host development tools for DSP development

As mentioned earlier, there are many software challenges facing the real-time DSP developer:

- Life-cycle costs are rising
- Simple system debug is gone
- Software reuse is minimal
- Systems are increasingly complex

A robust set of tools to aid the DSP developer can help speed development time, reduce errors, and more effectively manage large projects, among other advantages. Integrating a number of different tools into one integrated environment is called creating an Integrated Development Environment (IDE). An IDE is a programming environment that has been packaged as an application program. A typical IDE consists of a code editor, a compiler, a debugger, and a graphical user interface (GUI) builder. IDEs provide a user-friendly framework for building complex applications. PC developers first had access to IDEs.

Now applications are large and complex enough to warrant the same development environment for DSP applications. Currently, DSP vendors have IDEs to support their development environments. DSP development environments support some but not all of the overall DSP application development life cycle. As shown in Figure 15-3, the IDE is mainly used after the initial concept exploration, systems engineering, and partitioning phases of the development project. Development of the DSP application from inside the IDE mainly addressed software architecture, algorithm design and coding, the entire project build phase and the debug and optimize phases.

A typical DSP IDE consists of several major components (Figure 15-4):

- Code generation (compile, assemble, link)
- Edit

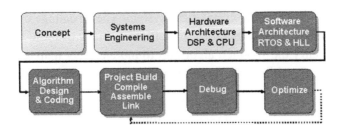

Figure 15-3:
The DSP IDE is useful for some, but not all, of the DSP development life cycle.

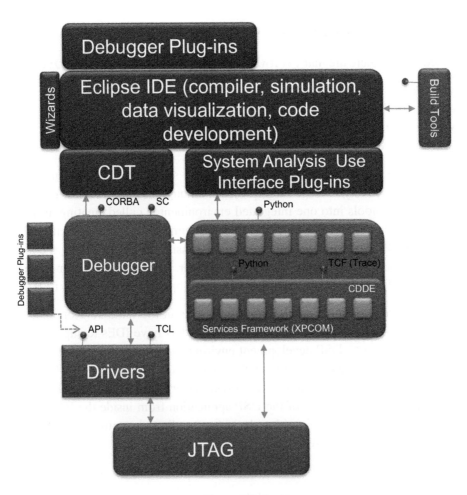

Figure 15-4:
DSP integrated development environment.

- Simulation
- Real-time analysis
- Debug and emulation
- Graphical user interface
- Other 'plug in' tools
- Efficient connection to a target system

A DSP IDE is optimized for DSP development. DSP development is different enough from other development to warrant a set of DSP-centric options within the IDE:

- Advanced real-time Debugging which includes advanced breakpoints, C-expression-based conditional breakpoints, and simultaneous view of source and dis-assembly
- Advanced watch window
- Multi-processor debug
- Global breakpoints
- Synchronized control over groups
- Probe points (advanced break points) provide oscilloscope-like functions
- File I/O with advanced triggering injects or extracts data signals

Data visualization, for example, allows the DSP developer to perform graphical signal analysis. This gives the developer the ability to view signals in native format and change variables on the fly to see their effects. You can read more about this topic in Chapter 17 on Developing and debugging DSP systems.

As DSP complexity grows and systems move from being cyclic executive to task based execution, more advanced tools are required to facilitate the integration and debug phase of development. DSP IDEs provide a robust set of 'dashboards' to help analyze and debug complex real-time applications. If more advanced task execution analysis is desired, a third party plug in capability can be used. For example, if the DSP developer needs to know *why* and *where* a task is blocked, not just *if* a task is blocked, rate monotonic analysis tools can be used to perform more detailed analysis.

DSP applications require real-time analysis of the system as it runs. DSP IDE's provide the ability to monitor the system in real-time with low overhead and interference of the application. Because of the variety of real-time applications, these analysis capabilities are user controlled and optimized. Analysis data is accumulated and sent to the host in the background (a low priority non-intrusive thread performs this function, sending data over the JTAG interface to the host). These real-time analysis capabilities act as a software logic analyzer, performing tasks that, in the past, were performed by hardware logic analyzers. The analysis capability can show CPU load percentage (useful for finding hot spots in the application), the task execution history (to show the sequencing of events in the real-time system), a rough estimate of DSP MIPS (by using an idle counter), and the ability to log best and worst case execution times. This data can help the DSP developer determine whether the system is operating within its design specification, and meeting performance targets, and whether there are any subtle timing problems in the run time model of the system[3].

System configuration tools allow the DSP developer to quickly prioritize system functions and perform what if analysis on different run time models.

The development tool flow starts from the most basic of requirements. The editor, assembler, and linker are, of course, are the most fundamental blocks. Once the linker has built an executable, there must be some way to load it into the target system. To do this, there must be run control of the target. The target can be a simulator (SIM), which is ideal for algorithm checkout when the hardware prototype is not available. The target can also be a starter kit (DSK) or an evaluation board (EVM) of some type. An evaluation board lets the developer run on real hardware, often with a degree of configurable I/O. Ultimately, the DSP developer will run the code on a prototype, and this requires an emulator.

Another important component of the IDE is the debugger. The debugger controls the simulator or emulator and gives the developer low level analysis and control of the program, memory, and registers in the target system. Because of the inherent nature of hardware/ software co-design, the DSP developer must develop at least part of the application using a simulator instead of the real hardware. This means the DSP developer must debug systems without complete interfaces, or ones with I/O, but that don't have real data available[4]. This is where file I/O helps the debug process. DSP debuggers generally have the ability to perform data capture as well as graphing capability in order to analyze the specific bits in the output file. Since DSP is all about code and application performance, it is also important to provide a way to measure code speed with the debugger (this is generally referred to as profiling). Finally, a real-time operating system capability allows the developer to build large complex applications and the Plug-In interface allows third parties (or the developer) to develop additional capabilities that integrate into the IDE.

A generic data flow example

This section will describe a simple example that brings together the various DSP development phases and tooling to produce a fully integrated DSP application. Figure 15-5 shows the simple model of a software system.

Input is processed or transformed and output. From a real-time system perspective, the input will come from some sensor in the analog environment and the output will control some

[3] If you can't see the problem, you can't fix the problem
[4] A case study of the hardware/software co-design as well as the analysis techniques for embedded DSP and micro devices is at the end of the chapter.

Figure 15-5:
A generic data flow example.

actuator in the analog environment. The next step will be to add data buffering to this model. Many real-time systems have a certain amount of input (and output) buffering to hold data while the CPU is busy processing a previous buffer of data. Figure 15-6 shows the example system with input and output data buffering added to the application.

A more detailed model for data buffering is a double buffer model as shown in Figure 15-7. A double buffer is necessary because, as the CPU is busy processing one buffer, data from the

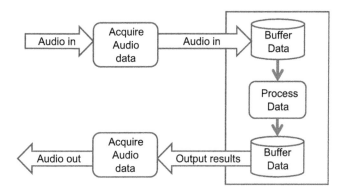

Figure 15-6:
A data buffering model for a DSP application.

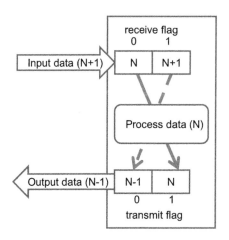

Figure 15-7:
A data buffering model for processing real-time samples from the environment.

external sensors must have a place to go and wait until the CPU has finished processing the current buffer of data. When the CPU finishes one buffer, it will begin processing the next buffer. The 'old' buffer (the buffer that was previously processed by the CPU) is now the current input buffer to hold data from the external sensors. When data is ready to be received (RCV 'ping' is empty, RCV flag = 1), RCV 'pong' buffer is emptied into the XMT 'ping' buffer (XMT flag = 0) for output & vice versa. This repeats continuously as data is input to and output from the system.

Analog data from the environment is input using the multi-channel buffered serial port interface. The external direct memory access controller (EDMA) manages the input of data into the DSP core and frees the CPU to perform other processing. The dual buffer implementation is realized in the on-chip data memory, which acts as the storage area. The CPU performs the processing on each of the buffers.

A combination of data in and data out views is shown in Figure 15-8. This is a system block diagram view shown mapped to a model of a DSP starter kit or evaluation board. The DSP starter kit in Figure 15-9 has a Codec which performs a transformation on the data before being sent to the buffered serial port and on into the DSP core using the DMA function. The DMA can be programmed to input data to either the Ping or Pong buffer and switch automatically so that the developer does not have to explicitly manage the dual buffer mechanism on the DSP.

In order to perform the coding for these peripheral processes, the DSP developer must perform the following steps:

- Direct DMA to continuously fill ping-pong buffers & interrupt CPU when full
- Select McBSP mode to match CODEC
- Select CODEC mode & start data flow via McBSP
- Route McBSP send & receive interrupts to DMA syncs

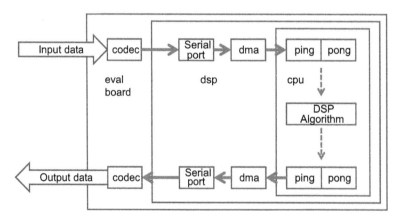

Figure 15-8:
System block diagram of a DSP application mapped to a DSP starter kit.

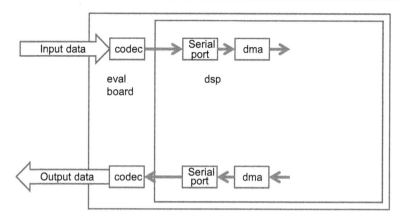

Figure 15-9:
The peripheral processes required to get data in and out of the DSP on-chip memory.

The main software initialization processes and flow include:

- Hardware Reset
- Reset Vector
- System Init
- Launch Main
- set up Application
- set up Interrupts
- enter infinite Loop
- If New Data In, then
  - Clear Flag
- Process Data
- Update Buffer Pointers
- Filter Data

The reset vectors operation performs the following tasks:

- Link hardware events to software response
- Establish relative priorities of hardware events

This is a device specific task that is manually managed.

Runtime support libraries are available to aid the DSP developer in developing DSP applications. This runtime software provides support for functions that are not part of the C language itself through inclusion of the following components:

- ANSI C standard library
- C I/O library − including printf()

- low-level support functions for I/O to the host operating system
- Intrinsic arithmetic routines
- System startup routine, _c_int00
- Functions and macros that allow C to access specific instructions

The main loop in this DSP application checks the buffer_data_ready_flag and calls the DSPBuffer() function when data is ready to be processed. The sample code for this is shown below.

```
void DSPBuffer(void)
void main( )
{
  initialize_application( );
  initialize_interrupts( );
  while( true )
{
    if(buffer_data_ready_flag)
    {
      buffer_data_ready_flag = 0;
      DSPBuffer( );
      Printf("loop count = %d\n", i++);
    }
  }
}
```

Figure 15-10 shows how the link process maps the input files for the application to the memory map available on the hardware platform. Memory consists of various types including vectors for interrupts, code, stack space, global declarations, and constants.

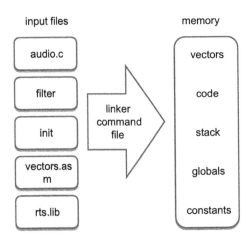

Figure 15-10:
Mapping the input files to the memory map for the hardware requires a link process.

The link command file, LNK.CMD, is used to define the hardware and usage of the hardware by the input files. This allocation is shown in Figure 15-11. The MEMORY section lists the memory that the target system contains and the address ranges for those sections. This code is DSP device specific. The SECTIONS directive defines where the software components are to be placed in memory. Some section types include:

- .text; program code
- .bss; global variables
- .stack; local variables
- .sysmem; heap

The next step in our process is to set up the debugger target. Modern DSP IDEs support multiple hardware boards (usually within the same vendor family). Most of the setup can be done easily with drag and drop within the IDE. DSP developers can configure both DSP and non-DSP devices easily using the IDE.

DSP IDEs allow projects to be visually managed. Component files are placed into the project easily using drag and drop. Dependency listings are also maintained automatically. Project management is supported in DSP IDEs and allow easy configuration and management of a large number of files in a DSP application.

The plug in capability provided by DSP IDEs allows the DSP developer to customize the development environment to the specific needs of the application and developer. Within the development environment, the developer can customize the input and output devices to be

```
MEMORY
{
        vecs:       o = 00000000h      l = 00000200h
        IRAM:       o = 00000200h      l = 0000FE00h
}
SECTIONS
{
        "vectors"   >     vecs
        .cinit      >     IRAM
        .text       >     IRAM
        .stack      >     IRAM
        .bss        >     IRAM
        .const      >     IRAM
        .data       >     IRAM
        .far        >     IRAM
        .switch     >     IRAM
        .sysmem     >     IRAM
        .tables     >     IRAM
        .cio        >     IRAM
```

Figure 15-11:
Sample code for the link command file mapping the input files to the target configuration.

used to input and analyze data, respectively. Block diagram tools and custom editors and build tools can be used with the DSP IDE to provide valuable extensions to the development environment.

Debug — verifying code performance

DSP IDEs also support the debug stage of the software development life cycle. During the phase, the first goal is to verify the system is logically correct. In the example being discussed, this phase is used to insure that the filter operates properly on the audio input signal. During this phase, the following steps can be performed in the IDE to set up and run the system to verify the logical correctness of the system:

- Load program to the target DSP
- Run to 'main()' or other functions
- Open Watches on key data
- Set and run to Breakpoints to halt execution at any point
- Display data in Graphs for visual analysis of signal data
- Specify Probe Points to identify when to observe data
- Continuously run to breakpoint via Animate

The debug phase is also used to verify that the temporal or real-time goals of the system are being met. During this phase, the developer determines if the code is running as efficiently as possible and if the execution time overhead is low enough without sacrificing the algorithm fidelity. Probe points inserted in the tool can be used in conjunction with visual graphing tools; this is helpful during this phase to verify both the functional as well as temporal aspects of the system. The visual graphs are updated each time the probe point is encountered in the code.

Code tuning and optimization

One of the main differentiators between developers of non-real-time systems and real-time systems is the phase of code tuning and optimization. It is during this phase that the DSP developer looks for 'hot spots' or inefficient code segments and attempts to optimize those segments. Code in real-time DSP systems are often optimized for speed, memory size, or power. DSP code build tools (compilers, assemblers, and linkers) are improving to the point where developers can write a majority, if not all, of their application in high level language like C or C++.

Nevertheless, the developer must provide help and guidance to the compiler in order to get the technology entitlement from the DSP architecture. DSP compilers perform architecture specific optimizations and provide the developer with feedback on the decisions and assumptions that were made during the compile process. The developer must iterate in this phase to address the decisions and assumptions made during the build process until the performance goals are met. DSP developers can give the DSP compiler specific instructions using a number of compiler options. These options direct the compiler as to the level of aggressiveness to use when compiling the code, whether to focus on code speed or size, whether to compile with advanced debug information, and many other options.

Given the potentially large number of degrees of freedom in compile options and optimization axes (speed, size, power), the number of tradeoffs available during the optimization phase can be enormous (especially considering that every function or file in an application can be compiled with different options). Profile based optimization can be used to graph a summary of code size versus speed options. The developer can choose the option that meets the goals for speed and power and have the application automatically built with the options that yield the selected size/speed tradeoff selected.

Typical DSP development flow

DSP developers follow a development flow that takes them through several phases:

- Application definition; it is during this phase that the developer begins to focus on the end goals for performance, power, and cost
- Architecture design; during this phase, the application is designed at a systems level using block diagrams and signal flow tools if the application is large enough to justify using these tools
- Hardware / software mapping; in this phase a target decision is made for each block and signal in the architecture design
- Code creation; this phase is where the initial development is done, prototypes are developed and mockups of the system are performed
- Validate / debug; functional correctness is verified during this phase
- Tuning / optimization; this is the phase where the developer's goal is to meet the performance goals of the system
- Production and deployment; release to market
- Field testing

Developing a well tuned and optimized application involves several iterations between the validate phase and the optimize phase. Each time through the validate phase the developer will edit and build the modified application, run it on a target or simulator, and analyze the results for functional correctness. Once the application is functionally correct, the developer will begin a phase of optimization on the functionally correct code. This involves tuning the application towards the performance goals of the system (speed, memory, power, for example), running the tuned code on the target or simulator to measure the performance, and evaluation where the

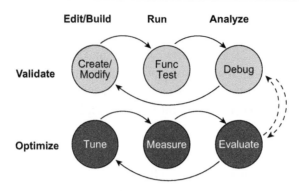

Figure 15-12:
DSP developers iterate through a series of optimize and validate steps until the goals for performance are achieved.

developer will analyze the remaining 'hot spots' or areas of concern that have not yet been addressed, or are still outside the goals of performance for that particular area (Figure 15-12).

Once the evaluation is complete, the developer will go back to the validate phase where the new, more optimized code is run to verify functional correctness has not been compromised. If not, and the performance of the application is within acceptable goals for the developer, the process stops. If a particular optimization has broken the functional correctness of the code, the developer will debug the system to determine what has been broken, fix the problem, and continue with another round of optimization. Optimizing DSP applications inherently leads to more complex code, and the likelihood of breaking something that used to work increases, the more aggressively the developer optimizes the application. There can be many cycles in this process, continuing until the performance goals have been met for the system.

Generally, a DSP application will initially be developed without much optimization. During this early period, the DSP developer is primarily concerned with functional correctness of the application. Therefore, the 'out of box' experience from a performance perspective is not that impressive, even when using the more aggressive levels of optimization in the DSP compiler.

This initial view can be termed the 'pessimistic' view in the sense that there are no aggressive assumptions made in the compiled output, there is no aggressive mapping of the application to the specific DSP architecture, and there is no aggressive algorithmic transformation to allow the application to run more efficiently on the target DSP.

Significant performance improvements can come quickly by focusing on a few critical areas of the application:

- Key tight loops in the code with many iterations
- Ensuring critical resources are in on-chip memory
- Unrolling key loops where applicable

The techniques to perform these optimizations were discussed in the chapter on optimizing DSP software. If these few key optimizations are performed, the overall system performance goes up significantly. As Figure 15-13 shows, a few key optimizations on a small percentage of the code early, leads to significant performance improvements. Additional phases of optimization get more and more difficult as the optimization opportunities are reduced as well as the cost/benefit of each additional optimization.

The goal of the DSP developer must be to continue to optimize the application until the performance goals of the system are met, not until the application is running at its theoretical peak performance. The cost / benefit does not justify this approach.

After each optimization, the profiled application can be analyzed for where the majority of cycles or memory is being consumed by the application. DSP IDEs provide advanced profiling capabilities that allow the DSP developer to profile the application and display useful information about the application such as code size, total cycle count, number of times the algorithm looped through a particular function, etc. This information can then be analyzed to determine which functions are optimization candidates.

An optimal approach to profiling and tuning a DSP application is to attack the right areas first. These represent those areas where the most performance improvement can be gained with the smallest effort. A Pareto ranking of the biggest performance areas (Figure 15-14) will guide the DSP developer towards those areas where the most performance can be gained.

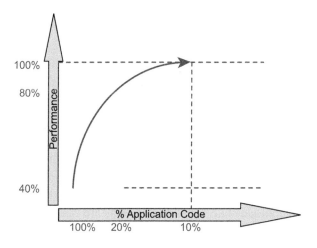

Figure 15-13:
Optimizing DSP code takes time and effort to reach the desired performance goals.

Per function relative to baseline

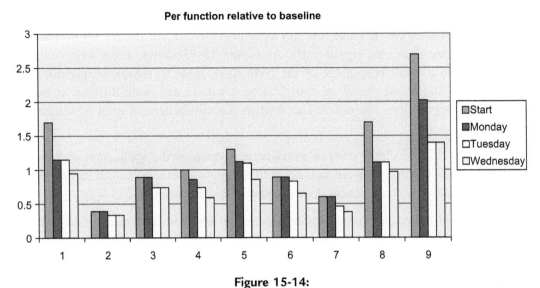

Figure 15-14:
A pareto analysis of a DSP function allows the DSP developer to focus on the most important areas first.

Getting started

A DSP starter kit is easy to install and allows the developer to get started writing code very quickly. The starter kits usually come with a daughter card expansion slots, the target hardware, software development tools, a parallel port interface for debug, a power supply, and the appropriate cables.

An evaluation module is more complex and is used for more in depth analysis of an application space. Evaluation modules have more advanced hardware and software to support this analysis and evaluation.

Putting it all together

Figure 15-15 shows the entire DSP development flow. There are five major stages to the DSP development process:

- System concept and requirements; this phase includes the elicitation of the system level functional and non-functional (sometimes called 'quality') requirements. Power requirements, Quality of Service (QoS), performance, and other system level requirements are elicited. Modeling techniques like signal flow graphs are constructed to examine the major building blocks of the system.
- System algorithm research and experimentation; during this phase, the detailed algorithms are developed based on the given performance and accuracy requirements. Analysis is first

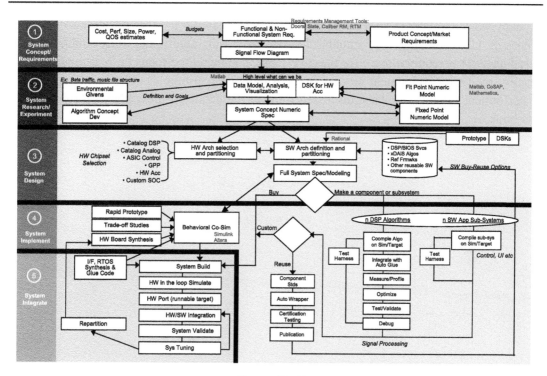

Figure 15-15:
The DSP development flow.

done on floating point development systems to determine if these performance and accuracy requirements can be met. These system are then ported, if necessary, to fixed point processors for cost reasons. Inexpensive evaluation boards are used for this analysis.

- System design; during the design phase, the hardware and software blocks for the system are selected and/or developed. These systems are analyzed using prototyping and simulation to determine if the right partitioning has been performed and whether the performance goals can be realized using the given hardware and software components. Software components can be custom developed or reused, depending on the application.

- System implementation; during the system implementation phase, inputs from system prototyping, trade off studies, and hardware synthesis options are used to develop a full system co-simulation model. Software algorithms and components are used to develop the software system. Combinations of signal processing algorithms and control frameworks are used to develop the system.

- System integration; during the system integration phase, the system is built, validated, tuned if necessary and executed in a simulated environment or in a hardware in the loop simulation environment. The scheduled system is analyzed and potentially re-partitioned if performance goals are not being met.

In many ways, the DSP system development process is similar to other development processes. Given the increased amount of signal processing algorithms, early simulation-based analysis is required more for these systems. The increased focus on performance requires the DSP development process to focus more on real-time deadlines and iterations of performance tuning. We have discussed some of the details of these phases throughout this book.

CHAPTER 16

Multicore Software Development for DSP

Michael Kardonik, Akshitij Malik

Chapter Outline

Introduction

Symmetrical multi-processing (SMP) is generally used to refer to a hardware system in which identical processors have equal access to the same memory subsystem. In this context, asymmetrical multi-processing (AMP) refers to a system in which different processors, such as a DSP and RISC processor, have access to non-equal memory systems.

However, the SMP and AMP concepts can also be applied to application software and to operating systems. When referring to application level software in a multicore environment, an application with a symmetrical software model can have functionally identical software processes executed by each of the cores in a multicore environment. Conversely, an asymmetrical application would have functionally different tasks executing in each processor.

DSP for Embedded and Real-Time Systems. DOI: 10.1016/B978-0-12-386535-9.00016-0

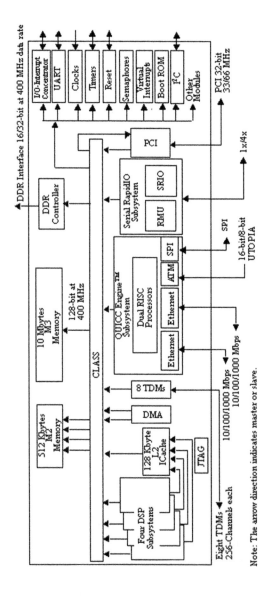

Figure 16-1:
MSC8144 block diagram.

This chapter discusses symmetrical multi-processing devices, such as the Freescale MSC8144 DSP. This device consists of four identical StarCore SC3400 cores and a memory subsystem equally accessible to each core as shown in Figure 16-1.

Each SC3400 core is a high-performance, general-purpose, fixed-point processor capable of 4 million multiply-accumulate operations per second (MMACS) for each MHz of clock frequency.

In addition to the SC3400 DSP cores, the MSC8144 includes the following memory system components:

- 16 Kbytes of Level 1 (L1) instruction cache and 32 Kbytes of Level 1 data cache per core offering zero-wait state access by each core
- 128 Kbytes of Level 2 (L2) instruction cache shared among the SC3400 cores
- 512 KBytes of SRAM M2 memory with four interleaved banks for up to four simultaneous 128-bit wide accesses at 400 MHz
- 10 MBytes of M3 memory implemented on a second die in a multi-chip-module (MCM) using embedded DRAM

The chip-level arbitration and switching system (CLASS) is an interconnect fabric that provides the system master devices with an access to slave resources, such as memory and device peripherals. The DMA controller on the MSC8144 enables data movement and rearrangement in parallel with the SC3400 core processing.

The MSC8144 is a true SMP device because all four SC3400 cores are identical processors and all four have access to the full memory subsystem in the device. This chapter uses the terms processor and core interchangeably.

Multcore programming models

Two general multicore processing models applicable to an SMP device such as the MSC144 are introduced in this section:

- Multiple-single-cores in which the cores in an SMP environment execute an application independent of each other
- True-multiple-cores in which the cores in an SMP environment cooperate in some fashion to accomplish the task

This chapter will focus on the three areas, listed in Table 16-1 below. These topics are important when designing an application using either of the multicore models presented here.

In addition to the items listed in Table 16-1, there are other important areas of consideration when porting applications to a multicore environment that are not addressed by this chapter.

Table 16-1: Multicore considerations.

Areas to Consider	Description
Scheduling	The scheduling methodology for an application allocates the resources in the multicore system, primarily by managing the processing of the cores to meet the timing and functional requirements of the application most effectively.
Inter-core communications	The interaction between cores in a multicore environment is largely used for passing messages between cores and for sharing common system resources such as peripherals, buffers, and queues. In general, the OS includes services for message passing and resource sharing.
Input and output	The management of input and output data. Defines the partitioning and allocation of input data among the cores for processing and the gathering of output data after processing for transfer out of the device.

The following sections describe two programming models for a multicore device along with a discussion of some of the points that typically create challenges when implementing applications using these concepts.

Multiple-single-cores

In a multiple-single-cores software model, all cores in the system execute their application independently of each other. The applications running on each core can be identical or different.

This model is the simplest way to port an application to a multicore environment, because the individual processors are not required to interact. Thus, porting basically involves replicating the single-core application on each of the corresponding cores on the multicore system. Thus, the developer basically replicates the single-core application on each of the cores in the multicore system such that their processing does not interfere with each other.

Figure 16-2 shows a multiple single core system. This example uses a Media Gateway for a voice over IP (VoIP) system on the MSC8144 DSP. Each core executes independently from the other cores in the DSP using data streams corresponding to distinct user channels. QUICC Engine™ is a network subsystem that is used to perform parsing, classification, and distribution for incoming network traffic.

Advantages of multiple-single-cores

One broad benefit of the multiple-single-cores model is that scaling the system by the addition of cores to the system, or by increasing the functional complexity of the application executing on each core can be as straightforward as making the same change on a single-core system, assuming other system constraints such as bus throughput, memory, and I/O can support the increased demand.

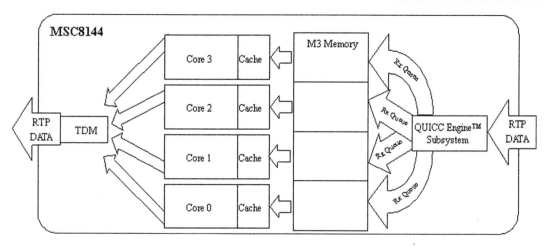

Figure 16-2:
Multiple-single-core system example.

Table 16-2: Advantages of multiple-single-cores.

Areas to Consider	Advantage
Scheduling	The lack of intentional interaction between cores precludes the need for task scheduling and load balancing between cores. Consequently, the associated complexity and overhead is eliminated which results in a more predictable system that is easier to maintain and debug.
Inter-core communications	Independent core operation eliminates the need for inter-core communication and its resulting overhead. This also minimizes data coherency issues between cores.
Input and output	The cores are not involved in partitioning or distributing the input or output data.
	Although the input data for a device may arrive through a single peripheral device or DMA controller, the data is partitioned into independent 'streams' for each core by the hardware peripheral. Thus no software intervention by the core is required to determine which portion of the incoming data belongs to it.
	The same applies to output data. It can be reassembled into a single output stream by the peripheral hardware from the independent data streams coming from each core, and then be transmitted over the appropriate output port(s).

There are other worthwhile advantages of using this model for an application, as indicated in Table 16-2.

Disadvantages of multiple-single-cores

The multiple-single-core model has inherent drawbacks, as listed in Table 16-3.

Table 16-3: Disadvantages of multiple-single-cores.

Areas to Consider	Disadvantage
Scheduling	Applications using the model may have cases in which some cores are overloaded while others are minimally loaded or even idle. This occurs simply because the system does not have a way of scheduling the processing tasks among the cores; each core must process the data assigned to it.
Inter-core communications	Inability to communicate or dynamically assign tasks between cores.
Input and output	The I/O peripherals must be capable of partitioning the data into independent streams for each core. In the example shown in Figure 16-2, the MSC8144 QUICC Engine subsystem supports the multiple I/O ports necessary to interface to the IP network. In addition, the operating system or framework used to execute the application must provide adequate services to manage the I/O devices.

Characteristics of multiple-single-cores applications

The following list describes the general characteristics of an application suitable for the multiple-single-cores model:

- A single core in the multicore system is capable of meeting the requirements of the application using the corresponding portion of the system resources associated with that core (memory, bus bandwidth, IO, and so forth).
- The I/O for the application must be assignable to each core with no runtime intervention. The assignment of data to a core occurs at compile time, at system-startup, or by an entity outside the multicore device.
- The multiple-single-core model supports more predicable execution because the application executes on a single core without any dependence on or interaction with other cores. Thus, applications that have a complicated control path or very strict real-time constraints are better suited to a multiple-single-cores implementation.
- The application has a data processing path consisting of tasks or functional modules that efficiently use the caches on the device. An application that has processing modules that do not use cache efficiently may require partitioning among multiple cores so the caches do not thrash.

True-multiple-cores

In the true-multiple-cores model, the cores in a multicore environment cooperate with each other and thus better utilize the system resources available for the application. For some applications, the true multiple cores is the only option because the application is too complex or large to process using the multiple single core model.

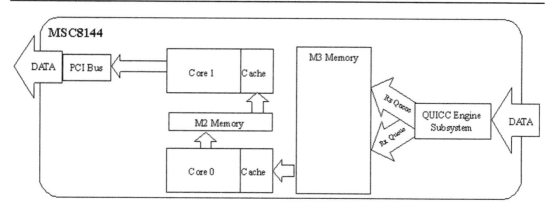

Figure 16-3:
"True" multiple-cores model example.

In a true-multiple-core system the cores do not generally perform identical tasks because the processing is partitioned among the cores, either at the application level, the scheduling, the I/O or in some other manner. The example in Figure 16-3 shows an application in which Core 0 and Core 1 each perform different portions of the application processing. The input stream is used by Core 0 and the output data is generated by Core 1. Intermediate results are passed between Core 0 and Core 1.

Advantages of true-multiple-cores

Table 16-4 lists the advantages of using the true-multiple-cores model in achieving design goals.

Disadvantages of true-multiple-cores

Use of the true-multiple-cores model is limited by a point of diminishing returns beyond which the application complexity simply requires too much overhead or renders the system less than deterministic. The complexity is largely due to the required scheduling, inter-core communication, and I/O activities, all of which impose overhead onto the basic application processing. Table 16-5 summarizes the details associated with each of these areas.

Porting guidelines

This section provides guidelines for porting a single-core application to a true-multiple-core model with a master-slave approach. The motion JPEG example presented illustrates the principles presented.

Design considerations

The first major activity in porting an application from a single to a multicore environment is to identify the threads or tasks that can execute concurrently by the

Table 16-4: Advantages of true-multiple-cores.

Areas to Consider	Advantage
Scheduling	The scheduler for true-multiple-cores has the ability to dynamically manage the system resources as a whole. This scheduling can be implemented in different ways. The scheduling intelligence can be centralized in a single core which assigns tasks to the remaining cores in the system, or distributed among multiple cores in the system with each core deciding which tasks to perform. In either situation, the system resources are better utilized and thus the performance of the application is maximized.
Inter-core communications	The nature of this communication is application specific and generally involves the passing of control and status information between the cores. The messages can be to a specific core or broadcast to all the cores. In general, the OS provides the necessary mechanisms for the inter-core communications through an API. The communication between cores allows the cores to cooperate with each other.
Input and output	In true-multiple-cores, it is possible to have a centralized entity, such as a core, manage the I/O for the application. This has an advantage that it can reduce the overall overhead associated with managing the I/O.

Table 16-5: Disadvantages of true-multiple-cores.

Areas to Consider	Disadvantage
Scheduling	The by the scheduler incurs overhead in the system, which can adversely affect the real-time requirements of the application. This must be offset by the increase in performance obtained from the cores cooperating with each other.
Inter-core communications	The overhead due to messages going between the cores can also negatively affect the performance of the application. Furthermore, the dependencies between tasks executing in different cores will also affect the performance of the system as a whole.

multiple cores. Use the following general guidelines when determining and evaluating these tasks:

- Choose tasks with real-time characteristics.
- The tasks in a multicore environment should have clearly defined real-time characteristics just as they do in a single core application.
- Avoid tasks that are too short.
- The overhead associated with short tasks is proportionally more significant than for larger tasks. Over-partitioning an application with the aim of providing flexibility and concurrency will generally create a large number of tasks and priorities spread out over several

cores complicating the scheduler, increasing overhead, and making it harder to implement and debug.

- Minimize the dependencies between cores. Over-designing the tasks and their interaction complicates the application and makes the system more difficult to implement and debug. Inter-core dependencies also incur an overhead.
- Task execution in a single core device forces tasks to execute sequentially. In a multicore environment, the same tasks can execute concurrently and tasks do not necessarily complete in the same order as in a single core. A multicore environment can expose dependencies that are hidden in a single core environment.

Note

Consider a simple application with three tasks A, B, and C that execute at the same priority. Task C can execute only after task A and B have completed. In a single core environment the application can be written such that task A triggers B which then triggers C. In a multicore device, task A and B can execute simultaneously on separate cores, so task C must now wait for both tasks A and B to begin execution, not just task B as in the single core environment. Though this is a simple example, consider a more realistic situation in which several cores execute many prioritized tasks whose execution time may change at runtime due to dependencies on the data being processed.

Motion JPEG case study

This case study demonstrates how a motion JPEG application is ported to the MSC8144 DSP using the true multiple cores model. This application was selected because it illustrates most of the major discussion points for this application note. The input data stream consists of a real-time sequence of raw video images (frames) that arrive over the DSP network connection. The DSP cores all cooperate in the processing of the video stream by encoding a portion of the current frame using the JPEG compression algorithm. The output data stream is then reassembled into the same order in which it was received and it is sent back over the network connection in this encoded format to a personal computer (PC) where it is decoded (uncompressed) and displayed real-time.

JPEG encoding

The JPEG encoding process consists of the five steps shown in Figure 16-4. The decoding process consists of the same steps inversed and performed in the reverse order.

Input data

The input video stream consists of continuous raw digital images partitioned into blocks of pixels called Minimum Coded Units (MCUs). An MCU consists of a 16×16 array of 8-bit pixels composed of 256 bytes of luminance information (Y) and 256 bytes of chrominance

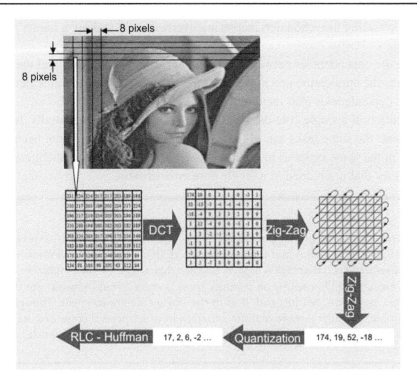

Figure 16-4:
The JPEG encoding process.

(Cb and Cr) information. Luminance is provided for every pixel in the image and chrominance is provided as a medium value for a 2×2 block of pixels.

The 512-byte MCU is partitioned into four 8×8 pixels blocks that serve as inputs to the Discrete Cosine Transfer processing block. There is no relation between any two MCU blocks.

Discrete cosine transfer

The purpose of the discrete cosine transfer (DCT) is to convert the information in the original raw pixels blocks into a spatial frequency representation of the block. These spatial frequencies represent the level of detail in the image. Thus, a block with a lot of detail in it has many high spatial frequency components while blocks with low detail are represented by a majority of low frequency components. Typically, most of the information is concentrated in a few low-frequency components.

The DCT is applied to an 8×8 block of pixels from left to right and from top to bottom of an image frame. The result is a new 8×8 block of integers (called DCT coefficients) that are then reordered using a zig-zag pattern.

Zig-zag reordering

The 8×8 block of DCT coefficients is traversed using a zig-zag pattern as shown in Figure 16-5.

The result of this reordering is a vector of 64 elements (0 to 63) arranged from lowest to highest frequency components. The first value in the vector (0) is called the DC component and represents the lowest frequency component. The other coefficients in the vector (1 to 63) are called the AC coefficients. The 64-item vector is then passed to the quantization block for processing.

Quantization

In this step, each value in the 64-coefficient vector resulting from the zig-zag reordering step is divided by a predefined value and rounded to the nearest integer. The quantization step removes the high frequency components (greater detail) of the input vector because the human eye is more sensitive to lower frequency components than higher frequency components. This is done by dividing the higher frequency coefficients in the vector by larger values than those used to divide the lower frequencies. This action forces the higher frequency components to have more zeros.

Run-length coding

This run-length coding (RLC) exploits the fact that we have consecutive zeros for the higher frequency components of the input vector by providing a pair of integers indicating the number of consecutive zeros in a run followed by the value of the non-zero number following the zeros. For example, consider the run of coefficients: 45, 33, 0, 0, 0, 12, 0, 0, 0, 0, 0, 0, 0, 0,

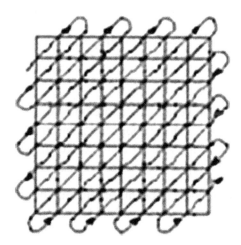

Figure 16-5:
Zig-zag pattern used to traverse DCT coefficients.

5. The zero run-length code becomes (0,45), (0,33), (3,12), (7,5). There are special situations that are not addressed here.

Huffman coding

This process uses a variable-length code table to map the right integer in each numbered pair generated in the previous coding step with another bit string that uses minimal space to represent the original information. This is advantageous because the variable-length code table is carefully designed to represent the most common input data patterns with shorter bit strings than for the less common input values. The result is a string of bits that is smaller in size than the original input data.

Design considerations

The overall requirements for processing the JPEG algorithm require only a small portion of the MSC8144 processing capabilities. Therefore, several JPEG encoder tasks can execute on each core on multiple input video data streams. This section discusses several characteristics of the motion JPEG (MJPEG) application and how these influenced the decisions made in the process of porting to the multicore MSC8144 DSP.

Input

The input data stream for a motion JPEG (MJPEG) encoder consists of a contiguous flow of raw digital images (frames). The frames are sent to the DSP at a particular frame rate determined on the PC. When a frame is sent to the DSP, the frame is partitioned into blocks of MCUs and then transmitted to an IP address defined on the network interface of the DSP. The rate at which the MCU blocks are transmitted to the DSP is predetermined and does not change regardless of the video frame rate.

For this application, it is sufficient (and simpler) for a single core to manage the QUICC Engine™ subsystem and service the resulting interrupts and then partition the input data block for processing by the other cores.

Scheduling

This is a soft real-time application because there are no hard real-time constraints. There are no imposed output frame rates at which the application must transmit the output video stream; the PC simply stalls the display of the video stream received from the DSP if the output frame rate slows below the expected rate (or stops all together).

Similarly, latency is of no consequence for this application because the DSPs do not have a fixed amount of time to complete the processing. However, we will require the DSP to process the incoming data blocks as they arrive over the network interface.

Because the MCU blocks in a frame are independent, there are no restrictions on which core processes a given block. Furthermore, it is best not to assign the incoming blocks to a specific core statically, because the performance of some of the tasks in the JPEG algorithm depends on the data being processed. Thus, the cores on the DSP are better utilized by dynamically allocating the data blocks to the cores based on the processing load of the core.

These characteristics, in conjunction with the input considerations discussed in Input, make the application a good candidate for the true multiple cores model using the master-and-slave approach to scheduling. The master core manages the incoming data stream and assigns tasks to the available cores, including itself, based on the available processing cycles.

Inter-core communication

The master core copies a pointer to the memory location of the next MCU data block to process into a global queue that is accessible by all the cores and then notifies the slave cores that there is data available for processing.

All non-idle cores, including the master core, then compete to process the input block. If a core is already processing a block, it ignores the message from the master core.

Output

The output video stream is sent over the network to a PC for decoding and display in real-time. Due to the data-dependent nature of the application JPEG algorithm, the encoded data blocks resulting from the core processing can potentially not be available in the same order as the input data blocks. This is an example of a hidden dependency on the flow order that would not exist if the application was executing on a single processor. Thus, the output data blocks must be placed in order before transmitting back to the PC. This process is called 'output serialization' and is assigned to the master core. The master core must pause the output data stream from the device until the next data block in the sequence is available.

Implementation details

One implementation of a true-multiple-cores model is referred as master-slave. In a master-and-slaves implementation, the control intelligence for the application resides in a master core. The other cores in the system become slave cores. The master core is responsible for scheduling the application processing and possibly managing the I/O. Figure 16-6 shows a system in which Core 0 is assigned the role of master core. Core 0 manages the I/O for the application and assigns tasks to the slave cores 1, 2, and 3 via a task queue in memory.

The remainder of the chapter focuses on a master-slave approach to a true-multiple-cores application.

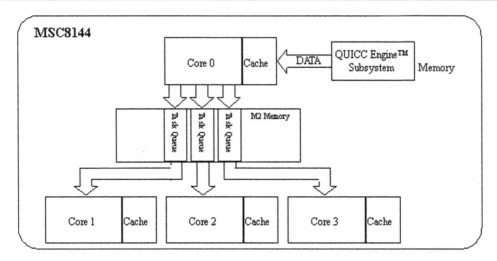

Figure 16-6:
Master and slaves systems.

The development tools for the MSC8144 used in the development process included the following:

- CodeWarrior® Development Studio.
  The development studio is a complete integrated development environment (IDE) that contains all of the tools needed to complete a major embedded development project, from hardware bring-up through programming and debugging embedded applications.
  Two useful features used from the CodeWarrior IDE were the Kernel Awareness plug-in module for visualization and debugging of the SmartDSP Operating System (SDOS) and a Profiler that allowed evaluation of the performance of various modules and interactions in the system.
- SmartDSP Operating System (SDOS).
  SDOS is a preemptable, real-time, priority-based operating system, specially designed for high-performance DSPs operating with tight memory requirements in an embedded environment. It includes support for multicore operations such as synchronization modules and inter-core messages to manage inter-core dependencies. A convenient and unified application program interface (API) supports various types of peripheral I/O devices and DMA controllers.

The steps we used to port the MJPEG application followed a straightforward approach as follows. First we ran a single instance of the MJPEG application on a single core of the MSC8144. Once this functionality was properly validated, we ran two or more instances of the MJPEG processing on the same core. This was helpful because the debugging process is simpler on one core than on multiple cores. During this process, we were also able to use the

MSC8144 functional simulator available with the CodeWarrior IDE, eliminating the need for the actual hardware for this portion of the development. After these initial steps, the MJPEG application was executed on more than one core without any inter-core communication. Lastly, we added the inter-core functionality.

Scheduling

In the master-and-slaves scheduling implemented on the MSC8144 for this application, the main scheduler functionality resides in core 0, the master core. Cores 1, 2, and 3 in the system become slave cores. The master core manages the I/O and processing for the application and the slave cores wait for tasks to be processed as shown in Figure 16-7.

Figure 16-7 shows that the incoming raw video images are received in blocks by the MSC8144 QUICC Engine network interface. The master core services the QUICC Engine interrupt and then sends a message to the slave cores with the information pertinent to the received block. The slave cores are notified by messages placed in this queue which then vie to access the message and be assigned the task of JPEG encoding the video data block. If a slave core is already processing a block, it de-queues a task until it completes the encoding of the current data block.

Note

Core 0, even though it is the master core, is also notified when messages are posted to the task queue and can perform the JPEG encoding of a block if it has available processing bandwidth.

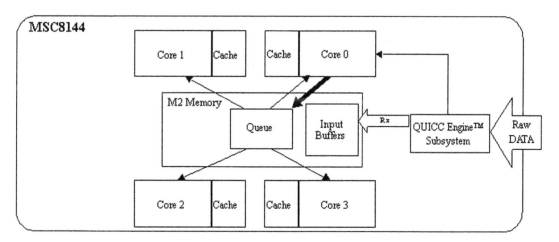

Figure 16-7:
Task scheduling.

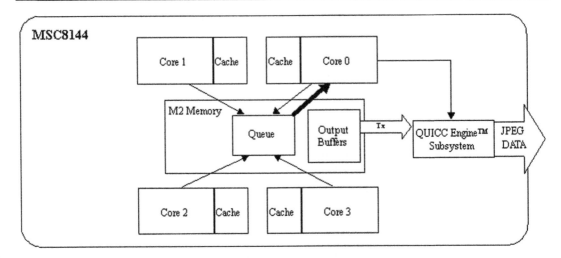

Figure 16-8:
Task completion.

Once a slave core finishes the JPEG encoding process, it notifies the master core via another message queue as shown in Figure 16-8. The master core de-queues the information for each encoded block and determines whether it is the next output block in the output stream (a process called serialization). If that is the case, the master core transmits the encoded block to the network using the QUICC Engine interface.

It also transmits any additional encoded blocks that are available in the serialized sequence.

If the master core performed the encoding task, it does not use the queue process, and it simply serializes the output and transmits the buffer (if possible).

The SDOS operating system provides the basic building blocks for the scheduling of the application. In other words, the master core implements the scheduling methodology by making calls to SDOS services through the operating system API. Similarly, the slaves respond to the master core using SDOS services.

The background task in SDOS is a user-defined function that executes when no other task in the application is required to execute. This task has the lowest priority and executes indefinitely in a loop until a higher priority task is enabled. The background task for this application places the corresponding core in the WAIT state by executing the **wait** instruction. The WAIT state is an intermediate power-saving mode used to minimize core utilization and reduce power consumption. In this case, the cores each remain in the WAIT state until a message arrives indicating a block of pixels is available for JPEG encoding, or, if it is the master core, an encoded block is ready to transmit.

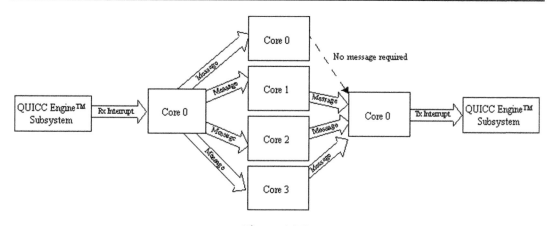

Figure 16-9:
Inter-core communication flow diagram.

Inter-core communication

The inter-core communication for the application consists primarily of the messages exchanged between the master core and the slave cores. These messages are implemented in this application using services provided by SDOS accessed through calls to the operating system API. The flow of inter-core communication for the application is shown in Figure 16-9.

The QUICC Engine subsystem interrupts the master core after several blocks of the raw video image are received. In the receive interrupt service routine (ISR), core 0 sends messages to all the slave cores and to itself to indicate there is data ready for JPEG encoding. After the encoding process, the slave cores send a message back to the master core indicating there are blocks of encoded data ready to transmit. Core 0 does not need to send a message to itself.

During the initialization process, each core creates queues used to send and receive messages as indicated in Table 16-6. The messages have two purposes. Messages from the master core (core 0) indicate that a block of raw video data was received and is available to encode. Messages from the slave cores (cores 1 through 3) indicate that a block has been encoded and

Table 16-6: Message queues defined during initialization.

Message	Location	Call-back Priority	Purpose
Core 0 to Cores 0–3	Core 0	6	Send/Receive messages indicating block ready to encode
	Cores 1–3	5	Receive messages indicating block ready to encode
Cores 1–3 to Core 0	Cores 0	3	Receive messages indicating JPEG encoding completed
	Core 1–3	N/A	Send messages indicating JPEG encoding completed

is ready to transmit. A block encoded by core 0 is serialized for transmission with no message generation.

Messages are implemented using MSC8144 virtual interrupts between the cores. Priorities associated with the user function called when a message is received by a core are indicated in Table 16-6.

Input and output

The I/O for the motion JPEG application also requires special consideration. The input of the raw video data arrives at the MSC8144 QUICC Engine interface from a PC over an IP network connection. As mentioned previously, the master core (core 0) initializes and services the interrupts for the QUICC Engine subsystem. Incoming data blocks are not copied. Instead, control information is passed to the slave cores in a message with a pointer, size, and other information needed to locate the data and complete the JPEG encoding processing.

The data output process is more involved. The output blocks must be sent back to the PC over the IP network in the same order in which they were received, but since the JPEG encoding processing is data dependent, the cores can complete the encoding process for a block out-of-order, which means the output blocks become available out of order. Thus, the output blocks must be buffered and placed back in order, a process called output serialization.

The master core executes the serialization by collecting pointers to the output buffers as they are made available by the slave cores. Encoded buffers are then transmitted to the PC only

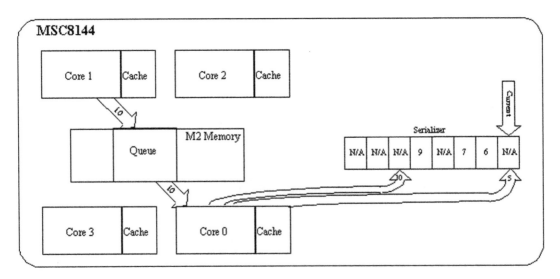

Figure 16-10:
Output serialization.

when the next blocks in the sequential output data stream are available. This process is depicted in Figure 16-10. In this example, the serializer shows that output blocks #6, #7, and #9 are available to be sent to the PC; however, the next block the PC is expecting is block #5 as indicated to by the Current pointer. Slave core 1 finishes encoding output block #10 and notifies the master core which then adds it to the serializer accordingly. Core 0 then provides output block #5 which then allows blocks #5 through #7 to be sent to the PC, after which the serializer must wait for block #8 in the sequence.

The serializer concept is similar to the jitter buffer used in voice over IP (VoIP) applications. The differences are that the jitter buffer in VoIP is located at the receiving end of the voice connection.

Conclusions

Designers of multicore applications must consider several factors including processing and task properties, hardware characteristics such as the cache and peripheral support, and the multicore support provided by the OS and tools. For example, the Kernel Awareness module of the CodeWarrior Development Studio proved invaluable in the implementation and debugging of the motion JPEG application.

Because a multicore system also exposes hidden dependencies and complexities, it is always a good idea to start with a simple scenario or subset of the application processing, and then improve form there. For example, it is generally impractical to try to obtain optimal task scheduling, inter-core communication, and I/O methodologies in the early development stages. It is better to begin with a single core using static scheduling and improve the design though testing and tuning.

To support effective application development in a multicore and multi-task environment, Freescale multi-processing offerings not only include multicore SMP devices such as the MSC8144 DSP, but also provide development environments like the CodeWarrior Development Studio, and hardware and software tool support that can help the designer develop application systems, measure their system performance and resource and processing utilization, and tune, test, and debug their functionality.

CHAPTER 17

Developing and Debugging
a DSP Application

Daniel Popa

Chapter Outline

Integrated Development Environment overview

The CodeWarrior IDE is a complete integrated development environment based on the latest
Eclipse platform, designed to accelerate the development of embedded applications.

On top of the Eclipse platform, the specific CodeWarrior plug-ins such as compiler, linker,
debugger, and software analysis engine, were implemented in order to provide access to the
specific DSP features.

Even more, being an Eclipse based development environment the user is encouraged to use
any third party software that can be integrated as a plug-in into the IDE to maximize the

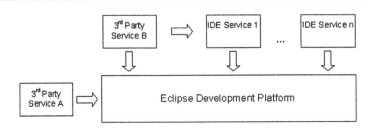

Figure 17-1:
CodeWarrior modular architecture overview.

user experience in such way that other tools cannot offer. For example, the user might choose to use his own plug-in for source versioning system or his own editor if this makes him more comfortable in using the tool.

The modular philosophy is depicted in Figure 17-1. Any other 3rd party service which is Eclipse platform compatible can also plug-in into the IDE and the user will reap the benefits of this.

In order to reuse parts of software created for other platforms the IDE specific tools are also created as plug-ins (e.g., the flash programmer is the same plug-in between Power Architecture and StarCore families).

In order to maintain a relative high performance and a good out-of-the-box user experience, some of the critical software components that are part of IDE are built-into Eclipse rather than delivered as standalone plug-ins.

Creating a new project

First time users, who are not familiar with the Eclipse terminology, will find themselves confronted with a new way of doing things compared with other classic development tools. The first thing that the IDE asks when it is opened is the path towards the *workspace* as depicted in Figure 17-2.

The *workspace* is a container that maintains all the 'things' the embedded software engineers need for editing, building, and debugging a project. For each user, the workspace takes the form of locally physical location where the IDE will place the .metadata structure that will hold the project history and configuration tools settings.

The usage of local workspace allows a better and more efficient collaboration between team members and easy exchange of code source through network shared repositories. Each team member can customize his own local workspace in order to accommodate his preferences related with the plug-ins used or the way in which the IDE displays the windows, widgets, text editors, and so forth. After the *workspace* is created the workbench opens and the user can perform the following actions from a rich set of predefined options.

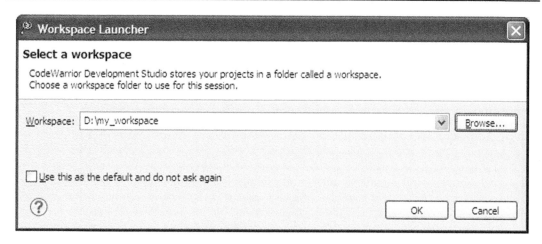

Figure 17-2:
Workspace dialogue

Enhancements can be made to the default Eclipse stock menus with intuitive actions such as:

- Create a new project with default initializations, DSP start-up code, and JTAG connection
- Import an existent project into workspace
- Migrate an older project type to newest IDE
- Create a new project based on a collection of source files
- Create and characterize a new target connection for debugging purposes
- Create new configurations files for peripherals configuration

Figure 17-3 shows all the options available in the New menu. To create a new project in a couple of simple steps the StarCore Project needs to be used. This menu will guide the user throughout the new project wizard which contains simple actions like:

- Set the project name
- Choose the target device for the project
- Choose the debugger connection and board for debugging

The main steps for a new project wizard are shown in Figure 17-4.

Apart of rich menus and options for supporting different features, the IDE provides a rich set of demos that covers the main hardware capacities. These demos are designed for shortening the development cycle by providing ready-to-run applications, drivers, and APIs.

For demonstration of the software development features in the following sections an application based on SmartDSP OS (SDOS) will be used.

The asymmetric_demo is based on a standard SDOS demo (that can be created via New menu), that has been modified to demonstrate the use of asymmetric memory model for each

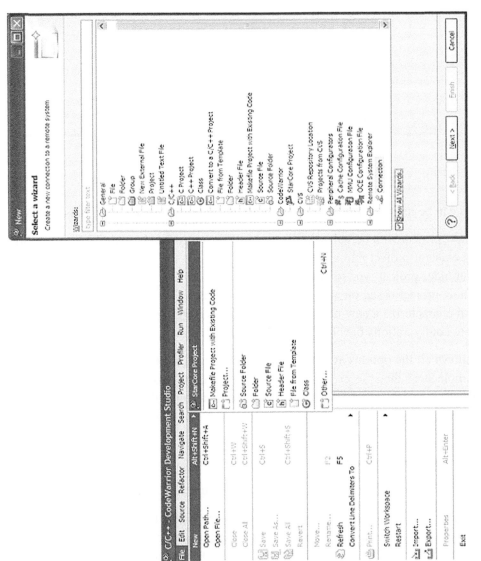

Figure 17-3:
New project menu.

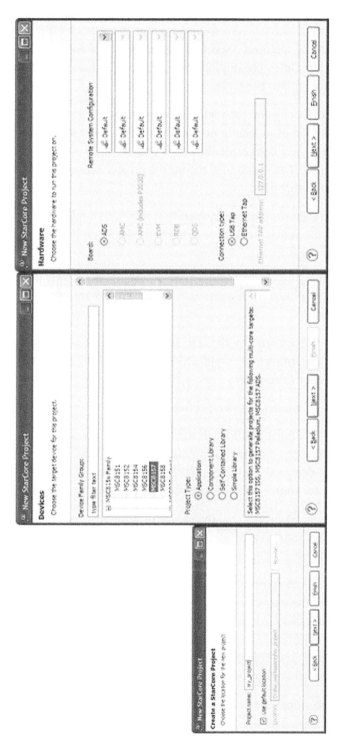

Figure 17-4:
New project wizard main steps.

of the DSP cores and thus enabling each core to run different code and data. The demo also illustrate how to link a multi-core system so that each of the cores have different implementations of the same methods. Additionaly, the usage of software interrupts, times, and specific SDOS functionalities are implemented. Figure 17-5 shows the steps for importing an existing project into the workspace.

The IDE asks the user to specify the path where the projects are located. Then, it automatically updates the list of available projects within the user specified folder. Within a folder, only one single project can reside. For each project there is a pair of two XML files called .project and .cproject. These two files record the project settings like file organization, tools settings, and project properties.

The Import dialogue also requires the user to specify if the IDE should copy the projects from the current location into the active workspace. This is useful when the original project version needs to be preserved.

After the project is imported successfully, the IDE automatically opens the default C/C++ perspective and offers complete overviews of the project resources as in Figure 17-6.

The term of perspective is borrowed by the IDE from the Eclipse platform and refers to a 'visual container' that determines user interactions with the IDE and in the same time offering a task oriented user experience.

The IDE is delivered with two default perspectives:

• C/C++ perspective is designed to offers all the tools needed for software development such as: text editor, project explorer, outline, build console and problem viewer, etc.
• Debug perspective is designed for debugging DSP targets by displaying the main features like stack frames, editors for source and disassembly, variable and register views, output console

Any of the perspectives can be customized by adding or removing the any of the views based on user requirements. The entire list of tools and features supported by the IDE can be accessed via menu Window/Show view/Other, as is shown in Figure 17-7.

On the left hand side, the *Project* panel displays all the files that belong to the asymmetric demo project. Within this panel, the user can control all the project settings and can easily navigate throughout the project files and folders. The user can choose between different modes of file organization within the project since the IDE offers a wide range of features like grouping files in virtual folders or groups (these folders do not exist on the physical drive) or adding links to real files. The IDE offers full control of project files and directories organization.

The overall project or individual file or folder properties can be easily accessible via a simple right click on each the items. The list of actions is displayed in Figure 17-8.

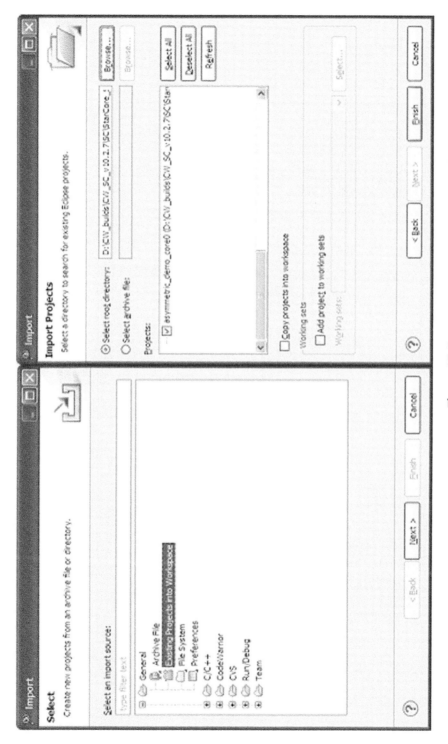

Figure 17-5:
Importing existing projects.

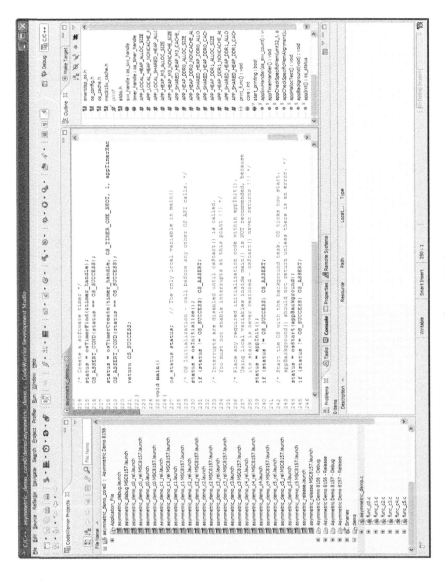

Figure 17-6:
Default C/C++ editor perspective.

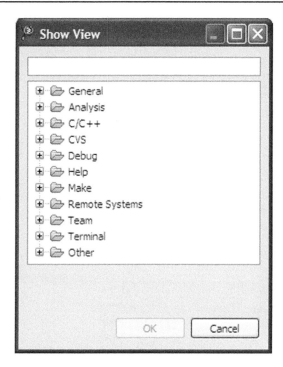

Figure 17-7:
List of all available tools.

Build and link the application for a multi-core DSP environment

Before starting to describe the building and linking process for the application considered in the previous chapter, a few details about the SDOS and the application memory map must be presented.

DSP (SDOS) Operating system

SDOS is a real-time operating system designed for DSP build on StarCore technology. It includes: royalty free source code, real-time responsiveness, C/C++/asm support, small memory footprint, support for SDPS family products, and full integration with the IDE.

The main features of the DSP OS include:

- Priority based event-driven scheduling triggered by SW of HW
- Dual-stack pointer for exception and task handling
- Inter-task communication using queues, semaphores, and event
- Designed as a asymmetric multi-processing system where each core runs its own OS instance
- Manages shared resources and supports inter-core communication

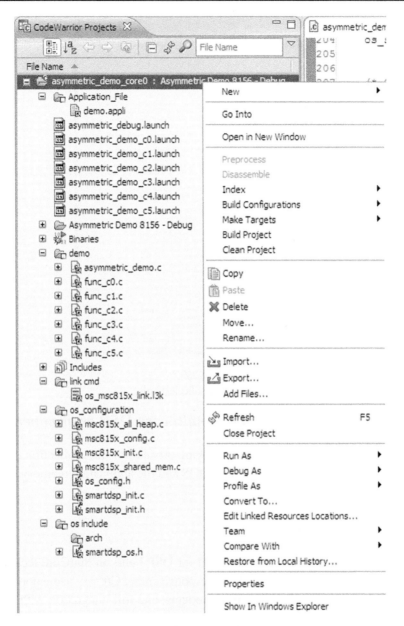

Figure 17-8:
List of project editing options.

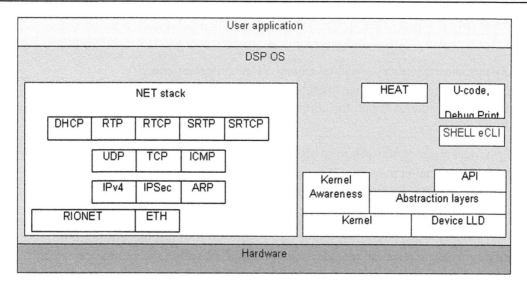

Figure 17-9:
DSP operating system architecture.

The DSP OS architecture is depicted in Figure 17-9. The operating system includes: kernel, peripherals drivers, the Ethernet stack (UDP/TCP/IP) and runtime enablement tools (Kernel awareness, HEAT, eCLI).

The SDOS, is integrated into the asymmetric demo project as a component library that needs to be linked with the user application. Apart of the library, SDOS configuration files and headers must be included into the project in order to initialize and configure the OS based on user requirements.

These configurations files are accessible from the `IDE Project` panel within the `os_configuration` and `os_include` virtual folders. The user is free to change any settings starting with the stack and heap sizes up to the settings associated with OS functionalities like number of running cores, cores synchronizations, cache policies, barriers.

Application memory map

The application, considerer in this analysis, covers the complex topic of the Multi Instruction and Multi Data model. In this case, each core is going to execute private code and handle private data.

This concept permits a complex software design to be broken down into smaller and simpler tasks. Therefore the MIMD design can divide the application processing requirements by distributing a portion of it across all the cores so that the DSP resources are used more efficiently.

The most critical aspect of the application is to divide the DSP resources properly among the tasks. Each memory type available on DSP has specific characteristics and purposes:

- System shared memory — shared among all the processor cores
- Symmetric memory — private to each core but the objects within this memory reside on the same virtual address for all the cores in the system
- Private memory

Beside virtual memory, the user must also choose the physical memory within the system to be used for application purposes. For example, on the Freescale MSC815x the available choices are: M2, M3, DDR0, and DDR1 (see Figure 17-10).

The system engineer needs to consider where each module of the application will be placed into the memory. Each physical resource (see Figure 17-10) must be partitioned by identifying which application modules need to be shared within the system and which modules are private to the cores. The steps required for code and data sections partitioning will be discussed in the next two sub-section that address the compiler and linker options. Specific examples with compiler and linker commands are provided for a better understanding.

In this case the modules that must be shared across all the cores are:

- ANSY libraries and runtime functions
- Start-up code
- SDOS

The modules that must be private to each core are:

- Stack and heap
- Address translation tables
- Private methods used in user application layer

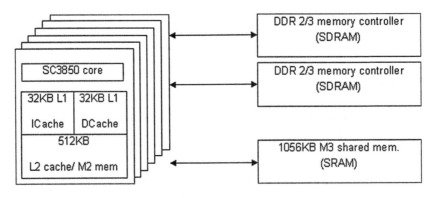

Figure 17-10:
Physical memory on MSC815x DSP.

Figure 17-11 shows the physical memory map of the asymmetric demo project. The user application layer is simulated via func_c(0-5).c modules. Inside these modules the user can insert the custom code that will be run indecently for each of the cores.

After the DSP platform is started-up, the initializations of SDOS and user application are performed. After all initializations are done, the operation system will start the default background job.

The user function is then called via a software timer created during the application initializations.

Compiler configuration for application

The IDE provides easy and straightforward methods to allocate the code and data so that the application requirements shown in Figure 17-11 are met. The application functions and variables are placed by the compiler in the appropriate sections via the application configuration file (*.appli).

Alternatively the placement can be done via compiler modifiers like pragma place and attribute to perform the setup. By using this method the code portability across different DSP platforms is affected since these code modifiers/keywords are recognized only by the DSP compiler.

The application configuration file is a compiler input file used to map the default compiler generated sections like .text, .data, .bss and .rom into user custom defined sections that will be later used by the linker to place the appropriate resources into the DSP memory.

Using custom sections makes the job of defining the mapping of the application elements easier in the context of multi-core environment.

Taking into account the nature of the application, the user specific sections must be placed into dedicated private sections.

In the demo.appli file the definition is implemented as:

```
section
      program = [
                Entry_c0_text            : ".private_text" core="c0",
                Entry_c1_text            : ".private_text" core="c1",
                Entry_c2_text            : ".private_text" core="c2",
                Entry_c3_text            : ".private_text" core="c3",
                Entry_c4_text            : ".private_text" core="c4",
                Entry_c5_text            : ".private_text" core="c5"
                ]
```

M2	Private	Core#0: 0x3000 0000 – 0x3003 FFFF (256KB) Core#1: 0x3100 0000 – 0x3103 FFFF (256KB) ... Core#5: 0x3500 0000 – 0x3503 FFFF (256KB)	0x3000 0000 0x3503 FFFF	OS stack, OS heap and Runtime Library heap for each core
M3	Private	Core#0: 0xC000 0000 – 0xC000 7FFF (32KB) Core#1: 0xC000 8000 – 0xC000 FFFF (32KB) ... Core#5: 0xC002 8000 – 0xC002 FFFF (32KB)	0xC000 0000 0xC002 FFFF	OS data and program segments that require being private
	Shared	Core#0, Core#1, ... , Core#5 (864KB)	0xC003 0000 0xC010 7FFF	OS shared data
DDR0	Private 0	Core#0: 0x4000 0000 – 0x40FF FFFF (16MB) Core#1: 0x4100 0000 – 0x41FF FFFF (16MB) ... Core#5: 0x4500 0000 – 0x45FF FFFF (16MB)	0x4000 0000 0x45FF FFFF	Address translation table
	Private 1	Core#0: 0x4600 0000 – 0x4600 7FFF (32KB) Core#1: 0x4600 8000 – 0x4600 FFFF (32KB) ... Core#5: 0x4602 8000 – 0x4602 FFFF (32KB)	0x4600 0000 0x4602 FFFF	User application code (program and data)
	Shared	Core#0, Core#1, ... , Core#5 (928MB)	0x4603 0000 0x7FFF FFF	OS Kernel run-time and initialization code
DDR1	Private	Core#0: 0x8000 0000 – 0x80FF FFFF (16MB) Core#1: 0x8100 0000 – 0x81FF FFFF (16MB) ... Core#5: 0x8500 0000 – 0x85FF FFFF (16MB)	0x8000 0000 0x85FF FFFF	Virtual trace buffer for trace collection
	Shared	Core#0, Core#1, ... , Core#5 (930MB)	0x8600 0000 0xBFFF FFFF	Free space

Figure 17-11:
Application memory map.

The user specific data must be placed in a private section too. The definition of data sections is done in `demo.appli` in a similar way as for the code sections:

```
section
      data =      [
                  Data0                         : ".data" core="c0",
                  Data1                         : ".data" core="c1",
                  Data2                         : ".data" core="c2",
                  Data3                         : ".data" core="c3",
                  Data4                         : ".data" core="c4",
                  Data5                         : ".data" core="c5"

                  ]
```

The last step is to instruct the compiler where to place the default sections generated for the user private code and data.

```
module "func_c0" [
                  program = Entry_c0_text
                  data= Data0
                  rom= Rom0
                  bss= Bss0

                  ]
```

For each user module, the compiler will rename the default sections into the newly created sections like `Entry_c0_text`, `Data0`, etc.

After the code is compiled the compiler will export towards the linker the newly created sections offering the user a clear image about the resources that needs to be placed into the appropriate memories. Otherwise, the process would be obfuscated and almost impossible to be handled by users when further changes are required.

The DSP compiler settings are all grouped into a single panel under: `Project properties/C/ C++ build/Settings` as is shown in Figure 17-12.

The user can control throughout the compiler settings panels how the compiler generates the code, reports the errors and warnings, and how it interacts with the user environments (build steps, make files, output files, logging, etc.).

The external configuration files can be added via the `Configuration Files` menu as shown in Figure 17-13.

The DSP compiler offers a wide range of optimizations from the standard C optimization up to the low level target specific optimizations. The user can choose between

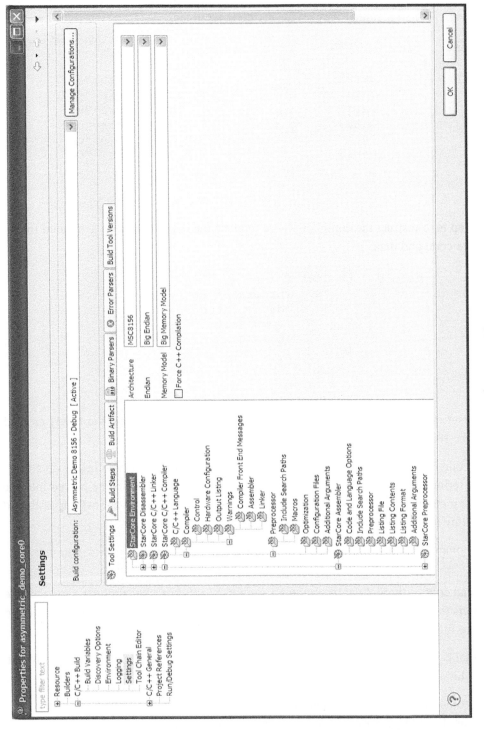

Figure 17-12:
DSP compiler settings panel.

Machine Configuration File		Browse...
Configuration View	Asymmetric_View	
Application Configuration File	..\..\..\Application_File\demo.appli	Browse...

Figure 17-13:
Adding custom settings and configuration for DSP compiler.

optimizations for speed or size or can choose to apply cross file optimization techniques (Figure 17-14).

DSP IDE's offer an interface that allows advanced users to interfere with compiler internal components and get optimal performance from the code. These options should be used only when the performance is a critical factor.

In order to benefit from the compiler optimization techniques, the user must be able to understand the internal structure of the compiler and how different components and compiler flags/switches and options interact with the user code.

A generic DSP compiler is depicted in Figure 17-15. The user invokes the compiler shell via the DSP IDE by specifying the C sources and assembly files that need to be processed with the

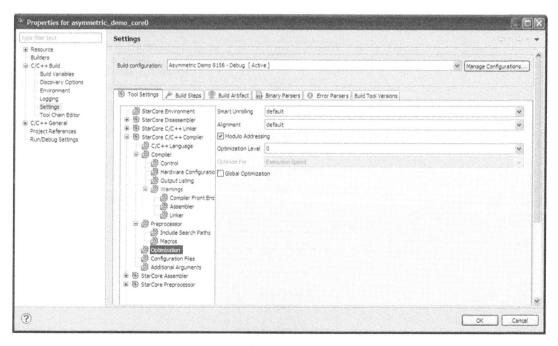

Figure 17-14:
Compiler optimization levels.

appropriate compiler options. The C Front-End indentifies the sources, creates the compilation unit by including all the headers, expands all the macro, and finally converts the file into an intermediate representation which is then passed to the optimizer.

Most of the commercial compilers for DSP have two optimization stages:

- High level optimizer which handles the platform independent optimization stage. Usually, these kinds of optimization are well known and common to most compilers.

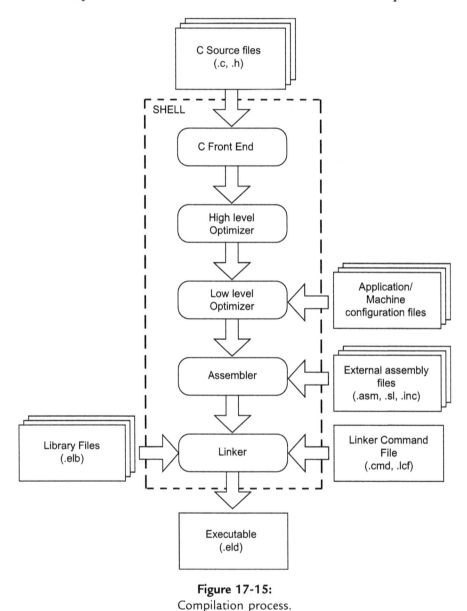

Figure 17-15:
Compilation process.

The optimizations done in this stage consists in: strength reduction (loop transformation), function in-lining, expression elimination, loop invariant code, constant folding and propagation, jump to jump elimination, dead storage and assignment elimination, etc. Also, depending on each vendor compiler technology, the high level optimizer can be instructed to generate specific optimization that includes additional information which is used later on other compiler internal components. At this point, the output is a linear assembly code.

- Low level optimizer which carries out specific target-specific optimizations. It performs instruction scheduling, register allocation, software pipelining, condition execution and prediction, speculative execution, post increment detection, peephole optimization, etc.

The low level optimizer is the most critical internal component, since it differentiates the different compiler technologies. It transforms the linear code into more optimal parallel assembly code. By default, to achieve this step, the low level optimizer uses traditional conservative assumptions.

Advanced users, who have a deep understanding of how the DSP works, can instruct the compiler at this stage to use a more aggressive approach for code optimization and parallelization.

The DSP IDE provides these means via the dedicated GUI as depicted in Figure 17-16.

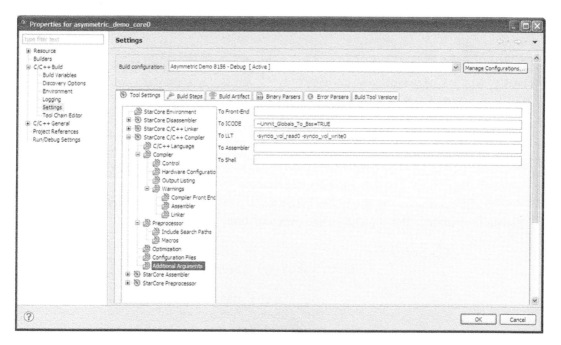

Figure 17-16:
Compiler options.

Linker configuration for application

Linking is the last step before getting the executable file. The linker job is to put together all the compiler object files, resolve all the symbols and, based on the allocation scheme, to generate the executable image that will be run on the target.

The DSP linker is based on GNU syntax and it has been enhanced with additional functionalities in order to match the multi-core DSP constraints. The linker is controlled via the linker command files *.13 k.

Most users find the writing of these files to be very difficult. The IDE tries to help such users by providing default physical memory configurations and predefined linker symbols.

The default physical memory is insured by using special designated linker functions such as arch() and number_of_cores(). These kinds of functions instruct the linker to generate the appropriate mapping for the physical memory layout and number of executable files that are needed for the application.

The process of linking all the object files is straightforward once the linker philosophy is correctly understood. In the simplest form the linker requires only three basic steps to correctly map the application:

- Define a virtual memory region where code and data segments need to be placed. This is achieved via the GNU style memory directive:

```
MEMORY {
  os_shared_data_descriptor              ("rw"): org = _SharedM3_b;
  shared_data_m3_descriptor              ("rw"): AFTER( os_shared_data_descriptor);
  shared_data_m3_cacheable_descriptor("rw"): AFTER( shared_data_m3_descriptor);
      }
```

The example above (*see the* os_msc815x_link.13 k from the asymmetric demo) defines three virtual memory segments called os_shared_data_descriptor, shared_data_m3_descriptor and shared_data_m3_cacheable_descriptor with read and write attribute (e.g. "rw"). The first memory segment os_shared_data_descriptor, starts at the virtual address pointed by the linker symbol _SharedM3_b while the other segments are placed in memory after the end of the previous one.

```
SECTIONS {
        descriptor_os_shared_data
        {
            .os_shared_data
            .os_shared_data_bss
            reserved_crt_mutex

        } > os_shared_data_descriptor;

    }
```

- Assigned the compiler generated output sections into linker output sections and place these sections into the appropriate virtual memory regions defined earlier.

  The compiler generated sections such as `.os_shared_data` are now grouped in a larger linker section called `descriptor_os_shared_data` that is placed into the virtual memory `os_shared_data_descriptor`.

```
address_translation (*) map11 {
    os_shared_data_descriptor              (SHARED_DATA_MMU_DEF): SHARED_M3;
    shared_data_m3_descriptor              (SHARED_DATA_MMU_DEF): SHARED_M3;
    shared_data_m3_cacheable_descriptor    (SYSTEM_DATA_MMU_DEF): SHARED_M3;
```

- The third step is to map the virtual memory into the physical memory using the address translation table in MMU (memory management unit).

  This example maps the virtual memory regions defines earlier over the M3 physical memory (defined here as SHARED_M3) with the MMU attributes SHARED_DATA_MMU_DEF. This symbol is defined in `sc3x00_mmu_link_map.13 k` file.

These three basic steps must be duplicated for shared and private memories definitions. In this example, the private code and data for the application that will run independently on each core will look like:

```
unit private (task_c0) {
    MEMORY {
            private_text_0                 ("rx"): org = _VirtPrivateDDR0_b;
            private_data_0                 ("rw"): AFTER(private_text_0);
    }
    SECTIONS {
            privateCode {
                "c0`.private_text"
            } > private_text_0;

            privateData {
                "c0`.data"
                "c0`.bss"
                "c0`.rom"
            } > private_data_0;
    }
}

address_translation (task_c0) {
    private_text_0                 (SYSTEM_PROG_MMU_DEF):PRIVATE_DDR0;
    private_data_0                 (SYSTEM_DATA_MMU_DEF):PRIVATE_DDR0;
}
```

The example shows how the virtual memory for code and data (`private_text_0` and `private_data_0`) are defined and how the compiler sections are then placed and mapped

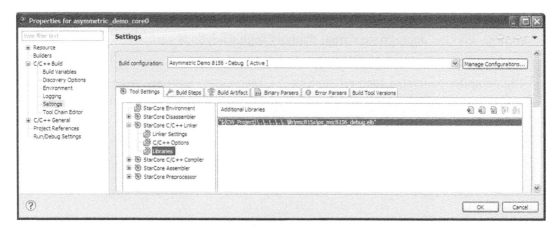

Figure 17-17:
Linker additional libraries.

into the proper private physical memory. This code is then duplicated for the rest of the DSP cores.

The last step before linking the application is to add the SDOS library. This is done via the linker Additional Libraries menu as is shown in Figure 17-17.

At this point all that remains is to build the project. The DSP IDE by default invokes the GNU Make to build the projects within the workspace. Using makefiles facilitates team work across the network and makes it possible to deploy and build the applications on other operating systems too. The maker settings can be set via the dialogues as shown in Figure 17-18.

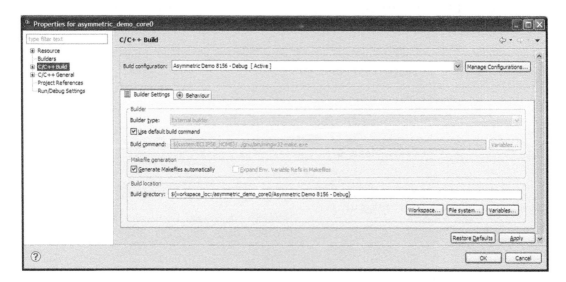

Figure 17-18:
Build settings.

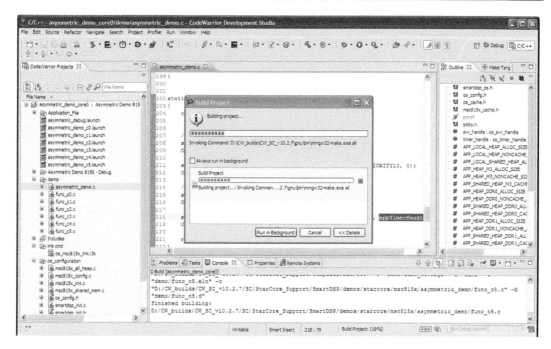

Figure 17-19:
DSP application building progress.

When the Project/Build Project option is selected, the IDE based on build settings generates automatically all the intermediate files and starts the building process as shown in Figure 17-19. During the process, the user can visualize, throughout the IDE console, the building progress.

Once the building process is finished the user has access to the executable files and can start to debug and instrument the code.

Execute and debug the application on multi-core DSP

The DSP IDE offers a very large set of features for DSP application debugging. The most relevant of them will be discussed in this chapter. As for the C/C++ editing part, the IDE provides a default, fully configurable debug perspective, shown in Figure 17-20.

The perspective provides information related to stack frames (upper left corner), variables, registers, cache and breakpoint views (upper right corner), C editor and disassembly (in the middle left and right) and miscellaneous views used for debugging (on lower part of the screen).

Throughout this chapter the main steps and features associated with multi-core debugging will be discussed.

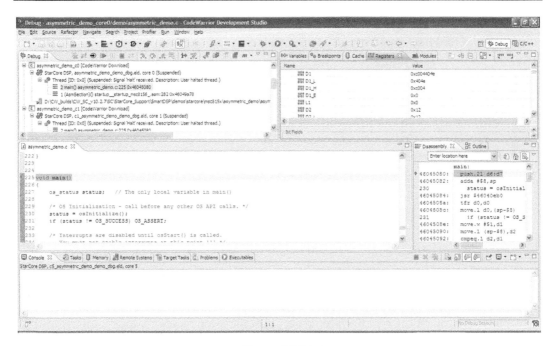

Figure 17-20:
IDE defaults debug perspective.

Create a new connection

Before downloading the code to the DSP memory the debugger connection must be defined. The CodeWarrior connection is based on Eclipse Remote System Explorer (RSE) wizard. This allows the user to create a new RSE system that will describe physical connection parameters. Further on, this system can be used to connect different applications with the DSP target without any additional steps.

DSP IDE's support tree types of connections, depending on the probe used to connect with the target. The physical connections can be done via USBTap or EthernetTap. A third type of connection is done for the simulator and in this case the connect type is CCSSIM2.

To create a new connection system, from the Remote Systems panel select New/Connection. The dialogue window that will appear should be configured as in Figure 17-21.

Once the connection type and Connection Server (CCS) port is selected the connection is ready. Apart from this basic setup, the CodeWarrior allows users to configure more complex scenarios like remote connection via a remote CCS running on a PC other than the one the DSP IDE is running on.

Also, different JTAG configuration chains and custom debugger memory configurations files and download methods can be selected from system properties (see Figure 17-22).

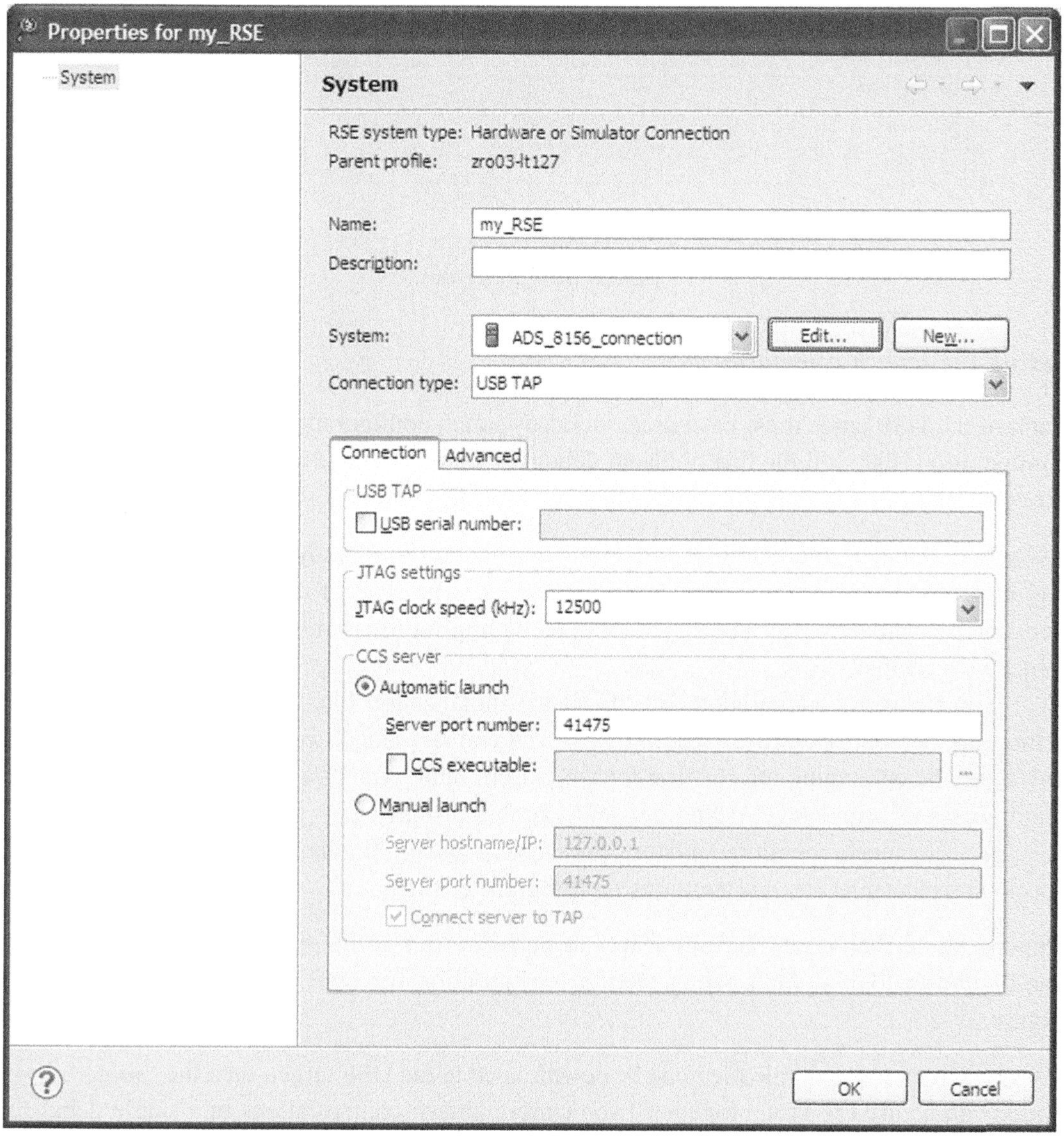

Figure 17-21:
RSE system.

After the connection is correctly configures, the RSE system created (in this case my_RSE) will be visible in the Remote System view and it can be reused in other projects that are built with the same DSP target in mind. The main advantage of this approach is that instead of having for each new project a new connection system, a single system that is kept in the workspace can be utilized for all the projects. If a modification of the RSE system is done, this will be immediately visible and available for all the projects within the workspace.

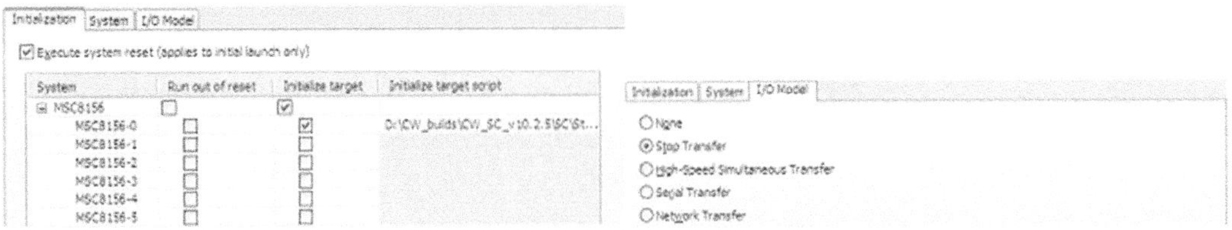

Figure 17-22:
Connection properties.

Set up the launch configuration

Each of the DSP cores must have an associated launch configuration. The launch configuration is an xml file that holds all debugger settings related to a particular executable file. CodeWarrior supports three types of launch configurations:

•   `Attach` launch configuration assumes that code is already running on the board and therefore does not run a target initialization file. The state of the running program is undisturbed. The debugger loads symbolic debugging information for the current build target's executable.

•   `Connect` launch configuration runs the target initialization file. This file is responsible for setting up the board before connecting to it. The connect launch configuration does not load any symbolic debugging information for the current build target's executable. Therefore there is no access to source-level debugging and variable display.

•   `Download` launch configuration stops the target, runs an initialization script, downloads the specified executable file, and modifies the program counter.

Depending on the action the user wants to perform, one of these launch configurations can be chosen. The default debug launch configuration for DSP core 0 is shown in Figure 17-23.

If more than one executable file must be downloaded to the DSP target, then the `Launch Group` option can be used to group together two or more launch configurations in a single debugger launch action. When this launch group is executed, the CodeWarrior will automatically load all the executable files and it will make all the appropriate settings with regard to the DSP target. For the asymmetric demo all six debug launch configuration files will be executed as a group (see Figure 17-24).

Debugger actions

The common debugger actions are listed on top of the `Debug` panel. Option such as: Reset, Run, Pause, Stop, Disconnect, Step In, Step Over, and Step Out are available for each Debug

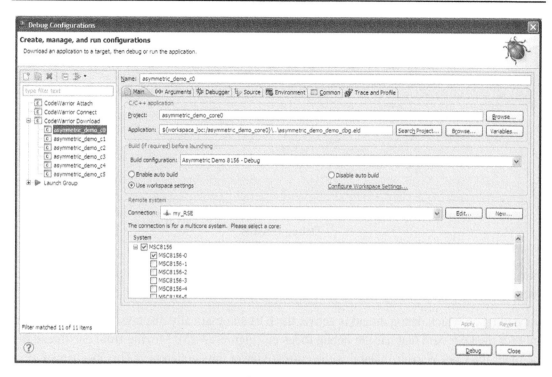

Figure 17-23:
Debug launch configuration for DSP core 0.

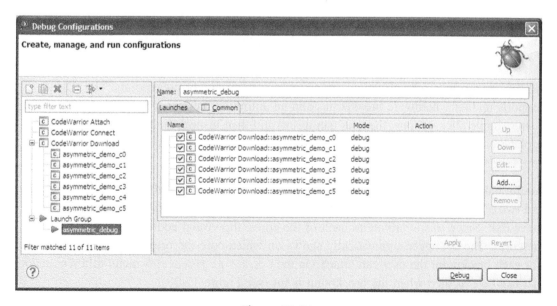

Figure 17-24:
Launch group.

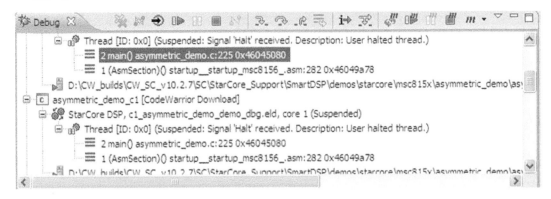

Figure 17-25:
Debug threads.

thread. These options are doubled by their counterpart multi-core options marked with the 'm' decorator.

Depending on which debug thread is active, the IDE automatically displays the entire context only for the DSP core that has the debug focus on (Figure 17-25). Moving from one thread to the other the CodeWarrior maintains the correct viewer data coherency. This mechanism is doubled by a caching algorithm whose scope is to reduce the communication with the target and to maintain a flawless fast change between the threads. The target data is read only if the debugger detects a change, otherwise the IDE communication with the target remains idle.

The Register view, shown in Figure 17-26, was enhanced to display all relevant information for the users. Each register that was updated since the last PC change is highlighted and a complete bit-field description for each part of the register is available.

No debug session can be effectively done without the usage of breakpoints. The IDE supports up to six hardware breakpoints and an unlimited number of software breakpoints. Using simple click actions, the user can install and then inspect the breakpoints in the application. Depending on breakpoint type, the IDE marks the breakpoint with the appropriate decorator as shown in Figure 17-27.

Additional information for each breakpoint is displayed. In this case the breakpoints (one HW and two SW) are installed on the same function print_func called from SDOS software timer task. Since this is private to each of the cores, the virtual address is the same for all the cores. The debugger automatically shows on which core the breakpoint was installed. For each breakpoint the user can attach via the Breakpoint Properties menu different actions that can be executed when the breakpoint is hit (see Figure 17-28).

Using the Breakpoint Properties the default behavior of the debugger in case of a breakpoint hit can be changed. A common functionality is conditional breakpoints (see Figure 17-29)

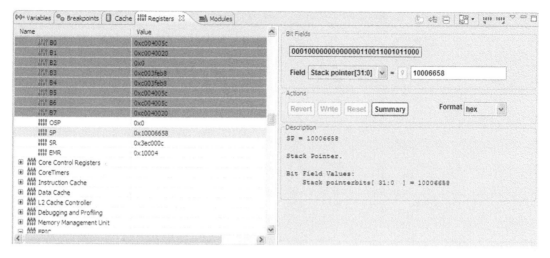

Figure 17-26:
Register view.

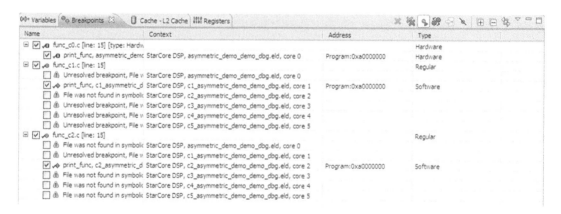

Figure 17-27:
Breakpoints view.

what can be used to control breakpoints behavior in case of debugging software loops. The user can instruct the debugger to trigger the breakpoint only when some conditions are met.

In the case where a breakpoint is hit or when stepping throughout the code, the debugger highlights the instructions executed and offering a visual trace of executed code as shown in Figure 17-30 where the last three assembly instructions are marked using a gradient color.

At this point it is very important for a user to understand exactly how a breakpoint interacts with the application and when and how to use a breakpoint in a multicore application.

The hardware breakpoints are implemented by a special processor unit — due to this reason they are limited as number. The logic circuit watches every bus cycle and when a match with

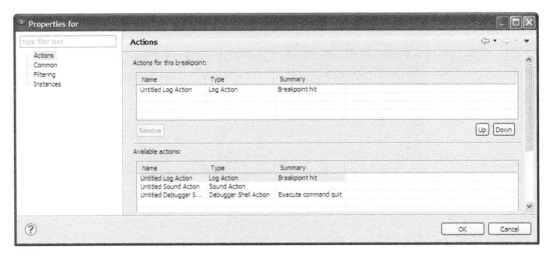

Figure 17-28:
Breakpoint actions.

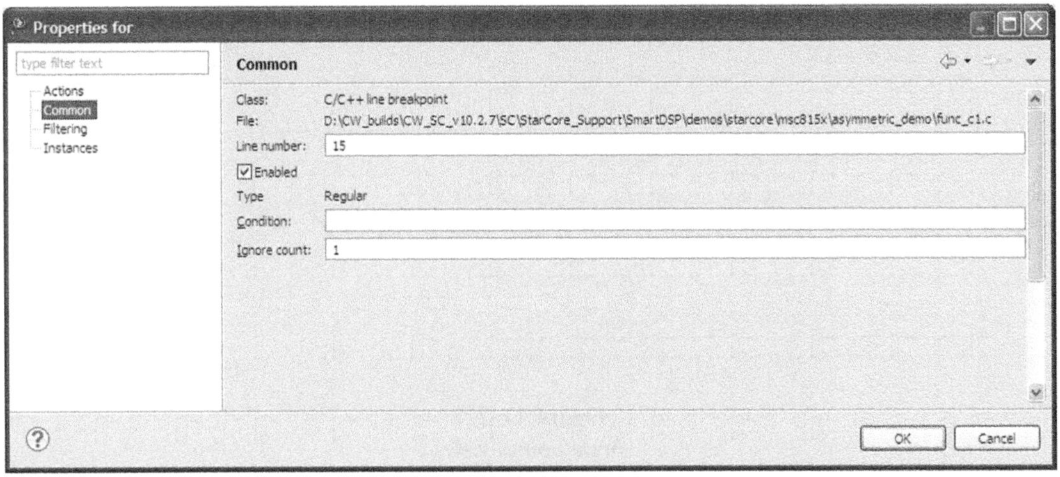

Figure 17-29:
Conditional breakpoints.

the address set for comparison occurs, it then stops the core. The hardware breakpoint does not modify the code, stack, or any target resource, being from this point of view completely non-intrusive.

Even further, it is advisable to use these kinds of breakpoint in a multicore application where the synchronization between threads at run-time is critical. Since these breakpoints do not interact with the program or the core, installation during run time can be done without any negative effects upon the application.

Figure 17-30:
Disassembly highlight of the latest executed instructions.

By contrast the software breakpoints always modify the code by inserting a small piece of code (usually the debugger inserts a special instruction that triggers an interrupt when it is executed). From this point of view software breakpoints are intrusive. After the debug interrupt is triggered the debugger stops the core and then replaces the code with the original code.

For DSPs that support variable length instructions sets (VLES) and depending on the type of breakpoint used for debugging, different results might be displayed. On hardware breakpoints case, the debugger stops immediately after the VLES that contains the matched address even if this was at the beginning or middle of the instruction set. Because of this the registers and memory contain the data for the rest of executed instructions within the VLES too.

In case of software breakpoints, the debugger rolls back with one instruction and the content of the registers and memory is actualized with the data before the real instruction execution.

The DPS IDE must provide the tools and method to limit the software breakpoint to the active debug context. Each of the software breakpoints can be set to act on all cores or on a single core. In case the user choose to debug a code section that is shared between multiple cores, then by using the functionality of limited software breakpoints to the active debug context, the debugger will control only core chosen by the user and leaves the other cores to run unaffected. Otherwise, being a shared code section, the debugger will control all the DSP cores.

While a debug session is active, the user can interact with the target via a comprehensive set of features to import or export memory like Memory monitors depicted in Figure 17-31, Target Tasks depicted in Figure 17-32, etc. The import/export memory can be done in various file formats depending on user requirements.

The IDE provides advance tools for target configuration and verification. For example the MMU configuration tool can be used to inspect the DSP target at any moment during debugging. This is a very useful tool for multi-core applications allowing users to modify on the fly the memory mapping (as is demonstrated in Figure 17-34) or to inspect the source of errors that might appear (as is shown in Figure 17-33 for an illegal memory access due to an initialized pointer).

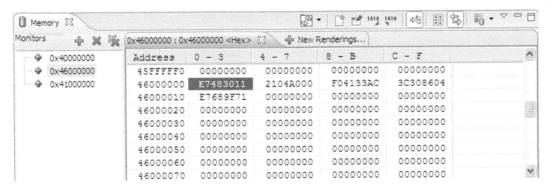

Figure 17-31:
Memory view.

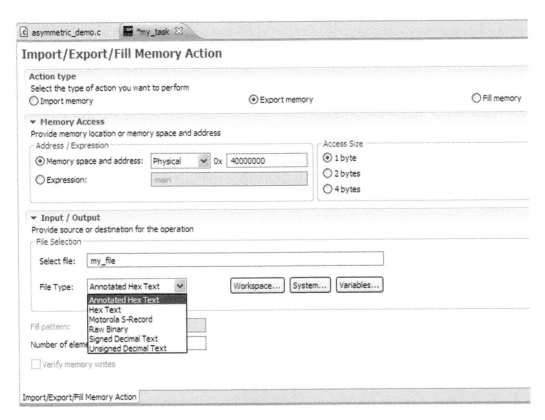

Figure 17-32:
Target tasks setup.

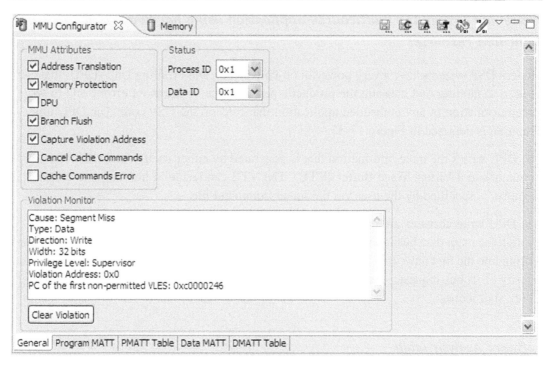

Figure 17-33:
MMU general view.

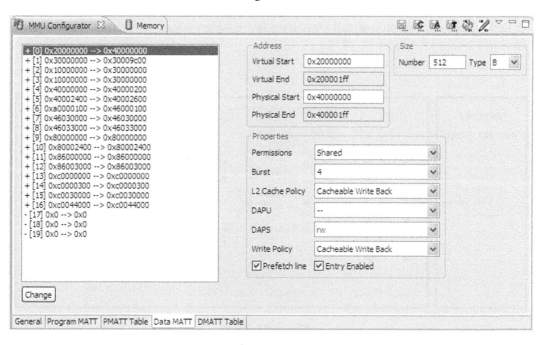

Figure 17-34:
MMU data memory translations.

Trace and Profile the multicourse application using hardware and software dedicated resources

Modern DSP systems have a very powerful on-chip Debug and Profiling Unit (DPU) that can be used to monitor and measure the product's performance, to diagnose errors, and to write trace information of any embedded application that runs on the DSP core. The DPU unit operation is depicted in Figure 17-35.

The DPU writes the trace information that is generated by either itself or the OCE (On Chip Emulator), to a Virtual Trace Buffer (VTB). The VTB can reside in internal or external memory, as specified by the user via the linker command file.

The DPU write accesses are buffered in the Trace Write Buffer (TWB), and then written through the main data bus interface in bursts. The lowest priority bus requests are used when writing into the first unwritten address of the VTB. When the TWB becomes full, it raises the priority of its bus request. At the first cycle after this priority change, the DPU generates a core stall request.

Software Analysis setup

In order to configure and analyze the trace date generated by the DPU, the IDE provides the Software Analysis plug-in.

The DPU setup is done automatically based on few simple options selected by the user as is shown in Figures 17-36 and 17-37.

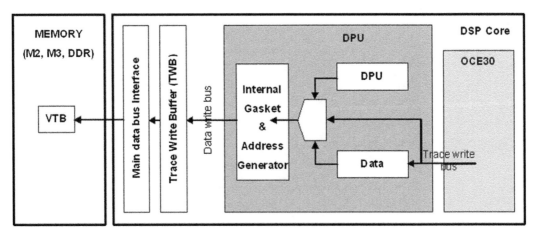

Figure 17-35:
DPU workflow.

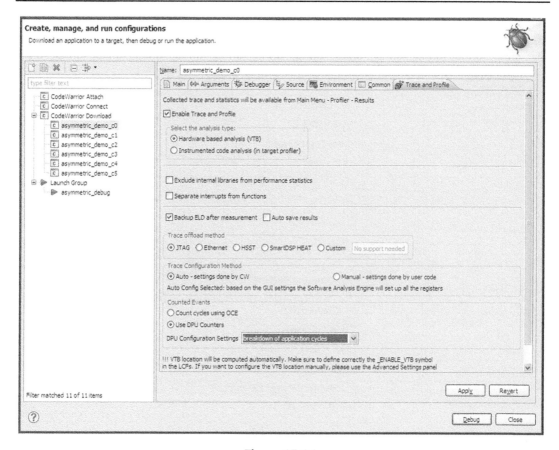

Figure 17-36:
Software Analysis main configuration dialogue.

For each of the debug configurations, the user can select to enable the trace. In this way one core or more can be instrumented. From the main configuration dialogue window, the user can select from a wide range of counting events. Using the default options provided by the IDE, the user can configure the DPU to measure the most representative events. Alternatively, the advanced options are available for individual DPU counters setup. Each of the six available counters within the DPU can be configured to measure any DPU supported event.

The VTB location is automatically placed into the appropriate memory via the linker command files. Alternatively the user can choose to place this buffer manually at his discretion into any free memory region on the target. The only constrain regarding the VTB placement is for multi-core application, where due to the nature of data, this buffer must have private privileges for each of the DSP cores. Otherwise the data might be polluted by the other cores events.

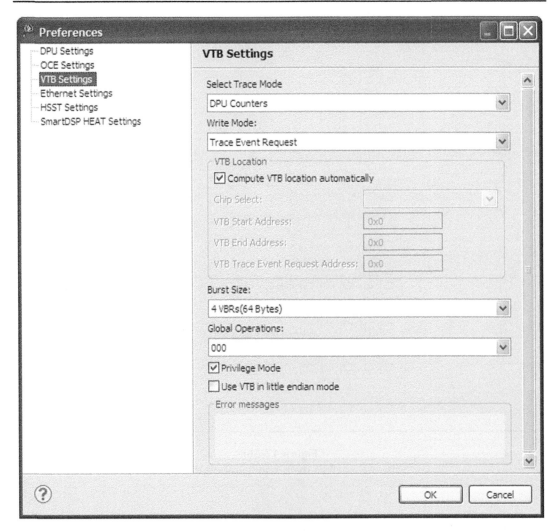

Figure 17-37:
VTB settings.

Once the VTB is defined at different addresses which do not overlap between the DSP cores, the data management is done automatically by the DPU unit without any additional interaction from the user side.

On the host PC, the DSP IDE handles the data from each VTB separately. In this way, each core can be examined separately from the others.

If the user wants to perform the appropriate DPU settings within the application and wants to utilize the Software Analysis only for results interpretation and display, then the Trace Configuration Method should be set to manual. In this mode, the CodeWarrior will not perform any DPU initializations and will interpret the results based on the DPU registers values found on the target. In this case it is the user's responsibility to do the correct DPU setup.

Since some applications might dump a lot of trace data into the internal VTB, CodeWarrior provides four methods for downloading the trace from the target to the host. Depending on user hardware and software setup, the CodeWarrior can be instructed to get the trace data via JTAG or Ethernet ports. The fastest method is to use the Ethernet port, but this method requires some additional libraries to be linked with the user application.

The code instrumentation is carried out automatically when a new debug session is launched. The trace results are available immediately when the DSP core is suspended or when the debug session is terminated.

The results for each application are available throughout the associated Trace and Profile Results panel as is shown in Figure 17-38.

For each of the debug configurations, there are five submenus available. Each of these submenu display specific analysis that was performed on the generic trace data.

Trace

The `Trace` submenu shows in the form of a list the raw DPU trace results (see Figure 17-39). It automatically computes and matches the addresses generated by the hardware unit with the executable file symbolic information and appends the corresponding counter values for each

Figure 17-38:
Trace and Profile results view.

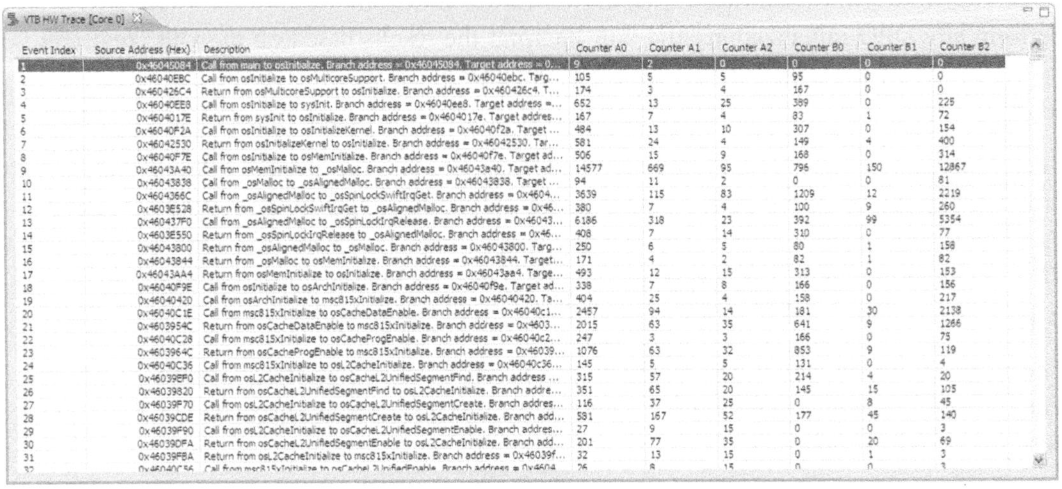

Figure 17-39:
Trace results.

pair of events. With the results in this form, the user can perform different actions like exporting the trace or trace filtering in order to find a specific event.

Critical code

The `Critical code` submenu is the most important view of all. It helps the users to optimize the critical part of the applications. For each of the functions within the application, the critical code computes, based on the value of the DPU counters, the number of cycles spent for each block execution as is shown in Figure 17-40.

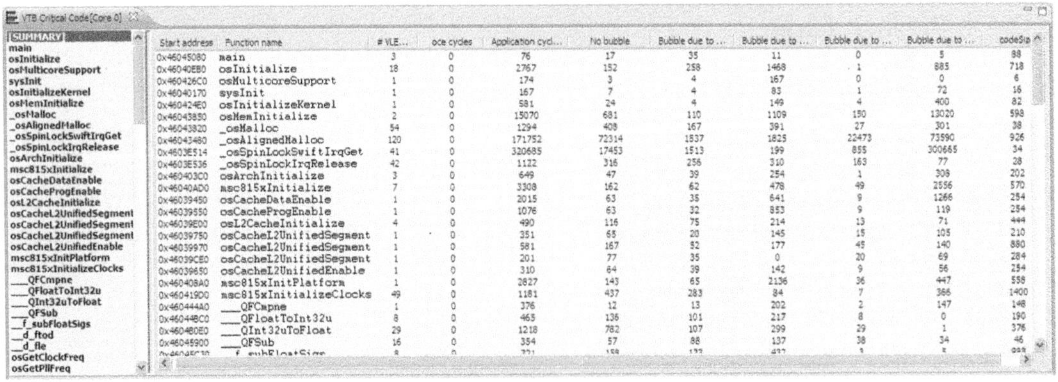

Figure 17-40:
Critical code view.

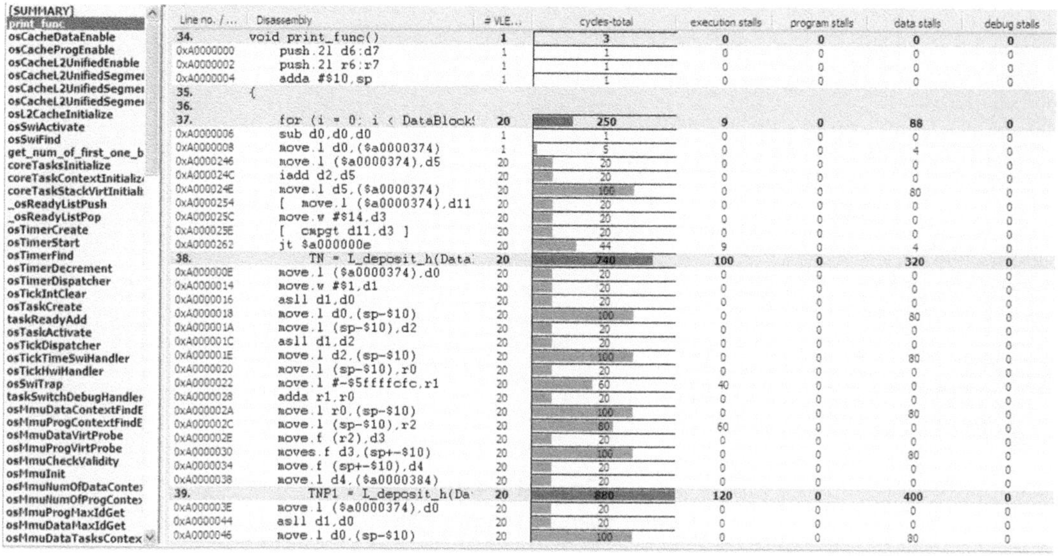

Figure 17-41:
Critical code function details level.

After the most critical functions have been identified the user can navigate one extra analysis level deep into the code. By selecting the appropriate function, the `Critical code` will display the instrumentation results for each code line. The viewer can be configured to display the results in different forms (C source only, assembly code only, or mixed).

As shown in Figure 17-41 the user can pinpoint the instructions where most of the processing cycles are spent and then can try to reduce the cycles by taking into account what caused the issue in the first place.

For example at line 38, the `Critical code` shows a high number of data stalls cycles need for `DataIn` buffer processing. This is a typical case of cache data miss, since the data is not yet in the cache at a time when the processing unit needs it.

Code Coverage

The Code Coverage view compares the actual application trace against the executable file code and produces the mapping of the program counter with the source code. An example is shown in Figure 17-42.

Usually this feature is used for checking if the algorithm flow is fully covered during the testing phase. Then the PC address found in the raw trace matches the one from the executable

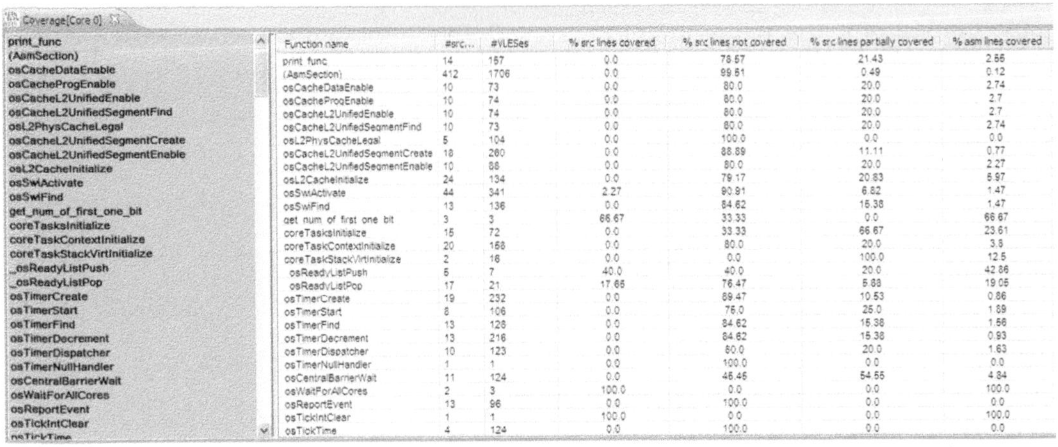

Function name	#src...	#VLESes	% src lines covered	% src lines not covered	% src lines partially covered	% asm lines covered
print func	14	157	0.0	78.57	21.43	2.55
(AsmSection)	412	1706	0.0	99.51	0.49	0.12
osCacheDataEnable	10	73	0.0	80.0	20.0	2.74
osCacheProgEnable	10	74	0.0	80.0	20.0	2.7
osCacheL2UnifiedEnable	10	74	0.0	80.0	20.0	2.7
osCacheL2UnifiedSegmentFind	10	73	0.0	80.0	20.0	2.74
osL2PhysCacheLegal	5	104	0.0	100.0	0.0	0.0
osCacheL2UnifiedSegmentCreate	18	260	0.0	88.89	11.11	0.77
osCacheL2UnifiedSegmentEnable	10	88	0.0	80.0	20.0	2.27
osL2CacheInitialize	24	134	0.0	79.17	20.83	5.97
osSwiActivate	44	341	2.27	90.91	6.82	1.47
osSwiFind	13	136	0.0	84.62	15.38	1.47
get num of first one bit	3	3	66.67	33.33	0.0	66.67
coreTasksInitialize	15	72	0.0	33.33	66.67	23.61
coreTaskContextInitialize	20	158	0.0	80.0	20.0	3.8
coreTaskStackVirtInitialize	2	16	0.0	0.0	100.0	12.5
osReadyListPush	5	7	40.0	40.0	20.0	42.86
osReadyListPop	17	21	17.65	76.47	5.88	19.05
osTimerCreate	19	232	0.0	89.47	10.53	0.86
osTimerStart	8	106	0.0	75.0	25.0	1.89
osTimerFind	13	128	0.0	84.62	15.38	1.56
osTimerDecrement	13	216	0.0	84.62	15.38	0.93
osTimerDispatcher	10	123	0.0	80.0	20.0	1.63
osTimerNullHandler	1	1	0.0	100.0	0.0	0.0
osCentralBarrierWait	11	124	0.0	45.45	54.55	4.84
osWaitForAllCores	2	3	100.0	0.0	0.0	100.0
osReportEvent	13	96	0.0	100.0	0.0	0.0
osTickIntClear	1	1	100.0	0.0	0.0	100.0
osTickTime	4	124	0.0	100.0	0.0	0.0

Figure 17-42:
Code Coverage summary view.

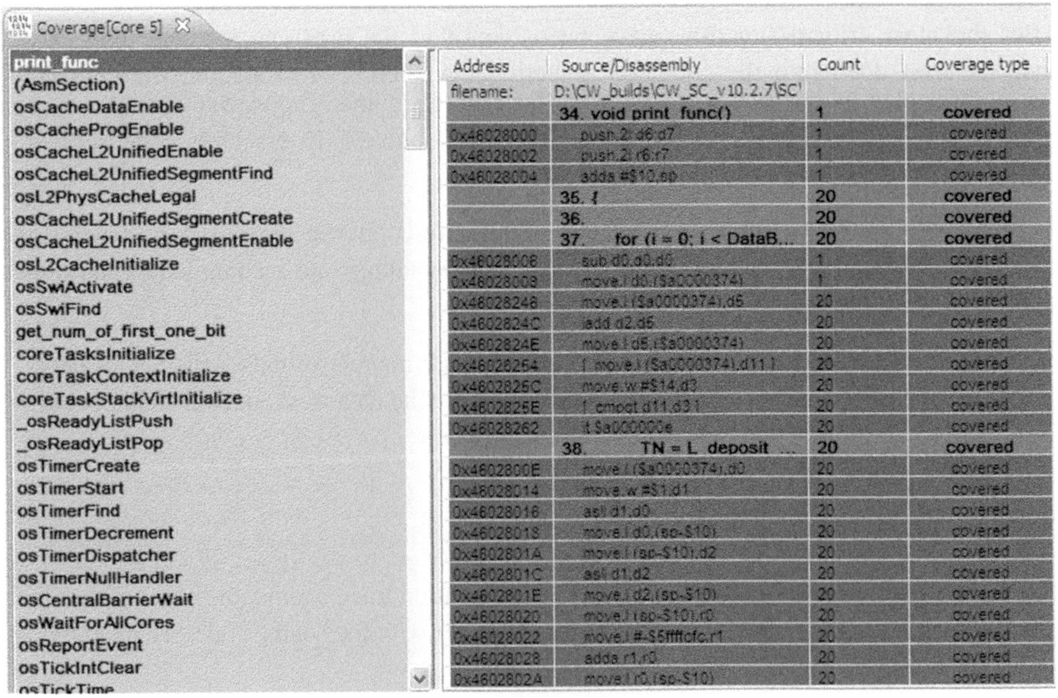

Address	Source/Disassembly	Count	Coverage type
filename:	D:\CW_builds\CW_SC_v10.2.7\SC'		
	34. void print_func()	1	covered
0x46028000	push.2l d6:d7	1	covered
0x46028002	push.2l r6:r7	1	covered
0x46028004	adda #$10,sp	1	covered
	35. {	20	covered
	36.	20	covered
	37. for (i = 0; i < DataB...	20	covered
0x46028006	sub d0,d0,d0	1	covered
0x46028008	move.l d0,($a0000374)	1	covered
0x46028248	move.l ($a0000374),d5	20	covered
0x4602824C	iadd d2,d5	20	covered
0x4602824E	move.l d5,($a0000374)	20	covered
0x46028254	[move.l ($a0000374),d11]	20	covered
0x4602825C	move.w #$14,d3	20	covered
0x4602825E	[cmpgt d11,d3]	20	covered
0x46028262	jt $a00000e	20	covered
	38. TN = L_deposit ...	20	covered
0x4602800E	move.l ($a0000374),d0	20	covered
0x46028014	move.w #$1,d1	20	covered
0x46028016	asl d1,d0	20	covered
0x46028018	move.l d0,(sp-$10)	20	covered
0x4602801A	move.l (sp-$10),d2	20	covered
0x4602801C	asl d1,d2	20	covered
0x4602801E	move.l d2,(sp-$10)	20	covered
0x46028020	move.l (sp-$10),r0	20	covered
0x46028022	move.l #-$5ffffcfc,r1	20	covered
0x46028028	adda r1,r0	20	covered
0x4602802A	move.l r0,(sp-$10)	20	covered

Figure 17-43:
Code Coverage function detailed view.

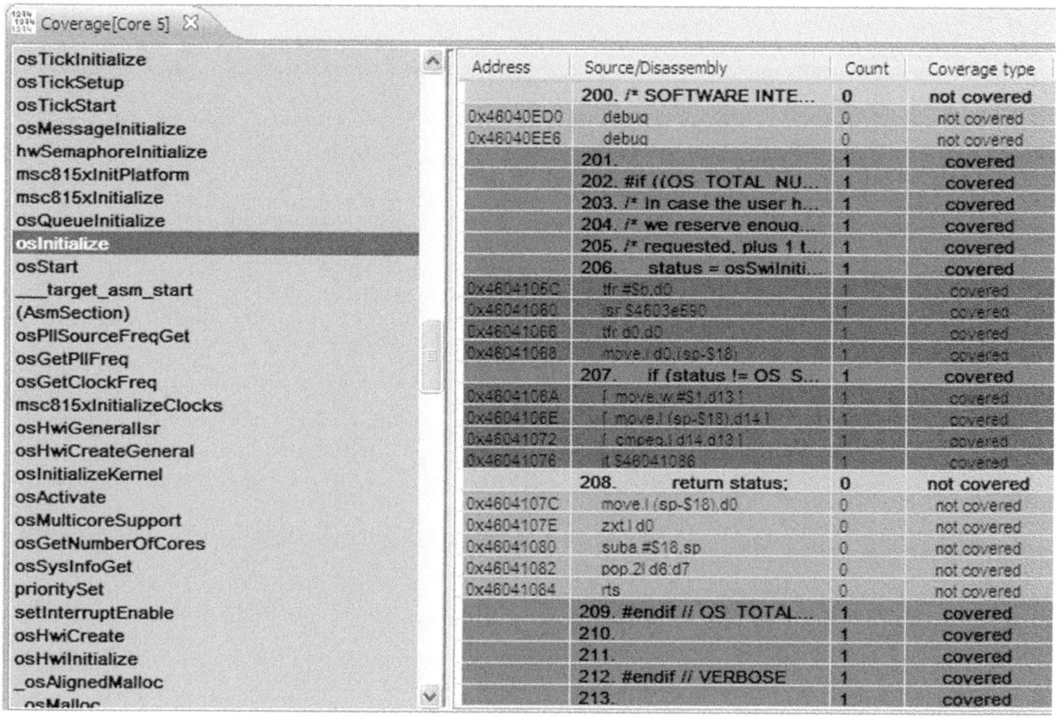

Figure 17-44:
Code Coverage details.

file; the code is displayed in green otherwise it is highlighted in magenta. Figures 17-43 and 17-44 show two of these cases of the asymmetric demo. (Please note the printed book does not show the color in these figures.)

If the entire code is highlighted as covered, the user can have the confidence that the tests carried out cover all the possible scenarios that might appear during the application life-time.

Performance

The Software Analysis functions would not be complete without a general summary of the application modules. The mail `Performance` view provides the application overview and this is demonstrated in Figure 17-45.

This view links together all the functions within the application and computes the statistics for both the caller and callee. Based on the raw trace, the IDE computes exactly how many cycles are spend for each function main code (*exclusive*), or the total cycles measured between a call and return (*inclusive*). By matching the trace against the executable file the code size can also be computed. In order to help the user to identify the most costly function, the results are aggregated in form of pie that show the overall performance.

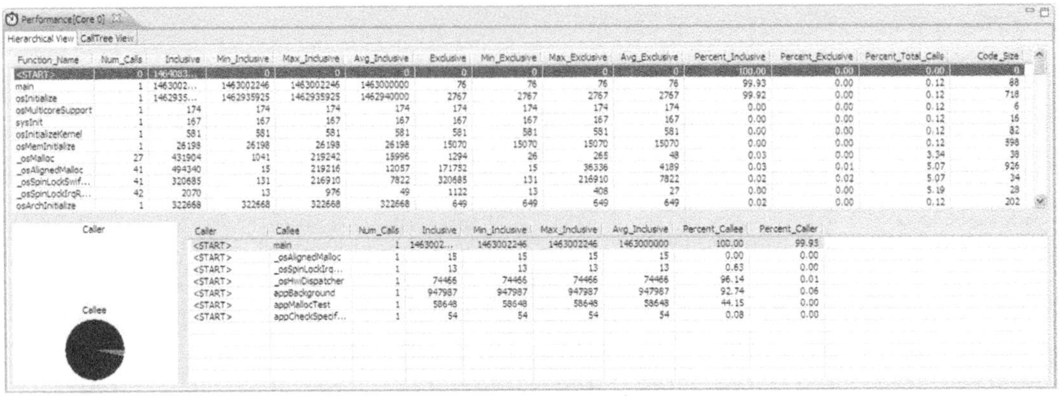

Figure 17-45:
Performance statistics.

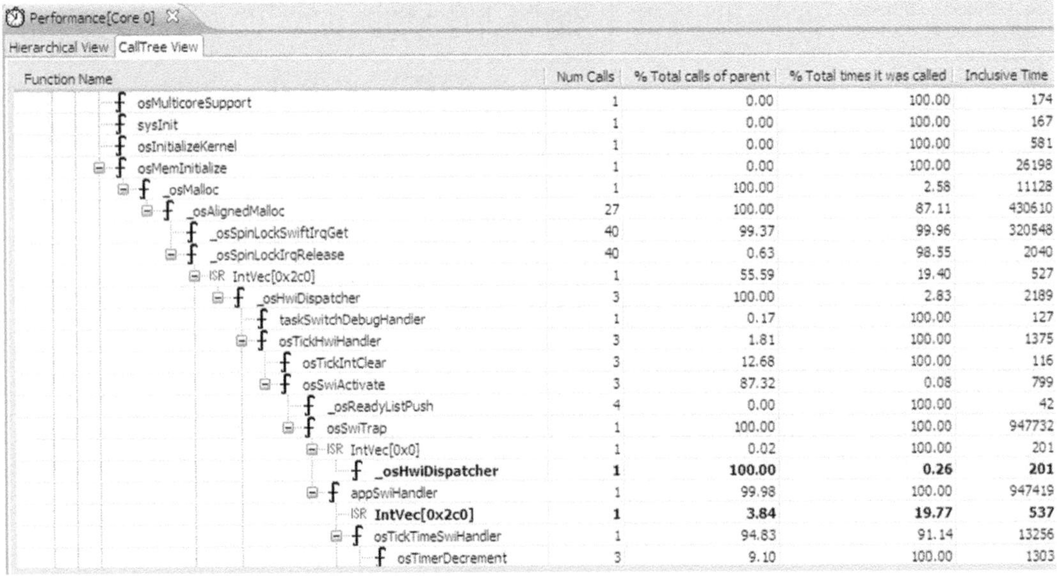

Figure 17-46:
Call tree view.

The results can be exported direct to Microsoft Excel CSV format. If needed the exported data can be formatted via the Excel in order to meet the user requirements.

Beside the statistics the `Performance` view assembles the application call tree. The entire map of functions calls for asymmetric demo, shown in Figure 17-46. This feature allows a better understanding of application and operating system flow.

Case Study — LTE Baseband Software Design

Arokiasamy I, Vatsal Gaur and Nitin Jain

Introduction

Wireless communications technology has grown tremendously in the past few years. The first generation (1G) analog cellular systems like Advanced Mobile Phone Systems (AMPS), Total Access Communication Systems (TACS) etc. supported only voice communication with some roaming limitations. The second generation (2G) digital cellular systems promised higher capacity, national and international roaming and better speech quality than analog cellular systems with Enhanced Full Rate (EFR) and Adaptive Multi-Rate (AMR) codec. The two popularly deployed second generation (2G) cellular systems are GSM (Groupe Special Mobile later commonly known as Global System for Mobile communications) and CDMA (Code Division Multiple Access). The 2G systems were primarily designed to support voice communication while lower data rate of up to 14.4 kbps data transmission were introduced in the later releases of 2G standards.

The International Telecommunication Union (ITU) initiated IMT-2000 (International Mobile Telecommunications-2000) paved the way for 3G. A set of requirements such as a peak data rate of 2 Mbps and support for vehicular mobility was published under IMT-2000 initiative. The GSM migrations are based on 3rd generation partnership project (3GPP) and CDMA migrations are based on 3GPP2.

High-data-rate, low-latency and packet-optimized radio access technology supporting flexible bandwidth deployments beyond 3G technologies are the main goals of LTE.

A new network architecture is designed with a goal to support packet-switched traffic with seamless mobility, quality of service and minimal latency. The air-interface related attributes of the LTE system are summarized in Table 1.

LTE architecture

LTE has been designed to support only packet-switched services, in contrast to the circuit-switched model of previous cellular systems. It aims to provide seamless Internet Protocol (IP) connectivity between User Equipment (UE) and the Packet Data Network (PDN),

Table 1: LTE features.

Bandwidth	1.25 − 20 MHz
Duplexing	FDD, TDD, half duplex FDD
Mobility	350 Km/h
Multiple access	Downlink: OFDMA
	Uplink: SC-FDMA
MIMO	Downlink − 2 x 2, 4 x 2, 4 x 4
	Uplink − 1 x 2, 1 x 4
Peak data rate in 20 MHz	Downlink − 173 Mbps with 2 x 2 MIMO and 326 Mbps with 4 x 4 MIMO
	Uplink − 86 Mbps with 1 x 2 antenna configuration
Modulation	QPSK, 16-QAM, 64-QAM
Channel coding	Turbo code
Other techniques	Link adaptation, Channel sensitive scheduling, Power control, Hybrid ARQ

without any disruption to the end users' applications during mobility. While the term 'LTE' encompasses the evolution of the radio access through the Evolved-UTRAN (E-UTRAN), it is accompanied by an evolution of the non-radio aspects under the term 'System Architecture Evolution' (SAE) which includes the Evolved Packet Core (EPC) network. Together LTE and SAE comprise the Evolved Packet System (EPS). EPS uses the concept of EPS bearers to route IP traffic from a gateway in the PDN to the UE. A bearer is an IP packet flow with a defined Quality of Service (QoS) between the gateway and the UE. The E-UTRAN and EPC together setup and release bearers as required by applications.

It consists of several different types of nodes, some of which are briefly illustrated in Figure 1. The eNodeB is connected to the EPC by means of the S1 interface, more specifically to the

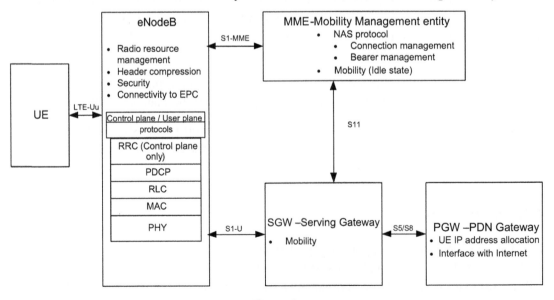

Figure 1:
LTE architecture

S-GW by means of the S1 user-plane part, S1-u, and to the MME by means of the S1 control-plane part, S1-MME. One eNodeB can be connected to multiple MMEs/S-GWs for the purpose of load sharing and redundancy.

Evolved system components and design

As the complexity and real time processing requirements of the system have grown over a period of time, the components of the LTE system have also evolved simultaneously. A discussion regarding the prime components and associated design considerations is followed here.

Multi-core digital signal processors

While clock speeds can only be increased to a certain point, multi-core processors have become preferred choice for implementation of complex software like LTE physical layer. In order for the applications to make the most of the evolving multi-core architectures, they must be designed from the beginning with multithreading in mind that allows an application to separate itself in many smaller tasks that can be run on different cores at the same time.

Challenges in multi-core design

Deadlock prevention and data protection

Use Case 1: System initialization

In a multi-core environment, when data is kept in a shared memory that is designed to be used by different processes on different cores then this data needs protection to ensure data integrity. This can be achieved through an effective implementation of locks.

It is also, sometimes, essential to ensure the synchronization in time when one process inherently assumes the completion of the work being done by another process. This can be achieved through usage of barriers.

Barriers are an extremely useful mechanism for regulating program execution, especially when multiple cores have to be brought to a common point of execution. These can be implemented with different scope. *Centralized* barrier ensure that all the cores come to the single point of execution while *partial* barrier ensure that only some cores are programmed to come to a common control function or label before execution can continue.

In a classical software design on a multi-core platform, the system initialization is undertaken by one core. This ensures centralized 'record keeping of initialization activities' and an ease of debugging. An example of a key system initialization task is the initialization of the task inventory structure that defines the various tasks existing in the system. Every task when it is created, dynamically or statically, is required to register itself with the system along with the associated handler function with the help of the task inventory structure. This structure, hence, maintains an inventory of various tasks in the system. This structure has to be available in shared memory to enable any core to trigger a task on itself or another core. In a multi-core system, different cores can come out of the reset at different times. In such a case, it is possible that even before the task inventory structure is initialized by the core assigned to perform the structure initialization, another core which has come out of the reset early has registered a new task with the system using this task inventory structure.

This design problem can be resolved by using the barrier capability, which will force all the cores to come synchronize at a common point of execution. After this synchronization point is reached, the structure will be initialized by the core assigned to do that task. Only after this step is complete, further execution by individual cores will continue. The initialization using barrier synchronization is illustrated in Figure 2.

A key method of data protection in software is the usage of locks. Using locks, the software designer can share the same resource across different processes effectively. This can be achieved through spinlocks where each core keeps on checking on the availability of the resource. An alternative implementation method is through the usage of *semaphore* logic, where the task goes into a queue if the resource is not available.

Inter-core communication

Use Case 2: Triggering of sequential and parallel processes

Software design on multi-core requires an efficient mechanism for inter-core communication. This communication mechanism should ensure that a process running on one core can trigger

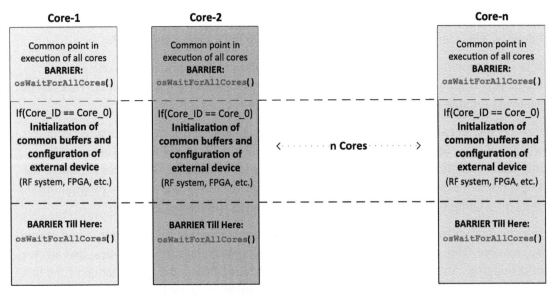

Figure 2:
Initialization and synchronization using barrier.

another task on the same core or a different core. Two types of inter-core communication mechanism are generally necessary: point to point transfer of messages and point to multipoint transfer of messages.

Point to point transfer of messages can be employed when one core after finishing its task triggers another task on a different core. Point-to-point message posting usually involves interrupt based message handling.

An example of this is demonstrated in Figure 3 where one core triggers a task on another core.

Point to multipoint transfer of messages is best suited for a master-slave model where all the slave cores will get interrupted or have to regularly poll for a message from the master. The message from the master will trigger different types of tasks on each slave core. In order to prevent any message getting lost or corrupted, spinlocks can be used. This ensures that the buffer space used by the message cannot be used for posting another message as long as the destination core has not read the message and cleared the spinlock (Figure 4).

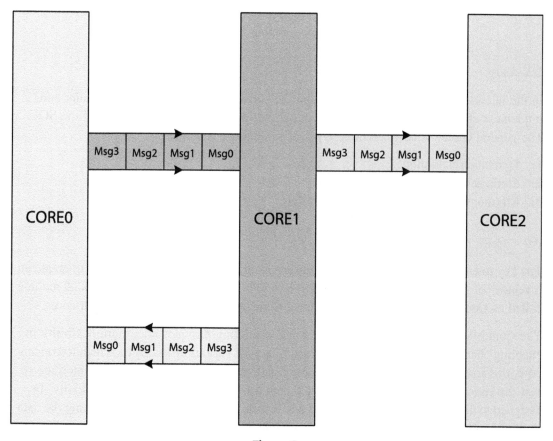

Figure 3:
Point to point queue.

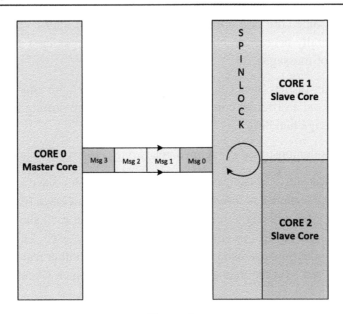

Figure 4:
Point to multipoint queue.

Scheduling — Dynamic and static

In the interest of an efficient system design the software designers have to consider and implement effective mechanism for scheduling various tasks on various cores of the SOC. The general choice is between three types of scheduling strategies:

1) Dynamic scheduling
2) Static scheduling
3) Mixture of static and dynamic scheduling

Use Case 3: Dynamic scheduling

The key to enabling any form of scheduling mechanism in a design is the ability to create an inventory of various tasks and control their execution sequence and dependencies. A module called as Controller has been designed to schedule specific tasks across different cores.

The Controller module in context of dynamic scheduling provides maximum flexibility in enabling changing parameters and scenarios. Since the various tasks can be scheduled on any core, this form of Controller capability necessitates usage of a shared memory interface so that the various cores have access to all the data for all the tasks that can 'potentially' be assigned to them. A shared memory interface, in turn, leads to the need for protecting the data through locks. Additionally, the Controller module running on a particular core depends on an effective point to point and point to multipoint communication mechanism so that various tasks can be offloaded to cores and the results of execution are available back to the Controller.

In order to effectively schedule various tasks on various cores the Controller software module requires the following three types of information:

1) The inventory of tasks that need to be processed in the system
2) The order of processing of each task such that for every task there is a list of dependencies. The Controller module should explicitly contain information on these dependencies so that it does not schedule a task unless all its dependencies have been resolved. This can be done through a dependency matrix.
3) A method to dynamically calculate and store the computation cost of each executing task. This information can be leveraged to ensure that all the tasks are distributed efficiently so that all the cores are loaded efficiently.

This design mechanism can be illustrated through UL Symbol Processing in Layer 1. Here, the inventory consists of four tasks:

1) Symbol receive
2) DMRS generation
3) Channel estimation
4) Equalization

Based on the inventory of the tasks that have been created a sample dependency matrix is shown below.

This matrix shows that tasks J4 can only be executed after completion of task J0 and J1. Similarly, equalization defined as task J6 can only be undertaken after the completion of tasks J4 and J5.

Task ID	Task	Task Description	Task Dependency
J0	DMRS generation	Generation of reference symbol for slot 0	After the sub frame configuration is available from L2
			Must be executed before channel estimation
J1	Symbol receive	Reference symbol in slot 0	After the symbol is available from Antenna interface
J4	Channel estimation	For slot 0	J0, J1
J2	DMRS generation	Generation of reference symbol for slot 1	After the sub frame configuration is available from L2
			Must be executed before channel estimation
J3	Symbol receive	Reference symbol in slot 1	Same as above
J5	Channel estimation	For slot 1	J2, J3
J6	Equalization	For interpolated	J4, J5 (for interpolated channel estimates)

Finally, the Controller software module needs to keep track of the computation cost of each task. This can be averaged data gathered and stored by the Controller for a historical set of processing of the same task. Through this method the Controller can ensure that if a task execution is dependent on two tasks to complete their execution then it is able to estimate on which cores all these tasks should be executed so that the idle waiting cycle for the dependent task is minimized.

As an example: Task J4 is dependent on J0 and J1. The Controller needs to have an estimate on the time required by J0 and J1 for successful execution. If time taken by J1 is greater than time taken by J0, and if J0 and J1 are executing on separate cores then unless J1 is complete the Controller cannot schedule J4. While, the task J4 is waiting for J1 to complete − the Controller can schedule task J2 on the core where the task J0 was executing earlier. This form of run time information allows the Controller to plan the task scheduling (Figure 5).

Using this form of approach the designer can expand the inventory of tasks through adding additional tasks and granularizing existing tasks through a logical division of functionality to

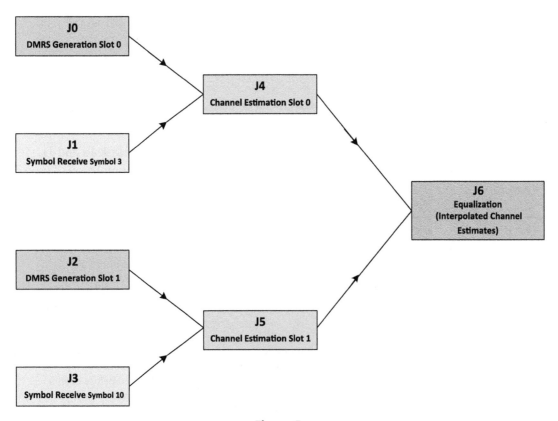

Figure 5:
Job dependency matrix.

provide more alternatives to the Controller for scheduling these tasks separately. For example, the symbol receive functionality can be sub divided into three sub-tasks: Cyclic prefix removal, cyclic shift and FFT. If these additional sub-tasks are also captured in the dependency matrix discussed earlier then the Controller module can schedule them for execution based on their dependency graph and computation requirements. While doing this micro level task break up so that each task can be exercised independently based on the dependency graph, the designer needs to be conscious about performance degradation due to higher fragmentation of tasks. This form of fragmentation can lead to higher inter-core communication as multiple tasks will have to communicate with each other for results and data dependencies. This will also necessitate extensive usage of shared memory in lieu of internal memory. The higher usage of shared memory can lead to higher data latencies as typically the shared memory has higher wait states than the internal memory.

Use Case 4: Static scheduling

The complexity of designing a Controller module that can dynamically schedule various tasks can be very high. The need for an exhaustive and intelligent dynamic scheduling mechanism may not be required when the use case is well defined and bounded. In such a case, a Controller module that is configured statically is a feasible option. This approach lends items itself to simplicity of implementation in Controller logic, as the task partitioning across various cores is estimated statically. This approach also yields to easier debugging against problems as the system behavior is more deterministic. Since the planning for task scheduling is done statically, the system is typically dimensioned for worst case scenario which can, in some cases, lead to non-optimum performance.

The example deliberated in Use Case 3 can be effectively implemented using the static scheduling approach vis-à-vis dynamic scheduling.

In such a case, the Controller module has to be made responsible for initialization of the various system resources and tasks in the processing chain. It also is responsible for maintaining a statically defined table that will identify the different cores on which different tasks will execute. This type of static binding of tasks to the cores imparts a sense of deterministic behavior in the system and simplifies the Controller logic.

In order to effectively schedule various tasks on various cores the Controller software module requires the following three types of information:

1) The inventory of tasks that need to be processed in the system.
2) The order of processing of each task such that for every task there is a list of dependencies. The Controller module should explicitly contain information on these dependencies so that it does not schedule a task unless all its dependencies have been resolved. This can be done through a dependency matrix.
3) Association and binding of the task to a core.

This design mechanism can be illustrated through UL Symbol Processing in Layer 1. Here, the inventory consists of four tasks:

1) Symbol receive
2) DMRS generation
3) Channel estimation
4) Equalization

Based on the inventory of the tasks that have been created a sample dependency matrix is shown below.

This matrix shows that tasks J4 can only be executed after completion of task J0 and J1. Similarly, equalization defined as task J6 can only be undertaken after the completion of tasks J4 and J5. This type of dependency information logic should be built into the Controller logic.

Task ID	Task	Task Description	Task Dependency	Core Number
J0	DMRS generation	Generation of reference symbol for slot 0	After the sub frame configuration is available from L2 Must be executed before channel estimation	1
J1	Symbol receive	Reference symbol in slot 0	After the symbol is available from Antenna interface	0
J4	Channel estimation	For slot 0	J0, J1	1
J2	DMRS generation	Generation of reference symbol for slot 1	After the sub frame configuration is available from L2 Must be executed before channel estimation	1
J3	Symbol receive	Reference symbol in slot 1	Same as above	0
J5	Channel estimation	For slot 1	J2, J3	1
J6	Equalization	For interpolated	J4, J5 (for interpolated channel estimates)	2

This form of task partitioning shows that at every 1 ms tick from the system, the symbol receive functionality is processed on Core 0 for all the slots. Similarly as soon as the sub frame information is available from the Layer 2, the DMRS generation is enabled on Core 1. Channel estimation is also undertaken on Core 1 and it depends on output result of DMRS generation and symbol receive functionality (Figure 6).

The aforesaid Controller design for supporting static scheduling has tightly coupled the various tasks to a given core. This logic can be extended, to impart a certain decision making capability to Controller. One approach for imparting higher intelligence in the Controller logic is to provide guidelines on binding tasks on a core after taking into consideration

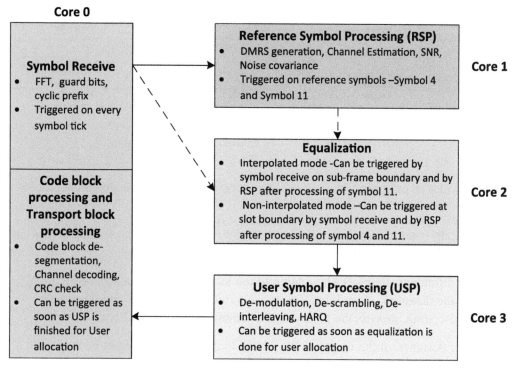

Figure 6:
Static scheduling design.

a limited set of system parameters. These parameters can be, for example, the number of resource blocks allocated to the user.

Parallelism and pipelining

Multi-core systems naturally lend themselves to the software designer to exploit parallelism in the underlying algorithm. Hence, one of the principal challenges for software designer is to modularize the software into independent and dependent modules that can be executed across the many cores of the system. This also requires identifying the dependency of various tasks on each other. This has been illustrated earlier in this chapter through Use Case 3 and Use Case 4. In these preceding examples we have discussed the task partitioning on various cores of the initial uplink tasks in Layer 1 software.

This section discusses additional techniques to derive parallelism in the Layer 1 processing thereby improving the overall system performance.

Use Case 5: Parallelism in downlink chain

The downlink physical layer data chain lends itself to high parallelism. The software designer can take advantage of this inherent parallelism by choosing architecture specific techniques.

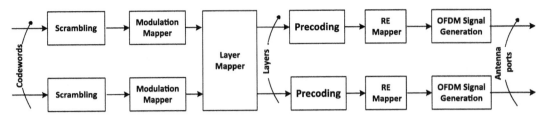

Figure 7:
Inherent parallelism in LTE downlink processing chain.

Figure 7 provides an example of the various key processing blocks in the downlink processing chain.

The software designer has three primary methods for choosing task partitioning across cores to leverage higher system performance:

1) Process different code blocks belonging to the same user simultaneously
2) Process the downlink chain for different users independently, but simultaneously
3) After pre-coding, process the Antenna data simultaneously across cores

The software designer, in the downlink can chose to process the different code blocks belonging to the same user simultaneously for the optimum system performance.

When the task partitioning is done across cores as defined in Figure 8, then the parallelism with respect to code block processing is not completely leveraged because even when there is more than one code block the code block processing is done on Core 1, in a sequential manner. This can add latency to the result of the code block processing and this can ultimately lead to a situation where the processing overshoots the 1 ms hard real time window.

This scenario can be avoided by ensuring that the code block processing is distributed across both the available cores. The division of code block processing across two cores can be done through different decision making processes. The example illustrated below, in Figure 8, does this such that only the odd number of code block is processed on Core 1. This was experimentally found to create a balanced system as the Core-1 is also used to perform the downlink control channel processing.

Use Case 6: Parallelism in uplink chain

In Use Case 3 and 4 we discussed the first three tasks in uplink processing, namely: Symbol receive, DMRS generation and channel estimation. After this processing the major tasks in the uplink are:

1) Equalization
2) User Symbol Processing

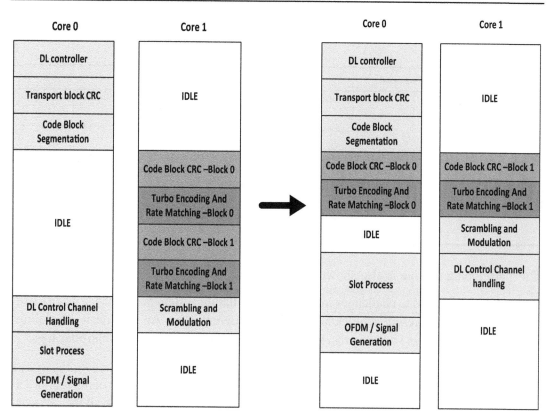

Figure 8:
Parallel processing of code blocks to improve core usage and performance.

3) Code block processing
4) Transport block processing

These four tasks are executed sequentially and thus straight forward parallelism is not evident.

Hence, parallelism can be gauged through two different ways:

1) Parallelism per user
2) Parallelism across users

In order to leverage task parallelism for every independent user, the tasks are distributed such that while all the four tasks are executed sequentially, for each user they are executed on different cores simultaneously. Thus all the four tasks, for every user, are executed on the same core. So if there are four cores available for the uplink processing then the Core-0 is allocated to User-1; Core-1 to User-2; Core-2 to

User-3; Core-3 to User-4. Thereafter, the user number wraps around to the core number.

Core Number	User Number	Tasks on Core
0	1	Equalization → User Symbol Processing → Code block processing → Transport block processing
1	2	Equalization → User Symbol Processing → Code block processing → Transport block processing
2	3	Equalization → User Symbol Processing → Code block processing → Transport block processing
3	4	Equalization → User Symbol Processing → Code block processing → Transport block processing
0	5	Equalization → User Symbol Processing → Code block processing → Transport block processing

This partitioning scheme is also shown in Figure 9.

This approach can lead to challenges in deriving an optimum system performance when there is a large variation in resource block allocations to different users. An intelligent scheduling logic can ensure that as soon as processing for one user is completed the Controller logic schedules another user on the idle core.

An alternative mechanism to derive parallelism across users is to assign each of the four tasks to a unique core. Data for all the users that need the task processing have to avail the services of the core on which the task is running.

Task	Core Number	User Number
Equalization	Core 0	All users
User Symbol Processing	Core 1	All users
Code block processing	Core 2	All users
Transport block processing	Core 3	All users

The equalization process is assigned to Core-0. After completion of the equalization process for User-1, the equalization for User-2 is computed on Core-0. This process proceeds for every user. The result of equalization process for User-1 is fed to Core-1 where the User Symbol Processing is undertaken. Over a period of time, a pipeline of processing tasks is built across all the four cores such that each core is doing its allocated task for a different user. So at some given instant — Core-0 is doing equalization for User-3; Core-1 is performing User Symbol Processing for User-2 and Core-2 is performing code block processing for User-1 (Figure 10).

This type of task partitioning leads to a very high inter-core communication as all the relevant data for every user has to be made available to the next core which is performing the next task in the processing chain.

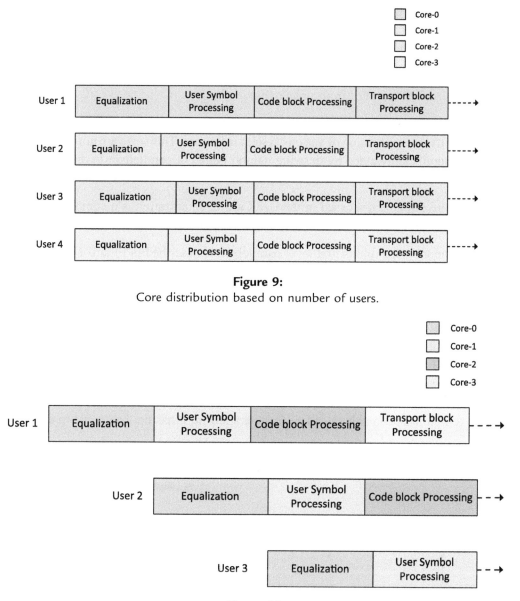

Figure 9:
Core distribution based on number of users.

Figure 10:
Core distribution based on individual tasks.

The second challenge with such an approach is that the amount of processing per user per task is dependent on key algorithm characteristics such as the number of resource blocks per user that need to be processed. If the number of the resource blocks for the first user is less than the number of resource blocks for the next user then the complexity of every task for the first user will get reduced as there is lesser data to process. This can lead to 'idle wasted cycles'. For example — if the equalization for the second user takes longer time vis-à-vis the User Symbol Processing for the first user; an idle slot is created on Core-1. This is because while the Core-1 has completed the User Symbol Processing for User-1, it cannot start the User Symbol Processing for User-2 as equalization process for User-2 is still ongoing on Core-0.

Use Case 7: Algorithm parallelism in uplink chain

Use Case 6 has demonstrated mechanisms through which tasks can be distributed across cores leading to an effective utilization of various system resources. This software portioning across cores can be significantly aided by deriving natural points of parallelism in the underlying algorithm.

As started in the example discussed in the Use Case 6; equalization is a computationally intensive task and it can be performed only after the channel estimation is complete.

A simplistic model that exploits first level parallelism is described in Figure 11. This shows that Core-0 is used to perform operations related to symbol receive functionality. Core-1 is used for Reference Symbol Processing based on the symbol information for third and tenth symbol received on Core-0. After the Reference Symbol Processing is complete (which includes channel estimation), equalization task is started on Core-2. The equalization for the current sub frame 'n' is triggered after the last symbol (symbol 13) is received in the given sub frame. Figure 11 shows that once the equalization task is started its processing will carry over into the next sub frame 'n + 1'. The processing required for computing the equalization for 12 symbols received in the sub frame 'n' requires an additional time equivalent of 4 symbols. While the design has been able to utilize task parallelism, it has not been able to extract

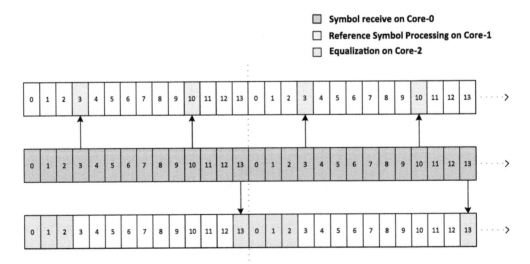

Figure 11:
Equalizer design with sub-frame granularity.

sufficient algorithmic parallelism. Ultimately this will lead to latency in obtaining results for the current sub frame 'n'. Also, this design approach does not utilize the Core-2 continuously and most of the time the Core-2 is 'idle'.

The aforesaid design approach can be modified to improve the algorithmic level parallelism. This can be done by ensuring that the granularity of the task processing is changed to symbol level. This means, as shown in Figure 12 that after Reference Symbol Processing is completed for symbol 3 the equalization is started for symbols 0–2. This processing is expected to take time equivalent of 1 symbol, as shown in Figure 12. As soon as all the symbols for slot 0 are received; the equalization is again triggered for symbols 0, 1, 2, 4, 5, and 6. This is found to take time equivalent to 2 symbols.

Similarly for slot 1; the equalization is done after Reference Symbol Processing of Symbol 10 and this takes time equivalent to 1 symbol. After all the symbols related to slot 1 are received, the equalization is again triggered for symbols 7, 8, 9, 11, 12, and 13. This is found to take time equivalent to 2 symbols, as shown in Figure 12.

There are two key advantages of this design. First, the result of equalization for the current sub frame 'n' is available in first symbol time of the next sub frame 'n + 1'. In the previous design, this was 4 symbol time of the next sub frame. Second, the Core-2 is more evenly loaded and hence better utilized.

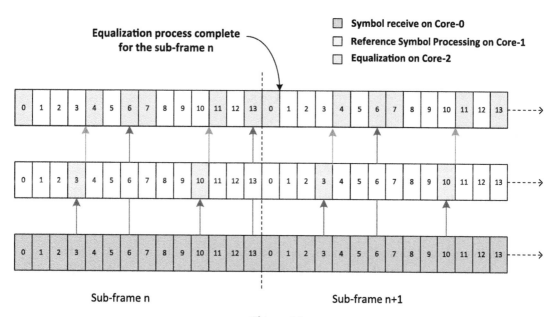

Figure 12:
Equalizer design with symbol level granularity.

The key challenge that still remain is that the equalization results for the current sub frame 'n' are only available in the next sub frame 'n + 1'. Hence this approach still adds to the overall system latency.

The aforesaid design approach can be further modified to ensure that the equalization result for the current sub frame 'n' does not get computed into the next sub frame 'n + 1'. This type of implementation requires a deeper deliberation on the equalization process.

Equalization ($\hat{s}$) using Interference Rejection Combining (IRC), includes an estimate of the transmitted symbols in the frequency domain according to the following equation

$$\hat{s} = (H^H \times P_n^{-1}H + C_x^{-1})^{-1} \times H^H \times P_n^{-1} \times r$$

In this equation the key input parameter are the Number of Receiver antennae (NRxAnt) and the Number of Transmitter antennae (NTxAnt). Based on the received data the following additional information is derived:

Equation Variable	Value Represented	Value derived from
H	Channel estimates	Matrix of NRxAnt rows and NTxAnt columns
r	Received symbols	Matrix of NTxAnt rows and NRxAnt columns
P_n	Noise covariance matrix	Matrix of NRxAnt rows and NTxAnt columns
C_x	Covariance matrix of transmitted signal	Matrix of NTxAnt rows and NRxAnt columns

The aforesaid equation can be represented as

$$\hat{s} = A \times r$$

In this equation,

$$A = (H^H \times P_n^{-1}H + C_x^{-1})^{-1} \times H^H \times P_n^{-1}$$

The most important property of the matrix 'A' is that it is dependent only on Reference Symbol Processing (i.e. channel estimates, transmit covariance and noise covariance). As there is only 1 reference symbol per slot, the matrix 'A' will remain same for a complete slot when non-interpolated channel estimates are used. As a result, the matrix 'A' can be computed immediately after the processing of reference symbol as shown in Figure 13.

In this example, immediately after the symbol 3 is received, the matrix 'A' is computed on Core-2. Along with this the equalization results are also computed for the symbol 0, 1, 2 on the Core-2.

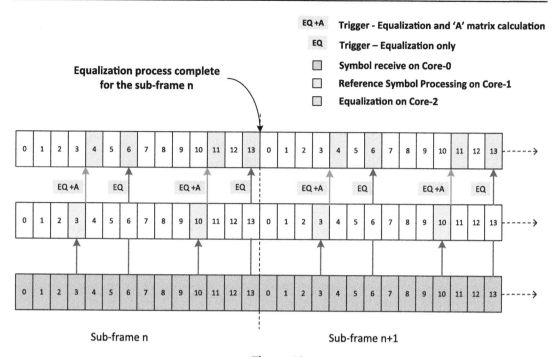

Figure 13:
Equalizer design with division of algorithmic computation into several tasks.

Now, as soon as Core-0 receives Symbol 6 the equalization is triggered on Core-2 for symbol 5, 6, 7. This uses the matrix 'A' calculated earlier. In this manner, the equalization for the last three data symbols is made even more simple without involving the re-calculation of matrix 'A' and hence, the overall system latency is improved by ensuring that the equalization processing results are available in the same slot of the current sub frame.

Use Case 8: Sub-frame pipelining in uplink processing

One of the key challenges in uplink processing is that the processing of the complete sub frame 'n' is dependent on receiving all the symbols of that sub frame. As discussed in the examples illustrated in Use Case 7, it is possible through efficient design techniques to ensure that for the current sub frame 'n' the symbol receive, Reference Symbol Processing and equalization of data symbols can be completed in the same sub frame 'n'. However, it is not feasible to complete the entire uplink chain processing in the current sub frame boundary. User Symbol Processing, code block processing and transport block processing are invariably carried over to the next sub frame.

Figure 14 shows that the code block processing is performed on the Core-3. For the current sub frame 'n', the code block processing completes toward slot 1 of the sub frame 'n + 1'. This creates a natural latency of one sub frame which is unavoidable, as explained earlier.

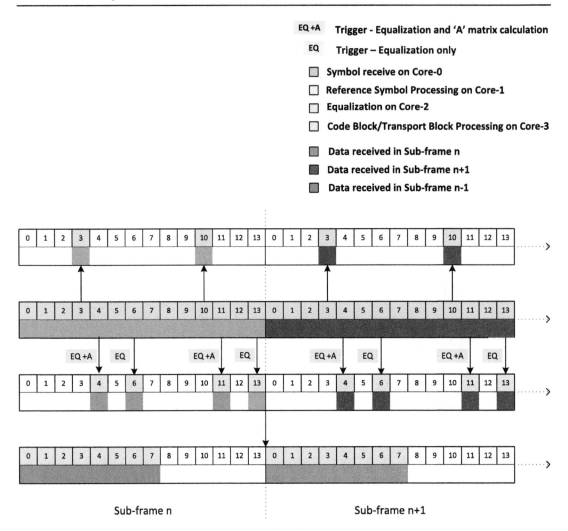

Figure 14:
Parallel pipeline of tasks between sub frames.

Overall, this approach demonstrates an extremely parallelized approach where two levels of parallelism have been exploited. At the first level the parallelism is attained by partitioning various uplink tasks across different cores. At the second level, this task partitioning is enhanced to also take into account algorithmic parallelization.

Load balancing

One of the key goals for an efficient software design on a multi-core architecture is an ability to create a system that is fully load balanced. A load balanced system ensures that all the cores are effectively used and the number of idle cycles is reduced to the minimum possible, for a finite number of use case scenarios. Indirectly, a fully load balanced system also ensures that the software designers have been able to exploit the process parallelism and data parallelism adequately.

The discussion in Use Case 3 and Use case 4 had introduced the concept of a Controller module that manages the scheduling of tasks across the system. The Controller module can itself be working on one core which is used as a master core to control the various tasks in the system.

This can be effectively done through a creation of the dependency matrix. This dependency matrix should contain an exhaustive list of all tasks that need to be performed along with information on the dependency of each task toward others tasks in the processing chain. The Controller logic can then ensure that unless all the dependencies of a task are executed, the task is not scheduled for execution. This form of basic intelligence in the Controller logic avoids idle core cycles by not scheduling tasks that are not ready for execution.

In order to assist in additional load balancing, the Controller can be imparted with higher intelligence by ensuring that the Controller is also aware of the amount of processing required for a particular task and the amount of processing cycles available across various cores. This form of information to the Controller ensures that the Controller is able to bind the tasks to the most appropriate core, based on the number of processing cycles available on that core.

The overall functionality of scheduler with load balancing can be summarized in Figure 15.

Hardware accelerators

Philosophy behind hardware acceleration in DSP systems

Advanced multi-core SoC are increasingly assisted by complex hardware accelerators to perform computationally complex activities. As a result, the overall system performance becomes dependent on partitioning of tasks across cores and an efficient usage of the hardware accelerator capability.

The hardware accelerators lend themselves to multiple advantages related to speed of software implementation and significant reduction in testing costs as the functionality is abstracted in hardware; offloading critical tasks from core to dedicated hardware which helps in making the system performance more deterministic and lower power consumption as the accelerators work at a lower clock frequency.

Hardware acceleration in baseband (LTE eNodeB) infrastructure

This section illustrates through a series of examples, usage of the hardware accelerator for Layer 1 processing. For the purpose of this discussion the hardware accelerator described in the examples is an advanced co-processor called Multi-Accelerator Platform Engine for

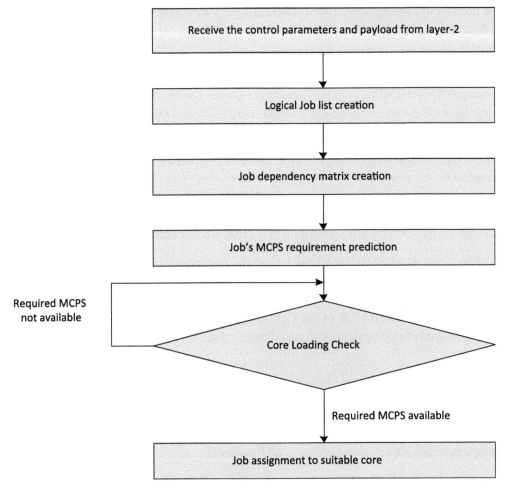

Figure 15:
Load balancing scheduler.

Baseband (MAPLE-B). MAPLE-B is available on Freescale's MSC8156 Multi-core processor. It consists of a programmable controller called Programmable-System-Interface (PSIF), a DMA and four processing elements:

- CRCPE (CRC Processing Element)
- TVPE (Turbo/Viterbi Processing-Element)
- FFTPE (FFT Processing-Element)
- DFTPE (DFT Processing-Element)

These processing elements (PEs) can be used in interrupt based or polling based modes. The MAPLE-B model avoids repetitive setup overhead on the various processing elements by allowing a single job preparation and trigger mechanism.

The Layer 1 software can leverage the MAPLE-B capabilities from the user space through a set of well defined APIs for different functionalities.

Use Case 9: Hardware—software partitioning for downlink shared channel processing

Figure 16, shows the complete downlink chain processing for an LTE Layer 1 implementation. In a pure per se multi-core architecture the implementation of the entire chain is required in software. However by using the hardware accelerators, such as MAPLE-B in case of MSC8156 multi-core, various computationally intensive blocks have been offloaded to the hardware accelerator.

The hatched blocks in Figure 16, depict the various functionalities that are mapped either fully or partially on to the hardware accelerators. These include:

1) The transport-block (TB) CRC computation is mapped on the CRCPE of MAPLE-B hardware accelerator.
2) The code-block (CB) CRC computation is also mapped on the CRCPE of MAPLE-B hardware accelerator.

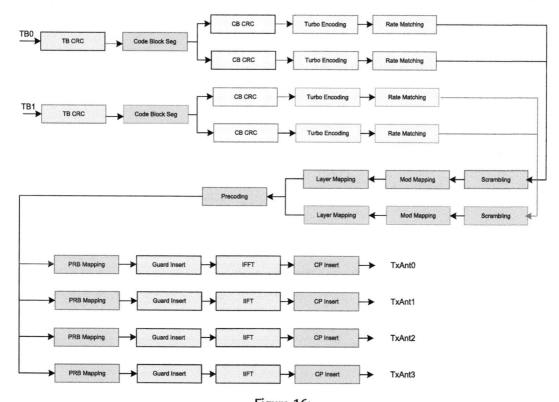

Figure 16:
LTE eNodeB shared data downlink processing chain.

3) At the time of generating the OFDM symbol the FFT operation is mapped to FFTPE of MAPLE-B hardware accelerator. Along with performing FFT, the FFTPE hardware accelerator is also used to insert the guard sub-carriers.

Use Case 10: Hardware–software partitioning for uplink shared channel processing

Figure 17, shows the complete uplink chain processing for an LTE Layer 1 implementation.

The hatched blocks in Figure 17, depict the various functionalities that are mapped either fully or partially on to the hardware accelerators. Hardware co-processor plays an important role in uplink processing in managing the high computation requirements for processing the reception of symbols and subsequent tasks for

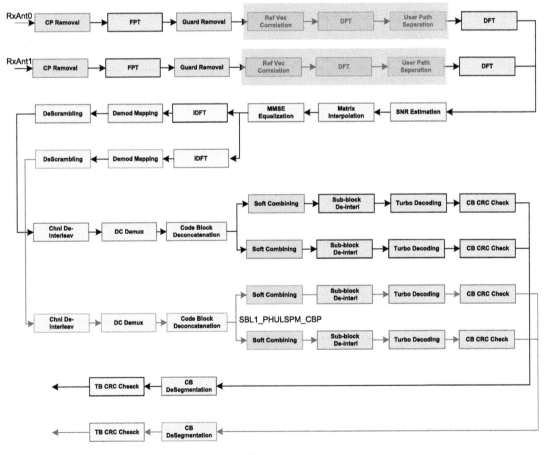

Figure 17:
LTE eNodeB shared data uplink processing chain.

hundreds of active LTE users. The MAPLE-B accelerator is hence leveraged for the following tasks:

1) After the OFDMA symbols are received — the Cyclic Prefix (CP) removal, FFT computation and guard removal is performed in the FFTPE
2) The time-domain channel estimation uses DFTPE for the IDFT and DFT operations
3) The IDFT functionality required to retrieve the modulated symbols from equalized user data is performed using the DFTPE
4) Viterbi decoding and Turbo decoding are computed using TVPE. TVPE internally uses CRCPE for Code Block CRC check
5) Final TB CRC check is computed using the CRCPE

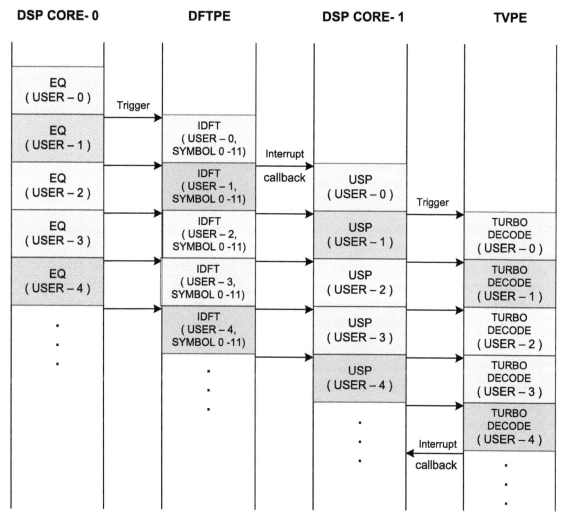

Figure 18:
Parallelism between processing in HW and SW.

Use Case 12: Pipelining and parallelism in uplink shared channel processing

Figure 18, demonstrates potential pipelining and parallelism in a system design involving multiple core and MAPLE-B hardware co-processor for LTE Layer 1 uplink shared processing. As explained earlier in the section, the MAPLE-B co-processor supports DFT/IDFT and Turbo decoding processing engines. They can run independently and simultaneously.

As shown in Figure 18, the IDFT operation is being executed on MAPLE-B in parallel to the Equalization operation being executed on core. After IDFT, the User Signal Processing (De-modulation mapping, De-scrambling, De-interleaving, De-multiplexing and code block de-concatenation) on core is run in parallel to Turbo decoding operation on MAPLE-B hardware co-processor. The above design clearly boosts the overall throughput performance by reducing the uplink execution latency.

Case Study Part 2: Wireless Baseband Software on Multi-core — Layer 2

Akshitij Malik, Umang Garg

Introduction

New generations of wireless protocols such as LTE lend themselves to the advantages of high throughput and improved spectrum utilization, resulting in improved user experience and improved Return on Investment (RoI) for service providers. These advanced protocols demand higher processing capabilities from the underlying processor architecture. The increased processing requirements can be efficiently managed through a new generation of multi-core architectures such as Freescale's P4080 System-On-Chip (SOC).

The adoption of advanced multi-core embedded platforms has also increased the traditional software development challenges to include emerging challenges such as: ability to use the multi-core devices effectively, adopting evolving tools, flexible software partitioning, deriving sequentially higher performance, and achieving an ever increasing software quality requirement. The ability to effectively enable the same software load on multi-core devices with different numbers of processing cores and with minimal changes to the software is another evolving challenge.

This discussion illustrates through a real life, wireless protocol software lifecycle, the art & science of navigating these challenges effectively.

Process and quality

Guiding principles

Software development on multi-core platforms is an evolving paradigm. The foremost step towards a dependable software development on multi-core devices is the need for a *Quality* driven methodology to ensure that the challenges introduced by evolving technology leaps do not overwhelm the software development cycle and degrade the customer experience. The need for higher quality is greater than ever before.

The 5 key *Quality* principles that can assist in ensuring a successful development cycle are:

* *Agile* development practices
* Modular software design
* Refactoring
* Reuse
* Tools

Agile development practices

For a process purist, the science of *Agile* development methodology is a well documented subject. For a team focused on delivering complex software on complex hardware within a definite business critical timeline, A*gile* practices are a means to maximize productivity, enhance delivery effectiveness, and create a positive ROI. *Agile* development practices are well suited to evolving multi-core platforms where teams are challenged to satisfy customer needs for delivering complex software on the new devices while they are still learning about the intricacies of the new generation of silicon platforms. A key qualitative requirement for the methodology to succeed is a close, intra-team collaboration while a key quantitative requirement is the adoption of project discipline through metrics such as: Phase Containment Effectiveness, Root Cause Analysis, Monthly Defect Reports, Cyclomatic Complexity, and Test Coverage Data.

Modular software design

As this text shows later, when the software quality practices enforce a requirement for modular software design with well defined interfaces, clear tabulation of memory requirements and test modularity and then adapting the same software to different software portioning requirements becomes feasible with significantly lower overhead. Moreover, isolation and simplification of the application into small, simple modules automatically allows for integration of *Agile* methodologies in the development of such software.

Refactoring

The quality practices should enable the process of continuous design and code refactoring to support an ever increasing understanding of the platform architecture. When the code is continually re-factored, the process is bound to introduce new defects and thereby reduce the software quality. A modular software design which can be tested modularly reduces this risk by ensuring that all interfaces for various defined use cases and configuration parameters are tested. Tests which validate the application at individual module-levels also help isolate the problem areas. Ultimately, the advantage accrued from refactoring has been found to improve system performance and advanced functionality exploitation. The advantages outweigh the short term challenges.

Reuse

The intensity of product development is invariably determined by the 'time to market' requirements. More often than not, a product development has to fit the time to market deadline, rather than the other way around. A deliberate choice is recommended when considering reuse of the software or parts of it. This decision making process is crucial when the software may not have been originally designed for the multi-core architectures and architectures which may contain 'many core and associated accelerators.' This study builds on examples where certain parts of the software were re-written completely and many parts were reused with minor modifications. An appropriate balance is necessary. The choice can be determined based on the nature of the changes required to migrate to a multi-core platform, the number of changes, and the impact of the changes on the existing collateral such as test cases and documentation. As a general guideline and as a heuristic exercise − if the number of changes is more than 50% lines of code and more than 50% lines in the associated collateral then a software designed from scratch is a better alternative to creating a maze of fixes on a legacy working software. If the number of changes is less than 20% of code and collateral then, it is advisable to reuse the existing software. If the number of changes is between 20% and 50% then it is better to identify sub-modules that can be reused with some patchwork and sub-modules that need to be rewritten.

Product quality software requires sustained focus on easy to use and up-to-date documentation on requirements, design, and test. Additionally, product quality software is an outcome of non trivial test focus that seeks to ensure that the software does what it is supposed to do.

In fact, more than 50% of the engineering time is found to be spent on activities related to development of test cases and documentation efforts related to requirements, design, and test. It is hence essential that while code refactoring and software reuse is practiced diligently, there is an effort to ensure that the documentation and test collaterals are generated in a way that their upkeep is made simpler. Tools such as Doxygen (refer to [1]) can be leveraged to generate accurate and relevant detailed design information. This can allow the design documents to be at a higher level of detail while low level details are documented through informative comments in the code which are captured through Doxygen. This allows the low level documentation to be dynamically updated as the code changes through refactoring.

Tools

Mature software tools for development are an essential part of the development process for high quality, complex software. When the development platforms are new and evolving, the software tools are also found to be in an evolutionary lifecycle of their own. This imparts an additional level of complexity to the software development cycle as a single tool suite cannot be adopted to enhance the development process. The development process has to hence enlist the support of multiple tools. Some of these tools may be open source and others maybe proprietary and hence essentially bundled with the SoC's enablement software. While all tools are important, for the purpose of this discussion we have put the tools in 5 broad categories:

- Category 1 tools essentially enable software compilation, assembling, and execution of the target platform. They must hence provide the aforesaid capabilities in a dependable way.
- Category 2 tools enable debugging and profiling. Most often the maturity cycle of these tools lag the category 1 tools. At the very least, these tools must provide minimal command capabilities for debugging, such as setting breakpoint, step wise execution, etc. More mature debugging tool set provide higher visualization, trace information, etc. The support for code profiling is essential to measure and understand the performance impact of the software functionality under various use case scenarios. The most essential profiling support is the ability to measure CPU core cycles, cache hits, cache misses. Profiling tools with higher maturity confer the ability to analyze a large number of registers available within the SoC. These registers/counters can enable improved software development through knowledge of hierarchical cache behavior, function wise profiling capability, etc.
- Category 3 tools enable development process through Simulation software. The simulation environment is typically a functionally accurate representation of the silicon environment and can hence enable pre-silicon development thereby supplementing the 'time to market' requirement. A more evolved simulation platform also provides a 'cycle accurate' environment.
- Category 4 tools are typically the general purpose software tools. These include static analysis tools that enable the development team to browse the code, gather statistics such as line count, and measure Cyclomatic Complexity etc. This category also includes tools such as bug management tools, static defect analysis tools, configuration management tools.
- Category 5 tools are the differentiator tools that can enable the development teams to evaluate the impact of their development platform on the software being development. These tools provide ability to analyze software partitioning choices, isolate concurrency issues, etc.

Software blocks and modules

The enhanced Node B (eNB) is a key network element in the LTE network (refer to [3]). The eNB functionalities include both Control Plane and Data Plane Processing. The data path of eNB has 2 primary interfaces to the rest of the LTE network:

- S1-U Interface. The S1 interface is an IP based interface and it is used to exchange data packets towards the rest of the LTE network using IP-based protocols.
- LTE-Uu air-interface. The LTE-Uu interface is used to exchange data with the actual Mobile Stations (UE) over the wireless medium. The LTE-Uu consists of specific protocols that are meant to handle the LTE's wireless transmission technology.

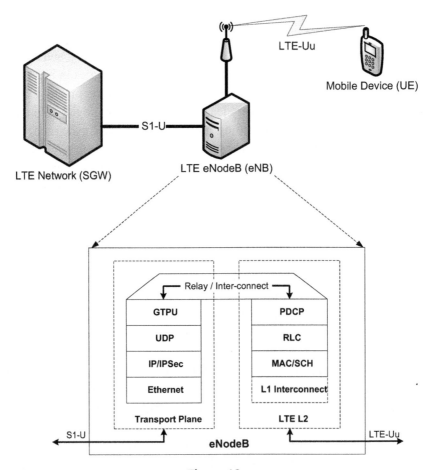

Figure 19:
LTE eNodeB protocol stack.

The S1 and LTE-Uu interfaces consist of multiple protocol sub-layers which form the protocol stack of the eNB. Each of these protocol sub-layers has its own independent functionality and thus is treated as a separate module/sub-modules.

Figure 19 gives a diagrammatic representation of the eNB data-path.

The challenges experienced in the migration of single-core software to multi-core and their resolutions can be explained through the example of eNB software. The eNB software is collectively termed eNB_App.

As part of the modularization exercise, the eNB_App has been logically separated into two software blocks: the Transport block and the Layer 2 block.

The Transport block has further been modularized on the basis of the protocol sub-layers that are implemented within it. It consists of two modules:

- IPSEC, and
- GTPU sub-layers

Similarly, the Layer 2 software block has been split into three modules:

- PDCP
- RLC, and
- MAC sub-layers

In addition to the protocol sub-layers of the Layer 2 software block, a simple Scheduler (SCH) sub-module has also been developed to carry out the scheduling of packets that are to be transmitted over the LTE-Uu interface. The SCH sub-module is a simple FCFS type of scheduler which has been integrated with the MAC sub-module.

The Transport software block along with the PDCP module is responsible for processing the packets received asynchronously at the S1 interface. The LTE-Uu interface requires frames to be exchanged between the eNB and the UEs at an accurate 1 ms timing. The synchronous processing of the LTE-Uu data takes place within the RLC and MAC software modules. The RLC and MAC modules, owing to their synchronous nature, have hard real time constraints and are required to perform processing within 1 ms.

Figure 20 gives a high-level representation of the arrangement of sub-modules within the eNB_App.

The IPSEC module is responsible for implementing the IPSec related functionality of the eNB. It processes the encrypted packets on ingress and applies the required decryption algorithms to extract and verify the payload. On the egress side, it identifies the security association of the outgoing packet and using that, it generates the encrypted IPSec packet that can subsequently be securely sent towards the destination.

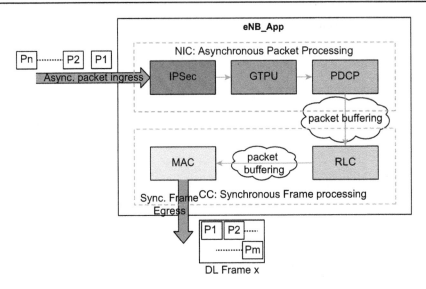

Figure 20:
eNB_App sub-modules.

The GTPU module receives the GTPU packets that were tunneled in the payload of the IPSec packets and identifies the data-transmission context towards the LTE L2 stack. It makes use of the addressing information along with the GTP Tunnel identifiers for this purpose. Similarly, during egress, the GTPU module receives the packets delivered by PDCP and identifies the outgoing GTPU context by building the GTPU headers containing the GTPU Tunnel identifiers, etc.

The PDCP module, along with the GTPU modules, acts as a relay between the Transport Stack and the LTE L2 Stack. The PDCP module receives the data of individual user-streams and (if configured) it applies the header compression algorithms to reduce the packet size which reduces the number of bits that need to be transferred on air. In addition, during handover the PDCP module forwards the buffered packets that have not yet been delivered to the UE towards the Target eNB. In the uplink (egress) direction, the PDCP module receives the data delivered by RLC and (if configured) it applies the header decompression algorithms to recover the header information that is relevant for the data-stream. The PDCP will then forward the packet with the uncompressed headers towards the GTPU along with the appropriate context identifier that would help GTPU identify the tunnel on which the packet must be sent out.

The RLC module is primarily responsible for carrying out link-level handling of UE data. In the case where the RLC is configured to operate in the Acknowledged Mode (AM Mode), it also performs ARQ functionality to ensure reliable delivery of the UE's data. The RLC module is also responsible for buffering data as it is scheduled for transmission and also when it is being re-constructed from multiple RLC PDUs during reception.

The MAC module (along with Scheduler functionality) is responsible for interfacing with the L1 layer of the eNB to send or receive UE's data over the air. The MAC module is responsible for selecting the UEs and their Logical Channel that transmit or receive data in each 1 ms frame. In addition, it is responsible for multiplexing data from multiple UEs (and their multiple logical channels) into a single frame for transmission and similarly, extraction of data from each of the UEs in the received frame. It also performs the task of HARQ processing to ensure efficient and reliable exchange of data between the eNB and the UE.

Single core application

SoC with high performance single cores have been the bedrock of embedded software development for a generation of engineers. These environments, through the natural evolution of their maturity, are accompanied by stable development tool chain, simplicity of programming choices, and early cycle tradeoffs on software capability (which itself is closely tied to use case requirement as well as the hardware capability).

Applications developed for single core platforms require minimal design choices on SW partitioning across cores. Depending on the expected number of core cycles and memory requirements for processing, the SW partitioning challenge is often resolved through usage of multiple SoC. This approach limits the flexibility for changing the software partitioning as it impacts software running on multiple SoC and the overall system architecture. Any changes to the SW partitioning also lead to an addition in the cost of SW Testing and changes to the associated collateral. Such solutions also face a limitation imposed by the data-transfer mechanisms that may be employed to exchange data across the SoC devices.

For the identified software blocks and modules in **Software Blocks and Modules**, when single core SoCs are available, the two plausible partitioning approaches may be:

a) Transport and L2 software running on the same SoC

This approach simplifies the design and partitioning choices by executing the complete stack on one single core SoC (Figure 21).

Following such an approach, all the blocks of the eNB_App execute on a single single-core SoC. While such an approach appears to be the simplest, however, the designer needs to be careful that the hard real-time requirements of the 1 ms frame processing are met under all circumstances. This adds the burden of dimensioning each of the modules such that each

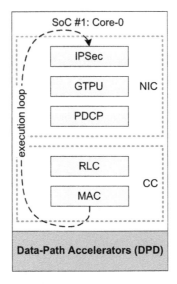

Figure 21:
eNB_App — single-SoC design.

module completes its processing well within a fraction of the 1 ms frame tick. Additionally, since the system dimensioning has to be done considering the worst-case processing times of each of the modules, it may not be possible to maximize the system utilization (since sufficient headroom must be maintained).

This approach also reduces the capability of the software to impart scalability for (future) use case requirements due to a finite processing capability:

b) Figure 22 shows transport software running on one SoC and the L2 software running on a second SoC

This approach allows the developers to implement the application more flexibly than with the Single-SoC approach.

The asynchronous, packet-triggered network processing modules (NIC) can be run on one thread associated with the first single core SoC, while the synchronous, timing sensitive part (CC) can run as an independent thread on a different single core SoC.

Compared to the previous approach in **Transport and L2 software running on the same SoC**, this design allows each set of sub-modules within the NIC and CC to maximize their respective performance. It may be possible to even use different SoCs for the NIC and CC — each having their specialized accelerators to further enhance performance.

However, this form of multiple SoC architecture leads to a higher bill of material cost, increased power consumption due to multiple SoC, and increased complexity of overall

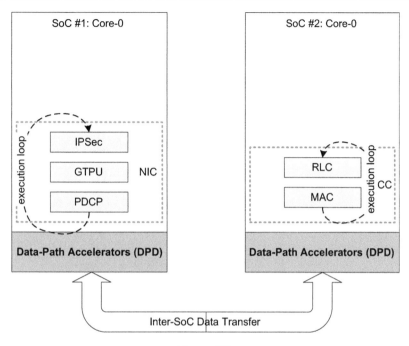

Figure 22:
eNB_App — multi-SoC design.

architecture. Additionally, the performance gained due to the reasons mentioned above may get offset by the latencies introduced due to limitations in the data transfer mechanism between the two SoCs. Finally, as this is a specific fixed arrangement of the SoCs, the scalability of the overall solution is limited.

An example of a multi-core SoC: P4080

The P4080 (shown in Figure 23) SoC is a high performance multi-core device. It consists of eight e500mc cores and multiple hardware accelerators that are designed to simplify the development of networking applications and yield high performance. The P4080 core usage is flexible and allows the eight cores to be combined as a fully-symmetric, multi-processing SoC, or they can be operated with varying degrees of independence to perform asymmetric multi-processing. Each of the cores may be used to execute independent systems. This allows the P4080 to be used with significant flexibility in partitioning between control, data-path, and applications processing. Additionally, the quantum of processing capability provided by the cores and the associated hardware accelerators allows consolidation of functions previously spread across multiple discrete processors onto a single device.

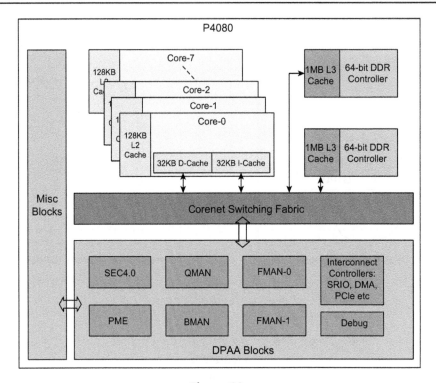

Figure 23:
P4080 High-level block diagram

Figure 24 shows a high level block diagram of the P4080 multi-core SoC. Hierarchical cache architecture and numerous accelerators on the device are a key differentiating factor resulting in improved software solution performance. The primary accelerators include:

- Security Engine (SEC), which can be leveraged for encryption and decryption
- Buffer Manager (BMAN) for memory allocation and management
- Queue Manager (QMAN) for intra system communication
- Frame Manager (FMAN) for high speed exchange using the Ethernet interface
- Pattern Matching Engine

For most end-users, it is not practical to write software to directly interface with the SoC hardware. Typically, a hardware abstraction layer is used which allows the higher layer applications to interpret the hardware using more generic APIs. For the P4080, the Light-Weight Executive (LWE) is one of the hardware abstraction layers. The LWE consists of software APIs which allow configuration of the hardware such as the DPAA modules — allowing the Data-path (DP) functionality to exercise the DPAA modules that support the DP functions.

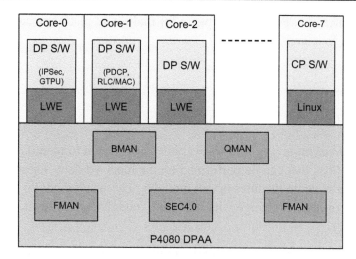

Figure 24:
Enablement software on P4080 - LWE.

Migrating to multi-core

Advantage

For the same deployment scenario and as compared to a mish mash of single core SoC, multi-core embedded architectures are a natural choice due to their inherent advantages such as lower bill of materials cost as well overall cost of ownership, lower power consumption, higher (future) scalability, and higher performance. Multi-core embedded SoCs are relatively new phenomena and hence committed engineering effort is required to realize the various advantages offered by the platform.

In the following sections, using the eNB_App as an example, we shall highlight the steps that may be followed to adapt current software applications from the single-core to the multi-core environment.

In order to directly develop applications for the multi-core environment, a similar series of steps may be followed.

Multi-core considerations

In addition to the standard design issues of single-core applications, the following issues have to be considered when designing applications for multi-core environment:

- Concurrency
- Information sharing
- Load sharing and distribution

The issues experienced in developing software for multi-core platforms can be addressed through a series of four steps:

1. Atomize
2. Serialize
3. Distribute
4. Balance

As a part of the *Atomization* step, the software designers need to identify various independent operations that can be performed on the received data. These operations can then be placed into separate modules or sub modules. These sub modules should have well defined interfaces for data exchange with other sub modules so that each sub module can work independently.

Advanced multi-core devices such as P4080 provide easy to use mechanisms for the (implicit) distribution of sub-modules across the cores. These mechanisms along with relevant examples have been explained in detail in **Advanced Accelerator Usage on P4080: Practical Examples**.

The next step is *Serialization*. As a part of the Serialization process, the software designer should identify the sequence in which the various modules and sub modules are required to execute. As a part of the serialization the designers need to take into account the method for sharing the packet context across various sub modules in an optimized manner for improved system performance. While sharing the packet context it is also essential to ensure that the concurrency issues are taken into account during the redesign phase.

In architectures such as P4080, the concurrency challenges associated with sharing the packet context can be simplified due to the underlying nature of the platform. As one possible approach, this can be undertaken by associating each context with a different Frame Queue. In situations where two or more modules executing on independent cores may access the same resource, then these resources must be protected using suitable mechanisms such as spin-locks, mutexes, etc.

The next step is the *Distribution* of the atomized and serialized modules on the available cores. This distribution is a function of the use-case required to be addressed, the estimated performance requirements of the modules/sub-modules, and any legacy design preferences.

Finally, the complete software requires *Balancing* for optimum system performance and for ensuring future scalability. This step may require modification of the distribution scheme undertaken in the previous step. It is recommended that for purpose of future scalability, *none* of the cores should be loaded more than 70% under maximum load conditions as defined by the given use case.

The procedure explained above was used to migrate a Single-Core eNB_App to a Multi-Core solution using the P4080 platform as the basis. Figure 25 highlights the work-flow as

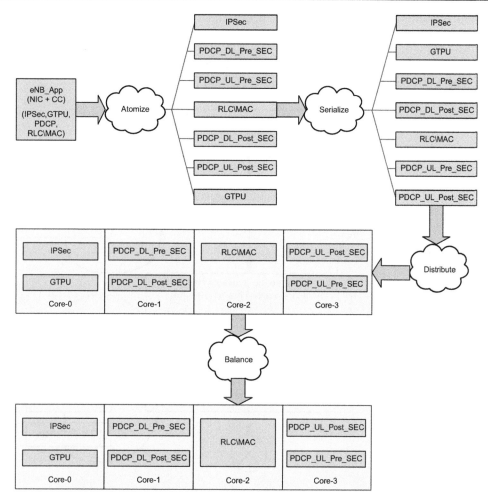

Figure 25:
Refactoring the eNB_App for Multi-Core P4080.

the eNB_App was re-factored using the steps — atomize, serialize, distribute, and balance. The detailed procedure that was followed in the migration is explained in the following sections.

Single Core to Multi-core

As discussed earlier, multi-core SoCs can be treated as a multi-threaded environment. So migrating from single-core to multi-core SoCs may be treated as an exercise of parallelizing a single-threaded application. The point of consideration is that there may be restrictions on the number of "threads" that can be spawned in the multi-core SoC environment. Additionally, since multi-core SoCs often also come with a rich set of accelerators, additional effort is needed to incorporate the use of the available accelerators within the application.

The migration of existing software from a single-core SoC to a multi-core SoC can be accomplished through the sequence of steps identified in Figure 26 below:

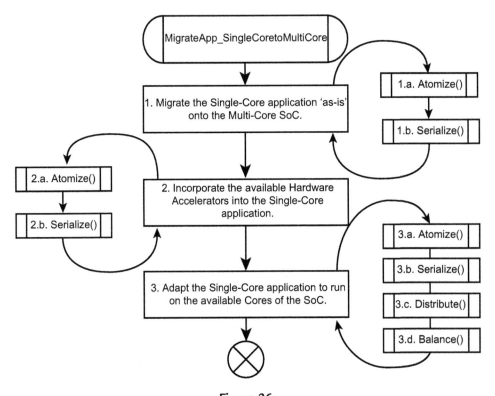

Figure 26:
Steps to migrate to Multi-Core application.

Step 1: Transplanting the single-core application to the multi-core SoC

The *first step* towards adopting a multi-core SoC is migrating (existing) single-core SoC application software to a single core of the multi-core architecture. This method allows early adoption of the new platform and is typically expected to lead to minimal software changes.

In the case of the eNB_App, this implies that we directly just port the Single SoC application to execute on a single core of the P4080. The execution loop of the eNB_App is run on any one core of the P4080 while the remaining cores are idle (Figure 27).

As can be seen in Figure 28, when adapting an existing single-core application to run on a multi-core SoC, the focus is on keeping the transition simple and minimalistic. In the code

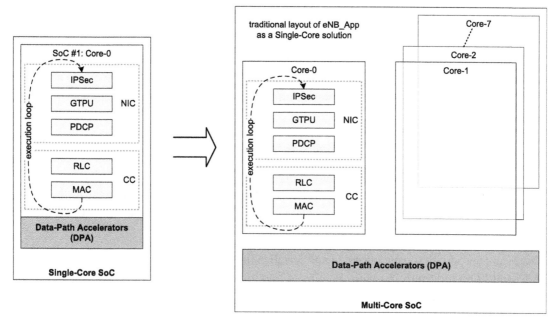

Figure 27:
Transition for single-core SoC to multi-core SoC.

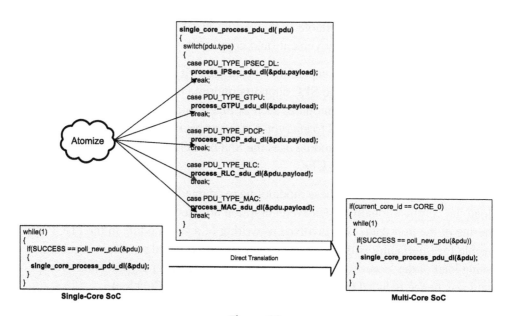

Figure 28:
Direct translation of single-core code to multi-core.

above, the only changes made to the existing application are to ensure that it executes only on one (CORE_0) of the available cores of the multi-core SoC.

In addition, it is useful to atomize and serialize (Steps 1.a. and 1.b. in Figure 26) the existing modules of the application as part of enhancing the adaptability of the software on the new platform. At this stage, only a bare minimum effort should be spent on atomization and serialization.

Step 2: Leveraging the hardware accelerators on the multi-core SoC

The *second step* towards adopting a multi-core architecture is to leverage the hardware accelerators supported on the hardware platform. This step will require changes to the software for leveraging the hardware accelerators. Hence, an initial understanding on the capability of the hardware and its usage is essential.

This is an appropriate stage to atomize and serialize the sub-modules of the application (Steps '2.a.' and '2.b.' in Figure 26) with the aim of maximizing the utilization of the available hardware accelerators.

The eNB_App first leveraged the Buffer Manager (BMAN) capabilities available on the P4080. A robust Buffer Management is a key functionality in any embedded software. By offloading the buffer management responsibilities such as allocation, de-allocation and memory management to the BMAN it is possible to reduce the software logic that would otherwise be required for a complex software buffer management sub-module.

The BMAN allows data to be available to the various cores of the SOC and its various accelerators, with minimal data movement (and copy) and without the need for translation.

The eNB_App then leverages the capabilities of the Security Engine (SEC) available on P4080, as shown in Figure 23. The SEC supports multiple encryption and decryption algorithms which are required by modules such as PDCP and IPSec. The IPSec module uses algorithms such as the DES/TDES to encrypt & decrypt packets exchanged with other nodes in the LTE network. The PDCP module uses the SNOW-3G F8-F9 algorithms to cipher and decipher data exchanged with the UEs.

As a part of the *second step* both these modules were adapted to interface with the SEC block to offload the computationally expensive encryption and decryption and cipher and deciphering of the user data. This involved re-writing of a sub-module in the PDCP that was responsible for interfacing with the encryption and decryption algorithms. Significant processing load of the 'cores' that were used in performing the encryption and decryption of user-data was hence reduced by offloading these tasks to the SEC block.

The next IP that was used was the Frame Manager (FMAN) on P4080, as shown in Figure 23. FMAN provides capabilities to parse the protocol header of the received packets and subsequently using the user configured policy rules, it can distribute these packets. As a result, the software sub-module that was responsible for the identification of the incoming IPSec packets and their Security Association (SA) to FMAN was deprecated and the work was offloaded to the FMAN. This step required minor changes in the implementation of the IPSEC module.

Subsequently, the FMAN block was used to perform checksum calculation and validation which led to further saving in the core cycles.

Finally, the Queue Manager (QMAN) provides a message-queue kind of IPC mechanism along with advanced capabilities such as prioritization of messages and message order guarantee, etc. The QMAN was thus used to uniformly replace the IPC used across all the modules and sub-modules of the eNB_App. Since the QMAN is 'aware' of the multi-core environment of the P4080, it contains support to handle concurrency related issues. This important capability provided by the QMAN has been heavily used in the next phase.

As a result of following Steps 2.a. and 2.b. in Figure 26, the PDCP module was further atomized into 2 sub-modules — pre-SEC and post-SEC. As part of the serialization, the arrangement of modules should be such that PDCP pre-SEC execute occurs first, followed by delivery of the *cleartext data* to the SEC block for ciphering. The ciphered payload is received back by the PDCP post-SEC sub-module to complete the PDCP to SEC offload. A similar approach was followed for the IPSec module.

As can be seen in Figure 29, in the downlink (DL) direction, the PDCP performs five primary tasks — Buffering of the received SDUs, Compression of network headers within the SDU,

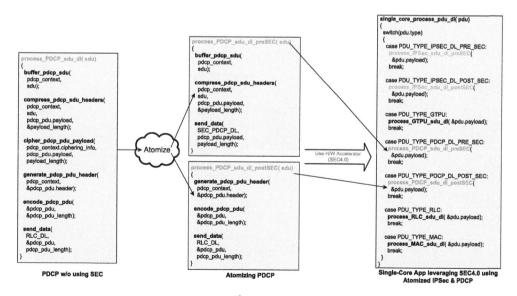

Figure 29:
Adapting the single-core app to leverage SEC4.0.

and ciphering of the PDCP PDU payload before its header is created and a PDU is formed and sent towards the RLC. To offload the task of ciphering to the SEC block, the `process_PDCP_sdu_dl` function that processed the PDCP SDUs in the DL direction was atomized into two atomic functions — `process_PDCP_sdu_dl_preSEC` and `process_PDCP_sdu_dl_postSEC`. Serialization of the operations across the two atomized functions was ensured since these are connected to the SEC block using QMAN which guarantees in-sequence delivery of queued data.

Figure 30 shows the eNB_App using the SEC block for encryption and decryption while the FMAN block is being used to offload the IP Header pre-processing and routing of incoming IPSec packets.

The indirect effect of using the hardware accelerators was the improvement of the overall quality of the system. This was accomplished as result of:

- Higher degree of 'reuse' since the hardware accelerators are commonly used across various modules
- Reduced code size which results in greater focus on the task at hand
  - Less software code statistically implies fewer defects
  - Less software code can be reviewed, tested, and covered even more thoroughly

Step 3: Distributing the application on the multi-core SoC

The *third step* for effectively leveraging the multi-core architecture is using the multiple cores along with the hardware accelerators, thereby realizing the complete potential of the architecture. This requires the software designers to analyze and determine the most

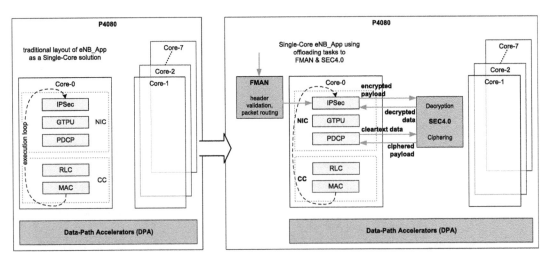

Figure 30:
Offloading processing to DPA on multi-core.

appropriate software partitioning approach that satisfies the use case and also facilitates a method for future scalability.

This phase of application refactoring primarily involves distribution of the modules and sub-modules across the available cores as well as balancing the load generated on each core so as to maximize the overall performance of the system. Steps 3.c. and 3.d. in Figure 26 target these tasks. In addition, as a result of distributing the sub-modules across multiple cores, or as a result of the need to balance the system, some further atomization and serialization of some sub-modules may be needed.

After following Steps 3.c. and 3.d. in Figure 26, in the first iteration we can distribute the NIC and CC functionality across two of the cores of the P4080. As a result, Core-0 of the P4080 is used for processing the modules associated with the transport stack and Core-1 is chosen for processing the sub-modules associated with the LTE L2 stack. However, since the timing requirement of the PDCP sub-module is different from the RLC and MAC sub-modules, as part of the balancing of the system, the PDCP module is also executed on Core-0.

To move to the two-core solution proposed above, the DL SDU processing function for the single-core solution – `single_core_process_sdu_dl`, was first atomized into two logically independent functions – `process_sdu_NIC` and `process_sdu_CC`. These functions could then easily be distributed across the two cores of the P4080 by adapting the main processing loop to use only the specific SDU processing functions on each of the cores (Figures 31 and 32).

A performance analysis of each of the sub-modules was done to assess their CPU Core-Cycle requirements. Figure 33 depicts an approximate distribution of the Core-Cycles that were consumed by each of the modules when running on a single Core. By offloading the encryption/decryption functionality to SEC, the processing requirements of the IPSEC module reduced significantly. The RLC and MAC modules consumed the highest proportion of Core-Cycles. In case of the PDCP module, it was found that the additional processing required by RoHC created a system imbalance and reduced the system throughput. As a result the eNB_App was re-factored and PDCP module was further atomized into Uplink (UL) and Downlink (DL) execution legs. These execution legs were then distributed to independent cores.

Using the information on the Load Distribution analysis along with the logical function that each sub-module performed, the sub-modules were distributed in the following manner:

> Core-0: Transport functionality – IPSec + GTPU modules
> Core-1: PDCP-DL functionality – PDCP_DL_Pre_SEC + PDCP_DL_Post_SEC sub-modules
> Core-2: Frame-driven Sector functionality – RLC/MAC + PHY Sim sub-modules
> Core-3: PDCP-UL functionality – PDCP_UL_Pre_SEC + PDCP_UL_Post_SEC sub-modules

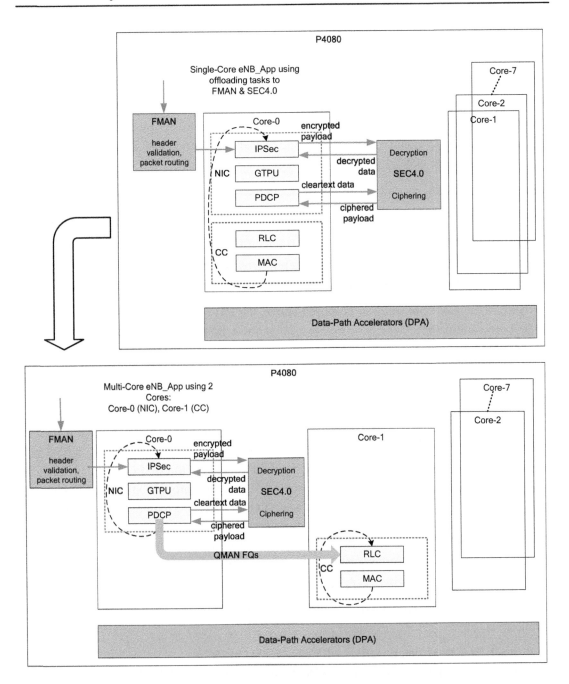

Figure 31:
Two-core distribution of eNB_App.

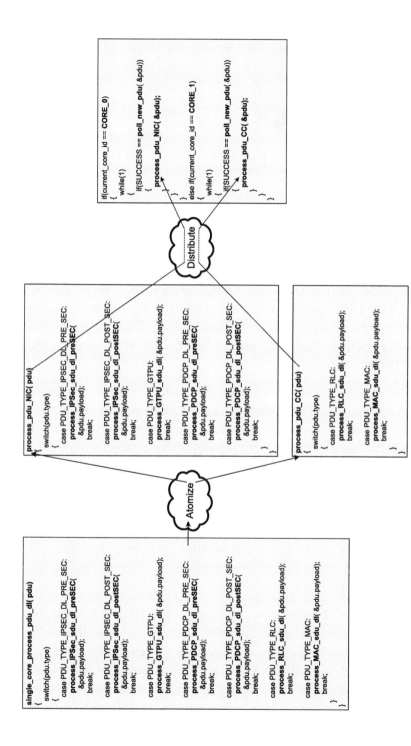

Figure 32:
Distributing the NIC and CC across two cores.

CPU Load %

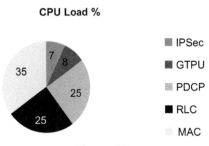

■ IPSec

■ GTPU

▨ PDCP

■ RLC

▨ MAC

Figure 33:
Load distribution for 1-sector on single-core application.

Scalability of the eNB_App to a 3-Sector solution was also kept in mind when distributing the modules. Figure 34 represents the eNB_App after the process of balancing the system.

To move to the 4-core solution proposed above, the processing load was balanced across the four cores of the P4080. The IPSec and GTPU functionalities were kept on Core-0, while the PDCP DL and UL functionalities were atomized into separate functions — `process_sdu_PDCP_dl` and `process_sdu_PDCP_ul`. Their execution was then transferred to Core-1 and Core-2 respectively. The execution of the RLC and MAC modules was directly moved to Core-3 (Figure 35).

The code modularity and the process of distribution of various sub-modules on cores can be improved through a framework. This framework among other module-specific information contains the binding of each sub-module with the specific Core on which it has to be executed. This greatly simplifies the task of rearranging the arrangement of sub-modules across cores.

As shown in Figure 36, a global list of sub-module initialization information is maintained in the Module Information Table. The designated Core and Pool Channel are maintained as parameters of the sub-module initialization information. During the Application initialization, the framework checks which sub-modules are to be executed on the current Core; the framework then requests QMAN to only dequeue data from FQs that are destined for the specific Pool Channel.

The code snippet in Figure 37 shows how the usage of the Module Information Tables simplifies the initialization routine of the software modules. Such an abstraction also brings in a desired amount of flexibility which permits the application to remain unchanged even when modules within the system are re-arranged or the underlying platform is changed. This greatly reduces the maintenance and adaptation burden which is common while developing software which is meant to be scalable and deployable across multiple SoC variants.

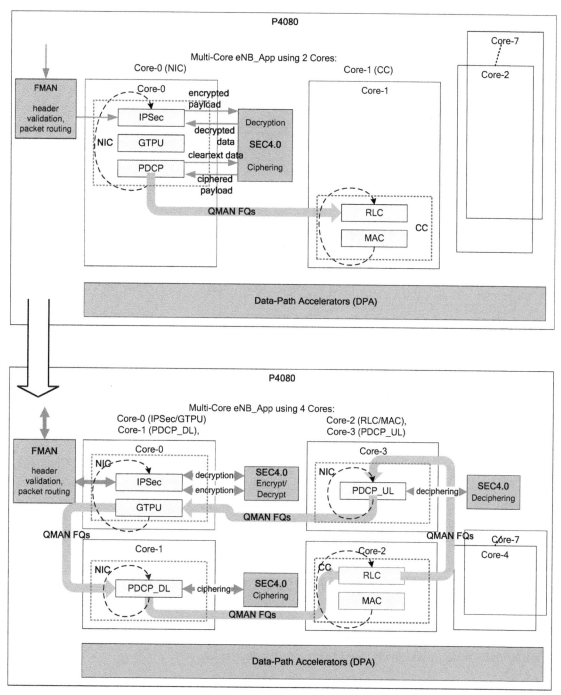

Figure 34:
4-core distribution of eNB_App.

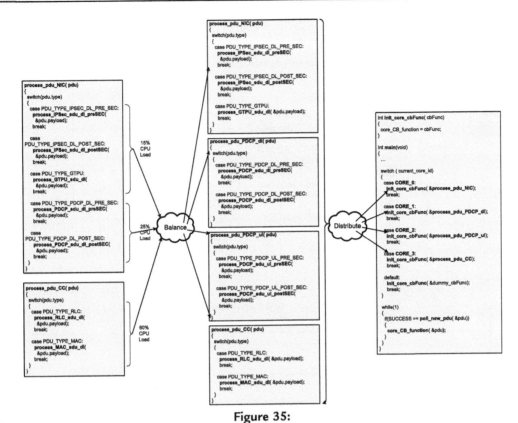

Figure 35:
Using atomized modules for load-distribution.

The actual binding of each sub-module to a specific Core is done using QMAN Pool Channels (the usage of QMAN for Communication is explained in greater detail in the following section). Each sub-module can be associated with a single Pool Channel, and it is possible for each Core to configure QMAN to de-queue FDs from only specific Pool Channels. Using this associative property, we can bind sub-modules to specific Cores.

Advanced accelerator usage on P4080: practical examples

As part of the migration of the eNB_App from single-core to multi-core (P4080), Step 2 and Step 3 involved leveraging the hardware accelerators available within the P4080 DPAA. The P4080 DPAA was extensively used for the following tasks:

FMAN — Parsing and distribution of Ingress packets, validation of common headers
QMAN — Inter-module communication, context-stashing
BMAN — memory management for data-buffers shared across all sub-modules
SEC — Offload of Ciphering/Deciphering operations

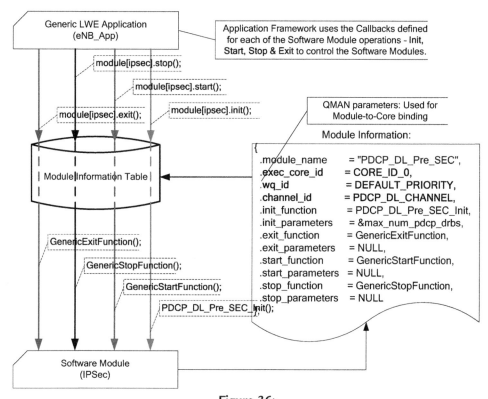

Figure 36:
Using the module information table for distribution.

The following are some of the examples of using the P4080 DPAA to offload and simplify the eNB_App software:

a) FMAN usage for IP-packet pre-processing

Figure 38 highlights the interaction between the IPSec module of eNB_App and the FMAN block of the P4080 DPAA.

The PCD (parse, classify, and distribute) functionality of the FMAN is utilized by the eNB_App to offload part of the packet processing (or pre-processing). Data packets received by the LTE eNodeB are first analyzed by the PCD functionality of the FMAN. The PCD functionality within the FMAN will first parse the incoming packet headers to identify relevant packets. For example, the FMAN will parse incoming IP packets to identify all the header fields of the IP packet. In fact, the parsing functionality will parse the received packet for all the common headers following the IP header as well (for example the UDP headers following the IP header in a UDP/IP packet).

```
int initialize_software_modules()
{
    int             ret;
    int             module_index;

    /*! Initialize the Modules specifically required by the Software Modules. */
    module_index = 0;
    while ( strlen( g_module_information_table[module_index].module_name))
    {
        if (g_current_core_id == g_module_information_table[module_index].exec_core_id)
        {
            /*! Call the Init-function of the Software Module. */
            ret = g_module_information_table[module_index].init_function(
                g_module_information_table[module_index].init_parameters);
            if (0 != ret)
            {
                APP_ERROR ("ERROR!!! Failed to initialize Module: %s",
                    g_module_information_table[module_index].module_name);
                return ret;
            }
            APP_INFO ("Initialized Module: %s",
                g_module_information_table[module_index].module_name);

        }
        module_index++;
    }
    APP_INFO ("Software Module-specific initializations done.");

    /* return SUCCESS */
    return 0;
}
```

Figure 37:
Initializing software modules defined in the module information table.

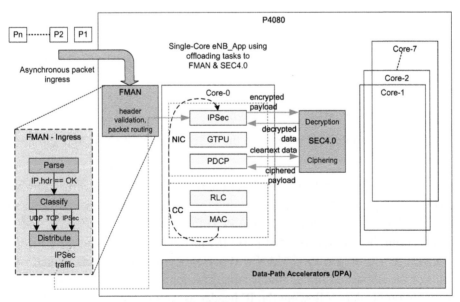

Figure 38:
FMAN Usage in packet pre-processing.

After the packet headers have been parsed, they will be classified based on the values of the packet headers. For example, IP packets whose 'protocol' header field has the value of 17 (0x11) are classified as UDP packets. Similarly, IP packets whose 'protocol' header field has the value of 50 (0x32) are classified as IPSec packets.

The classified packets can then be distributed according to special rules that can be applied for the header fields parsed earlier. Packet-distribution is done by enqueuing the selected data-packet onto a QMAN Frame Queue (FQ). The QMAN FQs provide a reliable and in-sequence delivery mechanism of the enqueued packets. The packet processing context (for example IPSec module) is the consumer of the enqueued Frames. It is possible to pre-process the packets even further by identifying specific streams within the class of packets.

The sample PCD file given in Figure 39, shows how FMAN is configured to parse incoming IPSec packets. The `ipsec_ingress_policy` first tries to distribute IPSec packets (using the `ipsec_in_distribution`). The `ipsec_in_distribution` forwards only the received ESP/IP packets (`protocolref name="ipsec_esp"`) for further classification (`ipsec_in_classification`) to identify the SA(s). If the received packet has an IP source & destination address pair of '192.168.100.11' and '192.168.100.10', then the packet is classified to belong to SA_1; this can be further classified on the basis of the SPI value of the ESP header. If the SPI value of the ESP/IP packet is 250 (`0xF0`) then the received packet would be placed on the Frame Queue with the FQ Id 0x4100 (for further processing). Any packet which does not have an SPI value of 250 is treated as an unknown packet and forwarded to an FQ reserved for 'garbage' data.

Thus, using FMAN, common tasks such as the routing of packets to appropriate modules of the eNB_App can be directly done and the modules as such can be simplified by removing the parsing and routing functionalities that are often associated as part of the SAP (Service Access Point) handling of the module.

b)   SEC4.0 usage for SNOW-3G offload

The ciphering/deciphering algorithms being repetitive and compute intensive are ideal candidates to be offloaded to a dedicated hardware block. The SEC4.0 block on the P4080 is capable of performing ciphering/deciphering based on SNOW-3G algorithms (which are used in LTE). In order to effectively use the SEC4.0 block for ciphering and deciphering the PDCP module was rewritten. This is in-line with the concept of code-reuse presented in **Reuse**.

In addition, the SEC4.0 block of the P4080 is also aware of its usage within the scope of LTE. The SEC4.0 block has advanced capabilities to perform 'protocol-aware' processing corresponding to the LTE PDCP sub-layer:

- parse the PDCP PDU header to extract information needed for deciphering, and
- build the PDCP PDU header after the successful completion of ciphering

```
<netpcd xmlns:xsi="http://www.w3.org/2001/XMLSchema-instance"
xsi:noNamespaceSchemaLocation="xmlProject/pcd.xsd" name="fman_test" description="FMAN Tester
configuration">

    <!-- Eth0 policy: This carries both dl & ul traffic -->
    <policy name="ipsec_ingress_policy">
       <dist_order>
          <distributionref name="ipsec_in_distribution"/>
          <distributionref name="..."/>
       </dist_order>
    </policy>


    <!-- ==================== -->
    <!-- START : IPSec Rules -->
    <!-- ==================== -->

    <!-- IPSec traffic; classify by SA. -->
    <distribution name="ipsec_in_distribution">
       <protocols>
          <protocolref name="ipsec_esp"/>
       </protocols>
       <action type="classification" name="ipsec_in_classification"/>
    </distribution>

    <!-- IPSec traffic: classification based on SA = { IP src, IP dst, IPSec SPI }
       Since different types of header extraction are used for IP and IPSec,
       need to do the classification in two stages -->
    <classification name="ipsec_in_classification">
       <key>
          <fieldref name="ipv4.src"/>
          <fieldref name="ipv4.dst"/>
       </key>
       <entry>
          <!-- SRC = 192.168.100.11; DST = 192.168.100.10 -->
          <data>0xC0A8640BC0A8640A</data>
          <action type="classification" name="ipsec_in_classification_sa_1"/>
       </entry>
    </classification>

    <classification name="ipsec_in_classification_sa_1">
       <key>
          <fieldref name="ipsec_esp.spi"/>
       </key>
       <entry>
          <!-- NOTE:- it is not necessary to have a mathematical relation
             between the SPI value and the FQID -->
          <data>0xF0</data>
          <queue base="0x4100"/>
       </entry>
       <action type="distribution" condition="on-miss" name="dl_ipsec_garbage_dist"/>
    </classification>


    ...

    <!-- Default queue for IPSec packets that didn't match any of the CC entries -->
    <distribution name="dl_ipsec_garbage_dist">
       <queue base="0x4006" count="1"/>
    </distribution>
</netpcd>
```

Figure 39:
Packet parsing automation using FMAN.

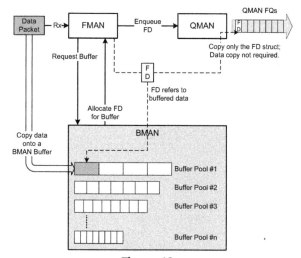

Figure 40:
BMAN buffers in FMAN, QMAN interaction.

The packets themselves are encapsulated within BMAN Buffers. The BMAN Frame-descriptors (FDs) provide references to the data stored within BMAN Buffers. The FDs allow the data packet to be processed without requiring actual data movement (Figure 40).

(Additionally, the FDs provide a uniform interface for both the software modules as well as all of the DPAA modules to access the data held within the BMAN Buffers.)

Figure 41 highlights the usage of the SEC4.0 block for deciphering PDCP payloads.

At the time of initialization of a deciphering context, the SEC4.0 block is configured with a 'descriptor' which contains the deciphering context. Additionally, a unique QMAN FQ (FQ-In) receiving the ciphered PDCP PDUs is associated with each aspect of the deciphering context along with a corresponding egress FQ (FQ-Out).

As shown in Figure 42, initializing a context within the SEC4.0 block consists of creating a Shared-Descriptor and a pair of QMAN FQs that are used to send and receive data from the SEC4.0 block. A memory area that is common to the Core as well as the SEC4.0 block is used to maintain the context within the Shared-Descriptor. The 'FQ-In' is initialized to contain references to both the Shared-Descriptor as well as 'FQ-Out'.

Figure 43 describes the algorithm used to initialize the SEC4.0 block in order to establish a 'security' context which would be used to cipher (or decipher) packets corresponding to the context. The SEC4.0 block within the P4080 maintains the ciphering context within a shared

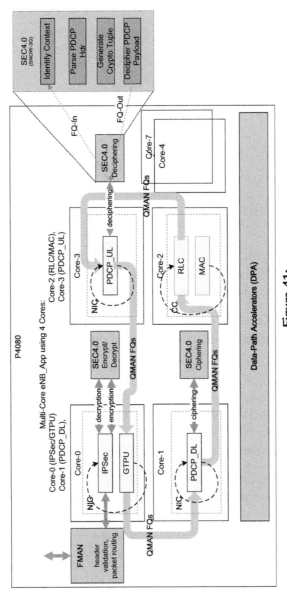

Figure 41:

Using the SEC block for SNOW-3G offload.

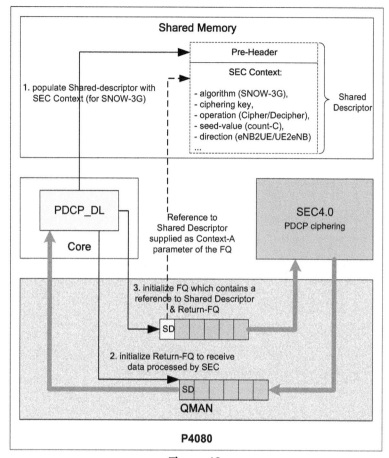

Figure 42:
Using the shared descriptor to initialize SEC4.0.

memory space — the Shared Descriptor. The sequence of steps followed to initialize the SEC context is as follows:

- Step 1 consists of allocating the memory that will be shared with the SEC4.0 block; a BMAN buffer is allocated for this purpose.
- In Step 2, the actual addresses of the start of the Shared-Descriptor and the Pre-Header Descriptors are identified. The Pre-Header Descriptor is used to pass parameters specifically used by the SEC4.0 block, while the Shared-Descriptor is used to pass the algorithm-specific parameters.
- In Step 3, a BSP API is called to construct the shared-descriptor corresponding to the ciphering algorithm (SNOW_F8). The parameters that are relevant to this specific algorithm are supplied to the BSP API which constructs the descriptor which is compatible with the SEC4.0 block.

```
struct bm_buffer        bman_buf;
uint32_t                ctxA_hi;
uint32_t                ctxA_lo;
uint32_t                *shared_desc    = NULL;
sec_descriptor_t        *prehdr_desc    = NULL;
...

STEP 1).
/*! Assign a buffer that can be shared with SEC4.0 Block. */
if( 0 >= dpa_allocator_get_buff( g_buff_allocator, sizeof(sec_descriptor_t), &bman_buf))

{
  APP_ERROR("Buffer allocation failed for SEC descriptor");
  return -ENOMEM;
}

STEP 2).
 prehdr_desc = (sec_descriptor_t *)BMAN_get_buffer_payload(bman_buf);;

 /*! Skip the Pre-Header space to point to the address of the Shared-descriptor. */
 shared_desc = (uint32_t *) ((uint8_t *)prehdr_desc + sizeof(struct preheader_s));

STEP 3).
/*! Construct the Shared-Descriptor for the PDCP Bearer's Ciphering context
 * in the SEC4.0 Block. */
if (0 > cnstr_shdsc_snow_f8(
      shared_desc,
      &shared_desc_len,
      ciphering_key,
      F8_KEY_LEN * BITS_PER_BYTE,
      DIR_ENCRYPT,
      count_c_value,
      pdcp_bearer_id,
      direction_eNB_to_UE, ...))
{
   APP_ERROR("Failed to Construct Shared-Descriptor.");
   ASSERT(0);
}

STEP 4).
/*! Now initialize the Pre-Header descriptor according to the Ciphering Algorithm. */
prehdr_desc->prehdr.hi.field. ... = XYZ;
prehdr_desc->prehdr.lo.field. ... = XYZ;

STEP 5).
/* Create an FQ to deliver unciphered packets to SEC.  */
fq = (struct qman_fq *)memalign(CACHE_LINE_SIZE, sizeof(struct qman_fq));
if (unlikely(NULL == fq))
{
  APP_ERROR( "malloc failed.");
  ASSERT(0);
}

/* Create an FQ to SEC which contains a reference to Shared-Descriptor memory. */
ctxA_hi = GET_HI_ORDER_PHYS_ADDR(prehdr_desc);
ctxA_lo = GET_LO_ORDER_PHYS_ADDR(prehdr_desc);
APP_DEBUG( "Initializing PDCP -> SEC DL FQ 0x%x.", FQID_PDCP_TO_SEC_DL);
ret = QMAN_initialiaze_frame_queue(
      fq              /* Pointer to the FQ structure */,
      FQID_PDCP_TO_SEC_DL  /* FQID for the FQ being created*/,
      qm_channel_caam     /* Channel ID for sending data to SEC*.,
      WQ_ID            /* Work-Queue ID giving default priority*/,
      ctxA_hi          /* High-order address of the Shared-Descriptor*/,
      ctxA_lo          /* Low-order address of the Shared-Descriptor*/,
      FQID_PDCP_FROM_SEC_DL /* ctxB == FQID of the FQ on which SEC will return the
                       ciphered data */,
      ...);
ASSERT(ret == 0);
...
```

Figure 43:
Initializing the SEC descriptor for SNOW-F8 algorithm.

- In Step 4, the Pre-Header descriptor is updated so that the SEC4.0 block can operate upon the information contained within the Shared-Descriptor.
- Finally, once the Shared-Descriptor has been updated with the ciphering context, a dedicated QMAN FQ is created in Step 5.

In addition to the default parameters of the function that creates the QMAN FQ, extra information, such as the High & Low-order addresses of the Shared-Descriptor, the specific Channel which is used to communicate with the SEC4.0 block (`qm_channel_caam`) and the FQ on which the ciphered data is to be returned (`FQID_PDCP_FROM_SEC_DL`), is provided to the FQ creation API.

Thus, when any data is sent to the SEC4.0 block using the QMAN FQ initialized here (`FQID_PDCP_TO_SEC_DL`), the SEC4.0 block will use the information stored in the Shared-Descriptor associated with this QMAN FQ, cipher the given data, and return the ciphered data on the return QMAN FQ (`FQID_PDCP_FROM_SEC_DL`).

When a ciphered PDCP PDU is received, it can be directly forwarded to the SEC4.0 block on the appropriate ingress FQ (FQ-In). The SEC4.0 block will retrieve the deciphering context based on the FQ on which it receives the PDCP PDU. To extract the run-time context information, the SEC4.0 block parses the header of the received PDCP PDU to directly generate the required deciphering context. The deciphered PDCP payload is then returned back to the PDCP sub-module which handles PDCP UL (Uplink) data.

NOTE: This 'insertion' of the SEC4.0 block into the PDCP UL processing path requires that the PDCP UL module be split into the pre-SEC and post-SEC functionalities. Thus, the PDCP UL module is further atomized into the PDCP_UL_Pre_SEC and PDCP_UL_Post_SEC sub-modules.

Similarly, in the DL (Downlink) direction the SEC4.0 block can directly receive the PDCP payload. The SEC4.0 block will cipher the PDCP payload based on the stored ciphering context (which gets automatically updated after the ciphering of each PDU). After the successful ciphering of the payload portion of the PDCP PDU, the SEC4.0 block can directly generate the PDCP header as well.

c) QMAN usage for communication

After FMAN processing, the received packet is forwarded to PDCP for the start of downlink (DL) processing in the LTE L2 DP. As part of the *atomization* activity the PDCP DL processing has been separated into the pre-SEC and post-SEC sub-modules.

In the third phase (Step 3) in the effort to optimize the eNB_App for the P4080 platform, the eNB_App was distributed across the cores of the P4080. This distribution was intended to balance the resource requirements of the various sub-modules of the eNB_App. In addition to this, a new mechanism (as shown in Figure 36) to flexibly configure the sub-module to core association was devised. Using this approach it is possible to support different configurations of the eNB_App with different arrangements of sub-modules across the available cores. This allowed the iterative process of balancing the eNB_App with the aim of maximizing its performance.

As mentioned in **Step 3: Distributing the application on the multi-core SoC**, QMAN FQs were the primary inter-module communication mechanism.

The use of QMAN FQs permitted a rapid conversion of the Single-core Application to a Multi-core Application.

Each FQ is associated with a Channel and a Work-Queue (WQ) within that Channel. The Channel associated with the FQ is used to define the consumer(s) of the FDs enqueued on the FQ while the WQ is used to assign a processing priority to the FQ.

There are two types of Channel — Direct Connect (DC) Portals and Pool Channels.

DC Portals are Channels that are associated with specific hardware (HW) blocks. Each HW block has one DC Portal associated with it. As a result only the HW block that owns the DC Portal can dequeue FDs that have been enqueued on an FQ associated with that DC Portal.

A Pool Channel, as the name suggests, permits the FQs associated with the Pool Channel to be consumed by a pool of consumers. Any of the Cores can be associated with a Pool Channel and a Core may be associated with multiple Pool Channels (Figure 44).

The flexibility imparted by the Pool Channels has been extensively used in refactoring the Application from Single-core to Multi-core.

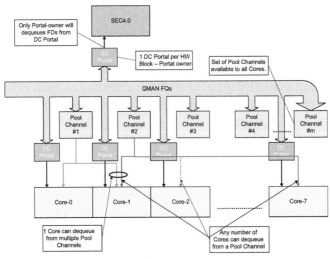

Figure 44:
Usage of QMAN for IPC.

The QMAN FQs can be initialized to carry out queue-based IPC between the various modules (hardware blocks and software modules) of the P4080 SoC. Each QMAN FQ that has to be used needs to be initialized before use. QMAN FQs can be used to either connect to a DC Portal, or a Pool Channel. DC Portals are hardwired and terminate on specific Hardware Blocks of the P4080 SoC — for example, the 'qm_channel_caam' is used to connect to the Security Engine block.

Setting up a QMAN FQ for IPC consists of two basic operations — qman_create_fq and qman_init_fq. When creating a QMAN FQ it is given a unique identifier — the FQ-Id, along with parameters which govern the FQ operation such as whether it terminates on a DC Portal or whether it will be used to receive BMAN buffers from a Hardware block, etc. Once a QMAN FQ is successfully created, its behavior and context can be initialized. The QMAN FQ context consists of information such as the Channel and Work-Queue with which the QMAN FQ is associated, any hardware or software contexts that are associated with the FQ, and whether the stashing feature is to be used or not (Figure 45).

```
int QMAN_initialiaze_frame_queue(
 *fq, fq_id, channel, wq_id, ctxA_hi, ctxA_lo, ctxB,
 <various_flags> ...)
{
 /*! Set the configuration & flags to be used when creating the FQ. */
 create_flags = ...;

 /*! Create the frame queue. */
 ret = qman_create_fq(fq_id, create_flags, fq);

 /*! Set the configuration & flags to be used when initializing the FQ. */
 init_flags = ...;

 /*! Update the S/W or H/W Contexts
  * that would be associated with this FQ. */
 /*! Use Context B to store reference of S/W Context. */
 if (is_sw_context_supplied)
   fq_opts.fqd.context_b = ...;

 /*! Use Context A to store reference of H/W Context. */
 if(is_hw_context_supplied)
   fq_opts.fqd.context_a = ...;

 /*! Enable Stashing (if required). */
 if (is_stashing_used)
   fq_opts.fqd.context_a.stashing = ...;

 /*! Initialize frame queue. */
 ret = qman_init_fq(fq, init_flags, &fq_opts);
}
```

Figure 45:
Initializing a QMAN frame queue.

As previously mentioned, the end-points of the QMAN FQ are decided by the Channel that it is associated with. While DC Portals are associated with hardwired end-points, the QMAN FQs associated with Pool Channels can be handled by any combination of cores. To process QMAN FQs associated with a Pool Channel on a specific Core, the Core must specifically instruct the QMAN to dequeue Frames that have been enqueued onto that particular Pool Channel. The qman_static_dequeue_add API is used to register the Pool Channels that would be de-queued on a particular Core (Figure 46).

Each of the software modules are associated with single and independent Pool Channels — QMAN_PC_NIC, QMAN_PC_PDCP_DL, QMAN_PC_PDCP_UL and QMAN_PC_CC. As shown in Figure 46, this allows the software modules to be easily distributed across Cores by:

1. Initializing the QMAN FQs that receive data for each software module to be associated with the specific Pool Channels assigned to that module.
2. Adding the specific Pool Channel (associated with the software module) to the list of channels that the QMAN will dequeue only on the Core assigned to run that software module.

As mentioned earlier in the section on multicore considerations, to modify the Single-core Application to run across multiple cores, it is essential to ensure the proper *Distribution & Balancing* of the sub-modules.

The example illustrated in Figure 47 shows the distribution and balancing of the PDCP_UL module across the cores of the P4080.

In this example using the three Cores, both PDCP_DL and PDCP_UL modules have been configured to execute on Core-1. This is made possible by configuring QMAN to dequeue FQs from pool-channel PDCP_UL_CHANNEL on Core-1. This module configuration related to the channel used by the module is maintained within the Module Information Table.

Figure 48 illustrates an example of distribution over four Cores, in which PDCP_DL is still configured to execute on Core-1, while, the PDCP_UL module has now been configured to execute on Core-3. This can be achieved by modifying the configuration of the PDCP_UL module in the Module Information Table. The configuration of the PDCP_UL module is modified so that QMAN FQs from pool-channel PDCP_UL_CHANNEL will now be de-queued on Core-3 (instead of Core-1 as shown in Figure 49).

Tips and tricks

1. Leverage Hardware

A majority of Multi-core SoCs contain various hardware accelerators which assist in implementing the target uses. Usage of these hardware accelerators not only improves the performance of the applications that run on them but also reduces the development and test effort to develop the application.

```
int add_IPC_channel_to_core( core_id, channel, ...)
{
 /*! Add the Pool Channels from which QMAN would
  * De-Queue Frames on this Core. */
 g_core_channel_map[core_id] |= channel;
 qman_static_dequeue_add( g_core_channel_map[core_id]);
}

int main(void)
{
  ...

 switch ( current_core_id)
 {
  case CORE_0:
   init_core_cbFunc( &process_pdu_NIC);

    /*! Specify the Channels from which IPC data would be read. */
    add_IPC_channel_to_core( current_core_id, QMAN_PC_NIC);
    break;

  case CORE_1:
   init_core_cbFunc( &process_pdu_PDCP_dl);

    /*! Specify the Channels from which IPC data would be read. */
    add_IPC_channel_to_core( current_core_id, QMAN_PC_PDCP_DL);
    break;

  case CORE_2:
   init_core_cbFunc( &process_pdu_PDCP_dl);

    /*! Specify the Channels from which IPC data would be read. */
    add_IPC_channel_to_core( current_core_id, QMAN_PC_PDCP_UL);
    break;

  case CORE_3:
   init_core_cbFunc( &process_pdu_CC);

    /*! Specify the Channels from which IPC data would be read. */
    add_IPC_channel_to_core( current_core_id, QMAN_PC_CC);
    break;

   ...
 }

 ...
}
```

Figure 46:
Setting up QMAN to restrict FQs per core.

```
int process_PDCP_sdu_dl_preSEC( ipc_pdu)
{
  PDCP_CONTEXT_t *pdcp_context;
  PDCP_SDU_t    *sdu = (PDCP_SDU_t *)ipc_pdu->payload;
  PDCP_PDU_t    *pdu;
  BMAN_BUFFER_t *frame_buffer;

  /*! Perform Error Checks. */
  if ((PDCP_SDU_SIZE_MIN > ipc_pdu->length) ||
     (PDCP_SDU_SIZE_MAX > ipc_pdu->length))
  {
    /* Error handling. */
    ...
  }

  /*! Use the Context-Stashing feature of QMAN FQs to retrieve PDCP Context. */
  pdcp_context = (PDCP_CONTEXT_t *)fq->stashed_data;

  /*! Perform Pre-SEC processing of the PDCP SDU. */
  buffer_pdcp_sdu( pdcp_context, sdu);

  frame_buffer = BMAN_get_buffer( PDCP_HEADER_SIZE + sdu->length);
  pdu       = (PDCP_PDU_t *)BMAN_get_buffer_payload( frame_buffer);
  pdu.length  = compress_pdcp_sdu_headers( pdcp_context, sdu,
                        pdu.payload, sdu->length);
  BMAN_set_payload_length( frame_buffer, pdu.length);

  qman_enqueue( FQID_PDCP_TO_SEC_DL, frame_buffer, ...);
}
```

Figure 47:
QMAN FQ being used for IPC.

A good example of the leveraging of such accelerators has been the leveraging of the SEC4.0 block described in the previous section.

2. Data-structure Access

While distributing software modules across the multiple Cores, care must be taken to ensure that data-structures are optimized for such an operation. The sharing of data-structures across multiple Core must be minimized as far as possible — this may require the original data-structures to be redesigned so that modules running on different Cores may concurrently access their respective data with minimal (if any) need for synchronization.

3. Scope of Variables

When moving from Single-Core to Multi-Core the use of global variables should be carefully analyzed. The use of global variables must be checked against race-conditions and synchronization errors.

A good practice is to minimize the usage of global variables. If they are at all required, they should be declared as volatile whenever the value of the variable is likely to change. For all other purposes, they should be treated as constant variables. If there are variables which have

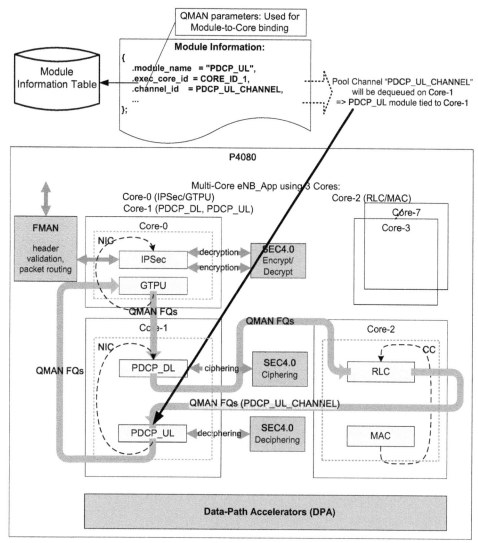

Figure 48:
QMAN Pool channel used to execute PDCP_UL on core-1.

scope that is global to each Core, then these could be declared as an array of global variables, with each index representing the Core-specific value of the global variable — access mechanisms to access Core-specific global variables may be defined for accessing such variables (Figure 50).

In the example shown in Figure 51, the scope of usage of the pointer to the currently active PDCP context has been modified so as to avoid race conditions in cases where multiple Cores

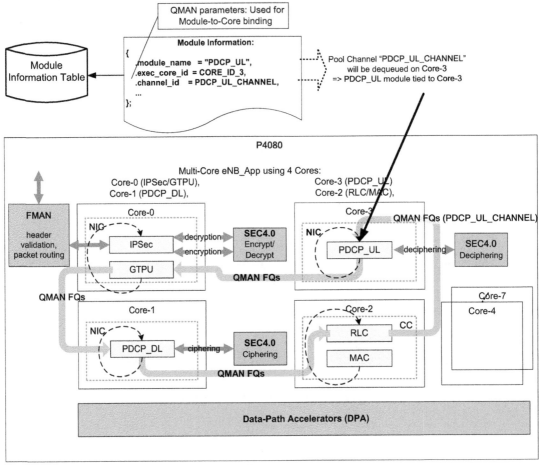

Figure 49:
QMAN pool channel used to execute PDCP_UL on Core-3.

could be used to simultaneously process PDCP SDUs. As can be seen in the example, the scope of the variable identifying the PDCP context has been reduced from global to local. Thus, when PDCP executes on multiple-cores, the race condition involving the correct PDCP context using which a PDCP SDU was processed is overcome.

4. Parameter Count

When calling nested functions, care must be taken to limit the number of parameters that are being passed as formal parameters.

A good programming practice is to limit the number of formal parameters passed to functions, as too many formal parameters will lead to data being pushed to the stack. The exceptions to this are the leaf-functions where the compiler may optimize the formal parameters by

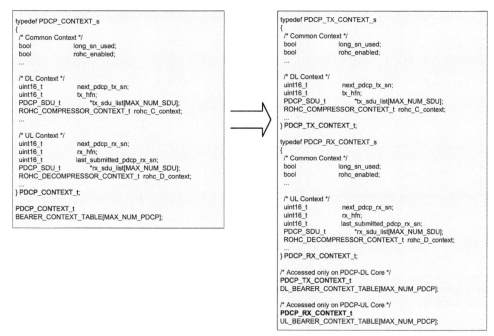

Figure 50:
Example of data distribution.

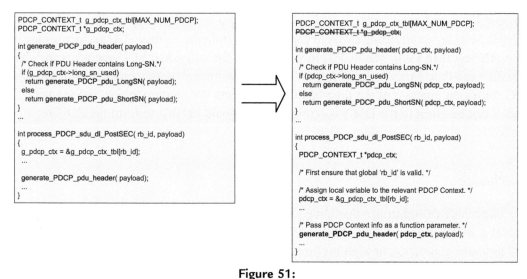

Figure 51:
Example of removing unnecessary use of global variables.

```
int foo_bar(void)
{
   int a = 5;
   int b = 3;
   int p;
   int q;

   . . .
   . . .

   p = a * 10;
   q = b * 5;
   return p+q;
}
```

```
int foo_bar(void)
{
   int a;
   int b;
   int p;
   int q;

   . . .
   . . .

   a = 5;
   p = a * 10;
   b = 7;
   q = b * 5;
   return p+q;
}
```

Figure 52:
Localization of variable usage.

retaining them as register variables. Even in such cases, excessive parameter counts have a detrimental impact on the performance.

5. Localization of Variable Usage

Variables that are used locally within functions should be initialized as close to the piece of code where they are to be used as possible. This allows the variables to be retained as register variables and the structures could be retained within the caches (Figure 52).

6. DMA copy vs. MEMCPY

In frame-based protocols (such as RLC/MAC) where a new PDU may be formed from multiple SDU segments, using DMA to copy the data from the various SDU segments may be more efficient than using MEMCPY. In such a case, the individual SDU segments may be identified and their memory locations and sized maintained in data-structures until the final PDU formation process. At the time of PDU formation, the memory offset and segment lengths can be given to the DMA controller which would be able to form the PDU more efficiently.

Conclusion

This case study demonstrates that a well defined and systematic development strategy enables a successful migration of embedded application solutions designed for single-core SoC to a new generation of high performance multi-core SoC. The migration path, discussed in Section 6 of this case study, is equally valid when the embedded applications are designed directly on the multi-core platforms. When the applications are directly being designed on the multi-core platform, the initial starting point can still be the single core of a multi-core SoC.

Thereafter, this solution can be incrementally scaled to use the various available accelerators and finally the various available cores for higher system performance. Lastly, the science of developing the software on high performance multi-core platforms is still evolving and the engineering teams need to adopt an incremental development approach that focuses on the highest quality of software to mitigate the risks introduced by a new technology paradigm.

Abbreviations

3GPP	3rd Generation Partnership Project
BMAN	Buffer MANager
CPU	Central Processing Unit
DL	DownLink
DPAA	Data-Path Acceleration Architecture
FMAN	Frame MANager
FQ	Frame Queue
GTPU	GPRS Tunneling Protocol − User plane
IPC	Inter-Process Communication
IPSec	IP Security protocol
LTE	Long Term Evolution
MAC	Medium Access Control
QMAN	Queue MANager
PDCP	Packet Data Convergence Protocol
RLC	Radio-Link Control
SEC	SECurity Engine
SoC	System on a Chip
UL	UpLink

References

[1] http://www.stack.nl/~dimitri/doxygen/
[2] http://www.freescale.com/webapp/sps/site/prod_summary.jsp?code=P4080
[3] http://www.3gpp.org/ftp/Specs/html-info/36300.htm

DSP for Medical Devices

Robert Krutsch

Medical imaging introduction

Medical imaging can be seen as the processes and methods used to create images of the human or animal body for clinical or medical science purposes. Considering that until 2010 about 5 billion medical imaging studies had been performed we can grasp the importance of the spread of use of medical imaging devices in our modern society.

Nowadays we have quite a few techniques for obtaining images that provide relevant information from the clinical point of view. The most common ones are MRI (magnetic resonance imaging), CT (computed tomography) and ultrasound imaging. CT imaging is based on ionizing radiation and is associated with health hazards. MRI is quite different and does not bring the same drawbacks in terms of health. Bought techniques (MRI and CT) involve large devices and significant power consumption and are not discussed in depth in this case study, since the focus is on portable devices. Ultrasound imaging is based on a phenomenon that is very familiar to a broad spectrum of people: echoes. While this technique may not provide all the anatomical details that CT and MRI yield, it has several advantages that make it a technique of choice:

- Very good tool to analyze moving structures
- No health hazards known or documented
- Low cost for device and scan procedure

The concepts of ultrasound imaging are very similar to those pertaining to sonar. The differences between the techniques come from the fact that, for sonar, single reflections from solid objects are used compared to ultrasound imaging, where reflections from tissue boundaries are recorded and displayed.

The ancestor of the modern ultrasound machine is the reflectoscope [6] that displayed the amplitude of the echoes versus time on the screen of the oscilloscope. This type of representation is known as A-Mode and it is the most basic representation in ultrasound imaging. Since then, advances in signal processing, acoustics, and electronic circuitry have brought significant advances in ultrasound imaging, not only in terms of image quality but also in terms of the use cases and areas of applications.

The main limitation of the ultrasound imaging technique has shifted during the years from the technological limitation to a more fundamental, physical obstacle, the speed of sound in the human body (this is more or less fixed to about 1540 m/s). Forming the image implies waiting for echoes to return from the region of interest, which means that the frame rate is limited by the speed of sound and by the depth of scan. While for slowly moving organs such as liver or kidney a low frame rate can be acceptable, for cardiac imaging the frame rate is of high importance for a qualitative assessment of the heart contractions.

Ultrasound basics

In this chapter the emphasis will be on deducing some of the formulas that will serve as rules of thumb for designing an ultrasound system. The approach simplifies the problem of medical ultrasound and tries to give intuitive answers to design engineers.

Let us consider a linear aperture of length 2A and a target point at location $(0, R)$ as in Figure 1. The time needed for the ultrasound wave that started at a specific aperture point to reach $(0, R)$ is proportional to the distance path from the specific aperture point to the target point.

If we consider that the speed of sound is a constant in the propagation medium and the above mentioned observation we can conclude that: studying the difference path from two aperture points to a target point is equivalent to studying the time delay between two waves starting in two points of the aperture and reaching the target point. In medical ultrasound the speed of sound is typically considered as being constant ($c = 1540$ m/s) even though in reality this is not the case. This assumption leads to possible artifacts in the image; artifacts are called aberrations in the literature.

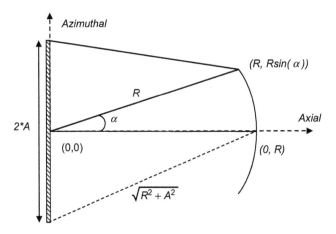

Figure 1:
Linear aperture of length 2A and a target point at location $(0, R)$.

We assume that our element of aperture 2A has a central frequency f_0. This implies a wavelength:

$$\lambda = \frac{c}{f_0} \tag{1}$$

The coordinate system will be considered in the center of the aperture. The direction along the aperture is called azimuthal direction and the direction normal to the aperture is called the axial direction.

Considering two points on the aperture, with coordinates (0, 0) and (−A, 0) we can calculate the distance delay, Δ, by using Pythagoras's theorem (2). Considering the delay of each aperture point with respect to the point (0, 0) we observe a maximum for (−A, 0) and (A, 0).

$$\Delta(A) = \sqrt{R^2 + A^2} - R \tag{2}$$

If we use the Taylor approximation in vicinity of zero we obtain:

$$\Delta(A) \cong \frac{A^2}{2R} \tag{3}$$

As can be observed from (3) the difference distance will determine when the target point is farther away from the aperture. It is desirable that the radiations from each aperture point arrive coherently at the target point and add constructively, so we would wish that Δ is very small.

Typically, Δ is considered negligible if it is less then $\frac{\lambda}{8}$. The distance $\overline{R}$ for Δ to be considered negligible is:

$$\overline{R} = \frac{A^2}{\lambda} \tag{4}$$

In the literature the region in which we can ignore the phase error obtained due to different propagation lengths (proportional to Δ) is called the far field or Fraunhofer region. The region where the phase error cannot be considered negligible is called the near field or Fresnel region.

In the use case of medical imaging we need to focus on more than a single point to form a complete image, so let's assume that our target points are on an arc as suggested in Figure 1. We can calculate the maximum delay by applying the cosine theorem:

$$\Delta(\alpha, A) = \sqrt{R^2 + A^2 - 2RA \sin \alpha} - R \tag{5}$$

After expanding $\Delta(\alpha, A)$ with the Taylor series, (6) is obtained:

$$\Delta \cong -A\alpha + \frac{A^2}{2R} \tag{6}$$

Under the assumption that we are in the far field we can ignore the second order term from (6) and obtain:

$$\Delta \cong -A\alpha \tag{7}$$

Destructive interferences appear if the phase error (φ) exceeds $\pi/2$.

$$\varphi = \frac{2\pi|\Delta|}{\lambda} \geq \frac{\pi}{2} \text{ or } |\Delta| \geq \frac{\lambda}{4} \tag{8}$$

From (7) and (8) we can deduce the angular interval such that destructive interference does not occur as being:

$$\alpha \in \left[-\frac{\lambda}{4A}, \frac{\lambda}{4A} \right] \tag{9}$$

It immediately follows that the angular extent or angular resolution is:

$$\bar{\alpha} = \frac{\lambda}{2A} \tag{10}$$

The azimuthal resolution can be calculated based on (10):

$$\bar{L} = R\bar{\alpha} = \frac{R\lambda}{2A} = F_{\#}\lambda \tag{11}$$

$F_{\#}$ (f-number) is defined as the ratio of range and aperture and is a very common parameter in many of the formulas in medical ultrasound. Equations (10) and (11) are some of the basic rules that help define ultrasound system parameters and help evaluate the overall complexity as we shall see later.

All the calculations that we have performed up to now are based on the assumption that we are in the far field, but this is not the case with medical ultrasound. To bring the approximation from far field into the near field we need to focus the aperture: this is done by using delays as suggested in Figure 2.

Considering the same physical parameters as before we will get the following error for a target point $(R, Rsin(\alpha))$:

$$\Delta(\alpha, A) = \sqrt{R^2 + A^2 - 2RA\sin\alpha} - \sqrt{R^2 + A^2} - (R - R) \tag{12}$$

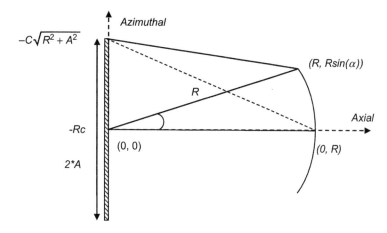

Figure 2:
Focusing the aperture using delays.

After expanding with the Taylor series in the vicinity of (0, 0) the following result will be obtained:

$$\Delta \cong - A\alpha \qquad (13)$$

As can be observed the angular resolution is no longer a function of the target depth and the azimuthal resolution is proportional to the target depth, similarly to the far field approximation case. This simplification does not come without a price tag in terms of computational prerequisites. If we deviate from the focus point as is suggested in Figure 3 the delay error will increase until it becomes unacceptable.

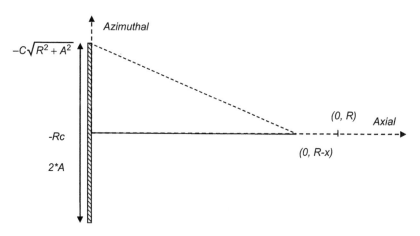

Figure 3:
A deviation from the focus point.

In the same way we calculate the error and expand with Taylor series in the vicinity of (0, 0):

$$\Delta(\alpha, A) = \sqrt{(R-x)^2 + A^2} - \sqrt{R^2 + A^2} - ((R-x) - R) \cong \frac{A^2 x}{2R^2} \tag{14}$$

As can be observed from (14) the error increases with the deviation x from the position of the focal point. The acceptable error is defined again as being:

$$|\Delta| \le \frac{\lambda}{8} \tag{15}$$

From equations (14) and (15) one can deduce the distance x for which we have acceptable errors:

$$|x| \le F_{\#}^2 \lambda \tag{16}$$

The range for which the error is acceptable is called focal depth and can be expressed based on (16) as:

$$\overline{x} = 2F_{\#}^2 \lambda \tag{17}$$

Another way of visualizing the effects of the delay and f-number is by using the ratio of the signal received at the target to the total excitation signal.

$$I_1 = \frac{\int_{-A}^{A} e^{2\pi j(f_0 t - \frac{\Delta}{\lambda})} da}{\int_{-A}^{A} e^{2\pi f_0 t} da}, \quad \Delta = -a\alpha = -\frac{aL}{R} \tag{18}$$

where L is the arc length, a is the variable that parses all points of the aperture of length $2A$.

After solving the integral for the case where x is 0 we obtain:

$$I_1(L) = sinc\left(\frac{L}{F_{\#}\lambda}\right) \tag{19}$$

The single processing chain used to form the B-Mode image implies a detection process (squared values of the echo envelope) and compression. This process has an interesting influence on the ratio I.

$$\left| I_2 \right| = \left| \frac{\int_{-A}^{A} e^{4\pi j\left(f_0 t - \frac{\Delta}{\lambda}\right)} da}{\int_{-A}^{A} e^{2\pi f_0 t} da} \right| = 2A sinc\left(\frac{2L}{F_{\#}\lambda}\right) \tag{20}$$

The first solution of the equations $|I_1| = 0$ *and* $|I_2| = 0$ are the azimuthal resolutions. It is observed that beam width decreases due to the detection process (Figure 4). This observation is helpful when defining the number of scan lines that would sufficiently sample a sector (practically, one needs to double the number of scan lines).

It is very interesting to observe that if the aperture is big the main lobe is narrow, and, practically, we can distinguish better objects that are closely spaced.

The geometrical approach is a good engineering tool that can give simple and intuitive responses to many of the questions in the design phase. Unfortunately the approach is limited and cannot give answers to many of the critical issues faced in medical ultrasound (such as understanding the transformation from object to image). Fourier analysis techniques will give a more in depth understanding and are used to explain some of the issues and challenges ([2],[6],[7]). A detailed explanation of these techniques is beyond the scope of this exposition which focuses on giving a more practical and engineer oriented perspective.

Ultrasound transducer

Ultrasound transducers contain one or more piezoelectric elements that contract/expand when they are stimulated with positive/negative voltages. The elements can so transform

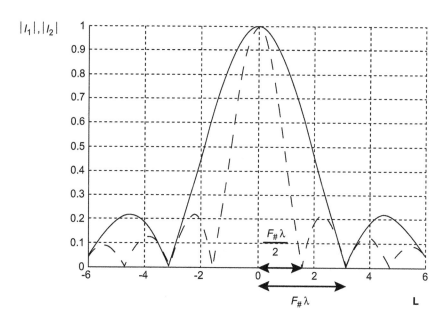

Figure 4:
Beam width decreases due to the detection process.

electric energy into acoustic energy and emit ultrasound waves at a desired frequency. The reverse is also possible with the same piezoelectric elements (transform acoustic energy intro electric energy). The basic 2D geometry of transducers is schematically represented in Figure 5.

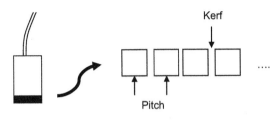

Figure 5:
Basic 2D geometry of transducers.

There are many types of transducers used, each bringing some gain for a certain type of scan. The most common are shown in Figure 6.

(a) Linear arrays: a wide span with large piezoelectric elements; usually no beam steering is employed and only a group of elements are used at the time

(b) Curved arrays: similar to linear arrays but provide a wider field of view due to curved geometry

(c) Phased arrays: smaller span than with linear arrays, with smaller piezoelectric elements. Scan lines have common origin on the transducers surface

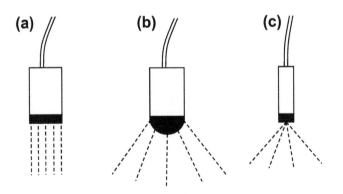

Figure 6:
Transducer types (a) Linear array (b) Curved array (c) Phased array.

Imaging modes

Today's technology permits the representation of various imaging modes that can be of interest for various use cases. We will concentrate on pulsed wave approaches and not on continuous wave since the former is more common and has several benefits, especially in the Doppler Effect based imaging modes [2]. The most common are:

*   A-Mode (Figure 7). This is the most basic representation, showing the logarithmic compressed magnitudes of the echoes against depth (time).

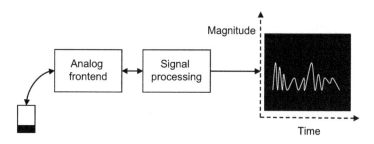

Figure 7:
Pulsed wave approach — A-mode.

*   M-Mode (Figure 8). The magnitudes of the echoes are represented through different brightness levels; on the x axis we have consecutive interrogations of the same direction and on the y axis is represented depth.

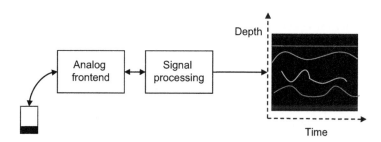

Figure 8:
Pulsed wave approach — M-mode.

*   B-Mode (Figure 9). The echo intensities are gray tone coded; on the y axis depth is represented and on the x axis we have the lateral position. The time dimension is introduced by updating the display with new image, creating a movie of the objects located in the scan sector.

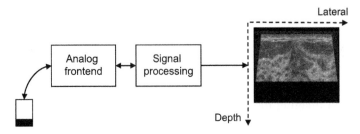

Figure 9:
Pulsed wave approach — B-mode.

- Color Doppler (Figure 10). Provides blood velocity and direction information, color coded and imposed over the B-Mode images. This method can give also information about blood turbulence.

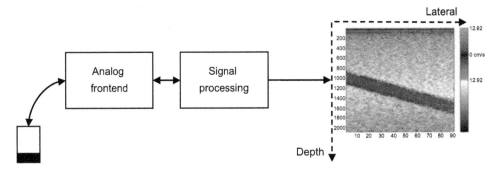

Figure 10:
Color Doppler.

- Power Doppler (Figure 11). Provides blood velocity information, color coded and imposed over B-Mode images. This method does not provide blood direction information but is very sensitive to low blood flows.

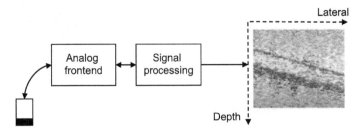

Figure 11:
Power Doppler.

- Spectral Doppler (Figure 12). Provides blood velocity estimates and direction information. It is localized and as such it has good temporal resolution. This method is good for analyzing the blood flow distribution. As one can imagine the blood flow velocity is not a constant value everywhere in the scan area.

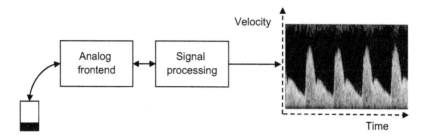

Figure 12:
Spectral Doppler.

Doppler effect basics

The idea behind the Doppler Effect is the change in frequency of a wave, as perceived by an observer that is moving relative to the source of the wave. In ultrasound imaging, the Doppler Effect is used to estimate the blood velocity. The ultrasound wave emitted by the transducer will hit the red blood cells (not only them) and get reflected back to the transducer with a frequency shift that is proportional to the blood flow velocity. When the direction of blood flow is towards the transducer the echoes will have a higher frequency and when the direction is away from the transducer the echoes will have a lower frequency (Figure 13). The difference in frequency between the transmitted wave and the received frequency off the echo is called Doppler Shift.

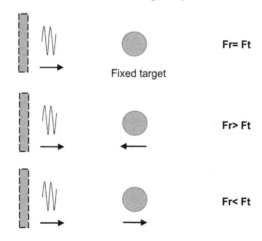

Figure 13:
Direction of blood flow to and from the transducer.

However, the practical case when the transducer is pointed head on at the blood vessel is so rare that we have to assume as almost certain that the waves will approach the target at an angle (Figure 14).

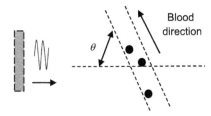

Figure 14:
Transducer is pointed head on at the blood vessel.

The frequency shift can be calculated by using the following formula:

$$\Delta F = F_r - F_t = \frac{2F_t v \cos\theta}{c} \tag{21}$$

where, F_r and F_t are the transmit and receive frequency, c is the speed of sound in the medium, v is the speed of the target and θ is the Doppler Angle.

As can be observed from (21), if the Doppler Angle comes very close to $\pi/2$ the Doppler shift will be almost zero. This can lead to the possible incorrect conclusion that there is no blood flow. Also, a miss estimation of the Doppler Angle can lead to incorrect scanning results. Usually sonographers perform a Doppler Angle correction and take care that the Doppler angle does not exceed 60–70 degrees (point from where angle corrections have a big impact due to the steep curve of the cosine function).

Since we are dealing with pulsed wave approaches (that means that we are transmitting a pulse with a frequency called pulse repetition frequency (PRF)) and blood velocities are calculated using the Doppler shift formula, an intuitive question can arise: "Will there be any aliasing if the speed of the blood is too high?" The answer is maybe; if the velocity of flow exceeds the Nyquist limit then aliasing can be observed. This can be easily identified in Color Doppler images, due to the mix of colors that correspond to positive and negative velocities.

Estimating the blood velocity magnitude and direction opens the door for parameter estimation in complex Gaussian processes. This will not be discussed in depth here, but the reader can find more detailed information in [1–4].

High level system overview

Figure 15 briefly illustrates the high level architecture of a diagnostic ultrasound system. The electrical signal controlled by the Tx part is converted by the transducer into acoustic waves. The wave propagates through the medium (in this case the human body) where a portion is reflected at the boundary of materials with different acoustic impedance (different types of tissue). The reflected part is captured by the transducer, amplified, and converted to a digital signal by the AFE. If we keep in mind that the transducer is formed from a number of

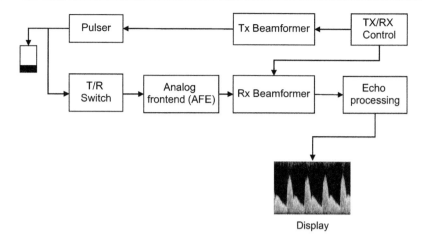

Figure 15:
High level architecture of diagnostic ultrasound system.

elements, the signal path just described has to be envisioned for all the elements of the transducer. The RX Beamforming part combines the digital signals captured by the transducer elements and forms the scan line. On the TX side the beamformer controls when and with what pulse the piezoelectric elements get excited. Finally, the scan lines are processed and the images are formed and sent to a display.

Parts like TX/RX switch, pulser, and the constituent parts of the AFE will not be discussed here; the reader can refer to the homepages of the semiconductor providers that offer such parts.

Overview of classic beamforming

Beamforming is the signal processing heart of any medical ultrasound machine. It is usually implemented in FPGAs and ASICS due to the large bandwidth and computational requirements. The recent advances in digital signal processors (highly parallel architecture with more than four cores, for example the six core DSP, MSC8156, from Freescale) opened the door for beamforming on general processing chips.

The medical ultrasound images are formed by first creating focused ultrasound beams in transmit and directivity patterns (in the same direction as for transmit) on receive. The area of interest is covered by adjacent beams with a spacing defined by the minimum resolution required. The ultrasound beams are formed by adding delays to the electrical pulses sent to each element with the purpose of controlling the radiation pattern. On the receive side, the approach is similar; the signals from each element are delayed by the same amount as on transmit and summed together. The delay phase insures that the waves/signals are all in phase and that no destructive adding takes place (Figure 16).

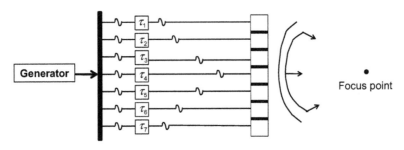

Figure 16:
Delay Phase, Transmit.

The beamforming process can be elegantly put into the following formula:

$$r(t) = \sum_{i=1}^{N} A_i(t) s_i(\tau_i(t)) \tag{22}$$

where N is the number of elements, A_i are the apodization coefficients (weights), s_i are the received echoes and τ_i are the delays. The apodization coefficients suppress the side lobes in the directivity pattern.

The delay coefficients calculation is based on the computation of time of flight of the wave. In Figure 17, the distance between the origin and the receiving element is denoted by x, the depth by R, the return path by d and the angle made by the scan line with the normal of the transducer surface by α. The full echo path can then be calculated based on the cosine theorem as being:

$$p = R + \sqrt{R^2 + x^2 - 2xR\sin(\alpha)} \tag{23}$$

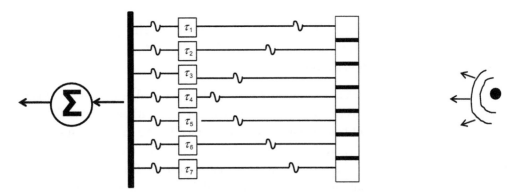

Figure 17:
Delay Phase, Receive.

By dividing equation (2) by the speed of sound we will get the time of arrival of the echo from point P. The coherent summation for two elements i and j will then imply adding

$$s_i\left(\frac{p_i}{c}\right), s_j\left(\frac{p_j}{c}\right)$$

As can be observed from equation (23), calculating the delays implies computation of the sin function and the square root; that can be cycle (for digital signal processors)/area (for FPGAs/ASICs) consuming. In the case of DSPs the delay calculation can be done offline and stored in memory but for FPGAs/ASICs the delays are usually calculated with some precision. In the literature one can find many approaches on how to approximate the delays ([2]·[6]·[7]). Computation of the full echo path is shown in Figure 18.

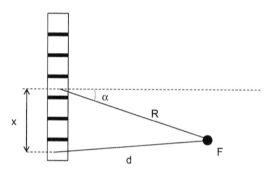

Figure 18:
Computation of the full echo path.

Design use case

Based on the rules of thumb deduced in the Ultrasound basics chapter we will go through a design example. First, let us assume the following:

- a probe of 30 elements is used, with 0.9 cm aperture for transmission and reception ($A_{tx} = A_{rx}$)
- the target depth is 17 cm (**R**)
- the area of interest is defined by a sector of angle 75 deg. (θ)
- sampling frequency of 32 MHz (**Fs**)
- a central frequency of 2 MHz (**F0**)
- the speed of sound in the human body is considered as 1540 m/s

As discussed, to form an image we have to acquisition multiple scan lines that cover the region of interest. To form a scan line, the ultrasound wave has to travel to the desired depth and get reflected back; this will give the time frame need to acquire a single scan line:

$$T = \frac{2R}{c} \tag{24}$$

This leads, in our use case, to *2.2078e-004* seconds needed per a scan line.

Based on the angular resolution (Equation 10) we can then calculate how many scan lines we need to sample the sector of interest:

$$\overline{\alpha} = \frac{\lambda}{A_{tx} + A_{rx}}, \lambda \overset{\text{def}}{=} \frac{c}{F_0} \tag{25}$$

$$N_{sl} = \frac{\theta}{\overline{\alpha}} \tag{26}$$

where $\overline{\alpha}$ is the angular resolution, N_{sl} is the number of scan lines and λ is the wavelength. In the angular resolution formula we introduced the aperture as the sum of the transmit and receive apertures because in the image context we have to take into account bought Tx and Rx direction.

As discussed in the chapter entitled Ultrasound basics, the process of forming the actual image implies also a detection phase where the actual resolution is degraded to about half of what we have considered. This implication leads to a required number of scan lines of:

$$N_{req} = 2N_{sl} \tag{27}$$

For our use case the number of scan lines required is:

$$N_{req} = 62$$

Based on (24) and (27) we can calculate the maximum frame rate achievable as being:

$$Fps_{max} = \frac{1}{TN_{req}} \tag{28}$$

This leads to about 73 frames/second as a maximum frame rate.

As can be observed from (28) the frame rate is very much limited by the speed of sound in the human body. In the literature one can find different approaches to increase the number of frames per second; the reader can refer to [6] for detailed explanations.

Once we have set our boundaries in terms of what is the maximum achievable we have to look at the hardware to see what is achievable from that perspective. As mentioned, beamforming was something only meant for FPGAs/ASICs, but nowadays DSPs have evolved and might offer a low cost alternative. There are two critical things to be evaluated:

1. Bandwidth
2. Cycle count

The MSC8156 is a six core DSP with two layers of cache per core. All cores can access an M3 (located on die) memory that has about 8 GB/s bandwidth and two DDRs with a total bandwidth of about 12.2 GB/s.

The IO interface that we will consider is Rapid IO since it can deliver up to 9.11 Gbps. The drawback of current DSPs or analog frontends (depending from where you are looking) is that they cannot interface one to another directly; perhaps in time some standard interface will emerge and get rid of this issue. Since one cannot interface directly, some sort of connector has to be designed. This connector can be an FPGA or an ASIC, and practically, a simplified version of the FPGA/ASIC that does the beamforming since it has only to collect data into FIFOs and give them to the DSP, which does all the work.

A pure software approach will give more flexibility and lower the NRE (non recurring engineering) since complicate hardware to generate delays and apodization coefficients is not needed; all can be pre-stored, pre-calculated, and easily updated if needed.

An important factor that influences the bandwidth is the directivity of the single element. An element will contribute to beamforming only after some depth that depends on the scan line orientation, element position, and the directivity pattern angle. If we consider that the directivity pattern angle is limited to

$$\left[-\frac{\pi}{4}, \frac{\pi}{4}\right]$$

and apply the sin theorem in triangle ABC (Figure 19) we obtain the following:

$$\frac{AC}{\sin\left(\frac{\pi}{4}\right)} = \frac{x}{\sin\left(\frac{3\pi}{4} + \varphi\right)} \tag{29}$$

$$AC = d_{start} = \frac{x}{\sqrt{2}\sin\left(\frac{3\pi}{4} + \varphi\right)} \tag{30}$$

For a scan line that makes an angle of 37.5 deg. with the normal (practically the last scan line in our use case) and an element at distance 0.44 cm, the depth where the element starts to contribute is about 2.4 cm.

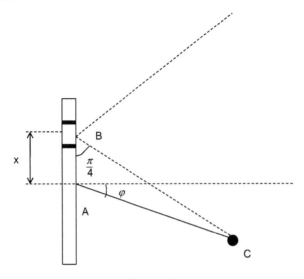

Figure 19:
Sin theorem in triangle ABC.

The overall system can be envisioned pipelined as shown in Figure 20.

While the DSP is processing the existing data a new acquisition and transfer is triggered. Beamforming and image forming happen in parallel with the transfer of new echo data. The time frame allocation, which finally defines the number of frames per second, can be envisioned as follows:

- Acquisition and transfer: 0.34375 ms
- Beamforming: 0.1385 ms
- IF: 0.205 ms

This leads to about 45 frames per second, considering 64 scan lines for acquisition.

The MSC8156 has six cores each capable of handling 1000 MCPS (million cycles per second); in the following estimation we will consider only 90% usage factor as possible, and

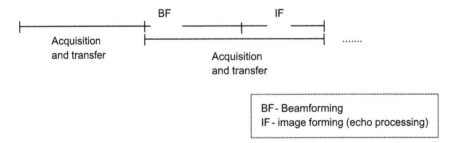

Figure 20:
Pipelined System.

leave room for operating system overheads. Also, the DDR and M3 bandwidths will be considered at maximum 60% so that the estimations are not close to the overload edge.

Considering that we have a depth range of 4 to 17 cm, 45 frames per second and about 64 scan lines, as was calculate previously, we will get the following bandwidth (based on (24)):

$$BW = \frac{32e6 \text{ Hz} \dfrac{0.17 \text{ m} + 0.13 \text{ m}}{1540 \text{ m/s}} \dfrac{16 \text{ bit}}{1024^3} 30 \text{ channels}}{T} \cong 8.1 \text{ Gbps} \tag{31}$$

$$T = 0.34375 \text{ ms}$$

The bandwidth can be handled by Rapid I/O on the DSP side and there is about 1 Gbps margin remaining ([8]).

The output of beamforming will be a scan line with about 6200 samples. Focusing and apodization coefficients should be updated at least as often as required by the focal dept limit (see the chapter titled Ultrasound basics, equation 17) . For the use case considered we have to have an update rate of at least 6.8 mm when the depth approaches 4 cm. We shall consider an update every 20 samples or about every 0.9625 mm (can be calculated based on speed of sound and sampling rate).

The beamforming module will update every 20 samples but should interpolate at each step. This will imply that we are pre-computing:

- Delay and apodization coefficients: two values on 8 bit
- An offset for updating: one value on 8 bit

It is fairly obvious that beamforming can not be handled by a single core so we have to split it in multiple tasks and parallelize it on multiple cores. We will show that five cores are sufficient for the task. The architecture is schematically shown in Figure 21. In *Phase 1* the five cores will process each six channels and in *Phase 2* the outputs of *Phase 1* will be added, and each core will process 20% of the data. The interpolation will happen completely in *Phase 1* so that in *Phase 2* a simple sum will be sufficient, since the vectors will be arranged coherently. The beamforming time frame will be allocated as follows:

- For *Phase 1*: 0.125 ms
- For *Phase 2*: 0.0135 ms

Considering that we have to transfer the same amount of data as calculated in (31), to perform beamforming, but in a shorter timeframe (0.125 ms), we can compute the bandwidth requirements for M3 to about 30 Gbps for the time timeframe of *Phase 2*; for the remainder of the time, M3 is used only by Rapid I/O.

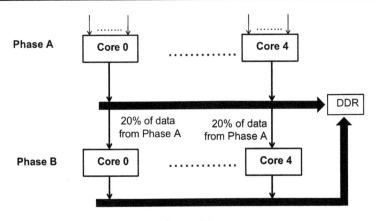

Figure 21:
Beamforming architecture split between cores.

Since we want to keep M3 mainly for I/O and considering the fact that the delay and apodization data is over 0.8 MB we have to keep this data in DDR. The bandwidth requirements needed to transfer this data in the time frame allocated for *Phase 1* can be calculated as:

$$BW_{DDR,\,coeff} = \frac{\dfrac{6200\ \text{samples}}{20} 24\ \text{bit}\ 30\ \text{channels}}{1024^3 T} \cong 1.66\ \text{Gbps} \tag{32}$$

$$T = 0.125\ \text{ms}$$

The data was considered on 16 bit, but since most frontends use 12 bit precision we will consider that the most significant four bits are 0. In *Phase 1* the cores will have to write to DDR five vectors of 6200 samples on 16 bit, since the maximum bit growth is four bits considering that we add together six values on 12 bit. This will lead to a bandwidth requirement from Table 1.

As can be observed, over all time frames, the DDR bandwidth is below 60% of the maximum. DDR bandwidth usage should not go over 60%. If it does, the estimation are not 100% reliable. The table shows that from the bandwidth point of view this is in the acceptable range where no special care and special evaluation is necessary.

Table 1: Bandwidth requirements

Memory	Phase 1		Phase 2	
	M3	DDR	M3	DDR
Required	30.04	5.3	8.1	33.9
Available	60	96	64	96

Each core of the MSC8156 has four data arithmetic logic units (DALU) and two address generation units (AGU). Also, there are two instructions that will decrease the cycle count significantly:

macd *Da,Db,Dn Dn + (Da.L * Db.L) + (Da.H * Db.H)* → *Dn*

mpyd *Da,Db,Dn (Da.L * Db.L) + (Da.H * Db.H)* → *Dn*

By using this instruction above, one core could perform up to four interpolations in a single clock cycle. If correctly pipelined, the coherent summation will take about 9 cycles (see pseudo code in Figure 22). Every 20 samples the delay and apodization coefficients have to be read; these require about 8 cycles, leading to about 9.4 cycles per output sample in *Phase 1* (see pseudo code in Figure 23). The total cycle count/core for *Phase 1* will be about 5.8e4 cycles without including the data read penalties. In *Phase 2* the five outputs of *Phase 1* have to be added together. In the simplest implementation, by computing two output samples in parallel, this could lead to about 2.5 cycles/output samples, leading to about 2.4e3 cycle/core (Figure 24).

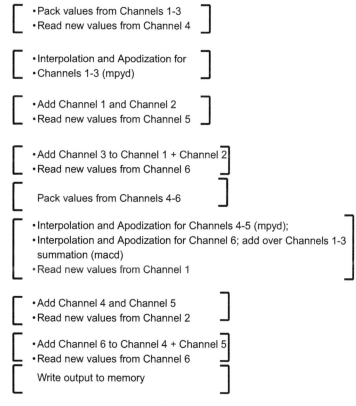

Figure 22:
Computation of cycles per core for *Phase 1*.

[Read offset for Channels 1,2]
[Read offset for Channel 3,4]
[Read offset for Channels 5,6]
[Add offsets for Channels 1,2]
[Add offsets for Channels 3,4]
[Add offset for Channels 5,6]
[Read interpolation and apodization coefficients for Channels 1-4]
[Read interpolation and apodization coefficients for Channels 5,6]

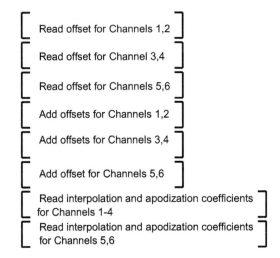

Figure 23:
Computation of cycles per core for *Phase 1*.

[
M = C0(i)+C1(i)
N = C2(i)+C3(i)
P = C0(i+1)+C1(i+1)
Q = C2(i+1)+C3(i+1)
Read new values for C0 and C1
]
[
A = M+N
B = P+Q
Read new values for C2 and C3
]
[
A +=C4(i)
B +=C4(i+1)
Read new values for C4
]
[
Write A and B
]

Figure 24:
Computation of cycles per core for entire path.

There are some rules of thumb when considering the penalties due to data transfers:

~80 cycles/128 byte when working with M3

~100 cycles/128 byte when working with DDR

This numbers can be decreased if one fetches data in advance, by using the dfetch instruction. Also it is to be mentioned that 80 cycles/128 bytes for M3 includes a big margin to cope with possible Rapid I/O bottlenecks.

The performance requirements for *Phase 1* and *Phase 2* are summarized in Table 2.

Table 2: Performance requirements for Phase 1 and Phase 2.

Phases	M3 penalty	DDR penalty	Core cycles
Phase 2	0	~9.6e3 Total Phase2: ~12e3 Available: ~13.5e3	~2.4e3
Phase 1	~4.6e4	~4500 Total Phase1 : ~10.8e4 Available: ~12.5e4	~5.8e4

The performance figures lead to about 86% core utilization for *Phase 1* and 89% core utilization for *Phase 2*; well within the bounds of the achievable margins.

The conclusion is that beamforming is no longer to be done only in FPGA/ASIC; for low cost and low end devices beamforming on DSP can be viable. Considering the fact that the limits of the DSP have not been reached one could evaluate further the feasibility of multiple line acquisition, but this is beyond the scope of this text.

Echo processing

The output of beamforming is RF data corresponding to scan lines. The echo processing phase that follows is responsible for forming the actual image from multiple scan line. The emphasis of this text will be on B-Mode image forming since it is a 'must have' for every ultrasound machine.

In Figure 25 two approaches of forming the B-Mode image are proposed. In Approach A, RF demodulation and decimation is used and in Approach B, a classic envelope detection algorithm, based on squaring the input signal.

The signal received from the transducer is a band pass signal with a typical bandwidth of 50–70% of the central frequency (Figure 26). This can be intuitively explained through the fact that transducer elements are sensitive to a frequency range.

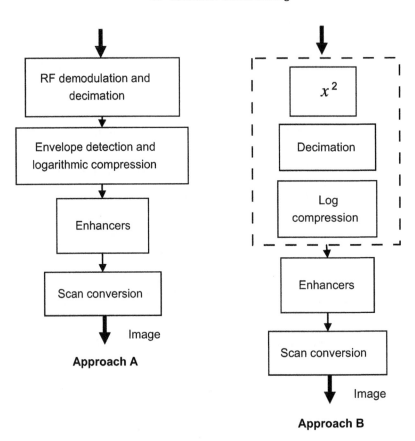

Figure 25:
Two approaches of forming the B-Mode image.

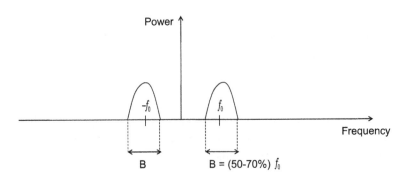

Figure 26:
Signal received from the transducer is a band pass signal with a typical bandwidth of 50–70% of the central frequency.

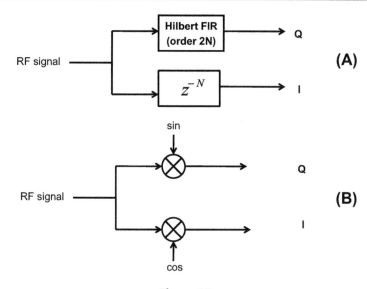

Figure 27:
RF demodulation methods.

For RF demodulation two methods are typically used:

a. based on Hilbert transform
b. baseband down conversion

For the first method, the in phase signal (I) is obtained by delaying the original signal, while the in quadrature-phase signal (Q) is obtained via Hilbert FIR filter (Figure 27(A)). A similar result can be obtained also if the Hilbert transform is applied in the frequency domain. The advantage of this method is that it is central frequency independent.

The second method is based on down-mixing with cos and sin as schematically pictured in Figure 27(B). The method is central frequency dependent, but can prove to be faster in terms of cycle count than the Hilbert transform based approach.

Independent of the method elected the next step is low pass filtering and decimation. This step is not only needed to reduce cycle count for the processing steps that follow but also due to the fact that displays have a limited resolution (e.g., 320×240, 640×480 etc).

Envelope detection and log compression is a simple algorithm that can be divided into the following steps:

$$Step\ 1 : O = \sqrt{I^2 + Q^2}$$
$$Step\ 2 : out = k + d * \log_{p}(O)$$

The second step is needed due to the large dynamic range between the specula and the scatter reflections. In order to overcome the dynamic range problem a gray level transform

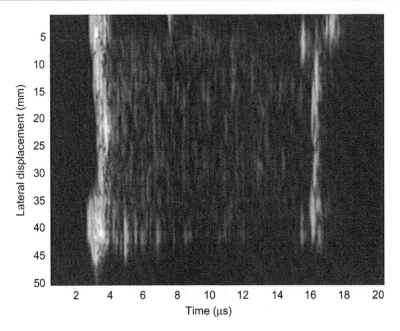

Figure 28:
Gray level transformation using logarithmic functions.

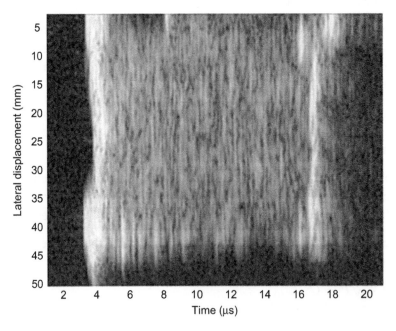

Figure 29:
Grey level transformation using logarithmic functions.

can be applied; typically this is done using a logarithmic function (see Figures 28 and 29). The parameters d, k, and p are estimated so that the dynamic range of the image fits in the maximum dynamic range of the human eye and also to adjust the desired brightness level.

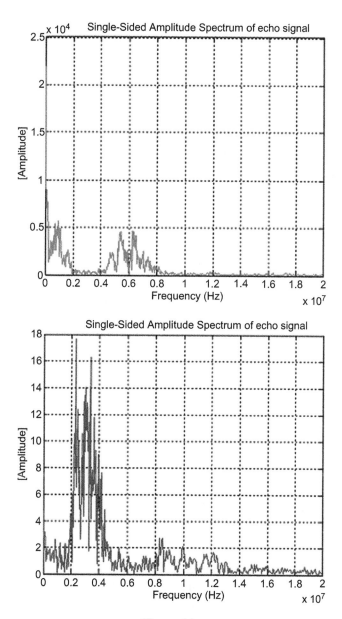

Figure 30:
Classic approach for Doppler imaging using approach B.

RF demodulation and decimation is an approach to be used for Doppler imaging modes, but for B-Mode, since we only need the signal envelope, it is too intensive from the cycle count point of view. The classic method, described in Approach B from Figure 27 is much more suitable. Squaring the input signal will shift the bandwidth close to 0 Hz and will bring a replica at twice the central frequency as can be seen in Figure 30.

The next step is to filter out the replica at twice the central frequency. This filtering can be incorporated into the low pass filter needed before decimation. Logarithmic compression follows after the decimation, to avoid spending cycles on samples that will be dropped anyway.

One cycle friendly method to perform decimation involves using a CIC filter and compensator approach. A detailed theoretical description of CIC filters can be found in [9] and is beyond the scope of this discussion. One interesting fact to be remembered about CIC filters is the bit growth. When the input is on B_{in} bits, the number of bits required by the filter is:

$$B = N \log_2(R) + B_{in} \tag{33}$$

where N is the number of stages and R is the decimation factor.

This can easily exceed the available bits in the registers and make the implementation more complicated. For example, in the case of an input on 19 bits (that could be obtained by summing 128 channels on 12 bits in the beamforming module), four filter stages, and a decimation factor of 10, the bit growth exceeds 32 bits. Some DSPs (like MSC8156 from Freescale) have a register on 40 bits, permitting efficient implementation of CIC filters with a larger decimation factor or a larger number of stages.

Scan conversion can be interpreted as transformation from one coordinate system (in this case the coordinate system in which we acquisition the image) to another coordinate system (in this case the coordinate system of the display). Typically, this is an operation of interpolation in a two dimensional space that might be performed with techniques like:

* nearest neighbor — fast but with less quality
* linear interpolation — fast but with less quality
* bilinear interpolation — usually a good tradeoff between quality and speed
* optimal interpolation — slow but with good quality

The interested reader can refer to [5] for an estimation of echo processing blocks on MSC8156. The discussion in [5] extends also to some classic image enhancing algorithms like median filtering and histogram equalization.

Conclusions

In this appendix we have presented a short theoretical background that will introduce engineers and application architects to the problems of medical ultrasound devices. Some of the classic image types are introduced and described.

The second part of this chapter details a typical use-case where the reader is introduced to methods of evaluating the capabilities of today's DSPs for medical ultrasound application. The focus is on beamforming due to the cycle intensive nature of the algorithm but the interested reader is directed to literature where echo processing modules are described and evaluated on modern high end DSPs.

References

[1] IEEE Transactions; May 1985; Chihiro Kasai, Koroku Namekawa; Real-Time Two-Dimensional Blood Flow Imaging using an Autocorrelation Technique.
[2] Thomas L. Szabo; Diagnostic ultrasound imaging inside out.
[3] US National Library of Medicine; Lind B, Nowak J; Analysis of temporal requirements for myocardial tissue velocity imaging.
[4] Jørgen Arendt Jensen; Estimation of blood velocity, using ultrasound.
[5] EETimes Design, 2/5/2011, Robert Krutsch, Taking a multicore DSP approach to medical ultrasound beamforming.
[6] Tore Gruner Bjastad, January 2009, High frame rate ultrasound imaging using parallel beamforming.
[7] Borislav Gueorguiev Tomov, January 2003, Compact beamforming for ultrasound scanners.
[8] Iulian Tapiga, Freescale Semiconductor Application Note, April 2010, Optimizing Serial Rapid IO Controller Performance in the MSC815x and MSC825x DSPs.
[9] Matthew P. Donadio, Iowegian, July 2000, CIC Filter Introduction.

Voice Over IP DSP Software System

Brief VoIP Domain Introduction

Wired TDM telecom network

The 'Plain Old Telephone Service' (or POTS) supplied the communication needs of the world for many decades during which it was refined and improved to a famous resilience known as '5-nines.' People relied on their Public Switched Telephone Network (PSTN) which provided services 99.999% of the time. Such high availability meant that the downtime per year was of 5.26 minutes. To achieve such high availability the network had to implement many control protocols and build fault tolerance within. The PSTN consists of a series of Central Offices (CO), also known as End Offices or Class 5 switches on the edge of the network that served the end users with analog or digital connectivity called 'local loop,' Tandem Offices in the network core linking the COs, a distribution digital network connecting the COs, and tandem offices and a signaling system to control these network elements.

Figure 1 depicts the PSTN architecture showing two types of transmission technologies: analog and digital. This combination of analog and digital links generated the most cost efficient way to reach the end-user and adaptation mechanisms that compensate for the artifacts are used to improve the conversations within this network with a quality that for a long time remained a benchmark denoted as 'toll quality.' Analog transmissions use continuously variable electrical waves representing the conversation. The digital transmission systems convert the analog electrical signals into discrete values represented by binary 0s and 1s for transmitting the information.

While the figure depicts a local loop made of twisted pair copper lines used for analog transmission, another technology called ISDN introduced digital lines to the Customer Premises Equipment (CPE), especially in the densely populated areas where the revenues could offset the higher deployment costs. If the local loop uses either analog or digital transmission techniques, the remainder of the network uses digital links to assure better transmission quality and efficiency.

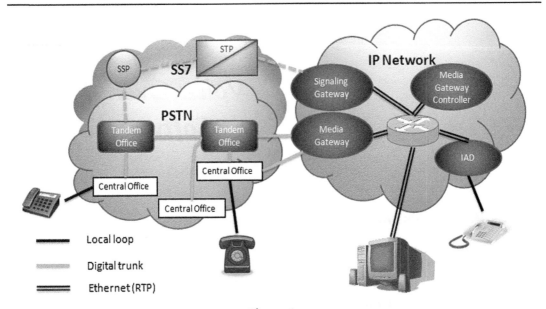

Figure 1:
PSTN Architecture.

In principle, digitization is done at the CO (Figure 2). However, there were many complications in deploying long (a few km) copper wires to distant locations from CO, such as electrical attenuation of the DC current used for subscriber signaling (e.g., off-hook), attenuation of the electrical voice signals, and the need to install amplifiers along the way which increase noise levels. The Subscriber Loop Carrier (SLC), also known as Digital Loop Carrier (DLC), was created to solve these problems and move the codec as close as possible to the customer premises, in the neighborhood of large buildings. The copper wirings connected the CPE to the DLC and digital trunks connected it to the CO, passing concentrated and digitized conversations upstream.

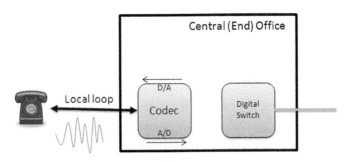

Figure 2:
Central Office.

With the voice conversations digitized, the same wires allow for more efficiency in the PSTN, namely sending voice and data (or signaling) using the same physical conduit. With the right protocol formats in place, bits can be packed and unpacked at a trunk's ends and the two types of information (voice and data) can be restored.

The most common modulation techniques are pulse code modulation (PCM) and adaptive differential pulse code modulation (ADPCM).

In the case of PCM, the bandwidth assumed is 4 kHz, although a bandpass filter of 3.1 kHz is actually used (from 300 Hz to 3,400 Hz), and each analog conversation is converted into a digital bit stream of 8000 samples × 8 bits/sample = 64,000 bits per second (64 kbps). ADPCM (realized by the mechanisms described in ITU-T Recommendation G.726) is a more technically advanced method for digitizing voice, using PCM-encoded digital input and analyzing its previous patterns to create a prediction of the next samples, relying on the inherent predictability of the human speech. It reduces PCM's 64 kbps to 40, 32, 24, or 16 kbps.

64 kbps PCM has been used in the PSTN for decades, a standard being released by ITU-T in Recommendation G.711. Since 1962, when it was employed by Bell Labs, the 64 kbps increments (referred to as DS0) have been used as the fundamental building blocks of digital transmission systems. In North America, the transmission system, called T-carrier, bundles 24 DS0 per T1 (also called DS1), for a total capacity of 1,544,000 bps. A larger capacity digital trunk is T3 with 28 T1s bundled within (also called DS3) for a total capacity of 45 Mbps. Multiple T3 channels can be bundled into an optical channel (OC) at even higher capacity, for which the Synchronous Optical Network (SONET) was developed. In Europe, a similar system named E-carrier (developed by European Conference of Postal and Telecommunications Administrations CEPT, the precursor of European Telecommunications Standards Institute ETSI, to improve the North American T-carrier) is used and combines 32 DS0s into a single bundle called E1 (out of which 30 timeslots are used for voice channels, one for framing and one for signaling), and multiple E1s are combined into E3s.

The pre-VoIP world was categorized as 'circuit-switched,' because all digital phone calls were sent over a time division multiplexed (TDM) link where resources were allocated throughout the network at the call setup, similar to multiple 'virtual circuits' being configured to create a continuous end-to-end voice path. This method benefits from an inherent quality of service because each transmission receives a dedicated time slot for the user's data or voice. That time slot cannot be over-run by another user's voice or data. As a drawback, the speculative usage of the temporarily idle resources is not possible, namely if the dedicated time slot is not used by the current subscriber's voice or data, the slot is wasted since it cannot be shared by another user.

Figure 3 below shows a T1 frame format. The framing bit was used for synchronization purposes and consecutives T1 193-bit frames were gathered into super-frame formats

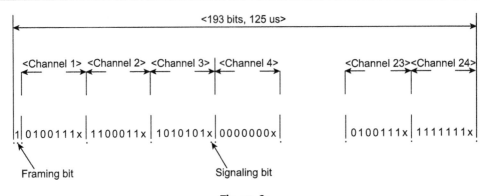

Figure 3:
T1 frame format.

(e.g., D4 or ESF). In early deployments, the least significant bit from the DS0 was used for signaling. Certain T1 frames from a super-frame had their signaling bit 'robbed,' for example every 6th and 12th in a D4 format, in which each super-frame was composed of 12 T1 frames. This effectively limited the throughput of a voice channel to 56 kbps in the early PSTN deployments.

To signal calls between users, the PSTN later employed a packet-based network called Signaling System 7 (SS7) in North America, or ITU-T's Common Channel Signaling System 7 (CCSS7) elsewhere, to determine the best call route, connect the callers, and provide call control.

Central offices provide basic features like dial tone, dialing (processing call set-up requests), and incoming call notification (electrical ringing current). Working together, the SS7 network and the CO provide more advanced features like caller ID and call forwarding.

Private voice network systems like private branch exchanges (PBXs) and key systems, their smaller scale realizations, also work with the PSTN to provide a hybrid public/private network. A PBX allows multiple users at a site to share incoming and outgoing PSTN trunk lines, so a site with PBX does not have to dedicate individual PSTN lines to each user of the phone station. The PBX also provides abbreviated number dialing to PBX extensions so internal callers can bypass the PSTN and save additional expense. In addition, the PBX offers features like call transfer, call pickup, and auto attendant. PBX and key systems also integrate with other value added applications and features like voice mail systems, unified messaging, and interactive voice response (IVR) systems. IVR systems interact with the caller to help direct calls based on the caller's selection of audio prompts.

Migration to IP based transport, market drivers, and technical changes

The goal of VoIP is to enable packet based Internet telephony users to communicate with each other and to regular PSTN telephone users. VoIP Gateways provide the interconnection between traditional telephony and the digital world of Internet telephony.

The primary function of the Gateway device is to provide translation services for the transmission formats, communications procedures, and audio codecs between the two domains.

The Gateway is a two-way interface between the telephone network and the IP-based network. It is optional when there is no need to interconnect with a regular public switched telephone network, such as in an enterprise LAN only architecture. It assumes responsibility for setting up and tearing down voice channels between H.323 or SIP endpoints and the PSTN network such as T1, B-ISDN, SS7, etc.

The Internet Engineering Task Force (IETF), in its Media Gateway Control Protocol (MGCP) from RFC2705, presented examples of Telephony Gateways which use specialized functions based on their application domains. The most common types of gateways are:

- Trunking Gateways, managing a large number of digital circuits,
- Access Gateways, providing support for analog circuits,
- Residential Gateways, with fewer capacity but supporting a larger variety of communication technologies found in people's homes

With the advent of the third-generation (3G) wireless network a fourth type was added:

- 3G Media Gateway

The Gateway model, adopted by both standardization bodies ETSI (TIPHON) and ITU-T, can be a single box that provides this interface but the Gateway model can be decomposed into three separate components and running on three different platforms:

1. Media gateway: is the critical inter-working element that translates media between networks of differing standards (e.g., Switched Circuit Network and IP/Ethernet). It provides conversion of streamed media formats — such as voice, data, or video — and manages the transfer of information between the different networks. As such, it is equivalent to either a local or a transit switch in the Switched Circuit Network. For instance, a PSTN/IP Media Gateway provides the voice traffic translation between an IP based G.723.1 at 6.3 kbps to G.711 at 64 kbps. From one side it is connected to a Local Area Network such as Ethernet 10/100BT. On the other side, it assumes the connection to the telephone network as a T1 trunk or ISDN line for video communication with H.320 compliant video equipment. This platform is required to remain permanently active to prevent any discontinuity of service between two endpoints. A high availability platform is required with a minimum downtime and allows maintenance operation while the system is operational. This node controls jitter, delay, echo cancellation or any other component that constitutes quality of service (QoS).

2. Media Gateway Controller provides the overall control of the gateway. It communicates with the Gatekeeper for database information regarding mapping between IP address and phone network.
3. Signaling gateway: responsible for the interface between SS7 signaling network and VoIP signaling such as H.323.

DSP role in VoIP applications

TDM-IP Media Gateway

A communications system like a PSTN to IP Media Gateway typically consists of five types of resources interconnected by internal backplane buses for transport of data and control signals:

1. A set of line cards to interface to the PSTN
2. Resource cards, typically loaded with DSP integrated circuits for specialized voice, fax, and modem processing
3. Network processor card with a general-purpose CPU and associated peripherals (as in Figure 4 and 5 but also with Ethernet connectivity; ATM, TDM; and DSP cards are separate)
4. Network interface cards for interfacing to the backbone packet network
5. Separate application processor(s), sometimes known as Host processors, in case a single CPU performing the network and application functions is not sufficient

The DSP subsystem in Media Gateway application (that can be seen in Figure 6 as complete system implemented using ATCA specifications as a Carrier Blade with four AMC cards,

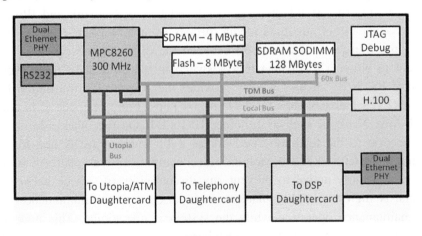

Figure 4:
Host Processor Baseboard (PDK).

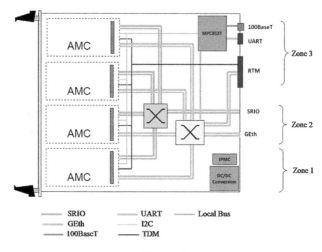

Figure 5:
Torridon 2, a low cost PTMC/AMC carrier

Figure 6:

from left to right, MPC8548 network processing card, MPC8641 host processing card, MSC8122 DSP processing card and T3 PSTN card) typically performs the data translation process for multiple conversations, or channels. Every conversation is full duplex, so every channel has both egress and ingress data to process (the data must be encoded from A to B, and decoded from B to A).

The complete voice processing of a full-duplex channel contains multiple signal processing and administrative tasks including speech encoding and decoding, voice compression, network echo cancellation, signaling, various telephony functions like dual tone multi-frequency (DTMF) generation and detection, and housekeeping functions (Figure 7). The type of speech coding algorithms employed will be dependent upon the application of the gateway system. For example, a wireless Media Gateway that interfaces the Universal Mobile Telephony System (UMTS) wireless system to a PSTN will include the Adaptive

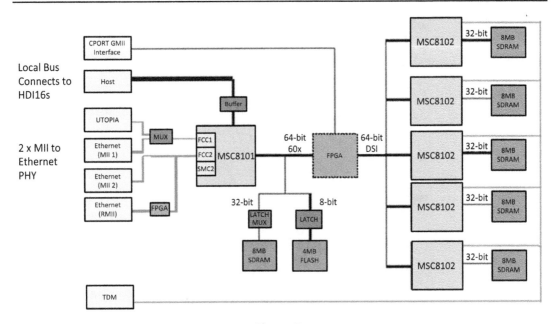

Figure 7:
DSP Daughter Card (8102PFC).

Multi-Rate (AMR) speech coder in addition to others required by the UMTS standard. Some channels may also include fax connections. Fax capability is generally associated with voice, but fax is only a small portion of the overall telephony traffic, being implemented by usage of the T.38/T.30 fax relay and associated data pumps.

A typical gateway will include a DSP subsystem, often consisting of multiple boards with multiple DSP processors, each one of which can handle multiple voice channels. The collective DSP subsystem is often called a DSP farm. When translating from the PSTN to the packet/cell network, the DSP processor receives 64 kbps data streams and puts out packets or ATM cells to the respective packet interface modules. The DSP performs the opposite processing steps when going from packet to PSTN. The DSP processors are usually controlled by a host processor or controller that is responsible for configuration and software download of the appropriate voice processing algorithms to the DSPs for load balancing between processors and in other network management functions.

Because the DSP performs the speech or fax processing required between these networks, it is assumed that the DSP I/O would support at least one packet interface and at least one Time Division Multiplex (TDM) interface that can support direct interfacing to Time Slot Interchange (TSI) devices.

The DSP farm cards ('farm' became a technical term for a system aggregating multiple sources of computational power) represented in Figures 7 and 8 concentrate the physical links

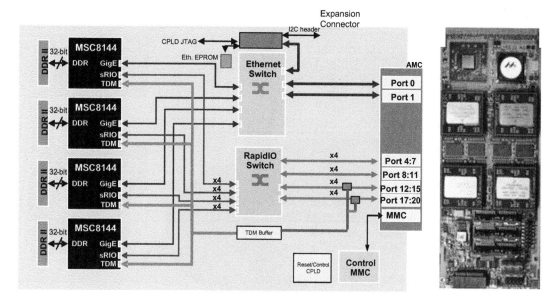

Figure 8:
MSC8144AMC-SA daughter card.

of the packet domain through a single focal point. This can be either a switch (Ethernet or RapidIO), or a CPU specialized in network processing. A network processor offers greater flexibility in configuring the data paths, assigning forwarding rules, offloading part of the network stack (e.g., the verification of IP header and UDP checksum of the incoming packets or the generation of the outgoing traffic), or even implementing system level features (e.g., Lawful Intercept, which is the ability to save the signals sent by all parties involved in a conversation that is tapped by legally authorized agencies).

A DSP Framework for the VoIP Applications

The DSP Framework provides comprehensive call processing engines covering the most dominant aspects of the modern telephony communications: voice calls, data communications (i.e., fax and modem). Auxiliary functionality for managing the signaling services is offered as well, compliant with the selected standards. This is shown in Figure 9.

The primary goal of the DSP Framework is creating a DSP enablement solution for the DSP Farm platform in support of the media-processing engine of the Media Gateway and IP-PBX type of products. The requirements for these two applications are a combination of the standardized features that customers expect (namely the list of capabilities supported, like codecs, tones), as well as the inherent collaborations with the other telecom applications (i.e., external systems part of, or interacting with, the telecommunication

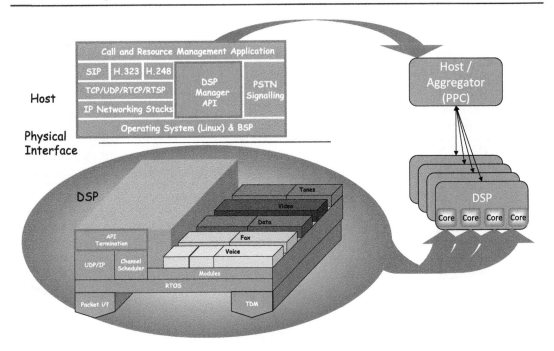

Figure 9:
High level architecture of VoIP Framework.

infrastructure in order to improve end-user experience). For example, a Customer Relationship Management application (CRM) might be connected to the telecom enterprise infrastructure to allow faster access to related information or to the same user's information communicated by phone. While some of these interactions with external actors use standardized structural mechanisms (known protocol formats allowing interoperability), a considerable number of them are driven by proprietary standards (e.g., IVR or voice mail with IP or TDM connections). Clearly, supporting these implementations is a key differentiation on the market.

DSP centric architectural details of a Media Gateway

High level architecture of a Media Gateway and the DSP VoIP application

The more media processing was needed, the more the DSP grew its role in the telephony infrastructure. It started to be used mainly for echo cancellation in long-haul telephone calls, as the longer the propagation delay through PSTN in the case of long distance or international calls, the more noticeable the echo at the far end speaker of its own speech was. The main cause of echo in telephone networks is the imperfection of certain network interface elements, called hybrid circuits, situated between four-wire and two-wire connections

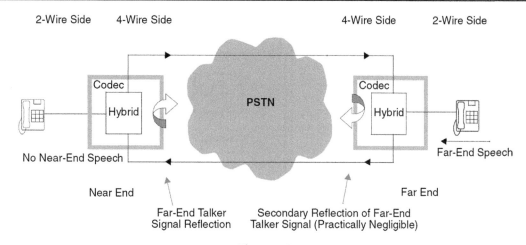

Figure 10:
Echo source.

(Figure 10). A hybrid circuit converts a four-wire physical medium (with two separate signal paths, one for the transmit direction and another one for the receive direction) into a two-wire physical connection that provides an electrical link for signals traveling in both directions at the same time[1].

After echo canceling, the DSP performed tone detection and generation and then other functionalities, so that its software codebase grew to contain multiple packages supporting different features.

Figure 11 shows a specific partitioning of DSP framework software. First of all, the partitioning takes into account the central API (DSPFWAPI) used by the Host Control Application running on the Application Processor (or Host). As in any API allowing remote control, there is a layer implemented on the CPU which is responsible for realizing the communication protocol and a resident component on the DSP that responds to queries and commands. The DSP image that contains it can also be structured in many forms. A factor to be taken into account is the containment of the features. The diagram represents three types:

- Component off the Shelf (COTS) is a convenient partition based on the assumption that its features are sufficiently well standardized or commonly available on the market to be purchased and integrated in the solution. Many providers offer these in binary or source form. Secondly, their performance at this standalone level gives a good metric to compare competing platforms. For example, one could determine the number of DSP cycles

---

[1] Reference: http://www.freescale.com/files/dsp/doc/app_note/AN2598.pdf, "Network Echo Cancellers and Freescale Solutions Using the StarCore™ SC140 Core" by Roman A. Dyba, Perry P. He, and Lúcio F. C. Pessoa

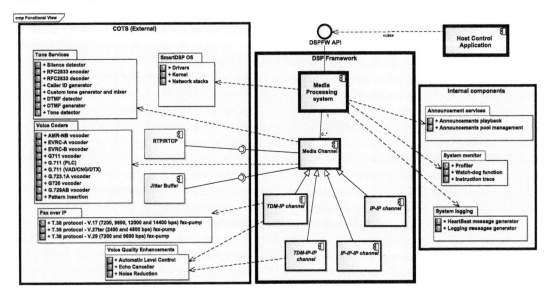

Figure 11:
Software architecture of VoIP framework.

required to process a codec frame and combine that with the core frequency, number of cores, and the chip cost to establish a first round of rankings.

• Media Channel is a software construct within the DSP framework which implements the control and data flow of the channel. Using the figure below (Figure 12) as a tentative data flow representation through a bidirectional TDM-IP voice channel, the software construct is represented by the integration code. Not only parameters and results handling, but also some business logic integrated into the decision flow are based on those results. As an example, the tone generation could be commanded to override or to mix with the current samples. As a result of this system level configuration, an additional mixer may be or not present in the data flow.

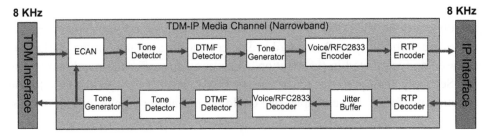

Figure 12:
Fundamentals of TDM-IP channel.

- Internal components/functions are part of the DSP framework as built-in features. They are tightly integrated within the Media Channel if they are associated with and local to a channel or with the Media Processing System in general, and they mostly follow proprietary conventions and formats.

System Software Functionalities

A high level representation of the dynamics of the Media Processing System is presented below (Figure 13). The main types of activities or services that should be developed starts with SoC boot customization. Every SoC has a ROM boot program that has configuration points which are modified to fit the needs. It will include options to load the actual application

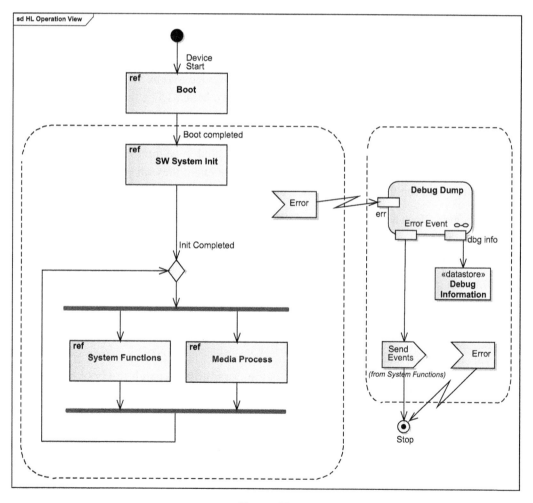

Figure 13:
High level operational view.

from an external support (either through a network link or from a storage device). Once the application finishes initializing the system and the software, it fundamentally runs in parallel two types of services: the system level services and the media processing. The 'parallel' in this context refers to what is possible based on the underlying support of the RTOS on which the application is based. It could mean 'preemptive' or 'cooperative' multitasking.

Sometimes neglected from the design process or specifications, the error handling to facilitate in-field diagnosis and possible on-the-fly recovery is of utmost importance. Placing such an area in the high level design diagrams from the start, forces the design team to investigate the error handling capabilities of the SoC right from the sourcing stage. It is rare though that in today's Request for Quotation forms that a potential client looking for sourcing DSP SoCs forwards to the suppliers on the market, room is made for requirements related to an on-chip hardware diagnosing subsystem that would allow for monitoring the bus errors or bottlenecks, triggering interrupts based on configurable thresholds, or maintaining on-the-fly statistics for the interconnect fabrics within SoC.

The boot process (Figure 14) is specific to each DSP device and as such the software initialization is tailored accordingly. For the Freescale's MSC8157[2], the boot program initializes the SoC after it completes a reset sequence. The MSC8157 can boot from an external host through the RapidIO interface or download a user boot program through the I2C, SPI, or Ethernet ports. The default boot code is located in an internal 96 KB ROM and is accessible to all cores. When cores finish the reset sequence, they all jump to the ROM starting address and run the boot code; then after that, they all jump to a user-defined address.

The boot program does not configure the DDR controller and this must be done by the user's firmware. In most cases, external memory is connected to the SoC, so the application must first configure the DDR controller to support the type of DDR memory in the system.

The ROM boot code mainly initializes the interrupt controller and the Vector Base Address that will point to the interrupt service routines. Finally, it sets up the microprocessor to execute the application code, enabling the core's caches, setting the stack pointer and jumping to the code. Depending on further configuration options, it will enable various peripherals needed to load the subsequent application code (I2C, Rapid I/O, or Ethernet controller). An interesting option available is the possibility to patch the boot code with user's code residing in I2C EEPROM. One can use this extension to save hardware diagnosing code that is run at start-up: verifying for example the external DDR memory, or the Ethernet connectivity.

The MSC8157 device can load files through the Ethernet port using DHCP (Dynamic Host Configuration Protocol) and TFTP (Trivial File Transfer Protocol) based on IPv4, as

---

[2] Freescale Semiconductor, MSC8157 Reference Manual.

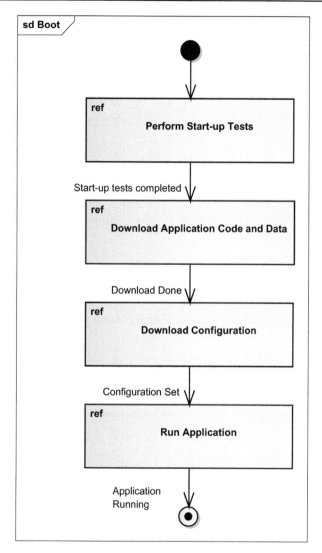

Figure 14:
Boot process.

represented in Figure 15. The MAC address of the SoC that is used in the first broadcast message DHCP DISCOVER of the client to find the server can be set in I2C EEPROM. Alternatively, it can be derived from Reset Configuration Word, in a combination of external pins that are sampled at power-on reset. Once the application code and data (including static configuration data that customizes the operation mode) are downloaded in the internal and external memory, the execution can start. As the final step in the boot process, all cores jump to the address written in 0xC000_0010 in M3 memory which should be defined in the boot image file.

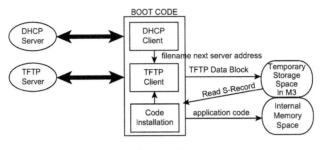

Figure 15:
DHCP boot.

Figure 16 above contains the typical system services required to be implemented by a communication system:

- Watchdog service for error detection and recovery
- System time keeping and transmission of various small metrics using the heartbeat message towards a remote control point
- Command handling received from the external control point and sending responses

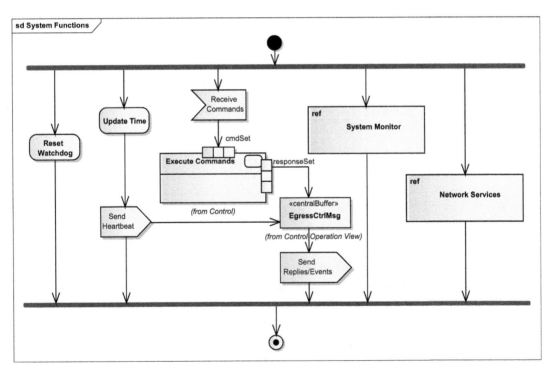

Figure 16:
System level functions.

- System monitoring function that gathers various statistics, for example the buffer status (how full or empty they are in order to signal potential overflow or underrun situations), the average processor load, the average task execution time, the number of real-time deadline misses in the task scheduler
- Network services include a selection of TCP/IP protocols (most common, IP, UDP, ICMP) as well as SNMP for network monitoring and management

A watchdog timer is a hardware counter that helps detecting and recovering from software failures. At the start of the application, the watchdog is initialized to count down a specified number of clock cycles and also is configured to a specific action needed to be executed upon expiration of the period, when the counter reaches zero (on MSC8157, the action can be either resetting the device, or generating a Machine Check interrupt to a specific core). It is frequent in various platforms that once started a watchdog can no longer be deactivated by the software, accidently or intentionally. The application services the watchdog by only resetting the counter to the initial value and making sure to return in time before the next expiration to reset it. Resetting the counter should be a difficult operation to be performed accidently, because resetting it means that the system is operating normally. A first step to making the operation more reliable is to fully decode the memory address of the access (considering the case where the address registers are wider than the SoC system bus, thus allowing accidental writes to watchdog location although the more significant address bits indicate an illegal space). An even powerful mechanism is to have the reset operation performed by back to back write operations. To be noted is that in the case of multicore devices there should be one watchdog timer per core to allow independent servicing by each core. One obvious reason is that each core runs different tasks and as such the timing requirements for error detection are different.

An example of a complex watchdog servicing routine is taken from the MSC8157 device. The software watchdog service sequence consists of two steps:

1. Write 0x556C to the System Watchdog Service Register
2. Write 0xAA39 to the System Watchdog Service Register

In this particular case, there can be any number of cycles passed between the two writes. This allows servicing the interrupts between the two write operations. Another approach used in devices from other semiconductor manufacturers imposes the condition that the writes must be back-to-back. This timing constraint requires that the two writes be protected from interrupts occurring between them. Proponents of the first approach argue that having interrupts disabled during the two write operations could cost tens, maybe couple of hundreds of cycles, depending on the bus hierarchy and the load conditions, reducing the system responsiveness to real time events. With or without interrupts disabled during the servicing sequence, the MSC8157 state machine is presented in Figure 17.

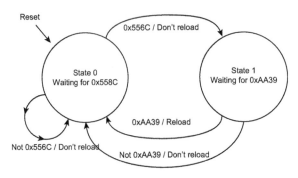

Figure 17:
Watchdog state

Another important aspect of watchdog operation is the identification of the cause of reset. It is intended that the application be able to identify the cause of reset when it restarts. At a minimum, it should determine if the start-up is a normal procedure, or it comes after watchdog expiration. A solution for this situation is provided by MSC8157 through a reset persistent register called Reset Status Register (RSR) that captures various reset events in the device. The fields in the register are sticky and can be cleared by writing 1. If not cleared, the events accumulate throughout multiple reset situations, except power-on-reset which restores the default value of the register. The RSR stores indications for events like:

- The identifier of the expired software watchdog timer that caused a reset (there are 8 counters)
- The identifier of the RapidI/O interface that received a reset event
- Indication whether there was a software hard reset, etc.

Configuring the duration of the watchdog requires a careful analysis of computation paths of the system. Setting a very relaxed watchdog (like a maximum 8.59 sec for a 500 MHz input clock for MSC8157) might not be sufficiently fast for the CPU to come back online and monitor the system, which might be dangerous for the appliance. Setting a too tight timeout might cause wrong expirations, or will force the application developers to reset the watchdog in many places in the code, causing less maintainability for the code and increasing the chance of missing a path that will later cause wrong timeouts. It is common to reset the watchdog in the background task loop and the duration of the interval must be larger than the duration of the biggest task plus the total of all interrupt service routines that could interrupt it.

The control traffic, represented in the System Functions figure, presents a design challenge: how to keep the system controllable, i.e., responsive to the interactions with the Host Control Application within the timing constraints, when the media traffic carrying RTP audio is significantly much larger and irregular (bursty) in its nature. It is not uncommon to have the control traffic exchanged through a different medium than the media traffic. We encounter both cases: same type of medium but different physical ports (e.g., two different Ethernet ports), or different types of medium (e.g., media carried through Ethernet or Serial RapidI/O, and control through PCI).

Often, because of simplicity of the hardware design, the same type of medium is preferred. Once the traffic arrives in the chip through a port, the platform must allow prioritization and queue separation between the two flows, control and media. A simple example of configurability can be given from MSC8122 DSP, a four core device. It includes an Ethernet controller that performs frame recognition in two ways: pattern matching or destination address. A frame can be rejected or accepted on the basis of the outcome of destination address filtering, pattern matching filtering, or both. Filtering of the received frames can be done based on any of the first 256 bytes of the frame and the frame is directed towards any of the four RX queues configurable. Being only four queues, the Symmetric Multiprocessing (SMP) model is harder to employ. Considering each core functionally identical, performing the same kind of tasks, it means each core will be identifiable and controllable by the Host Control Application. With only four incoming RX queues (that in the SMP model are allocated one per core), it implies that both media and control traffic will land in the same queue, potentially causing bottlenecks. The software design will be more cautious and will allocate more polling time for the control path aiming at checking with higher priority through the RX queue and pulling the control packets out of order, before the media packets, even though they arrive later. However, in the case of large flooding of media packets, the queues may have a hard time keeping the data, considering that the consumption rate is dictated by the TDM synchronous rate (in a TDM-IP channel). Consequently, with a particular queue full, the control traffic will be dropped, together with other media packets, by the Ethernet controller at the port level. Software architecture to work-around this potential problem is possible (e.g., dedicate RX queue for control on one core which can also execute media processing tasks but with packets taken from one or any of the other three queues left). But ideally the hardware platform should make the software simpler to design.

The subsequent generation of devices, MSC8144, another four core device, improves the configurability of the RX queues. By adding a more powerful QUICC Engine accelerator for the network traffic, the UCC Ethernet Controller (UEC) is capable of handling more queues, being increased to eight RX and TX, each with its own interrupt generation capabilities.

TDM to IP processing path in a Media Gateway

This section presents a functional diagram in Figure 18 of the TDM to IP processing path, with more concrete details about the engineering solutions used for integrating the DSP signal processing components into a field proven application for VoIP Media Gateway. More details

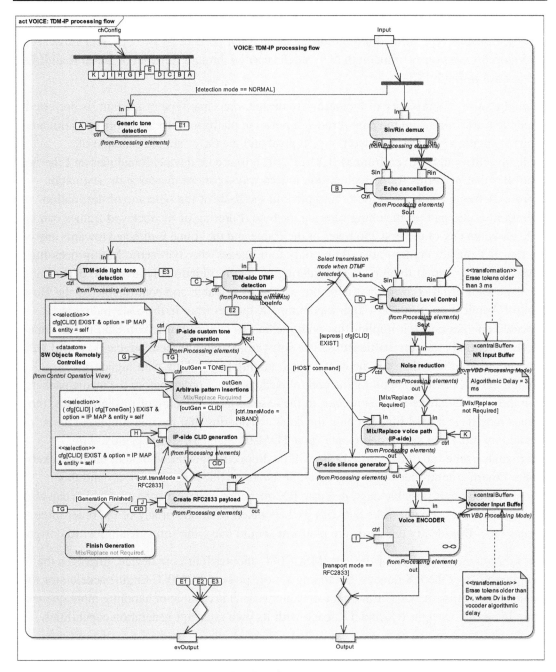

Figure 18:
Detailed design of TDM to IP path.

will be explored that cannot be depicted by a high level diagram such as Figure 12 'Fundamentals of TDM-IP channels,' above.

In the top left corner of the Figure 8 'Detailed design of TDM to IP path' diagram, an element called chConfig is represented. This is the database of all the channel's configurations (more than one hundred parameters) as well as its persistent data. Typically, a channel, like any software, needs three types of memory: stack memory, scratch memory (used temporarily for only one execution of the channel and does not need to be restored for the next execution), and persistent data (memory that holds critical state information and data processed by the channel in one execution and needed later in subsequent ones). This persistent data holds not only the inner computations of the software components seen as black boxes, but also the configuration parameters from the Host Control Application. For example, when the application configures the Voice Activity Detector (number of reflection coefficients = 10), that parameter ends up in the persistent data of the channel. The persistent memory is highly critical to manage because it may become a constraint of the system performance. Earlier DSPs did not have data cache. They relied on large internal M2 memory and assumed the software will bring using DMA the persistent data in M2 before executing the channel and swapping out also with DMA the data to a larger but slower memory off chip. Even in the newer DSPs with data cache, the persistent memory is often accessed throughout the execution of the channel (as it can be seen from all the labels from A to K that start from chConfig and link to an individual software component). Such random and frequent access can degrade the performance of the data cache.

The input of the TDM-IP channel is a TDM frame. In a synchronous system such as TDM-IP Media Gateway, the system is functioning at a clock that is based on the types of channels that will be required to run. For example, a G.729 codec can process 10 ms of PCM input. G.723.1A uses 30 ms frames. A common choice would be to configure the TDM system to gather 10 ms buffers and trigger an interrupt when buffers are full. That will represent the tick of the system. But if we add into the equation the G.711 channel which uses a codec that works on PCM samples, not on frames, then the tick period will decrease. It is common among equipments on the market that the G.711 channel will not be configured for less than a 5 ms packetization rate and will also be multiples of 5 ms (even though RTP standard allows for finer granularity). As such, based on these assumptions only, the tick of the system will be generated by the TDM subsystem interrupt after filling 5 ms worth of buffers. That is a reasonable compromise but will not be acceptable in premium quality applications. It is clear that the TDM subsystem will wait for 5 ms (starting from the moment the samples entered the SoC through the TDM serial line). This means adding 5 ms to the overall delay path from TDM to the far end receiver. Consider a complete path made of a TDM-IP gateway, through the IP cloud and then back to an IP-TDM gateway to an end user. The whole path is subjective to large delays (mid to high tens to hundreds of ms). A premium quality application will reach the lower end of the range and as such the two 5 ms buffering delays that are generated by our design decision in selection of the

TDM buffer length will cause concern. We conclude that the high quality VoIP Media Gateways work on 1 ms tick, thus with TDM buffering of 1 ms.

An interesting point to note in the diagram from Figure 17 is the many detection related components used in the channel. One is called Generic Tone Detection and it uses the signal directly from TDM. The others are TDM-side light tone detection and TDM-side DTMF detection. The first one is a superset of the signals that need to be detected on the TDM side and adjust the channel parameters. The light tone detection is optimized for signals generated by the fax/modems compliant with ITU-T V.8 using ANS (or ANS-AM) during handshake. The third is a DTMF detector to relay the digits pressed on the telephone in messages and relay to the far end over RTP using special packets RFC4733/RFC2833. As such, the audio signal that is identified as DTMF key is extracted (i.e., silence is generated to replace the original signal) from the audio media that is encoded using the codec and conveyed using notifications of start and stop using special packets. But, returning to the generic tone detection versus fax/modem tone detection, they allow flexibility in choosing the suitable algorithm for the necessary application. For example, a demanding VoIP Media Gateway application will require seamless transition for its customers on the copper wires to the new technology in the core network (IP based). Often households contain security devices (to counter burglars or fire/water accidents) that are connected to the telephone line with a monitoring center. Similarly, older ATM machines connected to the bank network through a secure telephone line. In some regions of the world, for telecom networks there are special regulations aimed at supporting the equipment used by people with disabilities (hearing or vision). All these three examples have in common, devices that use a variety of signals (many preserved for backwards compatibility and legacy interoperability) that need to be detected on the TDM side. This operation is more expensive in terms of processing power so if the application use case does not require it, it should be replaced with the more efficient standard fax/modem detection (note that there is also a large variety and legacy in the fax/modem world, but V.8 ANS-PR detection provides a reasonable but not complete coverage).

One of the parallel paths that the input signal takes is the Sin/Rin demultiplexing block. The purpose of this block is to provide the reference signal which the echo canceller searches for in the Sin input signal. This reference signal is one that the channel transmits in the opposite way, from IP to TDM, and as a consequence of the reflections of the output electric signal into the line hybrids converting from 4 to 2 wires (as described in Figure 9, 'Echo source'), there will be a distorted, attenuated instance of it returning from the hybrid into the incoming signal Sin. The purpose of this demultiplexing block is to provide timely access to the necessary Rin data from the TDM output buffers. After the Sin/Rin block, the input signal goes into three DSP components dealing with voice quality enhancements: the echo canceller, the automatic level control, and the noise reducer. These functions will be disabled if the tone detectors will detect transitions to other states like Voice Band Data or T.38 fax relay.

Several comments in the diagram also point to algorithmic delay that Noise Reduction and the Voice Encoder add to the channel. Two <<centralBuffer>> objects are used to depict circular buffers that are used to dynamically adjust the delay through the channel in the event of the reconfiguration of the structure. For example, when a command is received from the Host to disable the Noise Reduction, considering that it has 3 ms delay, in the next frame that will be processed without it, the output will suffer an audio distortion because of the missing 3 ms that were stored inside the Noise Reduction persistent memory and which were discarded. More details about the compensation mechanism are presented in the section, 'Support for Legacy Equipments.'

DSP VoIP Framework differentiators

The functionalities required by a VoIP Media Gateway, represented in the diagram 'Functional View', from Figure 10, 'The Software Architecture of VoIP framework'. are numerous and their implementation is open to innovation. Although many are standardized by ITU-T and IETF bodies, major industry players in both telecom and semiconductor industries play a significant role in committees as contributors as well as patent owners. The subsequent sections of this Appendix discuss two aspects in relation to engineering details.

The first is the product of long experience with testing the behavior of a variety of fax/modem equipment (either common end-user products or embedded in communication systems, like security systems or special communication equipment) over copper lines and VoIP migration of Access Gateways. The challenge was to reduce the distortions caused by VoIP signal processing such that the end-to-end call reliability is comparable with the case of tests over copper lines.

The second area for innovation and quality differentiation is DTMF detection and relay. The particular challenge is to have a DTMF key as short as 40 ms, for example, detected as fast as possible, extracted from the media stream with as little signal leakage as possible and reconstructed at the far end as accurately as possible.

Support for legacy equipment

In VoIP Media Gateways, the PSTN media is converted into RTP packets sent on the IP network. From the PSTN side (with its digital core, converts signal to digital G.711 64 kbps compression format from the analogue signal transported to the customer premises), the audio samples are gathered in frames and are encoded using various algorithms with higher or lower bandwidth usage. If the telephone call conveyed data transmission, dedicated detectors were used for monitoring the line and taking appropriate actions with the view of disabling as fast as possible the non-linear signal processing that might affect the frequency range of the communication. If a known signal characteristic for a data transmission is detected in the audio path, the parameters of the encoding path originally configured to process voice characteristics, change in order to avoid the distortion of the data (also known as voice-band data VBD mode). Most of the encoding schemes used for voice signals introduce 'algorithmic delays' in the audio stream. The different delays employed by various voice codecs initially active on the line when it was open, have to be handled properly when transitioning from voice to voice-band data mode because a signal loss causes a phase shift distortion on the original data signal. The most used codec for VBD mode is G.711 and it does not have algorithmic delay, thus when switching from voice to VBD mode, the encoding path suffers a delay variation. A supplemental configurable buffering mechanism is used to save audio frames to be consumed when switching from voice to VBD mode such that the overall processing delay from PSTN to IP path remains constant.

Phase distortions

The packet-switching technology was used to replace various network areas from the PSTN. Although initially the new media transport mechanism spread throughout the core of the network, mainly between digital, high traffic, telephone exchanges (e.g. Class 4 telephone switches), it was later deployed more aggressively towards the edge of the network. In some cases, it reached the households, but typical it stopped at the 'last mile', meaning that it reached the neighborhood of a residential area and from there to the phones in the home the line was preserved as analog two wire copper line. The PSTN could be depicted as a mixed map of areas of one of these two packet and circuit-switching technologies. Specific equipment to convert between the circuit-switched and packet switched domains is needed and this function is called Media Gateway. The Media Gateway function included signal processing procedures for transmission on the IP side which had to handle traffic characteristics: reducing bandwidth and taking into account the transit jitter. These functions were not compatible with data transmission devices and methods to change their parameters were derived to improve the data transmission as much as possible.

There are specifications for the attributes of the data VoIP link and for the events to be detected and to trigger the change of the characteristics of the path from voice to data (e.g., ITU-T V.152). The standards cover the parameters needed on VoIP path in order to support data transmission. The specific measures that are taken to switch between the 'voice' and VBD modes, are left implementation specific, being a differentiating quality aspect. One aspect that plays an important role is the algorithmic delay. The different delays employed by various voice codecs initially active on the line when it was open, have to be handled properly when transitioning to voice-band data mode because the signal loss causes a phase shift distortion on the original data signal. Occurrence of a phase shift distortion in certain data tones has a special meaning in telecom. For example, ITU-T G.168 states that a phase reversal in a 2100 Hz tone is an indication for disabling the Echo Canceller (ECAN) on the path, while a 2100 Hz without phase reversal keeps the ECAN active but without certain features in it.

The Media Gateway conveying data transmission over packet networks should avoid artificially introducing phase shift distortions as these degrade the quality of the data transmission.

It is common that the G.729AB codec is used as the initial voice codec. The algorithmic delay of this codec is 5 ms. In Figure 19, three probes are located at points A, B, and C. A is the output of the 'PCM media processing' phase, B is the input into the next phase, 'PCM encoding' and C is its output. The A and B signal lines are obviously identical. The C point represents the output frames resulting from the encoding phase being delayed compared to the input B line.

The Figure 20 shows a positive identification of a 2100 Hz tone. An ordinary implementation of a voice-to-VBD path switch would simply change the current voice codec (i.e., G.729AB) to the one used by VBD (i.e., G.711). In this simple implementation, a sub-frame of 5 ms was about to be encoded by the G.729AB and transmitted in the next 10 ms frame; however, because of the codec change, it will be discarded. This media loss represented by the 5 ms sub-frame has a different impact, depending on the frequency of the signal on which the loss occurred. In this example of a 2100 Hz tone, 5 ms is a multiple of the signal period plus a half a period. By losing 5 ms from the signal, the two segments that will be joined will be in anti-phase. This particular codec change on the 2100 Hz tone caused a 180 degrees phase shift (see the second frequency domain representation from Figure 20). This 180 degrees phase shift is an important event in the domain of data transmission. It was agreed in standards such

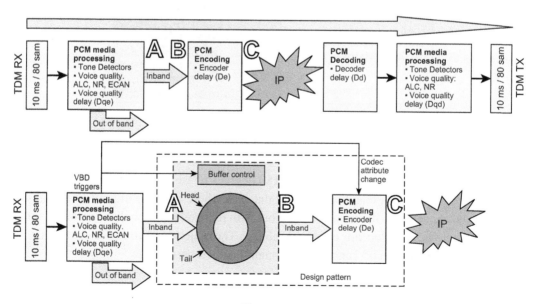

Figure 19:
Typical processing path PSTN-IP-PSTN.

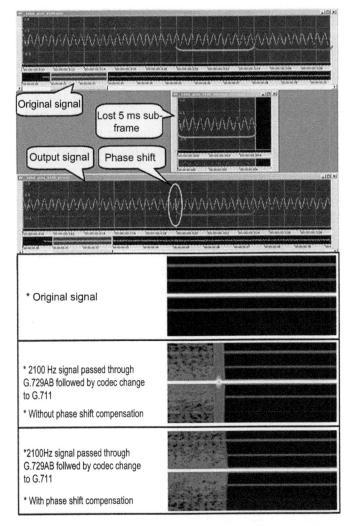

Figure 20:
Sub-frame loss when changing codecs from voice to VBD mode.

as ITU-T G.168, that a tone of 2100 Hz with 180 degrees phase shifts will be used to disable the echo cancellers along a telephone line. In the case where the tone did not show phase shifts, the echo cancellers will remain enabled but with reduced functionalities. A broad range of data equipment heavily depends on this capability. The majority of faxes (i.e., those that are part of Group 3) need an echo canceller to remain enabled on the line.

Delay Compensation Mechanism

In order to compensate for the delay variation and media loss caused by changing processing parameters in the VoIP channel, a software design method is employed which combines a media processing component in a VoIP path with a delay compensation mechanism in the form of a multi-purpose buffer (see diagram at the bottom of Figure 19) in order to serve as a supply source for signal when the processing component is disabled or its delay is reduced. The goals of this mechanism are:

- Keeping constant the algorithmic delay of a channel in order to preserve the accuracy of a potential data transmission conveyed through the channel.
- Reducing the algorithmic delay of a channel to the minimum possible by eliminating the delay compensation mechanism when a media loss is possible. Two examples are:
  Eliminate the buffering mechanism when the processing path has not detected a voice-band data communication and the signal power is low
  Eliminate the buffering mechanism when the voice band data communication ends

Every 'media processing element' in a VoIP channel that has algorithmic delay contribution must be accompanied by a delay compensation mechanism for the cases in which the element state changes (enabled, disabled, or reconfigured) in order to keep the delay constant through the channel (see the observation points A, B, and C in Figure 19).

The compensation mechanism for the media loss when changing from a delay De ms to 0 ms of a 'media processing element' is characterized by:

- Queue to storage of the incoming samples on the 'media processing element' that has a length capable of storing De ms of samples.
- 'Store' operation mode when P ms of input samples (in the figure P = 10 ms) are pushed in the queue, and the oldest samples are discarded, in order to keep only the newest De ms of samples in the queue.
- 'Compensate' operation mode when P ms of input samples are pushed in the queue and P ms of the oldest samples are retrieved from the queue, where the delay of each sample is De ms.

As a first example from Figure 21, we should assume a switch from codec G.723.1 (frame size Pa = 30 ms, algorithmic delay DeA = 7.5 ms) to G.711 (frame size Pb = 10 ms processing 80 samples at once, algorithmic delay DeB = 0 ms). In the initial state, the 30 ms frame represented by 6 pieces of 5 ms each, moves through the compensation mechanisms without delay; the last 7.5 ms (that is half of sub-frame 5 and sub-frame 6) are saved in the storage. At the probe point C, the encoder will output an encoded frame delayed by 7.5 ms; thus these samples will be saved in the internal channel data of the encoder to be included in the next encoded frame. At the next 30 ms frame, the currently saved 7.5 ms of samples (5 and 6 sub-frames) will be replaced by half of the sub-frame 11 and the sub-frame 12. At the moment of the switch from voice to VBD mode, the mechanism is used in compensation mode, meaning that it becomes a delay line of 7.5 ms which maintains constant the overall delay path across the encoding channel.

A more complicated example illustrates the usage of such mechanism in the case of switching complex codecs and preserving the integrity of the media signal. The case is represented in the

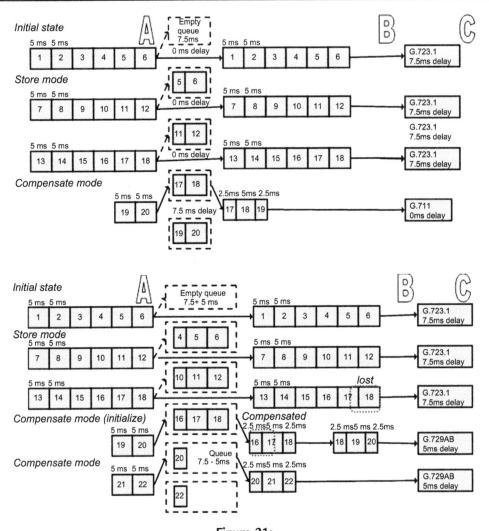

Figure 21:
Compensation mechanism in two examples.

bottom half of the Figure 20 'Compensation mechanism in two examples', and presents the switch from codec G.723.1 (frame size Pa = 30 ms, algorithmic delay DeA = 7.5 ms) to G.729AB (frame size Pb = 10 ms, algorithmic delay DeB = 5 ms). Unlike the previous example, the subsequent codec to which the switch is made does not have 0 ms algorithmic delay. The initial storage space for the compensation mechanism has to support DeA+DeB (i.e., 7.5+5 ms). After the codec switch has occurred, the compensation mechanism is able to provide enough elements to compensate for both the samples lost in the look-ahead buffer of the previous codec as well as the initial delay introduced by the new codec. It should be noted that the last half of sub-frame 16, sub-frame 17, and the first half of sub-frame 18 had been sent already to codec A. But, as it is highlighted in a dotted rectangle pointed by 'lost' label, those 7.5 ms were lost in the codec A history. At the first utilization of codec B, G.729AB, because of its algorithmic delay of 5 ms, this will encode 5 ms of silence added to the samples highlighted in a dotted rectangle indicated by 'compensated' label half of sub-frame 16, and half of sub-frame 17. The last half of sub-frame 17 and first half of sub-frame 18 will be preserved in the codec history. The first encoded frame by the new codec B (i.e., G.729AB in this case which takes 10 ms input signal) has a special purpose. It results from processing the uncompressed signal with a 10 ms length which is contained in the sub-frames: last 2.5 ms from 16, the entire 5 ms from 17 and the first half 2.5 ms from 18. The output of the encoding is not actually used to be transmitted to the destination but it is discarded. The encoding operation was performed only to allow G.729AB to properly initialize its internal history (i.e., the look-ahead buffer) in order to minimize the discontinuities caused by loosing fragments of the signal in the history of the previous codec A (i.e., G.723.1). Only the subsequent 10 ms samples, half of sub-frame 18 (2.5 ms), sub-frame 19 (5 ms), and first half of sub-frame 20 (2.5 ms) will be encoded and the resulted frame will be sent forward. After this phase, the storage space of the compensation mechanism will be reduced to DeA − DeB (i.e., 7.5 − 5 ms), such that the overall algorithmic delay in the encoding path remains as before.

One performance requirement of VoIP media gateways is minimal voice delay. While the buffering algorithm improves the quality of VBD transmissions, it may cause artificial delay increases for voice communications. The algorithm described below handles cases of 'voice-to-voice' codec changes, by implementing a time-out mechanism to drop the delay introduced by the buffering algorithm:

```
1. Start (buffering algorithm is active)
2. Initialize buffering time-out counter with 0
3. Process frame
4. If a switch to VBD mode has been detected
     a. While in VBD mode, keep the buffering algorithm active. The buffering is
     dropped when a switch back to voice is detected.
     b. End
5. If the current frame energy is above certain threshold reset time-out counter
6. If the current frame energy is below certain threshold increment time-out counter
     a. If buffering time-out counter reached a certain value, that is, a certain
     amount of time has passed with the signal energy below the threshold, drop the
     buffering.
     b. End
7. Repeat, starting from step 3
8. End
```

DSP VoIP framework differentiators

DTMF detection

Preserving backward compatibility and interfacing with PSTN services is one of the major requirements for any VoIP service provider. Every VoIP solution must be able to transparently provide connectivity, and call reliability and quality across a connection established between a PSTN subscriber and a VoIP user. DTMF detection and transport over packet based networks is part of this effort and the next paragraph explains how these three conditions can be achieved.

Sections

§1 DTMF specifications; Q.23 and Q.24 recommendations

§2 DTMF detection algorithm example

DTMF specifications; Q.23 and Q.24 recommendations

DTMF specifications DTMFs (Dual Tone Multi Frequency) represent signals that are sent through the analog telephone line composed of two frequencies with the purpose of exchanging information between the user equipment and the telephone exchange. Keypads from our day to day telephones are being used to generate over the line these kinds of audible signals which encode for example the phone number of the person we want to talk to. They were introduced in 1963 as a replacement for the old style of dialing — the pulse dialing — used mainly on telephones with rotary dial and analog exchanges. Soon, DTMFs became the most widely used method of interaction between the terminals (phones) and services provided by the switching centers.

ITU-T Q.23 recommendation A total of 16 DTMFs are defined by ITU-T recommendation Q.23 and each of these signals is built by combining two sets, each set containing 4 frequencies. Theoretically the frequencies are chosen in such a way to enable easy detection. The two frequency groups contain mutually exclusive values (see Figure 22 below).

Generally speaking the DTMF standard was not defined for transmitting data but rather control information; still, it can ensure a maximum of 50 bits/s transfer rate considering that a DTMF is being sent in a period of 80 milliseconds (40 milliseconds of DTMF and 40 milliseconds of pause after the signal), the minimum allowed by the Q.24 ITU-T recommendation and the amount of 4 bits/symbol, in this case the DTMF.

	1209Hz	1336Hz	1477Hz	1633Hz
697Hz	1	2	3	A
770Hz	4	5	6	B
852Hz	7	8	9	C
941Hz	*	0	#	D

Figure 22:
DTMF frequency allocation.

ITU-T Q.24 recommendation The Q.24 recommendation from ITU-T represents a set of physical characteristics that the DTMF signals are required to have at the receive point in order to enable reliable detection and compatibility with the sending equipment. Even if the recommendation is not intended to supersede any existing standard it was quickly accepted by the equipment manufacturers and it is now perceived by the industry as the de-facto standard in most of the exchange applications. Since DTMF detection is part of the receiver, Q.24 contain its main requirements and many compliance tests are being performed to ensure the rules from this recommendation are being followed.

The recommendation's requirements are:

1. No more than one DTMF is allowed at a specific moment; the detector should check for the presence of two or more DTMFs
2. The frequencies of each DTMF should be within specific tolerances
3. Since power level of the signal sent through telephony lines might vary for different frequencies, the difference between the power levels of each DTMF frequency should be within specific tolerances
4. In order to reduce the possibility of signal simulation by speech, DTMFs having the duration less than a specific threshold should be ignored
5. Double registration might occur when a digit is interrupted. To minimize this short-coming interruptions that are smaller than specific thresholds should be ignored and not detected as pauses
6. Signal velocity — the rate for DTMFs followed by pause — considering the two points from above, should have a specific threshold
7. The overall performance of the telephone network should not be perturbed by the signal simulation by speech (false detections on speech signals)
8. The DTMF reception should be immune to dial tones
9. The DTMF reception from long 4-wire transmissions should not be perturbed by echoes of milliseconds and the detector should be able to distinguish between the real signal and its echo
10. The DTMF reception should expect the occurrence of line noise and should not be perturbed by it

The recommendation also includes a table containing specific values for the characteristics discussed above from various administrations.

Due to its wide spread in the industry the current work will consider only the AT&T compliance.

Considering Table 1 one may conclude that the recommendation was written in order to define three detection areas: the white, the gray, and the black zone for DTMFs having the characteristics from the table. The white zone defines the value of each Q.24

Table 1: Q.24 table containing DTMF characteristics from various administrations

Parameters		Values				
		NTT	AT&T	Danish Administration [a]	Australian Administration	Brazilian Administration
Signal frequencies	Low Group	697,770,852,941 Hz	same as left column	same as left column	same as left column	same as left column
	High Group	1209,1336,1477,1633 Hz				
Frequency tolerance $|\Delta f|$	Operation	≤ 1.8%	≤ 1.5%	≤ (1.5%+2 Hz)	≤ (1.5%+4 Hz)	≤ 1.8%
	Non-Operation	≥ 3.0%	≥ 3.5%		≥ 7%	≥ 3%
Power levels per frequency	Operation	-3 to -25 dBm	0 to -25 dBm	(A+25) to A dBm	-5 to -27 dBm	-3 to -25 dBm
	Non-Operation	Max. -25 dBm	Max. -55 dBm	Max. (A-9) dBm (A=-27)	Max. -30 dBm	Max. -50 dBm
Power level difference between frequencies		Max. 5 dBm	+4 dB to -8 dB [b]	Max. 6dB	Max. 10 dB	Max. 9 dB
Signal reception duration timing	Signal Operation duration	Min. 40 ms	Min. 40 ms	Min. 40 ms	Min. 40 ms	Min. 40 ms
	Non-Operation	Max. 24 ms	Max. 23 ms	Max. 20 ms	Max. 25 ms	Max. 20 ms
	Pause duration	Min. 30 ms	Min. 40 ms	Min. 40 ms	Min. 70 ms	Min. 30 ms
	Signal interruption	Max. 10 ms [c]	Max. 10 ms	Max. 20 ms	Max. 12 ms	Max. 10 ms
	Signaling velocity	Min. 120 ms/digits	Min. 93 ms/digits	Min. 100 ms/digits	Min. 125 ms/digits	Min. 120 ms/digits
Signal simulation by speech		6 false/46 hours for speech with a level of -15 dBm	For the codes 0-9,1 false/3000 calls For the code 0-9,*,#,1 false/2000 calls For the codes 0-9,*,# A-D, 1 false/1500 calls	46 false/100 hours for speech with a mean level of -12 dBm		5 false/ for speech with a mean level of -13 dBm
Interference by echos			Should tolerate echoes delayed up to 20 ms and at least 10 dB down			

[a] Same characteristics are used by several European Administrations; values of a range from -22 to -30 to suit national conditions.
[b] The high group of frequency power level may be up to 4 dB more or 8 dB less than the low group frequency power level.
[c] For analogue multifrequency push-button receivers only.

parameter (characteristic) for which the component **must** guarantee a correct and reliable detection of the DTMF, the gray zone defines the value of each Q.24 parameter for which the component **might** detect the DTMF; if at least one DTMF parameter is situated in this zone, it is not guaranteed that the component detects the DTMF reliably and correctly, while having a parameter in the black zone the detector **mustn't** detect the current signal as a valid one. Just to give an example, the Q.24 AT&T specifies that signals having a duration of a minimum 40 milliseconds are in the 'Operation' zone, so they must be detected — this represents the white zone. In the meantime the signals that are below 23 milliseconds are in the 'Non-operation' zone, so they mustn't be detected — this represents the black zone. The question is what happens with signals that have a duration of between 23 and 40 milliseconds? Since the recommendation doesn't mention anything about this zone — 'the gray zone' — it was left to the implementer's decision about the detectability of DTMFs having these parameter values. In Table 2 is depicted the set of Q.24 parameters for which the white and gray zones apply.

Q.24 compliant DTMF detection algorithm example
 The design of this software module is based on the fact that any single-frequency tone signal having the general formula:

$$x(n) = A \cos(\Omega n + \phi)$$

is mapped to a constant value via the modified Teager-Keiser operator, as follows:

$$\Psi_k(x(n)) = x^2(n - k) - x(n)x(n - 2k) = A^2 \sin^2(k\Omega)$$

The modified TK energy operator is a special case f Volterra filter, which depends both on magnitude A and the normalized frequency Ω of the tone;

$$\Omega = 2f/f_s,$$

Table 2: Q.24 white and gray zones

Q.24 parameters	White zone	Gray zone
Frequency tolerance	≤ 1.5%	(1.5% .. 3.5%)
Power levels per frequency	0.. -25 dBm0	(-25.. -55) dBm0
Power level difference between tone frequencies	+4 dB to -8 dB	N/A
Signal duration	≥ 40 ms	23 − 40 ms
Signal interruption	≤ 10 ms	N/A
Signaling velocity	93 ms/digit	N/A
Interference by echos	Should tolerate echos delayed up to 20 ms and at least 10 dB down	N/A

where f is the tone frequency and f_s is the sampling frequency. Observe that

$$\Psi_k(x(n))$$

does not depend on the phase of the signal ϕ. The parameter k defines the underlying sub-rate processing (k = 1 in the original definition of the TK algorithm); notice that the effect of applying $\Psi_k(.)$ at a sampling rate f_s is equivalent to applying $\Psi_1(.)$ at a sampling rate of f_s/k. Depending on the range of frequencies of interest, sub-rate processing is a preferred approach because it reduces computational requirements.

Depending on the power of the signal (that is, the magnitude A of the tone), the energy operator generates different levels for the same normalized frequency Ω. Therefore, to estimate Ω, you must efficiently remove this magnitude dependency. This method (by using specific modules called LCUs) will be detailed in the following paragraphs. Once the dependency is removed, Ω is indirectly estimated by computing the following ratio (ρ_Ω) of energy operators:

$$\rho_\Omega = \frac{\Psi_k\left(\frac{1}{2}(x(n-1) + x(n-m))\right)}{\Psi_k(x(n))} = \cos^2\left\{\left(\frac{l-m}{2}\right)\Omega\right\}$$

This expression derives from the definition of the energy operators and the following trigonometric identity:

$$\frac{1}{2}[\cos(\alpha + \beta) + \cos(\alpha + \gamma)] = \cos\left(\frac{\beta - \gamma}{2}\right)\cos\left(\alpha + \frac{\beta + \gamma}{2}\right)$$

Therefore, selecting k = (1 − m)/2, the tone magnitude A is estimated by the following ratio (ρ_A):

$$\rho_A = \frac{\Psi_k(x(n))}{1 - \rho_\Omega} = A^2$$

Unfortunately when dealing with multi-frequency components like DTMFs the formulas above become too complicated to be handled easily therefore the preferred method in this case will be to separate the DTMFs into its frequency components and process each of them separately.

The first step in proper usage of the TK operator on each frequency after splitting the DTMF in its frequency components will be to eliminate the amplitude dependency. Special care should be taken in this case since the TK operator is very sensitive even to small variations of the signal level. Also, given the fact that TK is very sensitive to background noise special care shall be taken to eliminate the effect of noisy signals by using low-pass filters on the TK energy operator's output.

This section contains the details of Freescale's DTMF detection algorithm currently used in media gateway running on Freescale's StarCore DSP and also methods used to test the implementation against Q.24 rules and also under real-world conditions.

The software module is intended to work in real-time environments and to detect as quickly as possible any DTMF signal. This enables the media gateway to quickly react to the information that is being sent over the telephone line and also to perform specific optimizations, for example shut down specific modules that are not required in this mode of operation, etc. Thus it is assumed that the software module receives as input a multiple of 5 ms of data, in PCM format, 16 bit linear with the sampling frequency of 8 kHz containing the samples that are processed by the module in order to detect the presence of DTMF signals according to Q.24 specification and inform the caller function through an API, by using predefined events.

The detector (depicted in Figure 23) consists of several stages. The first stage comprises a coarse DTMF detection which is mainly used to set coefficients for the subsequent blocks and also to block irrelevant types of signals being processed by the entire chain. This stage consists of 8 Goertzel filters running in parallel; this should be a blocking stage, meaning that if the requirements have not been met the algorithm will exit. The second stage represents the signal conditioning stage, which is necessary to overcome the limitations of the core detection algorithm — the Teager-Keiser energy operator. As detailed in Figure 23, this stage consists of several peak and notch filters and also from the modules called LCU — level control units which will be defined in the next paragraph. The stage after the signal conditioning represents also a blocking stage: at this point, several checks are being performed to ensure that the DTMF is compliant with Q.24 rules and to restrict the signal traversing the core algorithm. This stage is necessary also to eliminate the possibility of false detection of DTMF (from voice signals for example) and it consists of a power estimation module. After all requirements are met, the signal is ready to be processed in the next stage representing the core detection algorithm, consisting of two Teager-Keiser filters (on for each frequency group). The last stage takes the final decision based on all the data available from the previous stages. The result consists of the DTMF code, its duration, and its level.

The Goertzel filters

The Goertzel algorithm provides an efficient means of calculating the DFT energy at a specific frequency. The transfer function of the process is:

$$H(z) = \frac{1}{1 - 2\cos(2\pi\omega)z^{-1} + z^{-2}}, \text{ where } \omega = 2\pi\frac{F_{interest}}{F_{sampling}}$$

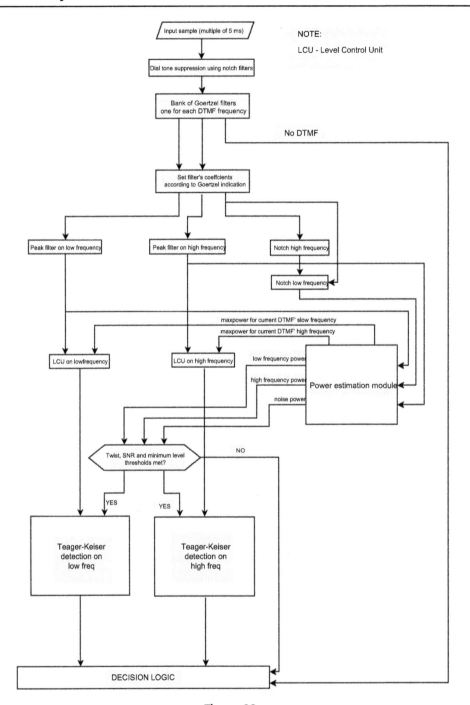

Figure 23:
DTMF detector overview.

The power of the frequency component at $F_{interest}$, using N samples, can be obtained by the following formula:

$$P = y(N - 2)^2 + y(N - 1)^2 - 2\cos(2\pi\omega)y(N - 2)y(N - 1),$$

where y is the output of the Goertzel filter.

The current bank of Goertzel filters is composed from 8 filter structures as defined above, each of them working on one of the 8 possible DTMF frequencies — 697, 770, 852, 941 Hz for low bank and 1209, 1336, 1477 and 1633 Hz for the high bank.

The main role of these filters is to provide a rough estimation regarding the presence of a DTMF in the current 5 ms frame and if so, to indicate the current DTMF frequencies. A secondary role is to prepare the proper filter coefficients used to extract both signals' frequencies. The filters used in this case are notch and peak filters and they will be discussed in the following text.

One important aspect of a Goertzel filter is its length (that is, number of samples). The more samples we have the better the frequency resolution. In contrast with this fact, having a high number of samples implies that the system will respond slower and will not be able to adapt to quick changes in signal dynamics resulting a poor time resolution. This is what is called in theory the uncertainty principle — the time-frequency resolution. For the current application it has been experimentally shown that using 5 ms (i.e., 40 samples @ 8kHz) of data is sufficient for the purpose of rough DTMF estimation. In Figure 5 was plotted the power level calculated for each DTMF frequency using Goertzel filters on 40 samples when using test signals comprising tones ranging from 400 to 2000 Hz with a step of 1 Hz.

After analyzing the plots the conclusion is that, considering only signals above a certain power threshold, there exists no risk of overlapping the DTMF frequency bins and thus eliminating the risk of false detections. The threshold resulting from this experiment proves to be sufficient to obey the Q.24 AT&T minimum level threshold (0...−25 dBm0 on white zone); moreover it even covers half of the white zone since experiments proved reliability in detection for signals as low as −30 dBm0.

Another important aspect resulting from Figure 24 is that Goertzel filters are inadequate for accurate Q.24 detection since the passband for each DTMF frequency (that is, the frequency band containing frequencies having levels higher than the minimum level threshold) is too large in respect to Q.24 rules. For example, Q.24 frequency tolerance is 1.5% maximum. Considering the first frequency from the high group of frequencies — 1209 Hz — the resulting tolerance is 1209 ± 18 Hz. This means that only DTMFs having a frequency in the range 1191−1227 Hz should be detected. Now going back to the results plotted in Figure 22, for the frequency considered above, the tolerance band is somewhere

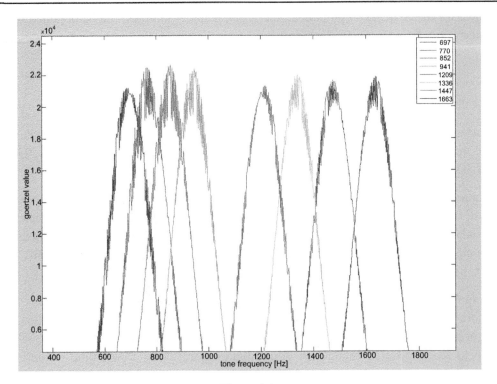

Figure 24:
Power computed using Goertzel filters using test signals with frequency ranging
400–2000 Hz.

around 1150–1290 Hz, which is much too wide to be used in order to achieve accuracy from Q.24.

The output of the Goertzel filter bank module consists of an indication of the DTMF's low bank frequency component *number_l* for the current frame, and one for the high bank frequency component *number_h*. The module also checks for the presence of more than two dominant frequencies and in this case it sends a non-DTMF indication, by setting the values of number_l and number_h with an out-of-range value. The purpose of this additional verification is to eliminate the detection of more than one DTMF signal at a time restriction that is imposed by Q.24.

Thanks to the structure of the StarCore DSP the filters can be run in parallel, four at a time and thus increasing the efficiency of the entire software solution.

The peak filters

These filters are used for extracting each DTMF frequency from the input signal, based on the coefficients set by the Goertzel filters.

The peak filters are second order IIR with a bandpass characteristic, with the following general transfer function:

$$b = \tan\left(\pi\,\frac{Bandwidth}{F_{sampling}}\right); A_{dB} = 3dB; A = 10^{-\frac{A_{dB}}{20}}$$

$$c = \cos\left(2\pi\,\frac{F_{interest}}{F_{sampling}}\right); \beta = \frac{A}{\sqrt{1-A^2}}B; g = \frac{1}{1+\beta}$$

$$H(z) = (1-g)\frac{1-z^{-1}}{1-2gcz^{-1}+(2g-1)z^{-2}}$$

- *Bandwidth*: represents the desired bandwidth at the A_{dB} level; in this case the bandwidth was set to $\pm 2\%$ from the targeted DTMF frequency, at the -3dB level. This tolerance was chosen to be enough for the Q.24 rules
- $F_{sampling}$: the sampling frequency, in this case 8 kHz
- $F_{interest}$: the frequency of interest for the peak filter: in this case the 8 possible DTMF frequencies

The filters' coefficients were pre-calculated and stored in tables. The filter is applied on a frame of 40 samples (5 ms). Figure 25 depicts a typical frequency response of a peak filter. Since it is an IIR filter it will present a slow response and the filter's decay tends to extend the duration of the signal. Both issues will be addressed in Figure 25.

Notch filters

One of the Q.24 requirements is that the DTMF should have a reasonable amount of background noise in order to be properly detected. Beyond this, the current algorithm also uses SNR (Signal to Noise Ratio) estimation to provide a good protection for false detections. In order to compute the background noise level, two cascaded notch filters are used, each being tuned on a DTMF frequency according to the indication provided by the Goertzel filters. They will eliminate the DTMF frequencies and obtain only the background noise. Having the level from the peak filters computed it is trivial in this case to obtain the SNR as detailed in the paragraph from below — *Power estimation module*.

The notch filters are second order IIR, with a bandstop characteristic, having the following transfer function:

$$H(z) = r\frac{1-bz^{-1}+z^{-2}}{1-rbz^{-1}+r^2z^{-2}}$$

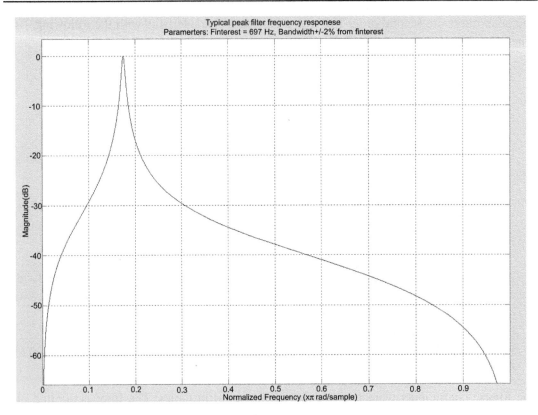

Figure 25:
The frequency response of the peak filters.

$$b = 2\cos\left(2\pi \frac{F_{interest}}{F_{sampling}}\right)$$

where: r is a parameter in the range [0;1] that controls the filter roll-off; in this case it was chosen at 0.93

The filters' coefficients were pre-calculated and stored in tables. The filter is applied on a frame of 40 samples (5 ms). Figure 26 shows a typical frequency response of a notch filter.

Power estimation module

This module is designed to estimate:

- the power of each DTMF frequency component − in dBm0
- the power of the DTMF signal − in dBm0
- the power of the background noise − in dBm0
- the SNR (Signal to Noise Ratio) − in dB

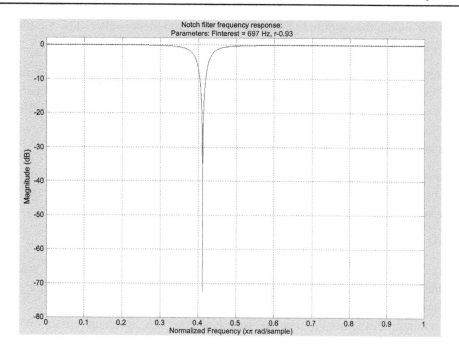

Figure 26:
The frequency response of the notch filter.

The module's inputs are the signals from the outputs of the peak and notch filters. The energy of the current frame is computed by passing the *squared signal* through a low-pass filter having the following transfer function:

$$H(z) = \frac{2^{-r}}{1 - (1 - 2^{-r})z^{-1}},$$

where parameter r is used to control the filter's transition band. In the current implementation, $r = 5$.

The energy is calculated using the formula above on a frame of 5 ms (40 samples) and after that it is converted in to dBm0.

The SNR is computed as:

$$SNR(dB) = P_{DTMF}(dBm0) - P_{noise}(dBm0).$$

A threshold was imposed for this measurement in order to avoid false detection of non-DTMF signals (e.g., speech). The threshold is 14 dB.

LCU (Level Control Unit) modules

DTMF detection is mainly based on filtering each of the DTMF's frequencies, and feeding it to the Teager-Kaiser algorithm. During the filtering stage, even if the DTMF frequency is kept as in the input signal, the amplitude of the frequency component is altered by the filter itself. This is a common effect that cannot be avoided by classical techniques.

It was theoretically proven and experimentally confirmed that feeding a signal with varying amplitude into TK-based algorithm for DTMF detection causes unwanted effects (slow adaptation, evaluation errors) that in the end are translated into limitations of DTMF detection for special cases like: short DTMFs (40 ms), frequency deviation, noisy environments; and this can also cause errors in DTMF duration reporting.

In order to address this problem, a special module was created (Level Control Unit − LCU) that receives as input the output of each peak filter; its main role is to normalize the signal to a constant amplitude, but also to match the input signal duration.

The method has four stages:

1. Build a discrete temporal envelope of the input signal's absolute values, by searching for local maximums as the points where the first derivative of the input signal changes its sign from positive to negative
2. Build a continuous temporal envelope by connecting two consecutive maximums with linear interpolated values
3. Divide the input signal by a scaled version of the continuous temporal envelope in order to obtain a target signal with constant amplitude
4. Limit the target signal to match the input signal duration

Below is an example of LCU functionality.

The test signal is considered to be a 1 kHz tonal signal; if multiplied by a Hanning window, its amplitude will vary as shown in Figure 27.

In stage 1, the LCU builds an array containing the absolute values of the input signal and then searches for local maxima inside this array. According to theory, when the first derivative of a continuous function changes its sign from negative to positive, it means that it passed through a local maximum. The same approach is used for first stage of the LCU. The local maxima are stored into a local array. Figure 28 shows the discrete temporal envelope for the signal above:

In stage 2, the envelope between two consecutive maxima is estimated using a *linear interpolator* which works by the following formula:

$$\text{current\_sample} = \text{previous\_sample} + \alpha,$$

$$\text{where} \, \alpha = \frac{\text{max}_2 - \text{max}_1}{\text{pos\_max}_2 - \text{pos\_max}_1}$$

After performing this step, a continuous temporal envelope is available (see Figure 29).

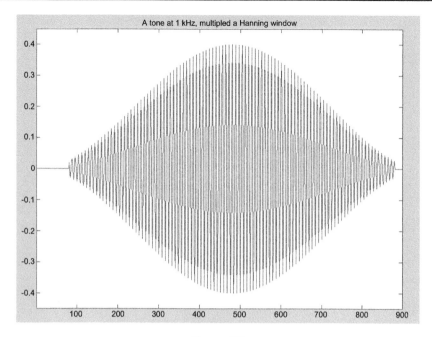

Figure 27:
1000 Hz test signal for LCU demonstration.

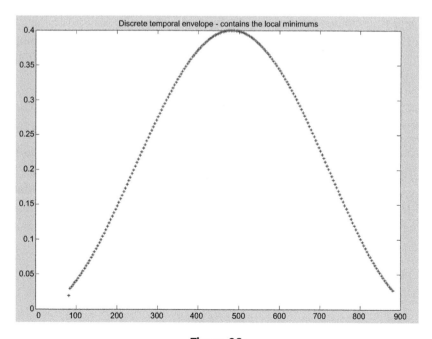

Figure 28:
Test signal maxima plot.

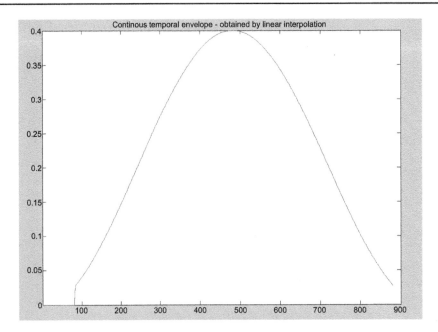

Figure 29:
Test signal envelope plot.

In stage 3, the continuous temporal envelope is first scaled (i.e., multiplied by two) and then the target signal is computed by dividing the input signal by the scaled version of the envelope. Freescale's StarCore DSP is a fixed point processor; scaling the envelope ensures that the target signal will not overflow; the resulting level will be continuous at approx. -8 dBm0 (see Figure 30).

The last stage (stage 4) is used to compensate the peak filters' tendency to extend the duration of the output signal with respect to the input, due to common causes like filter ring, etc.

The approach here is to keep in history the maximum energy of the current frequency component from the current DTMF (computed by the power estimation module, described above) and to cut (that is, set to zero) the target signal using a lookup table that matches each energy level to an amplitude value. The cut (zeroing) is performed in case the scaled continuous temporal envelope decreases below the threshold read from the lookup table using the current maximum energy as index.

On the StarCore DSP, for optimization purposes, some of the above operations are performed using approximations (e.g., division); also the signal is down-sampled to enable fast processing.

One drawback of the LCU is the additive noise caused by the imperfection of the interpolator and approximations used to optimize the performance. However, the noise is mostly of

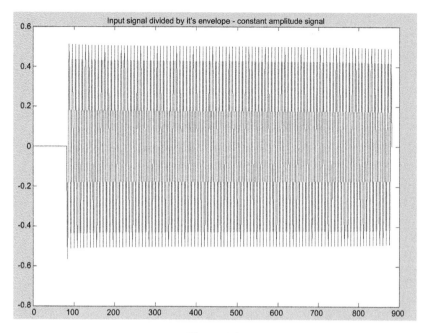

Figure 30:
Test signal shaped through the LCU.

white noise type and it can be efficiently eliminated by using a smoothing (that is, low-pass) filter over the TK output.

As an example of the LCUs necessity, the current detector was tested with LCU modules turned on and off. The results can be seen in Figures 31 and 32 below:

The test signal seen in Figure 33 represents the output of the peak filter. It can be observed that the input signal varies in amplitude by a small amount, except at the start and at the end, by less than 1 dB. If this signal is fed into the Teager-Keiser energy operator the result will be as depicted in Figure 34. Ideally the signal after applying TK would look like a steep valley with short 'grass' while the signal is active. Any destabilization of the trend would translate into signal interruption or termination. Unfortunately, as observed from Figure 30, the small variation in amplitude causes a high amount of error which, in certain conditions, makes the signal impossible to detect accurately.

In Figure 33 the test signal from Figure 31 was passed through the LCU as explained above. After that the signal is fed to the TK operator, which generates the output relating to Figure 32. As can be easily seen, the LCU compensates the error caused by signal level variation.

TK loops

Before feeding the signal to the TK, a final verification stage is required in order to assess the validity of Q.24 parameters. The level twist, minimum level per each frequency and SNR are

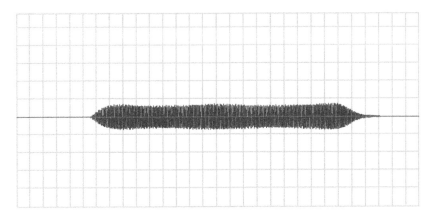

Figure 31:
Test signal after peak filtering.

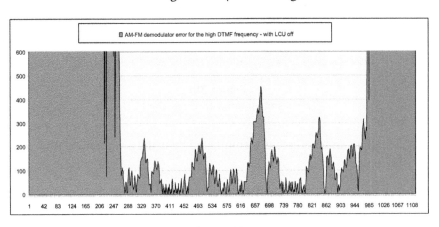

Figure 32:
The test signal fed directly to TK.

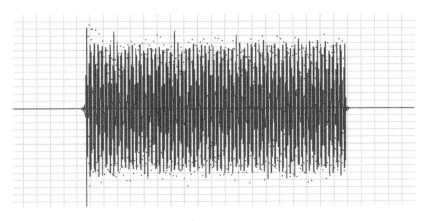

Figure 33:
Test signal obtained after LCU.

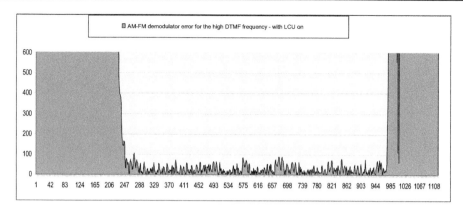

Figure 34:
The LCU-ed test signal fed to TK.

verified and then the target signals at the LCU outputs are fed to the Teager-Kaiser energy operator loops, which perform the operations as defined in the text above.

References

[1] AN2384. Generic Tone Detection using Teager-Kaiser energy operators on the StarCore SC140 core: http://www.freescale.com/files/dsp/doc/app_note/AN2384.pdf.

[2] ITU-T Recommendation Q.23 (11/88). Technical features of push-button telephone sets: http://www.itu.int/rec/dologin_pub.asp?lang=e&id=T-REC-Q.23-198811-I!!PDF-E&type=items.

[3] ITU-T Recommendation Q.24 (11/88). Multifrequency push-button signal reception: http://www.itu.int/rec/dologin_pub.asp?lang=e&id=T-REC-Q.24-198811-I!!PDF-E&type=items.

Software Performance Engineering of an Embedded System DSP Application

Robert Oshana

Based on 'Winning Teams; Performance Engineering Through Development' by Robert Oshana which appeared in IEEE Computer, June 2000, Vol 33, No 6. © 2000 IEEE.

Introduction and Project Description

Expensive disasters can be avoided when system performance evaluation takes place relatively early in the software development lifecycle. Applications will generally have better performance when alternative designs are evaluated prior to implementation. Software performance engineering (SPE) is a set of techniques for gathering data, constructing a system performance model, evaluating the performance model, managing risk of uncertainty, evaluating alternatives, and verifying the models and results. SPE also includes strategies for the effective use of these techniques. Software performance engineering concepts have been incorporated into a Raytheon Systems Company program developing a digital signal processing application concurrently with a next generation DSP-based array processor. Algorithmic performance and an efficient implementation were driving criteria for the program. As the processor was being developed concurrently with the software application a significant amount of the system and software development would be completed prior to the availability of physical hardware. This led to incorporation of SPE techniques into the development life-cycle. The techniques were incorporated cross-functionally into both the systems engineering organization responsible for developing the signal processing algorithms and the software and hardware engineering organizations responsible for implementing the algorithms in an embedded real-time system.

Consider the DSP-based system shown in Figure 1. The application is a large, distributed, multi-processing embedded system. One of the sub-systems consists of two large arrays of digital signal processors (DSP). These DSPs execute a host of signal processing algorithms (various size FFTs and digital filters, and other noise removing and signal enhancing algorithms). The algorithm stream being implemented includes both temporal decomposition of the processing steps as well as spatial decomposition of the data set. The array of mesh-connected DSPs is used because the spatial decomposition required maps well to the

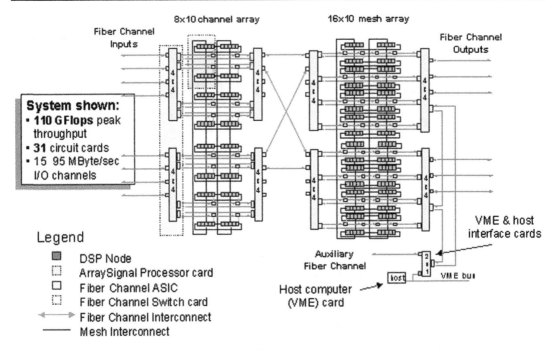

Figure 1:
DSP Array Architecture in Mesh Configuration

architecture. The required throughput of the system drives the size of the array. The system is a data driven application, using interrupts to signal the arrival of the next sample of data. This system is a 'hard' real-time system in the sense that missing one of the data processing deadlines results in a catastrophic loss of system performance.

This system was a hardware-software co-design effort. This involved the concurrent development of a new DSP-based array processor using high performance DSP devices. In this project, the risk of the delivered system not meeting performance requirements was a serious concern. To further complicate matters the algorithm stream was being enhanced and revised as part of the development effort. The incorporation of SPE techniques into the development processes of the various functional organizations was deemed critical to mitigating these risks.

The issue of performance was addressed from the inception of the program throughout its development phases. The main measures of performance are captured in three metrics:

*   processor throughput
*   memory utilization
*   I/O bandwidth utilization

These are the metrics of choice because monthly reporting of these metrics was a customer requirement for the program. Initial estimates of these metrics were made prior to the start of the program and updated monthly during the development effort. Uncertainties associated with key factors driving these estimates were identified. Plans for resolving these uncertainties during the development effort were developed and key dates identified. Updating the metrics and maintaining the associated risk mitigation plans was a cross-functional collaborative effort involving systems engineering, hardware engineering, and software engineering.

Initial Performance Estimates and Information Requirements

The information generally required for a SPE assessment are [1]:

Workload The expected use of the system and applicable performance scenarios. We chose performance scenarios that provided the array processors with the worst case data rates. These worst case scenarios were developed by interfacing with the users and our system engineers.

Performance objectives This represents the quantitative criteria for evaluating performance. We used CPU utilization, memory utilization, and I/O bandwidth because of the customer requirement that we report on these monthly.

Software characteristics This describes the processing steps for each of the performance scenarios and the order of the processing steps. We had accurate software characteristics due to an earlier prototype system using a similar algorithm stream. We also had an Algorithms Description document detailing the algorithmic requirements for each of the functions in the system. From this a discrete event simulation was developed to model the execution of the algorithms.

Execution environment This describes the platform on which the proposed system will execute. We had an accurate representation of the hardware platform due to involvement in the design of the I/O peripherals of the DSP as well as some of the DSP core features. The other hardware components were simulated by the hardware group.

Resource requirements This provides an estimate of the amount of service required for the key components of the system. Our key components were CPU, memory, and I/O bandwidth for each the DSP software functions.

Processing overhead This allows us to map software resources onto hardware or other device resources. The processing overhead is usually obtained by benchmarking typical functions (FFTs, filters) for each of the main performance scenarios.

CPU throughput utilization was the most difficult metric to estimate and achieve. Therefore, the rest of this paper will focus primarily on the methods we used to develop an accurate estimate for the CPU throughput utilization metric.

Developing the Initial Estimate

The process used to generate the initial performance metric estimates is shown in Figure 2. This flow was used throughout the development effort to update the metrics. The algorithm stream is

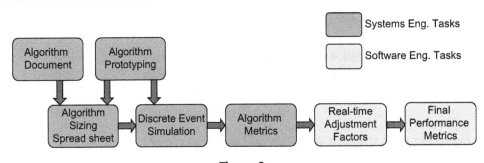

Figure 2:
Performance Metric Calculation Flow

documented in an algorithm document. From this document the systems engineering organization developed a static spreadsheet model of the algorithm stream which provided estimates of throughput and memory utilization for each of the algorithms in the algorithm requirements document. The spreadsheet includes allowances for operating system calls and inter-processor communication. The systems engineering organization used a current generation DSP processor to perform algorithm prototyping and investigation activities.

The results of this work influenced algorithm implementation decisions and were used to develop the discrete event simulations used to estimate the performance metrics. A discrete event simulation was used to model the dynamic performance of the algorithm stream. The simulation model included allowances for operating system task switches and associated calls. The initial algorithm spreadsheet of resource allocations for each algorithm and discrete event simulation processes provide the system engineering 'algorithm' performance metrics. At this point the metrics reflect the throughput, memory, and I/O bandwidth required to perform the algorithms defined in the algorithm document and implemented using the prototype implementations. The software engineering organization then updates the performance metrics to reflect the costs of embedding the algorithm stream in a robust, real-time system. These metric adjustments include the effects of system-level real-time control, built-in-test, formatting of input and output data, and other 'overhead' functions (processing overhead) required for the system to work. The results of this process are the reported processor throughput, memory utilization, and I/O utilization performance metrics.

Key factors in the spreadsheet that influence the processor throughput metric are:

- The quantity of algorithms to implement
- Elemental operation costs (measured in processor cycles)
- Sustained throughput to peak throughput efficiency
- Processor family speed-up

The quantity of algorithms to perform is derived from a straightforward measurement of the number of mathematical operations required by the functions in the algorithm stream. The number of data points to be processed is also included in this measurement. The elemental operation costs measure the number of processor cycles required to perform multiply accumulate operations, complex multiplies, transcendental functions, FFTs, etc.

The sustained throughput to peak throughput efficiency factor de-rates the 'marketing' processor throughput number to something achievable over the sustained period of time a real world code stream requires. This factor allows for processor stalls and resource conflicts encountered in operation. The processor family speed-up factor was used to adjust data gained from benchmarking on a current generation processor. This factor accounted for the increase in clock rate and number of processing pipelines in the next generation device compared to its current generation predecessor.

Key factors in the spreadsheet that influence the memory utilization metric are:

- Size and quantity of intermediate data products to be stored
- Dynamic nature of memory usage
- Bytes/data product
- Bytes/instruction
- size and quantity of input and output buffers based on worst case system scenarios (workloads)

The size and quantity of intermediate data products is derived from a straightforward analysis of the algorithm stream. A discrete event simulation was used to analyze memory usage patterns and establish high water marks. The bytes/data product and bytes/instruction were measures used to account for the number of data points being processed and storage requirement for the program load image.

All of these areas of uncertainty are the result of the target processor hardware being developed concurrently with the software and algorithm stream. While prototyping results were available from the current generation DSP array computer, translating these results to a new DSP architecture (superscalar architecture of the C40 versus the Very Long Instruction Word (VLIW) of the C67 DSP), different clock rate, and new memory device technology (synchronous DRAM versus DRAM) required the use of engineering judgment.

Tracking and Reporting the Metrics

The software development team is responsible for estimating and reporting metrics related to processor throughput and memory. These metrics are reported periodically to the customer, and are used for risk mitigation. Reserve requirements are also required to allow for future growth of functionality (our reserve requirement was 75% for CPU and memory). Throughout the development life cycle, these estimates varied widely based on the different modeling techniques used in the estimation and hardware design decisions which influenced the amount of hardware available to execute the suite of algorithms as well as measurement

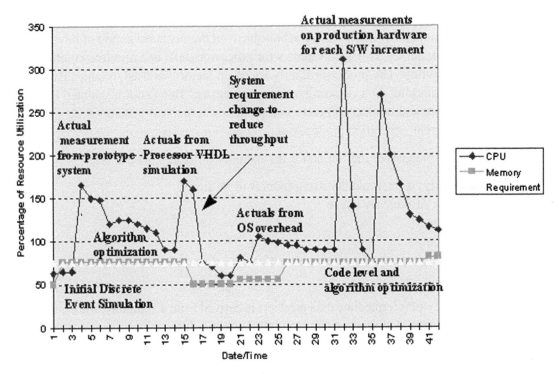

Figure 3:
Resource Utilization metric for Application 1

error. Figure 3 shows the metric history for throughput and memory for the first array processor application. There is a wide variability in the throughput throughout the life cycle, reflecting a series of attempts to lower the throughput estimate followed by large increases in the estimate due to newer information. In Figure 3, the annotations describe the increases and decreases in the estimate for the CPU throughput measurement. Table 1 describes the chronology of this estimate over the course of the project (not completed as of this writing).

The first large increase in the estimate came as a result of implementing the algorithm stream on a prototype current generation processor. These measurements were then scaled based on the anticipated performance of the next generation processor. An effort was then undertaken to optimize the implementation of the algorithm stream to lower the throughput estimate.

The next unexpected increase came from running representative benchmarks on the next generation cycle accurate simulator. This simulation allowed us to estimate the true cost of external memory accesses, pipeline stalls, and other processor characteristics that increased the cost of executing the algorithms. These results led the development teams to undertake

Table 1: Chronology of CPU throughput reduction for application 1

Increase or Decrease in Metric	Explanation
Initial discrete event simulation was used as the starting point for the metric estimation	Discrete event simulation was built using algorithm cycle estimations and first order modeling for task iterations due to context switching, etc.
Measurement on prototype C40 based array	Prototype code was ported to a C40 based DSP small scale array and measured. The measurement was then scaled based on the speedup of the C67 based DSP full scale array
Algorithm level optimization	Algorithms were made more efficient using algorithm re-structuring methods and reducing complexity in other areas of the algorithm stream.
Processor VHDL measurement	Big increase in throughput measurement was due to unexpected high cost of accessing data from external (off-chip) memory. Several benchmarks were performed and scaled to the entire application
System level requirement change	Project decision was made to change a system level parameter. This caused significant algorithm restructuring and was an unpopular decision with the customer
OS level overhead measured	Because the processor was new, the COTS OS was not immediately available. This point indicated the first time to run the application in a multi-tasking environment with the OS
Actuals on production hardware array of DSPs for each software increment	The production code was initially developed without code optimization techniques in place (make it work right and then make it work fast). Initial measurement for the full algorithm stream was not entirely optimized when we first took the measurement
Continued code and algorithm level optimization	Dedicated team in place to work code optimization and other algorithm transformation techniques to reduce CPU throughput (i.e., taking advantage of symmetry in the algorithms and innovative techniques to reduce communications between DSPs which were expensive)

another significant effort to optimize the algorithm stream for real-time operation. The main techniques undertaken during this phase included instrumentation of the Direct Memory Access (DMA) to stage data on and off chip, re-structuring of code to allow critical loops to pipeline, assembly language implementation of critical algorithm sections, and efficient use and management of on-chip memory where memory access time is much shorter.

The representative benchmarks showed us that we could reduce the throughput using code-level optimization techniques (use of on-chip memory, pipelining of important loops, etc.) but we were still in danger of not meeting our overall throughput requirement. It was at this time that a system requirement was modified to reduce throughput. Although a very unpopular decision with the customer (the change reduced data rate and performance of the algorithms), it allowed us to save money by not having to add additional hardware to the system (which is more cost per unit delivered). Algorithm studies also showed that we could still meet system performance by improvements in other areas of the system.

The third major increase came when we measured the full application on the target array of DSPs. The main reason for the increase was due to the fact that many of the algorithms were not optimized. Only a small percentage of algorithm was benchmarked on the processor VHDL simulator (representative samples of the most commonly used algorithms such as the FFTs and other algorithms called inside major loops in the code). The software group still needed to employ the same optimization techniques for the remaining code for each of the software increments being developed. By this time the optimization techniques were familiar to the group and the process went fairly fast.

The memory estimate, although not as severe as the throughput estimate, continued to grow throughout the development cycle. The main reasons for the increase in memory were:

- additional input and output buffers are required for a real-time system to operate
- additional memory was required for each section of code that is instrumented to use the DMA (although this does save on throughput cycles)
- additional memory is needed for code optimization techniques such as loop unrolling and software pipelining which cause the number of instructions to increase

The lifecycle throughput estimates for the second array processor application is shown in Figure 4. A similar pattern in the reported numbers is seen here due to the same basic issues. Table 2 shows the chronology of this CPU utilization estimation.

Once again the initial discrete event simulation proved to be inaccurate and the prototype system measurements were much higher than anticipated due to overly aggressive estimates of the CPU throughput, failure to account for realistic overhead constraints, etc. A long process of code and algorithm optimization was able to bring the estimate back down close to the goal before the VHDL simulation measurements uncovered some other areas that made us increase the estimate. The increase in the estimate in this application resulted in several risk management activities to be triggered;

The estimate in month 5 high enough and was made early enough in the program schedule that the program was able to add more hardware resources to reduce the algorithm distribution and lower the throughput estimate. This was made at the expense of more power and cooling requirements as well as more money for the hardware (no new designs were required, just more boards). These increases in power and cooling had to be offset by sacrifices elsewhere to maintain overall system requirements on these parameters.

The measurement in month 19 caused consternation among the managers as well as the technical staff. Although we felt continued optimization at the code level would reduce the number significantly, meeting the application requirement of 75% CPU throughput (25% reserved for growth) would be hard to accomplish.

CPU and Memory Utilization for Application 2

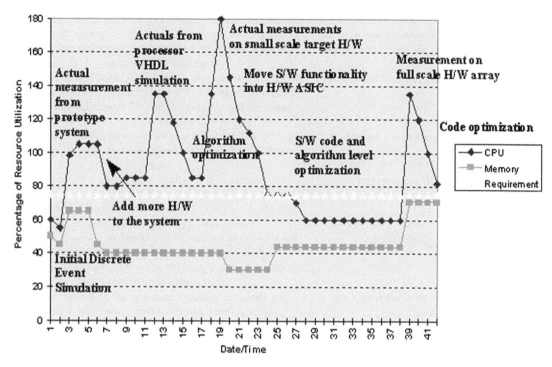

Figure 4:
Resource Utilization metric for Application 2

One contributor to the CPU throughput estimate increase was the result of an under-estimation of a worst case system scenario which led to an increase in data rate for the processing stream. This resulted in several algorithm loops being executed more frequently, which increased the overall CPU utilization.

The decision was made to move some of the software functionality being done in the DSPs into a hardware ASIC to reduce the throughput significantly (there were a sufficient number of unused gates in the ASIC to handle the increased functionality). With this decision coming so late in the development cycle, however, significant re-design and re-work of the ASIC and the interfaces were required, which was extremely expensive for the hardware effort as well as delays in the system integration and test phase.

The last increase in CPU utilization was a result of scaling the algorithms from the small (single node) DSP benchmark to the full array of DSPs. The increase was mainly due to a mis-estimation in the overhead associated with inter-processor communication. Once again, the development teams were faced with the difficult challenge of demonstrating real-time operation given these new parameters. At this late date in the development cycle, there are not

Table 2: Chronology of CPU throughput reduction for application 2

Increase or Decrease in Metric	Explanation
Initial discrete event simulation was used as the starting point for the metric estimation	Discrete event simulation was built using algorithm cycle estimations and first order modeling for task iterations due to context switching, etc.
Measurement on prototype C40 based array	Prototype code was ported to a C40 based DSP small scale array and measured. The measurement was then scaled based on the speedup of the C67 based DSP full scale array
Add more hardware to the system	Number of DSP nodes was increased by adding more DSP boards. Good hardware design made scalability relatively easy
Processor VHDL measurement	Big increase in throughput measurement was due to unexpected high cost of accessing data from external (off-chip) memory. Several benchmarks were performed and scaled to the entire application
Algorithm optimization	Because of the nature of the algorithms, we were able to significantly cut CPU throughput utilization by restructuring the algorithms to pipeline the major loops of the algorithm stream
Actual measurement on small scale target hardware	In our hardware/software co-design effort, we did not have full scale hardware until late in the cycle. Initial benchmarking for this application was performed on a single node prototype DSP card
Move software functionality into hardware ASIC	Decision was made for risk mitigation purposes to move part of the algorithm stream into a hardware ASIC in another sub-system, saving significant CPU cycles in the application software
Software code and algorithm level optimization	Dedicated team in place to work code optimization and other algorithm transformation techniques to reduce CPU throughput
Measurement on full scale hardware	Measuring the application CPU throughput on the full scale hardware showed that we had under estimated the overhead for communication among all the array nodes. We developed a tailored comm API to perform intra-node communications more quickly

many options left for the system designers. The main techniques used at this point to reduce the throughput estimate were additional code optimization, assembly language implementation of additional core algorithms, additional limited hardware support, and a significant restructuring of the algorithm control flow to circumvent the use of slow operating system functions. For example, we eliminated some of the robustness in the node to node communication API in order to save valuable CPU cycles.

It did not take long for management to realize that these 'spikes' in the CPU throughput utilization would continue until all of the application had been measured on the target system under worst case system loads. Rather than periodically being surprised by a new number

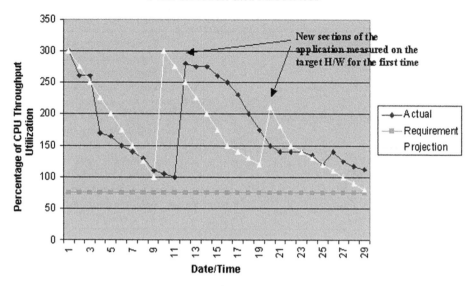

Figure 5:
Plan of Action and Milestones for Application 1

(we were optimizing the code in sections so every few months or so we would have actuals for a new part of the algorithm stream) we were asked to develop a Plan of Action and Milestones (POA&M) chart which predicted when we would have new numbers and the plan for reducing the throughput after each new measurement that would support completion by the program milestone. In the plan we predicted the remaining spikes in the estimate and the plan for working these numbers down (Figure 5). This new way of reporting showed management that we knew increases were coming and had a plan for completing.

Reducing the Measurement Error

The performance engineering plan detailed when hardware and software tools would become available which could be used to reduce the errors in the performance metrics. These availability dates when combined with the system development schedule provided decision points at which design trade-offs would be performed balancing algorithm, hardware, and software design points to yield a system that met cost and performance goals. Table 3 lists the tools identified and the error factors reduced by them.

As these tools became available, benchmark code was executed using them and the performance metrics updated accordingly. This data was used to support program level decision points to review the proposed computer design. This review included hardware resources in the computer, algorithmic functionality assigned to the computer, and the

Table 3: Tools identified in performance plan and errors resolved by them.

Tool	Error Factors Resolved
Code Generation Tools (Compiler, Assembler, Linker)	Compiler efficiency
	Quality of generated assembly code
	Size of load image
Instruction Level Processor Simulator	Utilization of dual processor pipelines
	Cycle counts for elemental operations
Cycle-accurate Device Level VHDL Model	Effect of external memory access times
	Instruction Caching effects
	Device resource contention between processor and DMA channels
Single DSP Test Card	Validate VHDL results
	Runtime interrupt effects
Multi-DSP Test Card	Inter-processor communication resource contention effects

proposed software architecture. At various decision points all of these areas were modified. The computer hardware resources were increased through the addition of more DSP processor nodes. The clock rate of the DSP was increased by 10%. Some algorithms were moved to other portions of the system. The software architecture was reworked to reduce overhead by eliminating extraneous interrupts and task switches. All aspects of the design were considered and adjusted as appropriate to meet the performance and cost objectives.

The performance plan also included the use of analytical tools to address the overall schedulability and large-scale performance of the array processor. We attempted to use Rate Monotonic Analysis (RMA) to validate the schedulability of the software architecture [3,4,5]. RMA is a mathematical approach to determining schedulability under worst case task phasings and allows the designer to determine ahead of time whether the system will meet its timing requirements. RMA has the advantage over discrete event simulation in that the model is easier to develop and change and the model provides a conservative answer that guarantees schedulability (using a simulation, it becomes hard to predict how long to execute the model before a certain set of task phasings causes the system to break). One powerful feature of RMA tools is the ability to identify blocking conditions. Blocking and preemption are the most common reasons for missing deadlines and are one of the main focuses of most RMA tools. We were interested in using RMA because the model could identify potential timing problems even before the building of the system. Alternative designs could be analyzed quickly before actually having to implement design. Our attempt at using RMA provided only a high level look at schedulability but not the details. The tools used did not scale well to large systems with thousands of task switch possibilities and non-preemptible sections (one of the compiler optimization techniques produced software pipelined loops which, because of the nature of the processor pipeline turns off interrupts during the pipelined loop, thereby creating a small non-preemptible section. Attempting to input and model thousands of these conditions proved to be too cumbersome for our application without becoming too abstract for our purposes).

As the array computer hardware was being developed concurrently with the software, the software team did not have target hardware available to them until late in the development life-cycle. To enable the team to functionally verify their software prior to the availability of hardware an environment was developed using networked Sun workstations running Solaris. Using features of the Solaris operating system, the environment enabled small portions of the array computer to be created with inter-processor communication channels logically modeled. The application code was linked to a special library that implemented the DSP operating system's API using Solaris features. This enabled the team to functionally verify the algorithms, including inter-task and inter-processor communications, prior to execution on the target hardware.

The fundamental approach was to make the application work correctly and then attempt to add efficiency to the code ('Make it work right — then make it work fast!'). We felt this was required for this application for the following reasons:

- Given the hardware/software co-design effort, the processor (and user documentation) was not available so the development team did not thoroughly understand the techniques required to optimize the algorithm stream.
- The algorithms themselves were complicated and hard to understand and this was seen as a risk by the development team. Making the algorithm stream run functionally correct was a big first step for a development team tackling a new area.
- Optimization of an algorithm stream should be performed based on the results of profiling the application. Only after the development team knows where the cycles are being spent can they effectively optimize the code. It does not make sense to optimize code that is executed infrequently. Removing a few cycles from a loop that executes thousands of times, however, can result in bigger savings at the bottom line.

Conclusions and Lessons Learned

Estimating throughput may not be exact science but active attention to it during life cycle phases can mitigate performance risks and enable time to work alternatives while meeting overall program schedules and performance objectives. This needs to be a collaborative effort across multiple disciplines. System performance is the responsibility of all parties involved. There are no winners on a losing team.

Processor CPU, memory, and I/O utilization are important metrics for a development effort. They give early indications as to problems and provide ample opportunity for the development teams to take mitigation actions early enough in the life cycle. These metrics also give management the information necessary to manage system risks and allocate reserve

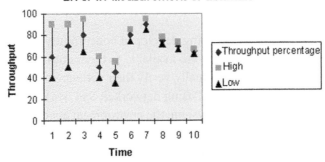

Figure 6:
Improvement in Accuracy in Estimate over time

resources (i.e., money and schedule) where needed. Often, one or more of these metrics will become an issue at some point during the development cycle. To obtain a system solution to the problem, sensitivity analysis is usually performed, examining various alternatives that trade off throughput, memory, I/O bandwidth, as well as cost, schedule, and risk. When performing this analysis, it is essential to understand the current accuracy in the metric estimates. Early in the life cycle the accuracy will be less than in later stages, where the measurements are much more aligned to the actual system due to the simple fact that more information is available (Figure 6).

There were several lessons learned in this experience:

- Prototype early in the development life cycle; several of the surprises we encountered could have been uncovered earlier if the proper level and type of prototyping was performed. Although prototyping was scheduled early in the life cycle, schedule pressures forced the development team to dedicate a limited amount of resources to this early in the development cycle.
- Benchmark; don't rely totally on the processor marketing information; Most processors will never achieve the throughput proposed in the literature. The numbers are often theoretical. In many cases, actual is much lower and very dependent on how the application maps to the processor architecture (DSPs run DSP-like algorithms very well but they are not very good at finite state machines and other 'control' software). Marketing information for processors will show how well they perform on the best mapped algorithms their processors support.
- Analyze the functions executing most often; these areas are where the hidden cycles can come back to haunt a development team. Eliminating just a few cycles from a function that executes many times will have significant impact on the overall throughput.
- Don't ignore the interfaces; real time systems carry an inherent 'overhead' that never seems to be accounted for in throughput estimates. Although the signal processing

algorithms may be where the main focus is from a system requirements and functionality point of view, real-time systems also need throughput for interrupt handling, data packing and unpacking, data extraction, error handling, and other management functions that are easily overlooked in throughput estimates. Many battles were fought over how much of the timeline should be devoted to overhead tasks.

- Benchmarks of discrete algorithms don't scale well to real-time systems; benchmarking an individual algorithm inherently implies that algorithm has complete control and use of all the processor resources including internal and external memory, the DMA controller, and other system resources. In reality, there may be other tasks competing for these same resources. Assumptions made when benchmarking individual algorithms may not apply when the system is put together and running under full system loads. Resource conflicts result in additional overhead that can easily be overlooked when forming throughput estimates.

- Keep management informed; as we approach the completion of the code level optimization effort, it appears the model we established early in the project was a relatively accurate estimate. However, it took a substantial amount of resources (schedule and budget) to accomplish this goal. Along the way, the estimate periodically rose and fell as we optimized and measured our algorithm stream. The reporting period for these metrics was short enough to catch these spikes which caused premature concern from management. A longer reporting interval may have 'smoothed' some of these spikes.

- Budget accordingly; the two pass approach of functional correctness followed by code optimization will take more time and more resources to accomplish. This needs to be planned. A one pass approach to code level optimization at the same time as the functionality is being developed should be attempted only by staffs experienced in the processor architecture and the algorithms.

References

[1] Connie U SmithConnie U Smith, Performance Engineering for Software Architectures, 21st Annual Computer Software and Applications Conference (1997), pp. 166–167.
[2] Michelle BakerMichelle Baker, Warren SmithWarren Smith, Performance Prototyping: A Simulation Methodology for Software Performance Engineering, Proceedings of the Computer Systems and Software Engineering (1992) 624–629.
[3] Oshana RobertOshana Robert, Rate Monotonic Analysis Keeps Real Time Systems On Track, EDN (September 1, 1997).
[4] C. LiuC. Liu, J. LaylandJ. Layland, Scheduling algorithms for multiprogramming in a hard real time environment, Journal of the Association for Computing Machinery (January 1973).
[5] Obenza RayObenza Ray, Rate monotonic analysis for real-time systems (March 1993).

Specifying Behavior of Embedded Systems

What makes a good requirement?

The criticality of correct, complete, testable requirements is a fundamental tenet in software engineering. Both functional and financial success is affected by the quality of requirements. So what is a requirement? It may range from a high-level abstract statement of a service or of a system constraint to a detailed mathematical functional specification. Requirements are needed for several reasons:

- Specify external system behavior
- Specify implementation constraints
- Serve as reference tool for maintenance
- Record forethought about the life cycle of the system, i.e., predict changes
- Characterize responses to unexpected events

The system designer must understand requirements and be able to organize them. A technical background and an understanding of the user are both required. Before design can start, each requirement must be understood in terms of significance and priority in the solution strategy. Because both the developer and the customer must understand the requirements, they are usually written in natural language. But natural language is a poor medium for communicating requirements. The use of natural language to specify complex requirements has at least two problems; those of ambiguity and inaccuracy. Many words and phrases have dual meanings and can be altered depending on the context in which they are used. A word that means one thing to one person can mean something entirely different to someone else. This is referred to as *semantic bypass*. For example the interpretation of the word 'bridge' can have completely different meanings depending on whether you are a dentist, a civil engineer, an electrical engineer, or someone who just retired early!

Other disciplines require a more exact language for communicating specifications. For example an architect blueprint or a circuit board schematic are formal specifications that define what is needed but not necessarily the implementation details of how to get there. Software requirements should be the same way. Software requirements specifications should not contain implementation details. These specifications should describe in sufficient detail what the software will do but not how. A good set of requirements has the following characteristics:

- correct; meets the need
- unambiguous; only one possible interpretation

- complete; covers all the requirements
- consistent; no conflicts between requirements
- ranked for importance
- verifiable; a test case can be written
- traceable; referring to requirements easy
- modifiable; easy to add new requirements

Trying to define a large multidimensional capability of a complex embedded system within the limitations of a linear two-dimensional structure of a document becomes almost impossible. At the other end of the scale, the use of a programming language is too detailed. This is nothing more than 'after the fact specification' which is just documenting what was implemented rather than what was required.

Developing a good set of requirements to specify a system can be hard. In many cases the stakeholders don't know what they really want. The various stakeholders also tend to express requirements in their own terms and different stakeholders may have conflicting requirements that need to be resolved. Organizational and political factors may also influence the system requirements and the requirements change during the analysis process as new stakeholders emerge. The term 'requirements engineering' has emerged that addresses the transformation by which vague and often unrelated customer requirements are transformed into detailed and precise requirements needed for system implementation.

Successful designs are usually the result of significant rethinking and reworking. Several iterations of design is the norm, rather than the exception. Design alternatives should be considered. In general, designs should get simpler rather than more difficult.

It can be difficult to specify the total behavior of a complex system because of the total number of possible uses of the system. But this is precisely what needs to be done in order to insure completeness and consistency in our designs. As James Kowal describes:

> 'If the systems planners and customer do not specify what is expected in all types of inter-actions with the system, i.e. the behavior of the system, someone else will.
> That someone else is most likely the programmer when he or she is coding the ELSE option of some IF statement. There is a very low probability that the programmer's guesses as to the expected behavior will be what the customer expects.'

The hard part is being able to determine all the possible types of interaction with the system. Use cases are used to explore and elicit requirements and can help determine certain types of interaction to expect between the actors (external stimuli to the system) and the system. But use cases may not always combine to describe *complete* and *consistent* behavior. Once use cases are used to explore the problem and for front end domain analysis, other techniques can then be used to fully specify the solution strategy. I want to explain one approach that has worked well in our embedded system project, called sequence enumeration.

Sequence Enumeration

Sequence enumeration is a way of specifying stimuli and responses of an embedded system. This approach considers all permutations of input stimuli. Sequence enumerations consist of a list of prior stimuli and current stimuli as well as a response for those particular stimuli given the prior history. Equivalent histories are used to map certain responses. This technique maps directly to a state machine implementation. The strength of sequence enumerations is that the technique requires the developer to consider the obscure sequences that are usually overlooked.

As an example, I will consider the cell phone shown in Figure 1. A very simplified set of natural language requirements for this system is shown in Table 1.

Assumptions

- All four-digit combinations are valid phone numbers.
- The cell phone will only work when the power has been applied. This is a prerequisite for the system and not a requirement. Because power must be applied for the system to be operational, it is not considered a stimulus to the system. System stimuli must impact the functioning system in order to produce a response.

Figure 1:
A cell phone.

Table 1: Simplified set of natural language requirements for the cell phone machine.

Tag Number	Requirement
1	When the 'POWER' button is activated, the display light comes on and the initial screen is displayed.
2	If the cell phone is on and the 'POWER' button is pressed, the cell phone turns off, and a 'good bye' message is displayed on the screen.
3	After each digit button is pressed, the character is echoed on the display.
4	If there is a valid phone number entered into the display, when the 'GO' button is pressed, a 'calling' message is displayed on the display and the number is dialed.
5	If there are less than four digits on the display, the 'GO' button is ignored.
6	If the phone is in a call, when the 'STOP' button is pressed, a 'terminating call' message is displayed on the screen and the call is terminated.
7	If the phone is not in a call but there are digits on the screen, when the 'STOP' button is pressed, the screen is cleared.
8	If there are no digits on the screen, the 'STOP' button is ignored.
9	This cell phone requires a 4-digit code for a phone number.

To start with, a use case can be developed that describes one type of interaction with the system. A use case is a story about the usage of a system told from an end users' perspective. It consists of:

- Preconditions; what must be available and done before the performance of the use case
- Description ; the story
- Exceptions; exceptions in the normal flow of the story
- Illustrations; figures to help understand the use case
- Postconditions; the state of the system and actors after the use case has been performed

The motivation for use cases are:

- tool to analyze and improve functional requirements
- tool to model the functionality of the system as soon as possible
- to allow the customer to understand the operation of the system

Use cases must specify the most important functional requirements. A use case depicts a typical way of using the system but nothing more. In general, a use case should not try to define all possible ways of performing a task. Certain important exception conditions are described in 'secondary' use cases or in exceptions within a use case. A simple use case for the coke machine could be as follows:

Precondition: The cell phone has been turned on and is displaying the default screen.
Description: A user decides to make a phone call using the cell phone. The user presses two of the numeric keys to dial the 4 digit number for this cell phone. The user makes an error on the second numeric digit, presses the Stop button to clear out the digits and

then proceeds to type in the correct 4 digit code. The user then presses the Go button to initiate the call, and, after finishing the call, presses the Stop button to terminate the call.
Exceptions: None
Postconditions: The cell phone is displaying the default screen

Other use cases can be developed depending on the various business cases for this system. Once we have a general idea of what the system should do (which use cases are useful for determining), we can then perform a sequence enumeration to more completely map all possible combinations of stimuli to a response. To start with, it's a good idea to represent the system as a 'black box,' showing the stimuli and the responses that are possible. The black box for the coke machine is shown in Figure 2. Table 2 summarizes the stimuli and responses for the system.

A sequence enumeration can be done in a simple Excel spreadsheet. Enumeration starts at evaluating a single stimuli to the system and determining the appropriate response for each stimulus. Table 3 shows the enumeration for the system. As the table shows, each stimulus is separately evaluated. Starting at the beginning, the enumeration shows that if the Power stimulus 'P' occurs (the power is applied and the system is initialized) the 'Phone On' is produced, as required by requirement 1. If any other stimulus occurs the response is null and it is considered an illegal sequence (you cannot use the cell phone if the power is not on, for example). The result from performing this first step 'P' is required before anything else can be done and we must remember that this event occurred. Therefore, the 'P' stimulus is used in the next level of enumeration.

The level 3 enumeration uses all stimuli sequences that were not illegal or did not have a shorter equivalent sequence in the previous enumeration level than the first stimulus in the sequence. So, as the length 2 section in Table 3 illustrates, the first stimulus is 'P' and the second stimulus is each of possible stimuli to the system (the simple cross product). As an example, the sequence P P shows, if the system receives a power on signal, followed by another power on signal, the system will toggle power and shut off. The sequence 'P D'

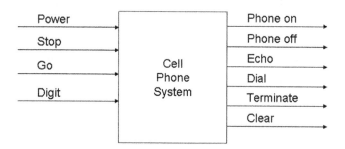

Figure 2:
Cell phone machine black box.

Table 2: Cell phone machine stimuli and responses.

Stimulus		
Stimulus	**Description**	**Req Trace**
POWER	Button pressed on the phone	1, 2
STOP	Button pressed on the phone	6, 7, 8
GO	Button pressed on the phone	4, 5
Digit	Digit entered from the keypad	3, 9
Response		
Response	**Description**	**Req Trace**
Phone on	Phone turns on, display light on, and initial screen displayed	1
Phone off	Phone turns off and 'good-bye' message on screen	2
Character echo	Digits displayed on screen	3
Dial	Number dialed and 'calling' message on screen	4
Terminate	Call ended and 'terminating call' message on screen	6
Clear	Screen cleared	7

describes that the sequence of Power on followed by the pressing of a single digit produces no response and has no equivalent sequence (another piece of behavior that the system needs to remember). In other words the system must remember the fact that a digit has been dialed. This means we will extend this specific sequence, and all others that do not have an equivalence and are not illegal to the sequence length n + 1. The sequence 'P D', for example, will get evaluated again at length 2 + 1 or level three, as shown in the table. Each sequence in this level that is not illegal or does not have a shorter equivalent sequence must be remembered and used in the next enumeration level.

This process continues until all behavior is eventually mapped to a shorter equivalent sequence (it must always end or the software system may never terminate!). Table 3 shows this process terminating in length 7.

The specific use case described earlier can be seen in the length 4 sequence 'P D D S' which effectively describes a case of powering the phone on, entering two digits, and then pressing the Stop button which takes the system back to an initial state. The second part of the user case is relected in the length 7 enumeration 'P D D D D G S,' which completed the call and returns the system to the initial state 'P' with the power on and the system in the initial state.

You may have noticed that the enumeration tables have a 'Requirement' column that is used to trace the requirement as we define the behavior. This is a simple and easy way to trace all requirements to behavior. You will also notice the derived requirements (indicated as D1, D2, etc.). A derived requirement is a lower level requirement that is determined to be necessary for a higher level requirement to be met. In general, a derived requirement is more specific

Table 3: Cell phone sequence enumeration.

Sequence Enumeration			
Sequence	**Response**	**Equivalence**	**Requirements Trace Number**
Length 0			
Empty	Null		D1: Phone is initially off
Length 1			
P	Phone on		1
S	Illegal		D1
G	Illegal		D1
D	Illegal		D1
Length 2			
P P	Phone off	Empty	2
P S	Null	P	8
P G	Null	P	5
P D	Character echo		3
Length 3			
P D P	Phone off	Empty	2
P D S	Clear	P	7
P D G	Null	P D	5
P D D	Character echo		3
Length 4			
P D D P	Phone off	Empty	2
P D D S	Clear	P	7
P D D G	Null	P D D	5
P D D D	Character echo		3
Length 5			
P D D D P	Phone off	Empty	2
P D D D S	Clear	P	7
P D D D G	Null	P D D D	5
P D D D D	Character echo		3
Length 6			
P D D D D P	Phone off	Empty	2
P D D D D S	Clear	P	7
P D D D D G	Dial		4
P D D D D D	Character echo	P D D D D	3
Length 7			
P D D D D G P	Phone off	Empty	2
P D D D D G S	Terminate	P	6
P D D D D G G	Null	P D D D D G	D2: GO while in a call is ignored
P D D D D G D	Null	P D D D D G	D3: Digit while in a call is ignored

and directed toward some sub-element of the project. Derived requirements often occur through the process of analysis.

There are a couple important conclusions to draw from this process. We can say that this process produced a set of behavior that is:

- *complete*; the enumeration considered, in effect, every possible combination of stimuli to the system for an infinite length of possible stimuli. This insures that we have considered every possible behavior condition. The requirement trace also ensures that we have covered all possible requirements.
- *consistent*; because we have considered every combination of stimuli only once, we have insured consistency. In other words we have not mapped a stimuli combination to more than one possible output sequence.
- *correct*; the mapping from stimuli to responses have been properly specified in the judgement of the domain experts.

Once the sequence enumeration is complete, it is a fairly simple step to develop the finite state machine that represents the behavior we have just specified. For one thing, I can determine how many states in the finite state machine there will be by simply counting the lines of enumeration that are not illegal and do not have equivalent sequences. These are referred as the *canonical* sequences. From Table 3, these are those sequences that are not 'illegal' and do not have an equivalence at that level of enumeration. These are listed in Table 4. State data can be invented that will encapsulate the behavior described in these canonical sequences. Table 5 shows the state data that has been invented to encapsulate this relevant system behavior. Table 6 shows how this state data is updated for each of the canonical sequences.

Table 4: Canonical sequences for the cell phone machine.

Canonical Sequence Analysis			
Canonical Sequence	**State Variables**	**Value before Current Stimulus**	**Value after Current Stimulus**
Empty			
P	Phone	OFF	ON
P D	Phone	ON	ON
	Phone_number	NONE	ONE
P D D	Phone	ON	ON
	Phone_number	ONE	TWO
P D D D	Phone	ON	ON
	Phone_number	TWO	THREE
P D D D D	Phone	ON	ON
	Phone_number	THREE	FOUR
P D D D D G	Phone	ON	ON
	Phone_number	FOUR	FOUR
	Status	NOT_CONNECTED	CONNECTED

Table 5: State data invention for each of the canonical sequences.

State Variables		
State Variable	**Range**	**Initial Value**
Phone	{ ON, OFF }	OFF
Number	{ NONE, ONE, TWO, THREE, FOUR }	NONE
Status	{ NOT_CONNECTED, CONNECTED }	NOT_CONNECTED

Based on the state data just invented and the behavior described by the enumeration sequences, the state machine for the cell phone is easily produced (Figure 3). Notice the seven states corresponding to the seven canonical states produced during the enumeration process.

It's really all just math!

The process described can be scaled to larger systems by using abstractions for the stimuli and responses. These abstractions can be decomposed and expanded at lower levels of behavior definition, as more and more details of the system are exposed. In effect, what we have done is define a rule for a function for this software system. This process is rooted in the mathematics of function theory, which maps a domain (valid inputs to the system) to a range (the correct results) within a co-domain (all *possible* results) (Figure 4). A relation, on the other hand, is

Table 6: State data for the cell phone machine and state data mapping.

State Mapping				
Tag #	**Current State**	**Response**	**State Update**	**Sequence Trace**
1	Phone = ON Number = NONE Status = NOT_CONNECTED	Phone on	Phone = ON	P
2	Phone = ON Number = ONE Status = NOT_CONNECTED	Character echo	Number = ONE	P D
3	Phone = ON Number = TWO Status = NOT_CONNECTED	Character echo	Number = TWO	P D D
4	Phone = ON Number = THREE Status = NOT_CONNECTED	Character echo	Number = THREE	P D D D
5	Phone = ON Number = FOUR Status = NOT_CONNECTED	Character echo	Number = FOUR	P D D D D
6	Phone = ON Number = FOUR Status = CONNECTED	Dial	Status = CONNECTED	P D D D D G

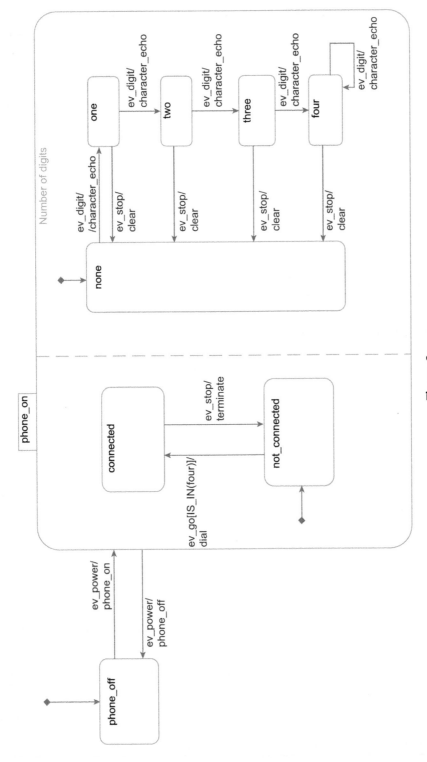

Figure 3:
State machine for the coke machine system.

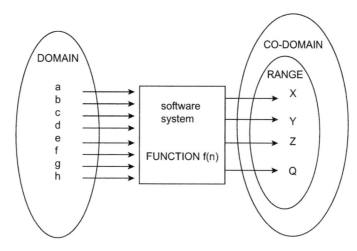

Figure 4:
A Function maps a domain to a range within a co-domain.

a more general mapping from domain to range and allows the same domain element to map to more than one range element (similar to what would happen in a software system with an un-initialized variable — different behavior for the same input sequence!).

Referring back to the use case described earlier, you will notice that we can map the behavior from this use case to parts, but not all, of the enumeration we just completed. There are certain parts of the enumeration that describe the story being told by the use case. The sequence enumeration goes a step further, defining the complete behavior that was originally described in the story told in the use case. This is where the synergy between use cases and sequence enumeration becomes apparent. Use cases and scenarios are an effective tool to decide what we have to do for the various stakeholders of the system. The sequence enumeration drives home the completeness and consistency that will give your developers a clear picture of what to do in all circumstances of use as well as resolve any inconsistencies between the stakeholders.

DSP for Software Defined Radio

Andrei Enescu

Introduction

The idea behind SDR (software defined radio equipment) is to have a multi-band, multi-standard terminal that can adaptively select the best communication stack available to establish a successful connection according to the QoS of the demanding services.

We can look at SDR from two perspectives: the mobile terminal and the base-station. The *mobile terminal* has multiple communication stacks implemented, but at any given time, it should use only one of these, according to a given criterion.

The criteria, depending on the application, may be:

- Best link
  - The terminal must provide measurements on each of the candidate bands
  - It then chooses the network that offers the best link conditions according to physical layer measurements: received signal power, signal to noise + interference ratio
  - As a boundary condition, the terminal must select the operating technology in a region when usually only one technology is available from a certain carrier
- Best capacity
  - The terminal should choose the network that provides the best throughput according to the individual link quality and to its throughput requirements.
- Predefined criteria
  - For example services cost (e.g., Always choose WiFi against 3G for data services)

In both cases, measurements are done at the physical layer level (L1), but decision is taken at an upper layer. In Figure 1 a mobile terminal monitors reference signals transmitted by one or several base stations on different bands, different technologies. Once the band is swept, L2 of the terminal takes the decision and chooses one available resource.

This is the case of a smart phone that can access the Internet via multiple available technologies: 2.5G EDGE, 3G UMTS, 3G+ HSPA, 4G LTE (soon to come), and WiFi. Only one of the available technologies is used at any given time for data transfers.

Right now, switching among these technologies is done at a higher layer in the protocol stack. SDR equipments tend to lower down the level of decision, into L2, according to fast L1 measurements. This will massively reduce the overhead of using the entire protocol stack.

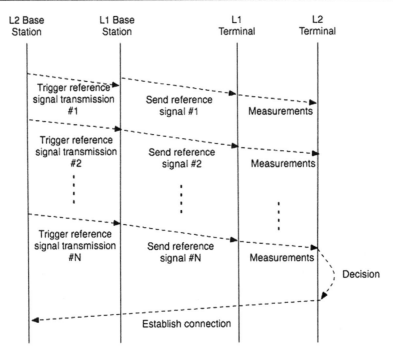

Figure 1:
Establishing connection at layer 1 for a SDR.

Also, the tendency is to use the same processor for all L1s, which is something most of the state-of-the-art smartphones do not currently support. They usually have different ASICs for different standards of communication (especially for WiFi and 3G). Even though at a given time, only one will be functional, area will be used to support all kinds of technologies. And most likely power consumption will not be optimized due to this. Using only one processor may lead to a significant resource save because of a very simple reason: *resource reuse*. A simple example can be stated. If we have a terminal that supports LTE, WiFi and WiMAX, then we can share the FFT resources, which are extensively used in all three cases.

On the base station side, we can have two cases. Consider a carrier network that supports any 3GPP communication technology from 2.5G to 4G (EDGE to LTE). This can be ensured at first glance by using several base stations, one for each supported technology. It is like having three different networks (one for EDGE, one for UMTS, one for LTE) and the mobile terminal will connect to any of these three. It is a rough solution to support all possible cases. An elegant solution would be to have one base station supporting all three standards, without considering additional network equipment, in a converging network architecture, using the same RNC (Radio Network Controller). Most of the standards are built so that time coherency is kept on the network architecture level. Therefore, support of a newer version of the specs can be done without changing the architecture.

We will focus in this chapter on the base station, since we find it a more general case. While the terminal has to support one technology at a time, the optimum one that fits the needs, the base station must operate with all sorts of stations simultaneously: some old terminals supporting only EDGE, some newer with UMTS capabilities, and some state of the art smartphones, equipped with LTE transceivers. Thus, designing a multi-standard base-station becomes a hot topic of today's telecom industry.

Functional architecture of a base station

General partition

The general partition of the protocol stack for software defined radio equipment is presented in Figure 2.

The common DFE is responsible for up and down conversion, up/down sampling, predistortion, crest factor reduction. The former two operations may strongly depend on the technology used and may not be part of the common block, according to design. This is mainly because of the sensitivity of multi-carrier transmissions to produce spikes in the time-domain signal and affect the PAPR (Peak to verage Power Ratio). To some extent, this is fixed in the LTE uplink transmission, where SC-FDMA is used instead of OFDM.

The equipment includes several waveform generators (WV), with appropriate Layer 1 (L1 WV #k) and Layer 2 (L2 WV #k) implementations. There are some measurement

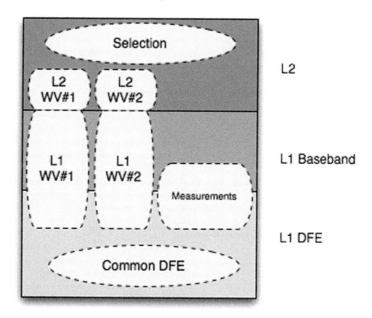

Figure 2:
Lower protocol stack partition for a SDR.

modules, so-called spectrum sensing in SDR terminology. These measurement modules monitor the energy of the received signals in different bands, on different standards.

Reports are forwarded to upper layers that eventually take the decision of initiating network entry on a specific technology and on a certain band. If already connected to a network, the upper layers may take the decision to switch to a new resource, in order to maximize either link or capacity, as stated previously.

LTE eNodeB

The functional block diagram of an LTE eNodeB transceiver is shown in Figure 3.

The transmitter processes independent bit streams, called codewords (CW). These will be transmitted using different virtual antennas. It is up to the configuration to map these virtual antennas onto physical antennas. Also, generalization of this principle may lead to a precoding that spreads the codeword energy on all the antennas, thus forming a virtual

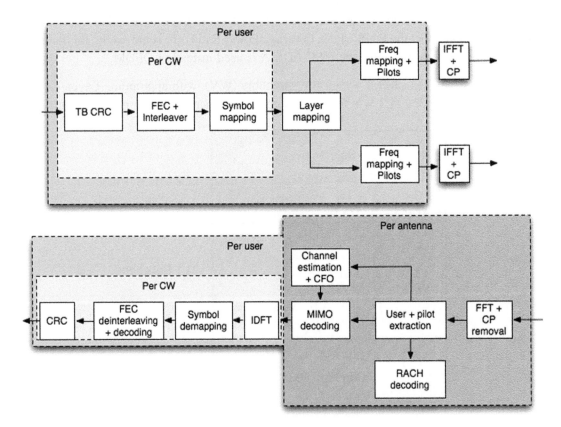

Figure 3:
Simplified LTE eNodeB transceiver.

antenna beam on which this CW is sent on air. Each codeword has its own bit processing part, consisting of CRC appending, FEC encoder (either convolutional or turbo), rate matching and interleaving. The so-formed bits map symbols from a constellation (QPSK, 16-QAM, 64-QAM) and are mapped onto the antennas, as specified, on specific time/frequency resources, called *resource elements* (RE). At this point, reference signals are added. These can be either dedicated or common per all users. OFDM modulation is further applied, using an IFFT transformer and a cyclic prefix (CP) padding.

On the receive side, the inverse operations are applied in general. OFDM decoding occurs, by removing the cyclic prefix and then using an FFT transformer. Each user is then extracted from the time-frequency grid. The reference signals are used to estimate the channel and any carrier frequency offset (CFO) per receive antenna. These channel estimates are further used for MIMO decoding, where a general ML decoding algorithm should be applied.
In practice, these receivers use either a quasi-ML approach (such as sphere decoding) or a sub-optimal receiver, as in the case of MMSE (Minimum Mean Squared Error).
Over uplink, LTE uses SC-FDMA (Single-Carrier Frequency Division Multiplexing). Data is not exactly mapped in frequency, as is, but rather spread at transmission, using a DFT transform. Hence, at the receiver, a despread IDFT operation occurs, followed by a symbol demapping, usually with soft bit outputs, called LLR (Log-Likelihood Ratio). Bit processing at the receiver then goes on in the exact manner as in the transmission, with de-interleaving, channel decoding, descrambling, and CRC check. If multiple retransmissions are employed, such as in the case of HARQ (Hybrid Automatic Repetition Request), the combining is done within the decoder.

UMTS and HSPA NodeB

3GPP WCDMA standard is backwards compatible, e.g., the same L1 implementation can host R99 (UMTS) and HSPA (and HSPA+) transport channel implementation, since they are both based on Direct Sequence − CDMA (DS-CDMA).

In Figure 4, we have a general block diagram of a UMTS/HSPA transceiver.

Over DL (downlink), each transport block is encoded and mapped onto symbols from constellation (QPSK, 16-QAM). Then more transport blocks are concatenated in a physical channel. This physical channel is spread with a channelization code and then scrambled using the cell primary scrambling code. On UL (uplink), each user has its own scrambling code. First, NodeB detects the fingers of the user, i.e., the propagation paths, in the attempt to coherently combine all propagation paths. This is done via a path searcher, which detects propagation paths, delays, and magnitudes. Each finger is descrambled and despread. Then all the fingers are coherently combined in a Rake receiver, using the channel coefficients, as provided by the channel estimator. The RACH channel is not explicitly shown here, but it uses the same architecture.

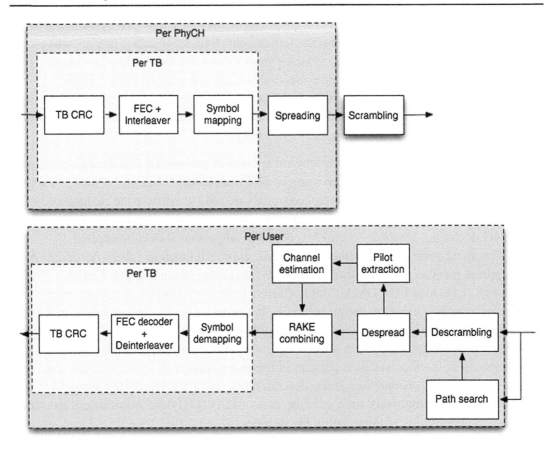

Figure 4:
General block diagram of a UMTS/HSPA transceiver.

Joint architecture

Solution for a hybrid (e)NodeB would be to use a baseband processor (Figure 5) with as much processing power, in order to be able to:

- carry the throughput from/to L2 for both standards
- carry the samples from/to RRH (Remote Radio Head) for both standards
- support enough processing power for both transceivers
- glue to a RRH or several RRHs

These requirements lead to constraints on the design of the platform. The need for concurrent tasks, corresponding to different and independent standards of communication, raises the idea of multicore processing. A single core cannot be shared among multiple real-time flows. Instead, multiple cores can do that. These cores work in parallel and raise the MIPS figure by a factor roughly equal to the number of processors. Increasing the number of cores leads to

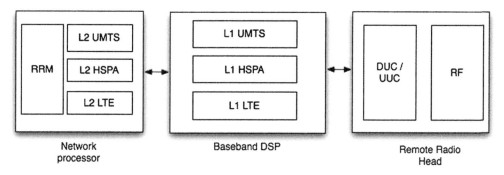

Figure 5:
Hybrid transceiver (L1 + L2 + RF).

increasing the overhead required to manage them (and possible bottlenecks to the shared resources (memory, I/O, accelerators)). The degree of improvement is therefore not exactly proportional to the number of cores. There are an optimum number of cores that provide optimum performance / cost + power ratio. Nowadays, it is mutually and generally agreed that this number is around 6.

Apart from multicore processing, processing power should come from hardware accelerators. It is bad design to block one core with repetitive operations, occurring on a symbol/chip/packet basis. Instead, if such operations occur on a periodic basis, a dedicated circuit can be used for that. Repetitive operations in the case of communication systems at the L1 level may consist of:

*   FFT operations for OFDM modulators or demodulators
*   DFT operations for SC-FDMA modulators (UE) or demodulators (NodeB)
*   Channel decoders (Convolutional, Turbo, Reed-Muller)
*   CRC detectors
*   Spreading / despreading for CDMA systems
*   Correlators for code detection

Processor

A case study for the baseband processor is Freescale's MSC8157. This is a six-core (SC3850) processor, whose unique powers derive from the HW accelerators on the chip. These are grouped into a single co-processor, called MAPLE. A block diagram is shown in Figure 6.

MAPLE comprises several hardware units, called Processing Elements (PE), specialized in typical operations or series of operations likely to be performed on a large scale in communication equipments.

A list of the PEs that may be used in a system and their use cases is presented in Table 1.

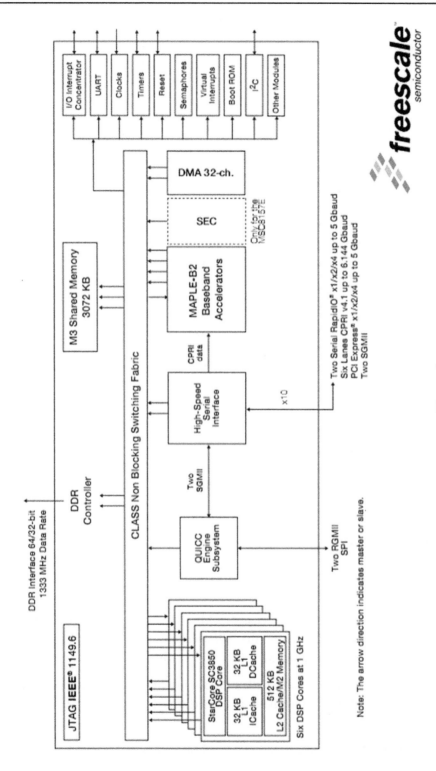

Figure 6:
MSC8157 block diagram.

Table 1: MAPLE coprocessor processing elements.

PE Name	Application	Use Case
CONV PE	LTE + UMTS + HSPA	Used for correlations (RACH channel, path searcher)
EQPE	LTE	Equalizer using either Zero Forcing, MMSE or Maximum Likelihood for MIMO
FTPE	LTE + UMTS + HSPA	FFT 128...2048 points used for OFDMA
		DFT up to 1200 points, used for SC-FDMA
		Used for fast correlations using CONV PE
CRC PE	LTE + UMTS + HSPA	CRC Check for uplink and CRC insertion for downlink
		Several polynomials used for CRC16, CRC24, CRC32, CRC16, CRC12, CRC18
eTVPE	LTE + UMTS + HSPA	Channel decoder for turbo and convolutional codes
		Soft or hard
		Several polynomials supported
DEPE	LTE + UMTS + HSPA	Turbo encoder
CRPE	UMTS + HSPA	Chip-rate processor
		Spreading and despreading with spreading factors up to 256.

Software architecture

When using multiple cores, care must be taken to design the software architecture in order to fully benefit from the processing power provided by the number of cores.

In Figure 7, the control plane is described. Each L1 accesses shared resources, such as cores, accelerators, and memory. Access to the cores is achieved through a real-time operating system. The scheduler of the operating system schedules and dispatches jobs to cores in different ways: dedicated or based on the core load at a given time. Triggering the job execution can be done in several ways, based on different criteria:

- Priority of the job
  - Some jobs should be executed once they are in the queue rather than the others.
  - Jobs that require reading from or writing to a buffer may need to be executed with higher priority to ensure that buffers are not filled up.
- Time triggered jobs
  - Some jobs need to be executed at precise times (synchronous operations).
  - Such jobs are the ones involving transactions at the synchronous interface (CPRI)
- Dependent jobs
  - Jobs should be executed only after previous jobs in the signal flow have been executed.
  - Example: in WCDMA, a chip-rate processing may only occur after each open and planned physical channel has been built at the symbol level.

An example of a job diagram for WCDMA case is depicted in Figure 8.

In Figure 8, TrCH#k coding are jobs that take care of encoding a transport channel. These are all command driven jobs. It means that their scheduling is the result of a previous command

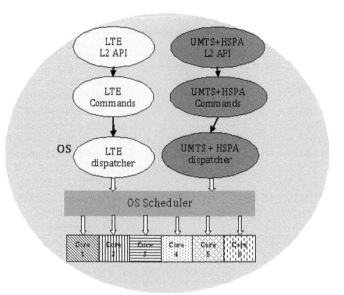

Figure 7:
Control plane of the SW architecture.

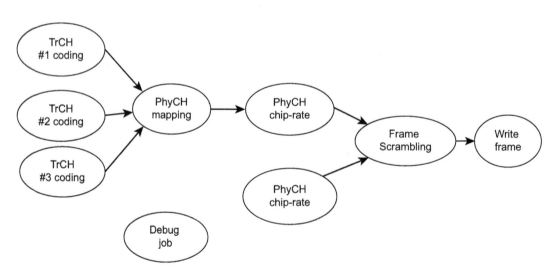

Figure 8:
Possible job dependence for WCDMA transmitter.

issued by a higher layer for the current frame. PhyCH mapping takes care of all the processing described in Figure 4 for a physical channel comprising three transport channels. This job should be executed as early as possible, but only after all TrCH #k jobs have been executed. Hence, this job has in fact 3 predecessors. Each physical channel is then processed as per

CDMA flow, by passing through a spreading operation. This is also dependent on the mapping job having been completed. Frame scrambling can occur only after all physical channels have been spread. This job then conditions the execution of the job of writing all the samples to a synchronous or asynchronous interface to radio. This job should also be triggered by a timer that indicates when these samples should be presented to the interface. No matter how early or fast frame scrambling has been performed, a subsequent job will also wait for the right time to send samples further. In addition, a debug job, logging different events can be introduced, usually with lower priority, as real time is not a real requirement for this. Scheduling should be done making sure that, in the worst case, in the highest load scenario, a frame can be completely processed once all commands have been issued, until the samples have to be pushed to the radio. Any of the jobs described above can be executed on the core or on an accelerator.

Such an approach ensures that a various number of L1 independent paths may access shared resources. The goal of the scheduler and of the high-level design is that the total combined processing power of the cores and accelerators covers the total needs of the arithmetic operations inside the L1 paths, minus a multicore scheduling penalty. Such a flexible and scalable approach ensures an easy design of a complex multi-standard and multicore architecture.

Conclusion

Digital signal processing has become a powerful premise for the world of wireless communications. All the latest technologies, including OFDM, CDMA, and SC-FDMA, that represent the main foundation of the 3G-4G networks, such as HSPA, LTE, or WiMAX, are now made possible by the high density of digital algorithms that can be squeezed into a small, low-power chip. On top of that, additional signal processing techniques, such as beamforming or spatial multiplexing, have contributed to the achievement of throughputs of hundreds of Mbps and spectral efficiencies of tens of b/s/Hz. Also, some other complex algorithms that can be found in channel estimators, equalizers, and bit decoders allow all these techniques to operate in a tough radio environment, including in non-line of sight communications with high mobility, even at large distance.

Large-scale integration allows a handheld device to include a multi-standard terminal, compliant with a large variety of wireless standards. Think about a smart phone that now conveniently chooses the best technology for data and voice transmission, according to its service needs and to the channel conditions. This is why your smart phone can connect to the BTS through GSM, EDGE, UMTS and soon, LTE. At the same time, it has capabilities for Wi-Fi, Bluetooth, and GPS connections, all these in a very small terminal, with good battery life. In the future, this service selection and multiplexing will only be defined through software. Without digital signal processing, relying solely on analog components, there would be no possible way to achieve such performance. Right now, the analog part in a transceiver is losing its pace in front of the digital part. More and more operations are performed in the digital part, including filtering, up/down conversion, and compensations of the distortions produced in the analog chain, at a smaller price and with better performances. Moreover, it is expected that the analog part will be completely displaced in the future, with the help of high-speed A/D and D/A converters, so that the digital transceiver will be glued directly to the antenna. This part is called *front-end* in a transceiver and while some 10 years ago it used to be completely analog, it is becoming more and more digital.

Index

Page numbers with "f" denote figures; "t" tables.

Printed and bound by CPI Group (UK) Ltd, Croydon, CR0 4YY

03/10/2024

01040317-0002